OXIDATION IN ORGANIC CHEMISTRY

Part D

This is Volume 5 of
ORGANIC CHEMISTRY
A series of monographs
Editor: HARRY H. WASSERMAN

A complete list of the books in this series appears at the end of the volume.

OXIDATION IN ORGANIC CHEMISTRY

Edited by *WALTER S. TRAHANOVSKY*

DEPARTMENT OF CHEMISTRY
IOWA STATE UNIVERSITY
AMES, IOWA

PART D

1982

ACADEMIC PRESS
A Subsidiary of Harcourt Brace Jovanovich, Publishers

New York London
Paris San Diego San Francisco São Paulo Sydney Tokyo Toronto

ACADEMIC PRESS, INC.
111 Fifth Avenue, New York, New York 10003

United Kingdom Edition published by
ACADEMIC PRESS, INC. (LONDON) LTD.
24/28 Oval Road, London NW1 7DX

Library of Congress Cataloging in Publication Data
Main entry under title:

Oxidation in organic chemistry.

(Organic chemistry, a series of monographs ; v. 5, pts. A-)
Pts. A- edited by Walter S. Trahanovsky.
Includes bibliographical references and indexes.
1. Oxidation. 2. Chemistry, Organic. I. Wiberg, Kenneth B. II. Trahanovsky, Walter S., Date.
QD281.O9W5 547'.23 65-26047
ISBN 0-12-697253-2 (Pt. D) AACR2

PRINTED IN THE UNITED STATES OF AMERICA

82 83 84 85 9 8 7 6 5 4 3 2 1

Contents

Contributors vii
Preface ix
Contents of Parts A, B, and C xi

CHAPTER I

Oxidations with Lead Tetraacetate

George M. Rubottom

I. Introduction 1
II. LTA Reactions with Hydroxyl Groups 2
III. LTA Reactions with Nitrogen-Containing Compounds 89
IV. LTA Reactions with Hydrocarbons 122
V. LTA Reactions with Organometallics 140

CHAPTER II

Phase Transfer Assisted Permanganate Oxidations

Donald G. Lee

I. Introduction 147
II. The Phase Transfer Process 152
III. Oxidation of Alkenes 168
IV. Oxidation of Alkynes 182
V. Oxidation of Arenes 186
VI. Oxidation of Alkanes 190
VII. Oxidation of Polycyclic Aromatic Hydrocarbons 192
VIII. Oxidation of Alcohols 193
IX. Oxidation of Phenols 196
X. Oxidation of Ethers 199
XI. Oxidation of Aldehydes 201
XII. Oxidation of Sulfur Compounds 201
XIII. Oxidation of Halides 203
XIV. Oxidation of Amines 204

CHAPTER III

Intramolecular Oxidative Coupling of Aromatic Substrates

Om Prakash Dhingra

I. Introduction . . . 207
II. Oxidative Coupling of Benzyltetrahydroisoquinolines . . . 209
III. Oxidative Coupling of Phenethyltetrahydroisoquinolines . . . 241
IV. Oxidative Coupling of *N*-Benzylphenethylamines . . . 257
V. Oxidative Coupling of Diarylalkanes . . . 264

CHAPTER IV

Oxidation of Coal

Ryoichi Hayatsu, Robert G. Scott, and Randall E. Winans

I. Introduction . . . 279
II. Coal Samples . . . 280
III. Extraction, Separation, and Identification of Oxidation Products . . . 284
IV. Oxidation . . . 291
V. Conclusion . . . 351
VI. Glossary . . . 352

Index . . . 355

Contributors

Numbers in parentheses indicate the pages on which the authors' contributions begin.

Om Prakash Dhingra, *Monsanto Agricultural Products Co., St. Louis, Missouri 63166* (*207*)

Ryoichi Hayatsu, *Chemistry Division, Argonne National Laboratory, Argonne, Illinois 60439* (*279*)

Donald G. Lee, *Department of Chemistry, University of Regina, Regina, Saskatchewan S4S 0A2, Canada* (*147*)

George M. Rubottom, *Department of Chemistry, University of Idaho, Moscow, Idaho 83843* (*1*)

Robert G. Scott, *Chemistry Division, Argonne National Laboratory, Argonne, Illinois 60439* (*279*)

Randall E. Winans, *Chemistry Division, Argonne National Laboratory, Argonne, Illinois 60439* (*279*)

Preface

Some of the most important and common reactions in organic chemistry involve oxidation and reduction. Because of the importance of this class of reactions, numerous reagents have been developed that will bring about certain oxidations or reductions selectively. This treatise is devoted to detailed discussions of specific oxidants or topics involving oxidation of organic compounds.

In this volume, the first chapter covers oxidations with lead tetraacetate, an oxidant that has been used widely for the selective oxidation of a variety of functional groups. The second chapter is concerned with an oxidant that has a long and extensive history—permanganate—but concentrates primarily on the recently studied phase transfer assisted permanganate oxidations. The third chapter discusses the intramolecular oxidative coupling of aromatic substrates and focuses on the utility of recently developed methods, such as vanadium oxytrifluoride oxidations, in the syntheses of complex natural products. The subject of the fourth chapter is the oxidation of coal with emphasis on the selective oxidative degradation procedures as tools for characterization of coal structure. Each chapter was written by an active researcher or researchers who have made important contributions in the area of chemistry being reviewed.

The level of all chapters is such that experts in these areas of research and students and researchers who wish a thorough and rigorous discussion of these topics should find them useful. In general, emphasis is on the scope and preparative use as well as the mechanistic aspects of the various oxidations.

WALTER S. TRAHANOVSKY

Contents of Part A

Edited by Kenneth B. Wiberg

ROSS STEWART, Oxidation by Permanganate

KENNETH B. WIBERG, Oxidation by Chromic Acid and Chromyl Compounds

W. A. WATERS AND J. S. LITTLER, Oxidation by Vanadium(V), Cobalt(III), and Manganese(III)

WILLIAM H. RICHARDSON, Ceric Ion Oxidation of Organic Compounds

RUDOLF CRIEGEE, Oxidations with Lead Tetraacetate

C. A. BUNTON, Glycol Cleavage and Related Reactions

AUTHOR INDEX—SUBJECT INDEX

Contents of Part B

Edited by Walter S. Trahanovsky

W. G. NIGH, Oxidation by Cupric Ion

PATRICK D. MCDONALD AND GORDON A. HAMILTON, Mechanisms of Phenolic Oxidative Coupling Reactions

ROBERT J. OUELLETTE, Oxidation by Thallium(III)

DONALD G. LEE AND MATTHIJS VAN DEN ENGH, The Oxidation of Organic Compounds by Ruthenium Tetroxide

AUTHOR INDEX—SUBJECT INDEX

Contents of Part C

Edited by Walter S. Trahanovsky

HANS J. REICH, Organoselenium Oxidations

ROY A. JOHNSON, Oxygenations with Microorganisms

BOŽO PLESNIČAR, Oxidations with Peroxy Acids and Other Peroxides

YOSHIRO OGATA, Oxidations with Nitric Acid or Nitrogen Oxides

SUJIT K. CHAKRABARTTY, Alkaline Hypohalite Oxidations

INDEX

OXIDATION IN ORGANIC CHEMISTRY

Part D

CHAPTER I

Oxidations with Lead Tetraacetate

GEORGE M. RUBOTTOM

I. Introduction 1
II. LTA Reactions with Hydroxyl Groups 2
A. Alcohols 2
B. 1,2-Diols and Related Systems 27
C. Enols and Related Systems 37
D. Phenols 55
E. Monocarboxylic Acids 61
F. Dicarboxylic Acids 81
III. LTA Reactions with Nitrogen-Containing Compounds 89
A. Amines and Related Compounds 89
B. Amides 98
C. Hydrazines and Related Compounds 101
D. Azomethines 118
IV. LTA Reactions with Hydrocarbons 122
A. Alkanes 122
B. Aromatic Hydrocarbons 127
C. Alkenes 131
V. LTA Reactions with Organometallics 140

I. Introduction

Lead tetraacetate [(LTA), mp 175°C] was first isolated in 1851[1] and has been used extensively as an oxidant by organic chemists since 1920. Although LTA can be produced in the laboratory by the reaction of red lead oxide (Pb_3O_4) with acetic acid in the presence of acetic anhydride,[2] the reagent is commercially available at moderate cost. The commercial LTA contains 10% acetic acid to retard decomposition due to reaction with water. This acetic acid is readily removed either by washing the LTA with anhydrous

[1] A. Jacquelain, *J. Prakt. Chem.* **53**, 151 (1851).
[2] J. C. Bailar, *Inorg. Synth.* **1**, 47 (1939).

Oxidation in Organic Chemistry, Part D

ISBN 0-12-697253-2

ether or by azeotropic distillation with benzene just prior to use. LTA is compatible with a number of common solvents including acetic acid, benzene, methylene chloride, chloroform, nitrobenzene, pyridine, DMF, and DMSO.

LTA exists as a monomer in the solid state and in both benzene and acetic acid solution.[3] A distorted cubic geometry has been proposed with both oxygen atoms of each acyloxy group coordinated with lead.[4] Infrared measurements support this arrangement.[5] The solubility and conductance properties of LTA suggest that the compound is covalent,[4] but in acetic acid or carboxylic acid solvents, in general, rapid ligand exchange occurs between LTA and the solvent.[6]

Oxidation of inorganic compounds by LTA is facilitated by the high redox potential, 1.6 V in perchloric acid, of the reagent.[7] With organic compounds LTA is normally reduced to lead(II) acetate in reactions involving both homolytic and heterolytic mechanisms. Photolysis of LTA affords lead(II) acetate and acetoxy radicals as primary products.[8]

The basic properties of LTA as an oxidizing agent have been reviewed[7] and several excellent general reviews on the reactions of LTA with organic substrates are extant.[9-12] Reviews concerning the behavior of LTA toward specific functional groups are cited where appropriate in the text. The primary literature included in this chapter covers mainly the period of 1970 through January, 1980.

II. LTA Reactions with Hydroxyl Groups

A. ALCOHOLS

The treatment of alcohols with LTA leads to the formation of acetates, oxidation to the corresponding aldehyde or ketone, fragmentation, and

[3] G. Rudakoff, *Z. Naturforsch. B. Anorg. Chem., Org. Chem.* **17**, 623 (1962).
[4] R. Partch and J. Monthony, *Tetrahedron Lett.*, 4427 (1967).
[5] K. Huesler and H. Loeliger, *Helv. Chim. Acta* **52,** 1495 (1969).
[6] E. A. Evans, J. L. Huston, and T. H. Norris, *J. Am. Chem. Soc.* **74**, 4985 (1952).
[7] J. Zýka, *Pure Appl. Chem.* **13**, 569 (1966).
V. Franzen and R. Edems, *Angew. Chem.* **73**, 579 (1961).
[9] R. N. Butler, *in* "Synthetic Reagents" (J. S. Pizey, ed.), p. 277, Ellis Horwood, Chichester, England, 1977.
[10] L. F. Fieser and M. Fieser, "Reagents for Organic Synthesis," p. 537; Vol. 2, p. 234; Vol. 3, p. 168; Vol. 4, p. 278; Vol. 5, p. 365; Vol. 6, p. 313; Vol. 7, p. 185. Wiley, New York, 1967.
[10a] G. W. Rotermund, *in* "Houben-Weyl, Methoden der Organischen Chemie" (E. Müller, ed.), 4th ed., Vol. IV/1b, p. 167. Thieme, Stuttgart, 1975.
[11] H. O. House, "Modern Synthetic Reactions," 2nd ed., p. 359. Benjamin, California, 1972.
[12] R. Criegee, *in* "Oxidation in Organic Chemistry" (K. B. Wiberg, ed.), Part A, p. 277. Academic Press, New York, 1965.

cyclization. The parameters leading to each mode of behavior have been thoroughly discussed with respect to substrate structure and solvent effects. Comprehensive lists of examples have also been tabulated.[13,14]

The formation of acetates is a process that accompanies most LTA oxidations. The retention of configuration encountered in the acetate has led to postulation of the generalized attack of the alcohol on the acetate carbonyl of a lead(IV) species and/or acetic acid (or anhydride) attack on an alkoxy lead(IV) intermediate.[13] Although the use of benzene seems to maximize acetate production (when cyclization is not possible) whereas

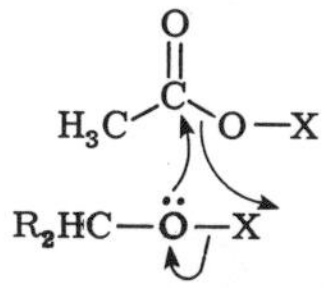

X = H, $Pb(OAc)_3$,
or $COCH_3$

pyridine minimizes it, generalizations are dangerous as shown by the production of **1**.[15]

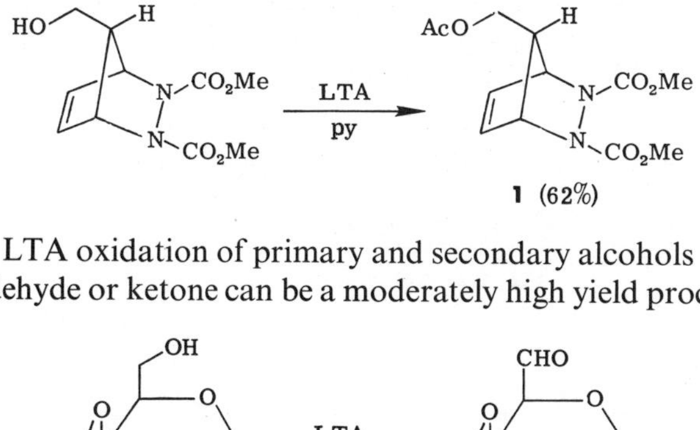

1 (62%)

(ref. 15)

The LTA oxidation of primary and secondary alcohols to the corresponding aldehyde or ketone can be a moderately high yield process.[16–18] With the

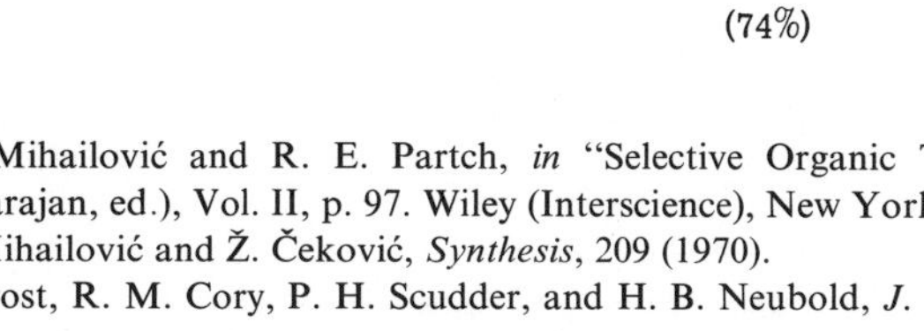

(74%)

(ref. 16)

[13] M. Lj. Mihailović and R. E. Partch, *in* "Selective Organic Transformations" (B. S. Thyagarajan, ed.), Vol. II, p. 97. Wiley (Interscience), New York, 1972.

[14] M. Lj. Mihailović and Ž. Čeković, *Synthesis*, 209 (1970).

[15] B. M. Trost, R. M. Cory, P. H. Scudder, and H. B. Neubold, *J. Am. Chem. Soc.* **95**, 7813 (1973).

[16] D. J. Ward, W. A. Szarek, and J. K. N. Jones, *Carbohydr. Res.* **21**, 305 (1972).

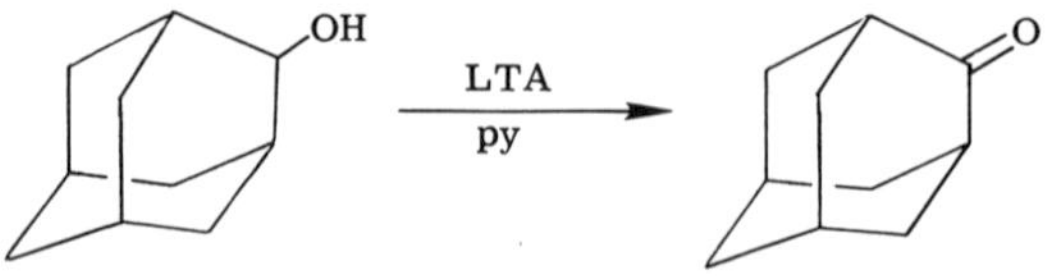

(ref. (17))

(100%)

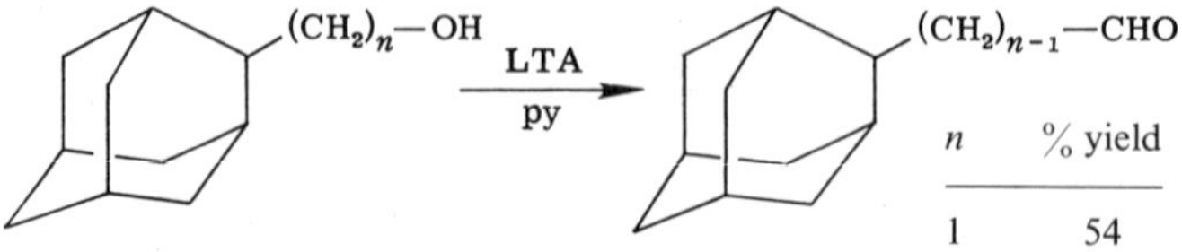

n	% yield
1	54
2	48
3	33
4	55

(ref. 18)

ω-(2-adamantyl)alkan-1-ols, the production of aldehyde was accompanied by acetate formation as well.[18] As noted by the examples, the use of pyridine gives maximum yields of carbonyl derivatives. This is in sharp contrast to the role of acidic or apolar solvents such as acetic acid or benzene wherein products of substitution, fragmentation, and cyclization become paramount.[17] This dramatic shift in reaction pathway can be attributed to the occurrence of a heterolytic mechanism in pyridine while homolytic routes prevail in both acetic acid and benzene.[13,14] It has been suggested that the carbonyl formation that does occur in the absence of pyridine arises from heterolytic disproportionation of the dialkoxylead(IV) derivative **2**[14,19–21]

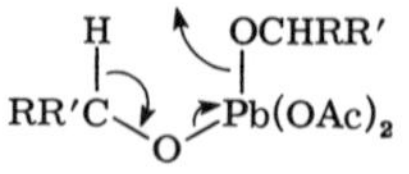

2

in line, for instance, with studies on the oxidation of benzyl alcohol and 1-phenylethanol in which the reaction is first order with respect to LTA and second order with respect to alcohol.[20] The role of pyridine is generally believed to involve [ROPb(OAc)$_3$ · py] formation,[14,20] and the suggestion has been made that Pb—N bond formation increases the electrophilicity of

---

[17] J. Burkhard, J. Jankû, V. Kubelka, J. Mitera, and S. Landa, *Collect. Czech. Chem. Commun.* **39**, 1083 (1974).

[18] J. Burkhard, J. Jankû, and S. Landa, *Collect. Czech. Chem. Commun.* **39**, 1072 (1974).

[19] Y. Pocker and B. C. Davis, *J. Chem. Soc. Chem. Commun.*, 803 (1974).

[20] K. K. Banerji, S. K. Banerjee, and R. Shanker, *Indian J. Chem. Sect. A* **15**, 702 (1977).

[21] K. K. Banerji, S. K. Banerjee, and R. Shanker, *Bull. Chem. Soc. Jpn.* **51**, 2153 (1978).

the complex and thus favors α-hydrogen removal over the homolysis which occurs when neutral or acidic conditions pertain.[20] The rate-determining nature of the proton removal step has been shown by isotope studies,[14,20,21] and transition state **3** has been proposed for the carbonyl forming oxidation of both benzyl alcohol and 1-phenylethanol.[20]

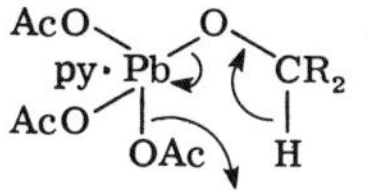

3

In cases where either 1,4- or 1,5-hydrogen transfer is possible, another mode of carbonyl formation becomes important when benzene (homolytic conditions) is used as solvent.[22,23]

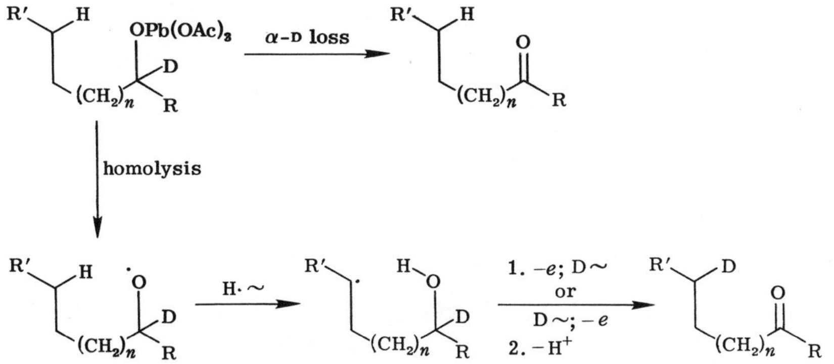

As noted above, although the reactions of alcohols with LTA to form esters and to undergo heterolysis in pyridine have received attention, a great deal more work has been devoted to the study of homolytic fragmentation and cyclization of alcohols by LTA in neutral and/or acidic media.[13,14] Scheme 1 indicates the generally accepted pathway leading to both cyclization and fragmentation.[14,23a] Fragmentation occurs best when substituents favor the stabilization of **4** and when ring strain can be released. Radical **4** then is oxidized to the corresponding carbocation **5** that undergoes typical

[22] D. Jeremić, S. Milosavljević, V. Andrejević, M. Jakovljević-Marinković, Ž. Čeković, and M. Lj. Mihailović, *J. Chem. Soc. Chem. Commun.*, 1612 (1971).

[23] S. Milosavljević, D. Jeremić, and M. Lj. Mihailović, *Tetrahedron* **29**, 3547 (1973).

[23a] W. S. Trahanovsky, *in* "Methods in Free-Radical Chemistry" (E. S. Huyser, ed.), Vol. 4, p. 133. M. Dekker, New York, 1973.

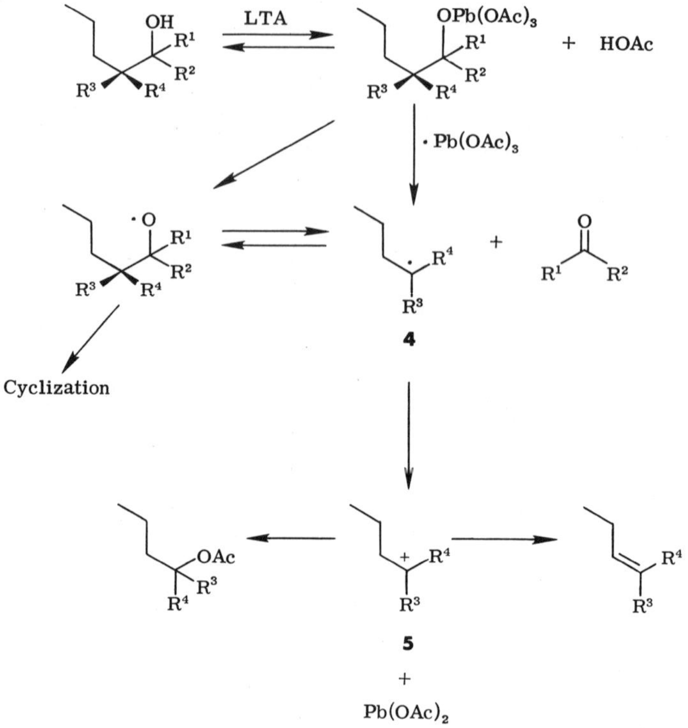

SCHEME 1

carbonium ion reactions.[14,23a] Production of **6**,[24] **7**,[25] and **8**[25a] reflects the importance of radical stabilization as does the increased amount of fragmentation noted from *trans*-2-methylcyclohexanol as opposed to cyclohexanol.[26] The preparative cleavage of both α- and β-amyrin to give the corresponding seco aldehyde is another case in point.[26a]

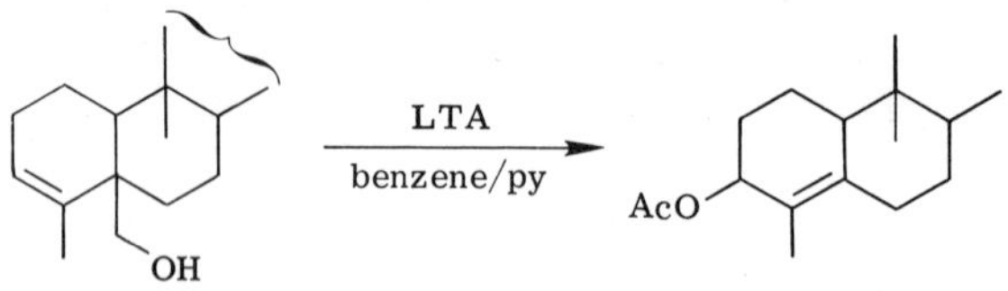

6 (99%)

(ref. 24)

[24] H. Kakisawa, T. Horie, and T. Kusumi, *Bull. Chem. Soc. Jpn.* **48**, 727 (1975).
[25] W. P. Schneider and R. A. Morge, *Tetrahedron Lett.*, 3283 (1976).
[25a] K. Shankaran and A. S. Rao, *Indian J. Chem. Sect. B* **18**, 507 (1979).
[26] J. Bošnjak, V. Andrejević, Ž. Čeković, and M. Lj. Mahailović, *Tetrahedron* **28**, 6031 (1972).
[26a] A. K. Devi, G. K. Trivedi, and S. C. Bhattacharyya, *Indian J. Chem. Sect. B* **16**, 8 (1978).

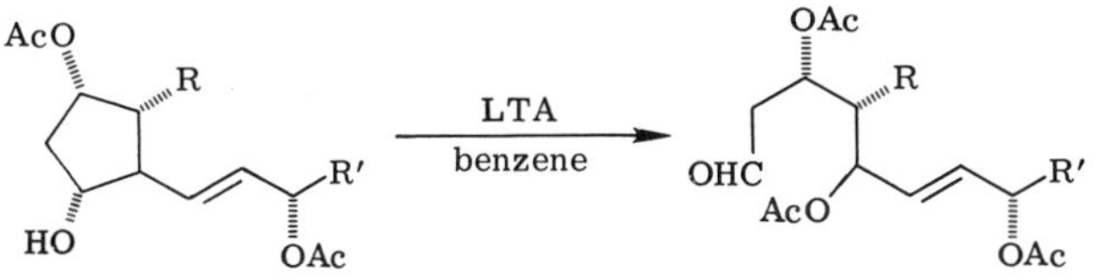

(ref. 25)

7

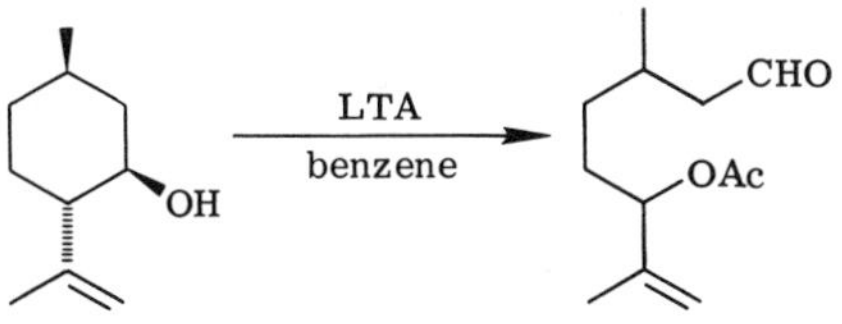

(ref. 25a)

8 (60%)

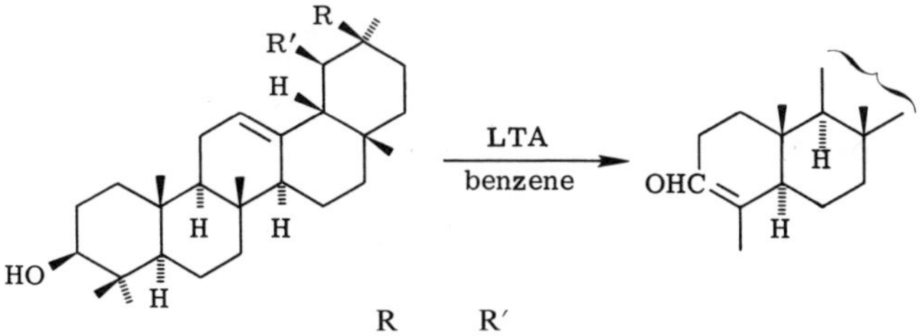

(ref. 26a)

	R	R′
α-amyrin	H	Me
β-amyrin	Me	H

The effect of relieving ring strain can be seen in the oxidation of a series of steroidal and triterpenoidal cyclobutanols to give either the corresponding hydroxyketones or hemiacetals.[27–29] Arguments involving substituent steric interactions have been advanced to explain the predominance of both types of product.[29] Patchouli alcohol fragments when treated with LTA in

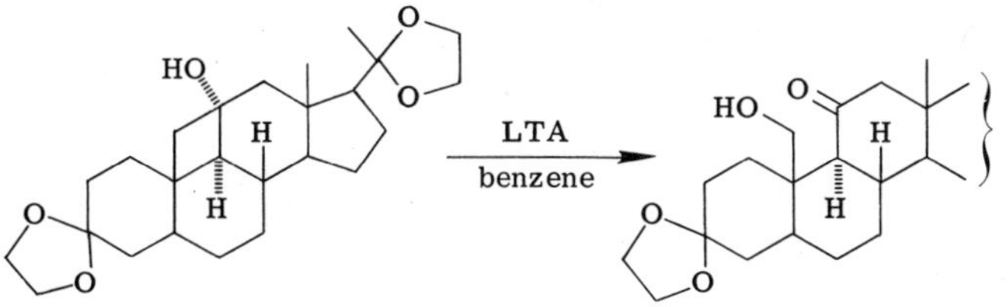

(ref. 27)

[27] H. Wehrli, M. S. Heller, K. Schaffner, and O. Jeger, *Helv. Chim. Acta* **44**, 2162 (1961).
[28] R. Imhof, W. Graf, H. Wehrli, and K. Schaffner, *J. Chem. Soc. Chem. Commun.*, 852 (1969).
[29] W. Herz and D. H. White *J. Org. Chem.* **39**, 1 (1974).

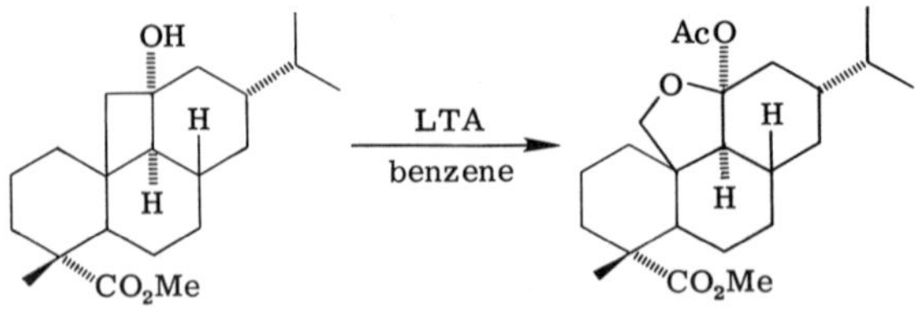

(ref. 29)

benzene/calcium carbonate, give a 40% yield of **9** in a reaction that may well be concerted.[30,30a] Fragmentation of bridged steroidal alcohols has been

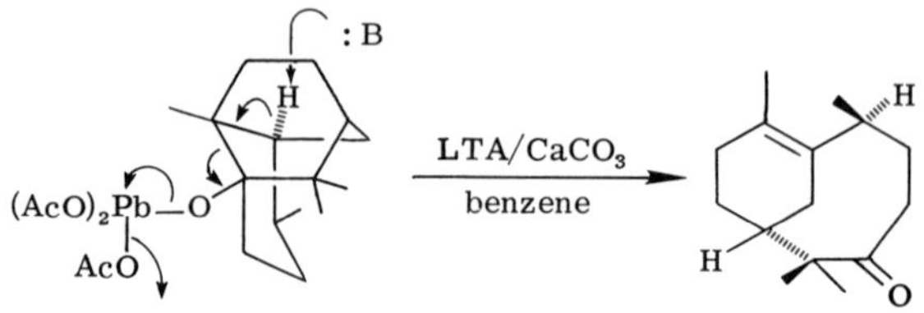

(refs. 30, 30a)

9 (40%)

used as part of a scheme directed toward the study of photolytically mediated remote oxidation.[31]

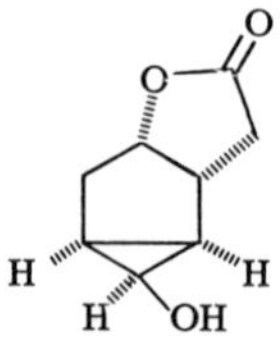

10

Although treatment of **10** with LTA led to no useful cleavage products,[32] recent studies have shown that, in general, cyclopropanols fragment readily when treated with LTA.[33,34] Cleavages of both **11** and **12** are stereospecific.[35]

Comparisons of the products obtained from LTA β-fragmentation of alcohols with those obtained from LTA decarboxylation of analogous carboxylic acids (Section II,E) point to the intervention of similar, if not

[30] A. F. Thomas and M. Ozainne, *J. Chem. Soc. Chem. Commun.*, 120 (1977).

[30a] A. F. Thomas and M. Ozainne, *Helv. Chim. Acta* **62,** 361 (1979).

[31] R. Breslow, S. Baldwin, T. Flechtner, P. Kalicky, S. Liu, and W. Washburn, *J. Am. Chem. Soc.* **95**, 3251 (1973).

[32] E. J. Corey, Z. Arnold, and J. Hutton, *Tetrahedron Lett.*, 307 (1970).

[33] G. M. Rubottom, R. Marrero, D. S. Krueger, and J. L. Schreiner, *Tetrahedron Lett.*, 4013 (1977).

[34] T. L. Macdonald, *Tetrahedron Lett.*, 4201 (1978).

[35] G. M. Rubottom, unpublished results.

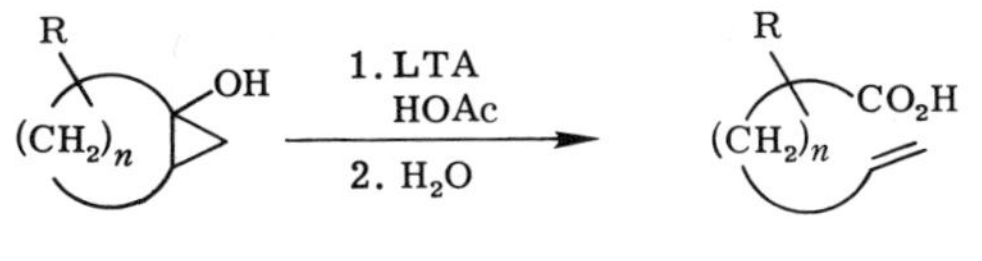

(62–92%) (ref. 33)

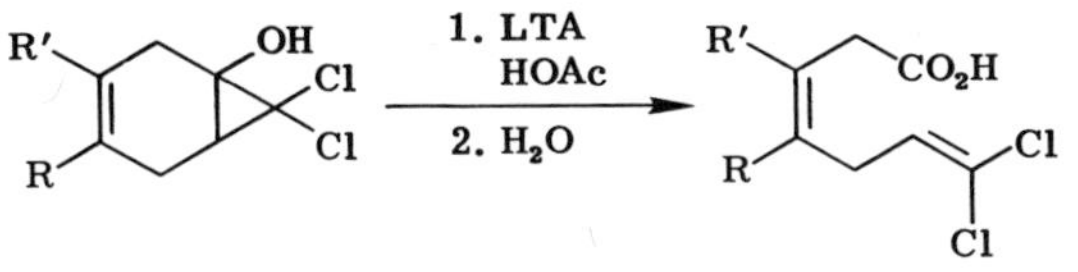

(65–85%) (ref. 34)

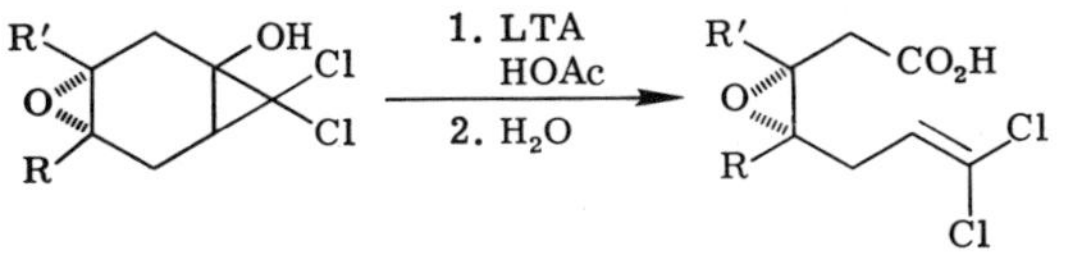

(59–69%) (ref. 34)

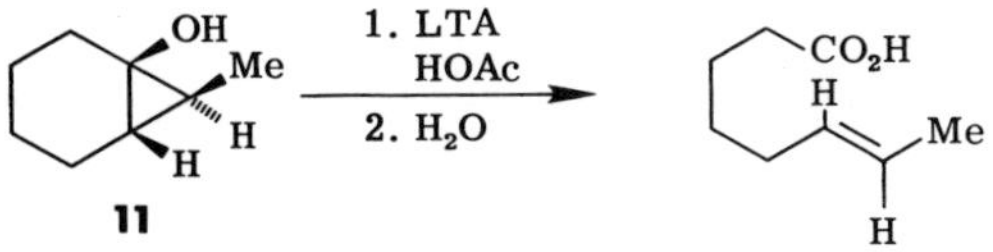

11

(ref. 35)

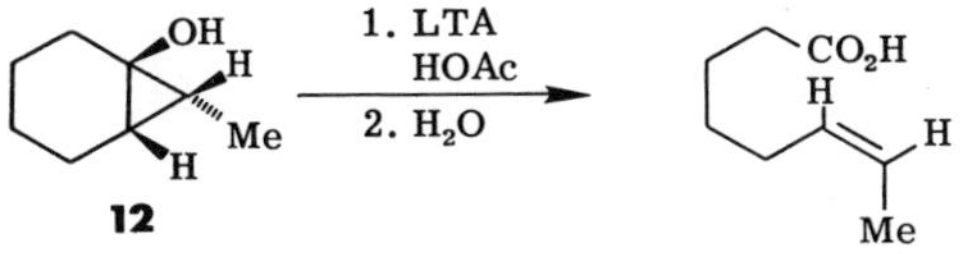

12

identical, reaction intermediates, presumably the alkyl radicals.[23a,36,37] The case presented[36] is illustrative and similar findings have been noted for systems leading to the cyclopropylcarbinyl/cyclobutyl interface.[37]

Another interesting aspect of the fragmentation reaction involves the reversibility of the cleavage process that can lead to stereochemical changes

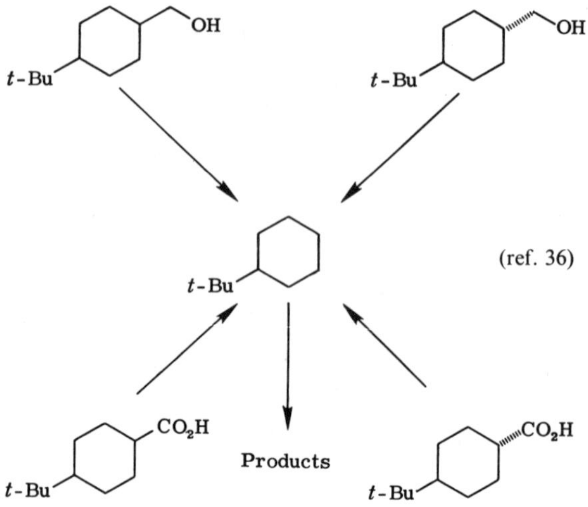

(ref. 36)

in **13**. The examples given indicate that the phenomenon is quite general,[26,29,38]

13

and the reader is directed to references 29 and 38 for more examples.

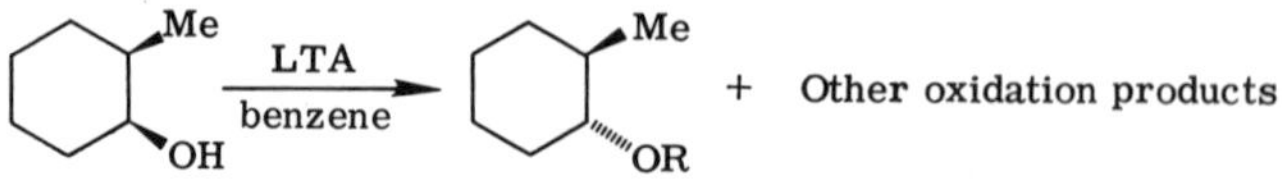

(ref. 26)

R	% yield
H	5.7
Ac	2.3

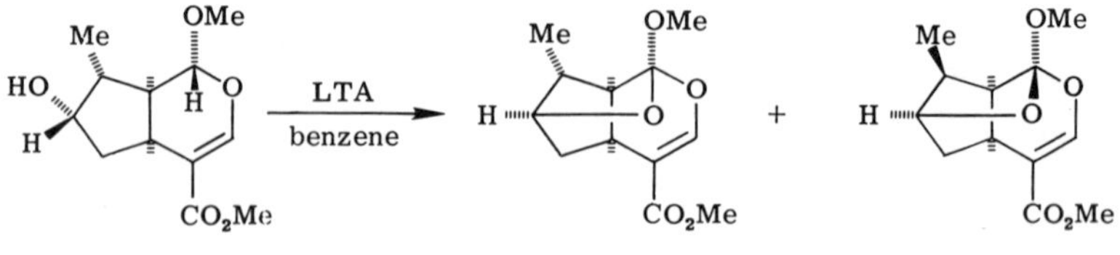

(49%) (3%)

(ref. 38)

[36] M. Lj. Mihailović, J. Bošnjak, and Ž. Čeković, *Helv. Chim. Acta* **57**, 1015 (1974).
[37] M. Lj. Mihailović, J. Bošnjak, and Ž. Čeković, *Helv. Chim. Acta* **59**, 475 (1976).
[38] J. J. Partridge, N. K. Chada, S. Faber, and M. R. Ushoković, *Synth. Commun.* **1**, 233 (1971).

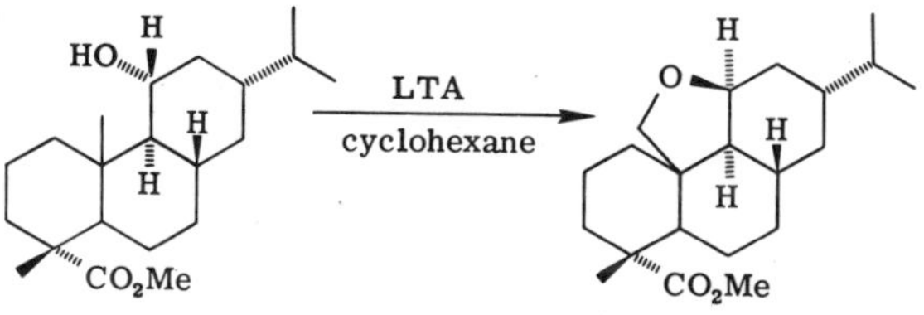

(ref. 29)

(40%)

The products shown above reflect the other major pathway noted in alcohol oxidation with LTA, cyclization.[13,14] Table I contains the results of other recent studies involving cyclic alcohols.[18,26,39–42]

The mechanism of cyclization has been extensively studied and these findings reviewed.[13,14] The generally accepted pathway is shown below.

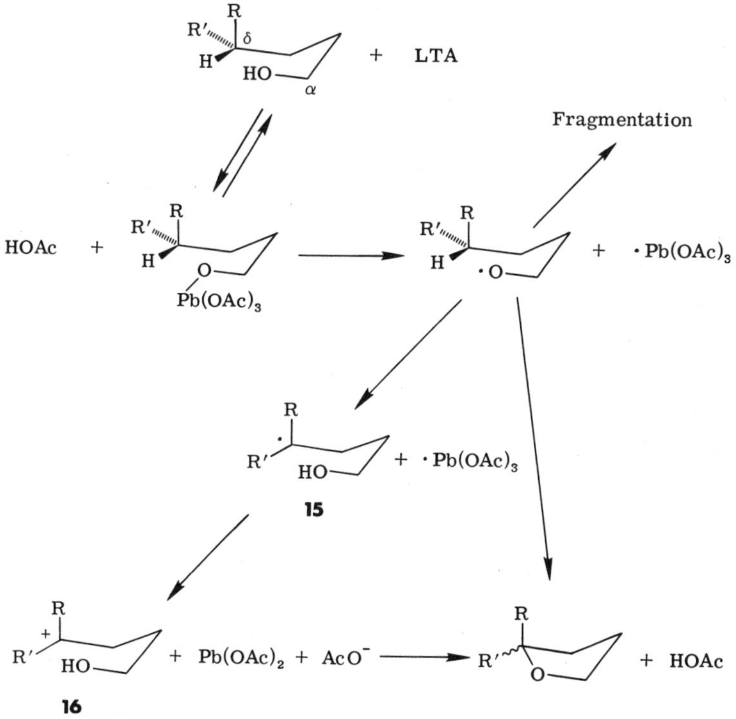

[39] W. T. Borden, V. Varma, M. Cabell, and T. Ravindranathan, *J. Am. Chem. Soc.* **93**, 3800 (1971).

[39a] A. B. Crow and W. T. Borden, *J. Am. Chem. Soc.* **101**, 6666 (1979).

[40] T. Kato, S. Kumazawa, C. Kabuto, T. Honda, and Y. Kitahara, *Tetrahedron Lett.*, 2319 (1975).

[41] P. Brun and B. Waegell, *Bull. Soc. Chim. Fr.*, 1825 (1972).

[42] P. Brun and B. Waegell, *Tetrahedron* **32**, 1137 (1976).

TABLE I

LTA-PROMOTED CYCLIZATION OF ALCOHOLS

Substrate	Conditions	Product	% Yield	Reference
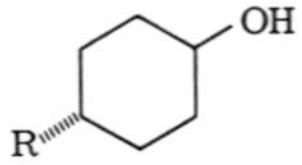	Benzene/$CaCO_3$	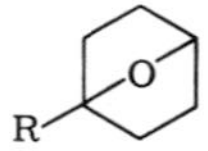	R H 0.8 *t*-Bu 1.5	26
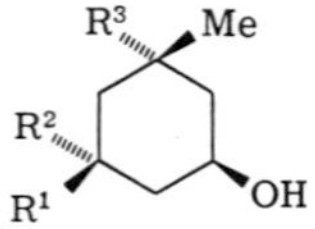	Benzene/$CaCO_3$	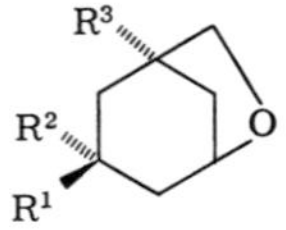	R^1 R^2 R^3 H H H 3 H H Me 9.5 H Me Me 55	26
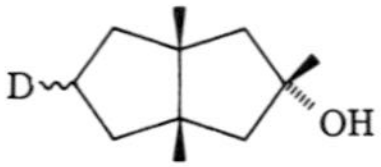	Benzene/$CaCO_3$	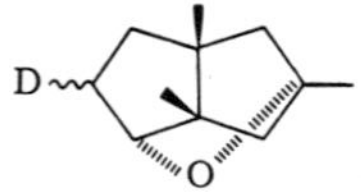	60	39, 39a
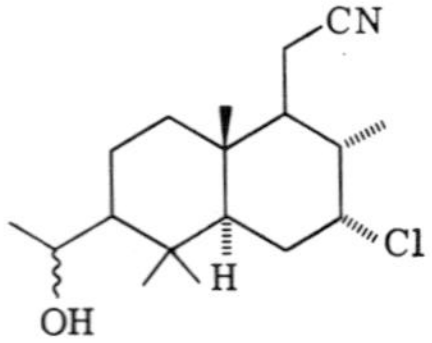	—	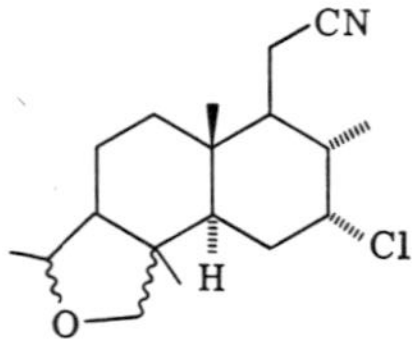	44	40
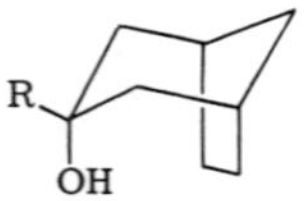	Benzene/$CaCO_3$	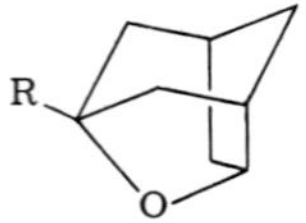	R H 25–28 Me 51	41
	Benzene/$CaCO_3$	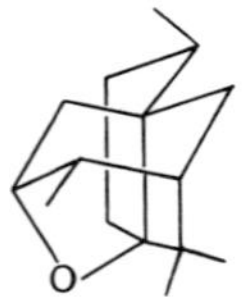	77	42

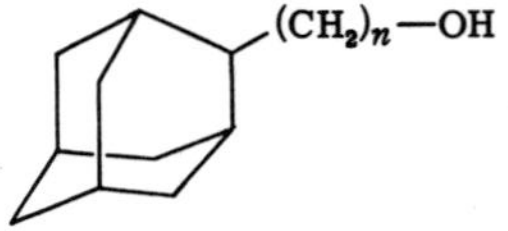	Benzene/$CaCO_3$	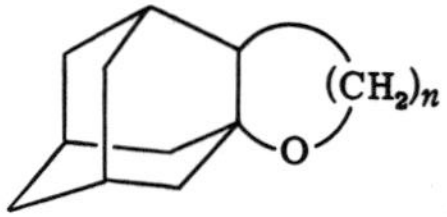	$\underline{n}$ 1 — 2 21	18
	—	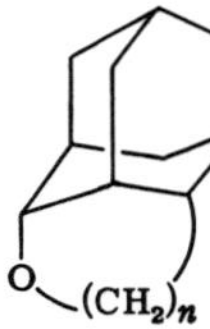	$\underline{n}$ 1 67 2 1	
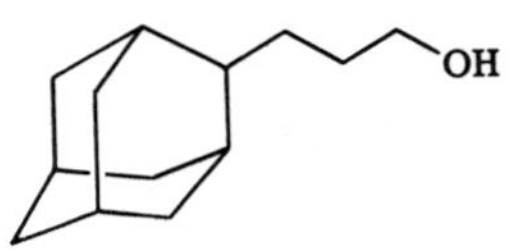	Benzene/$CaCO_3$	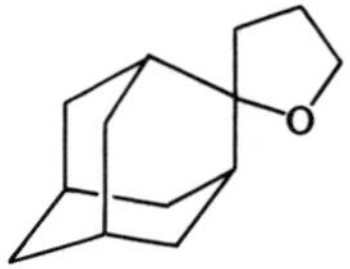	29	18
		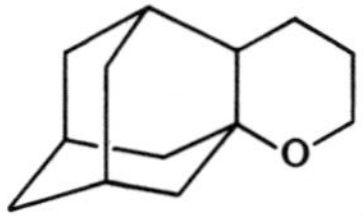	1	
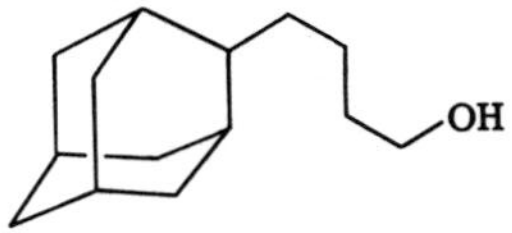	Benzene/$CaCO_3$	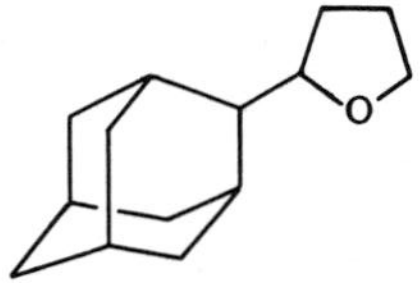	57	18
		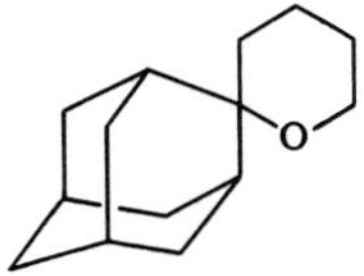	2	

As noted in Table I and the generalized mechanism, the product most commonly formed is a tetrahydrofuran derivative rather than a tetrahydropyran, a fact accounted for by the favorable transition state **14** for 1,5-hydrogen migration.[13,14] The optimal nonbonded distance between the

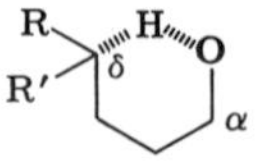

14

δ-carbon and oxygen has been determined to be 2.5–2.7 Å[13,14] In general, alkyl substitution at the δ-center aids abstraction due to stabilization of **14**, whereas substitution at the α-center slows the process. Since the radical **15** and/or carbonium ion **16** are intermediates, stereospecificity is not observed in the final cyclization step.[13,14,43] When rigid geometry offers close proximity to a remote hydrogen, formation of rings other than those containing five members can occur.[44–46] On the other hand, geometry that can remove

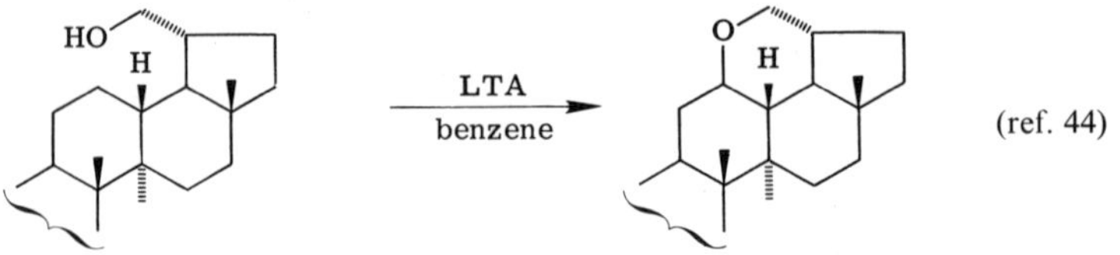

(49%)

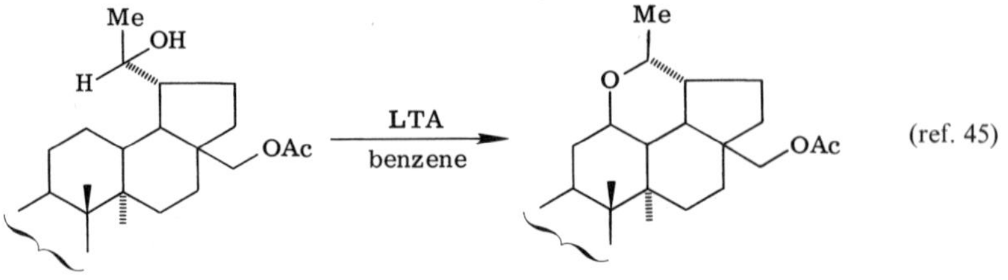

(50%)

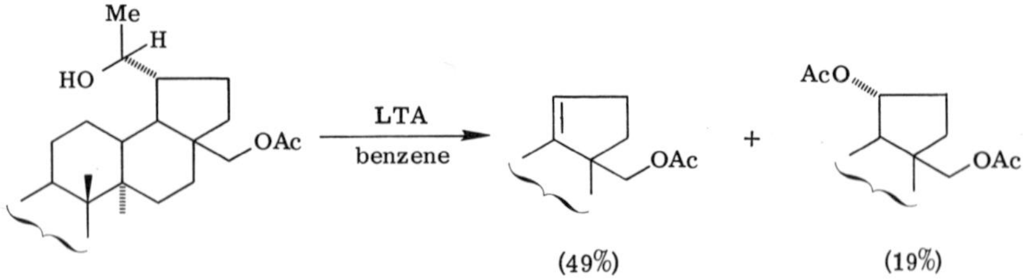

(49%) (19%)

[43] M. Lj. Mihailović, S. Gojković, and S. Konstantinović, *Tetrahedron* **29**, 3675 (1973).
[44] V. Pouzar and A. Vystrčil, *Collect. Czech. Chem. Commun.* **43**, 2190 (1978).

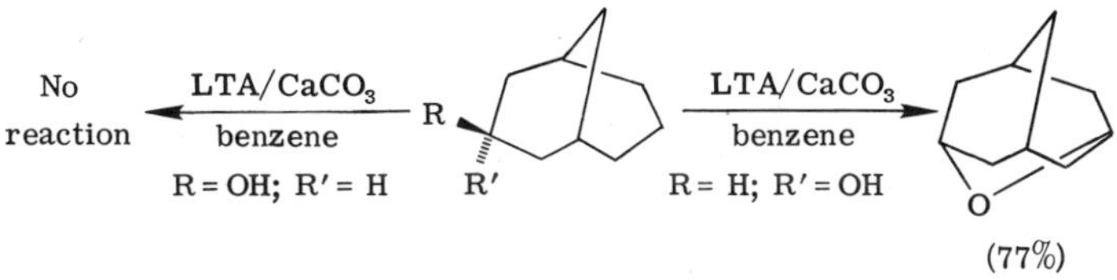

(ref. 46)

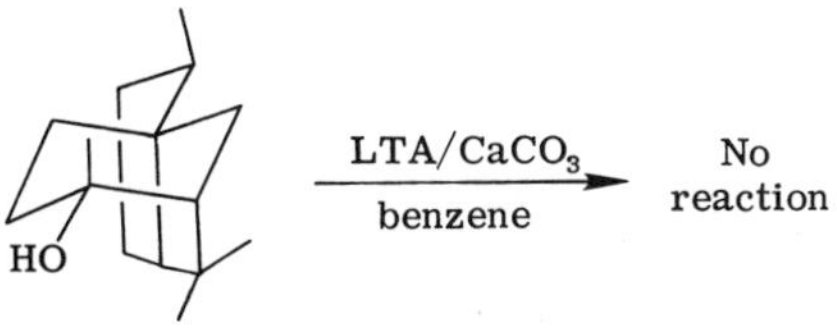

(ref. 42)

the oxygen from a remote hydrogen retards the reaction.[42,45,46] In one instance, rearrangement occurs prior to ether formation resulting, once again, in the production of a five-membered ring.[47]

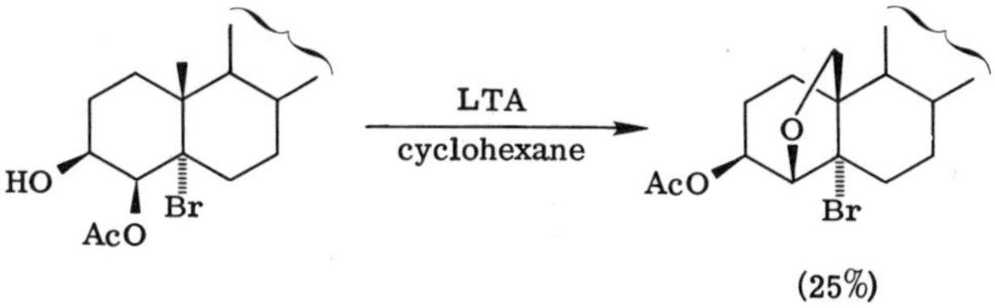

(ref. 47)

Oxidation of **17** with LTA to give **18** as the major product has been interpreted to involve carbonium ion **19** since oxidation with both HgO/Br_2 or Ag_2CO_3/Br_2 should not involve ionic intermediates.[48]

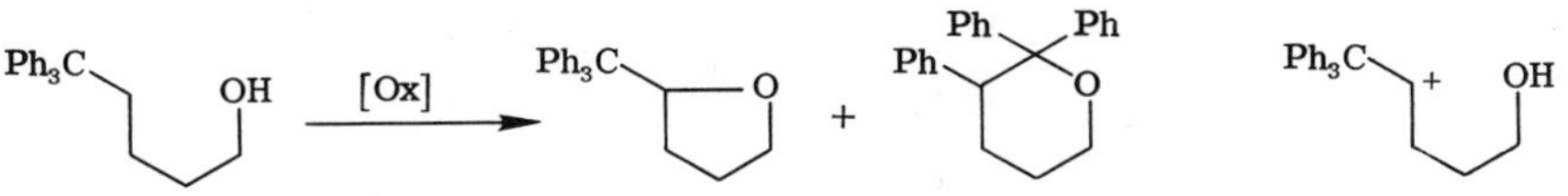

17 **18** **19**

(ref. 48)

LTA	5.5	59
HgO/Br_2	6	33
Ag_2CO_3/Br_2	4	19

---

[45] V. Pouzar, J. Protiva, E. Lisá, E. Klinotová, and A. Vystrčil, *Collect. Czech. Chem. Commun.* **40**, 3046 (1975).

[46] M. Fisch, S. Smallcombe, J. C. Gramain, M. A. McKervey, and J. E. Anderson, *J. Org. Chem.* **35**, 1886 (1970).

[47] P. Morand and A. Palakova-Paquet, *Can. J. Chem.* **51**, 4098 (1973).

[48] M. Lj. Mihailović, G. Misošević, and A. Milovanović, *Tetrahedron* **34**, 2587 (1978).

The cis/trans ratios obtained from the oxidation of **20** with LTA relative to those encountered with added $Fe(ClO_4)_3$ or $Cu(BF_4)_2$ indicate that cationic cyclization predominates with LTA, whereas cyclization in the presence of Cu(II) or Fe(III) proceeds via intermediates such as **21**.[49]

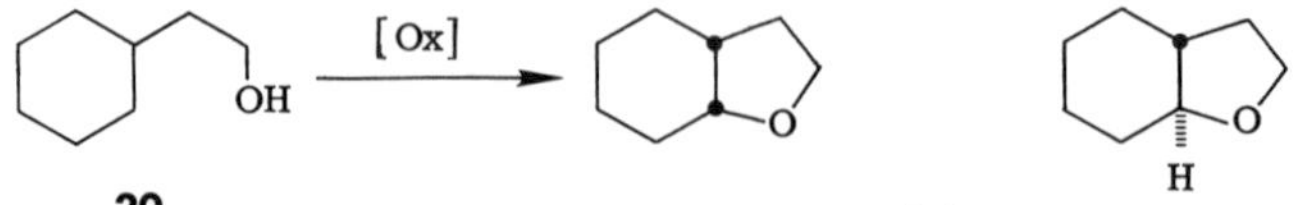

20

	cis/trans
LTA	3.9
LTA/Fe(III)	2.0
LTA/Cu(II)	0.2

(ref. 49)

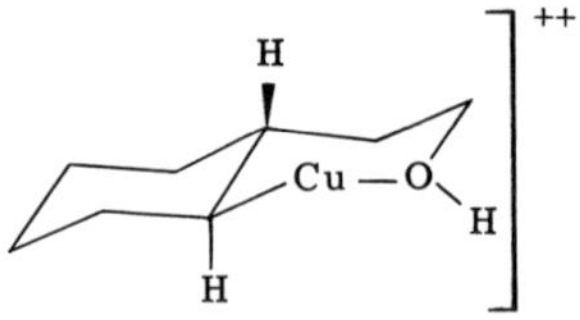

21

Isotope studies utilizing **22** and **23** have shown a 20-fold preference for *trans*-hydrogen abstraction via a chair (or "half-chair-like") transition

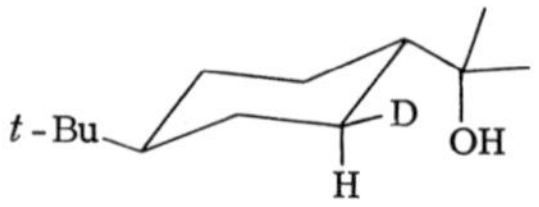

22

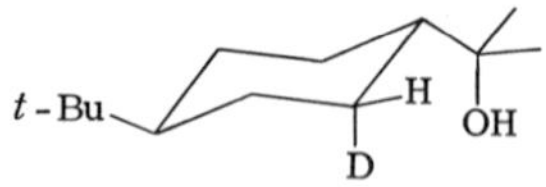

23

state.[50] The process has also been found to parallel closely abstraction reactions induced by electron impact.[50,51]

Introduction of remote double bonds into alcohols oxidized with LTA gives the opportunity for neighboring group participation.[13,52–55b] In these

[49] J. T. Groves, *Tetrahedron Lett.*, 3113 (1975).
[50] G. Eadon, *J. Am. Chem. Soc.* **98**, 7313 (1976).
[51] M. M. Green, J. G. McGrew, II, and J. M. Moldowan, *J. Am. Chem. Soc.* **93**, 6700 (1971).
[52] R. M. Moriarty *in* "Selective Organic Transformations" (B. S. Thyagarajan, ed.), Vol. II, p. 183. Wiley (Interscience), New York, 1972.
[53] T. Sasaki, S. Eguchi, and T. Kiriyama, *J. Org. Chem.* **38**, 2230 (1973).
[54] M. P. Zink, J. Ehrenfreund, and H. R. Wolf, *Helv. Chim. Acta* **57**, 1116 (1974).
[54a] N. Langlois and R. Z. Andriamialisoa, *J. Org. Chem.* **44**, 2468 (1979).
[55] P. K. Grant, R. T. Weavers, and C. Huntrakul, *Tetrahedron* **29**, 245 (1973).
[55a] M. P. Bertrand, J. M. Surzur, M. Boyer, and M. Lj. Mihailović, *Tetrahedron* **35**, 1365 (1979).
[55b] R. B. Gupta and R. N. Khanna, *Indian J. Chem. Sect. B* **16**, 35 (1978).

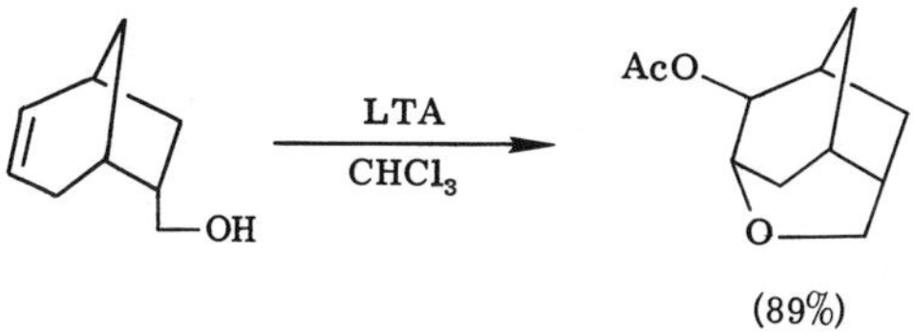

(ref. 53)

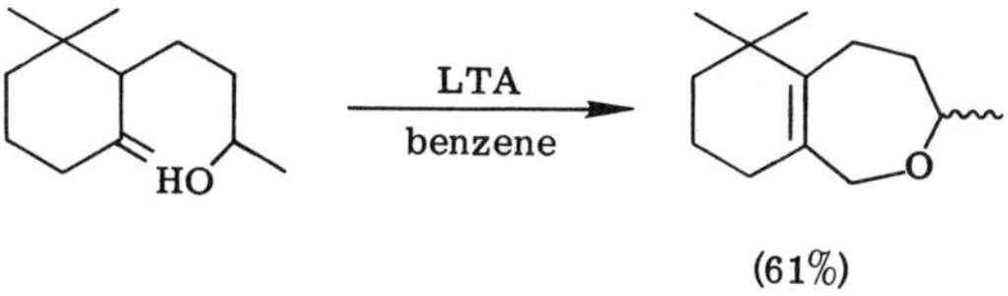

(ref. 54)

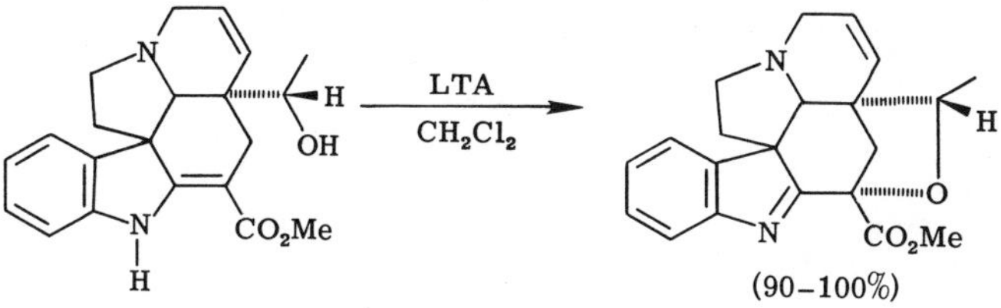

(ref. 54a)

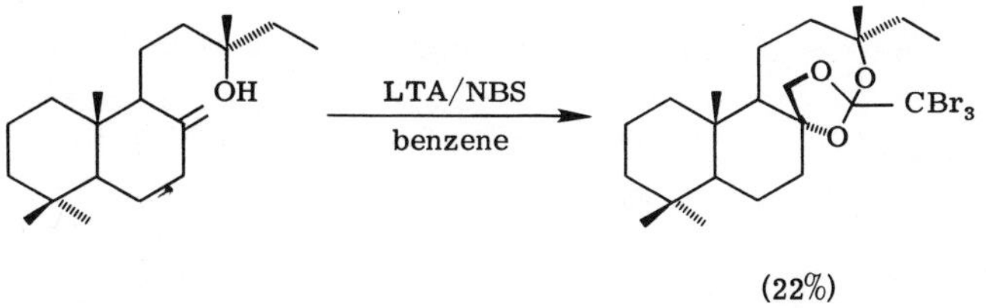

(ref. 55)

cases π-interaction with LTA can mediate products; however, "normal" alcohol fragmentation and cyclization can compete as well.[54,56–58]

An interesting example of epoxide formation occurs when the appropriate allylic alcohol is treated with either LTA[58] or $Pb(OCOCF_3)_4$.[59]

[56] M. Lj. Mahailović, Ž. Čeković, J. Stanković, S. Djokic-Mazinjanin, D. Marinković, and S. Konstantinović, *Glas. Hem. Drus., Beograd* **43**, 69 (1978).

[57] M. Lj. Mahailović, Ž. Čeković, J. Stanković, N. Pavlović, S. Konstantinović, and S. Djokić-Mazinjanin, *Helv. Chim. Acta* **56**, 3056 (1973).

[58] J. Ehrenfreund, M. P. Zink, and H. R. Wolf, *Helv. Chim. Acta* **57**, 1098 (1974).

[59] D. Westphal and E. Zbiral, *Liebigs Ann. Chem.*, 2038 (1975).

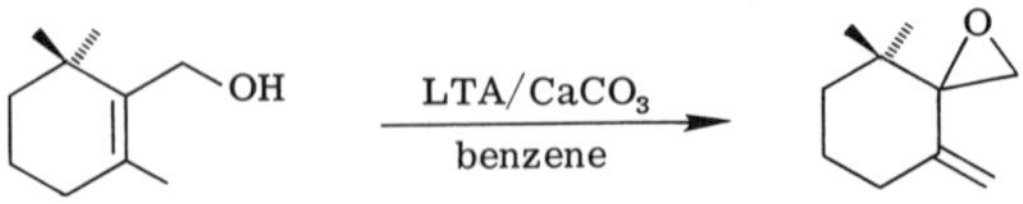

(40%) (ref. 58)

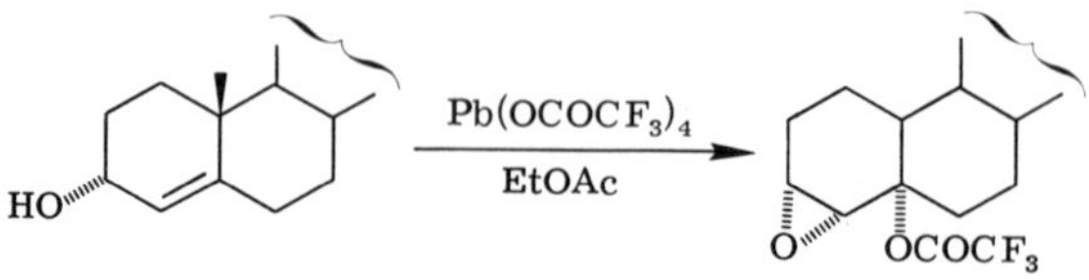

(46%) (ref. 59)

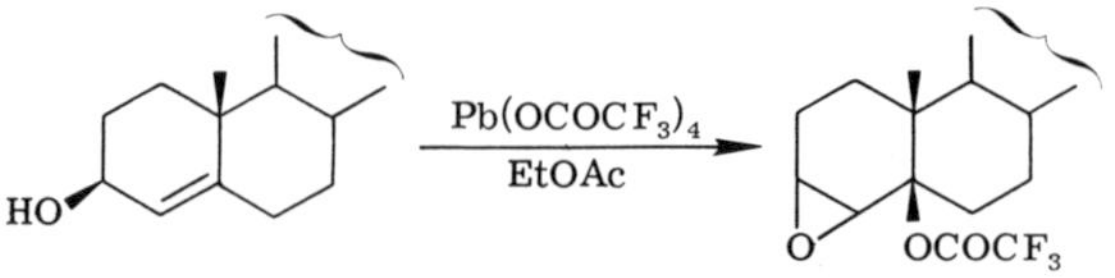

(20%) (ref. 59)

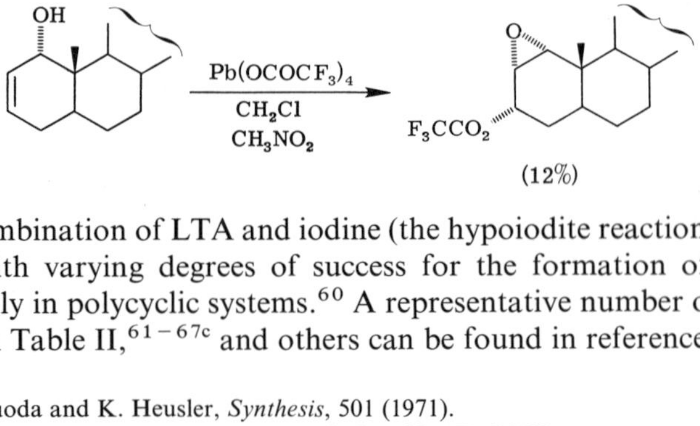

(12%) (ref. 59)

A combination of LTA and iodine (the hypoiodite reaction) has also been used with varying degrees of success for the formation of cyclic ethers, especially in polycyclic systems.[60] A representative number of examples are given in Table II,[61–67c] and others can be found in reference 60.

[60] J. Kaluoda and K. Heusler, *Synthesis*, 501 (1971).
[61] T. Nakano and A. K. Banerjee, *Tetrahedron* **28**, 471 (1972).
[62] T. Nakano and A. K. Banerjee, *Tetrahedron Lett.*, 165 (1971).
[63] H. Khan, A. Zaman, G. L. Chetty, A. S. Gupta, and S. Dev, *Tetrahedron Lett.*, 4443 (1971).
[63a] A. K. Banerjee, C. D. Ceballo, M. N. Vallejo, and E. H. Bolivar, *Bull. Chem. Soc. Jpn.* **52**, 608 (1979).
[63b] A. L. Campbell, H. N. Leader, C. L. Spencer, and J. D. McChesney, *J. Org. Chem.* **44**, 2746 (1979).
[64] J. Bull and C. Van Zyl, *Tetrahedron* **28**, 3957 (1972).
[65] J. Wicha, *Tetrahedron Lett.*, 2877 (1972).
[66] P. Roller, B. Tursch, and C. Djerassi, *J. Org. Chem.* **35**, 2585 (1970).
[67] M. Kaufman, P. Morand, and S. A. Samad, *J. Org. Chem.* **37**, 1067 (1972).
[67a] P. Kočovský and V. Černý, *Coll. Czech. Chem. Commun.* **44**, 2275 (1979).
[67b] H. Suginome, N. Sato, and T. Masamune, *Bull. Chem. Soc. Jpn.* **52**, 3043 (1979).
[67c] P. K. Jadhav and U. R. Nayak, *Indian J. Chem. Sect. B* **16**, 952 (1978).

TABLE II

LTA/I_2-Promoted Cyclization of Alcohols

Substrate	Conditions	Product	% Yield	Reference
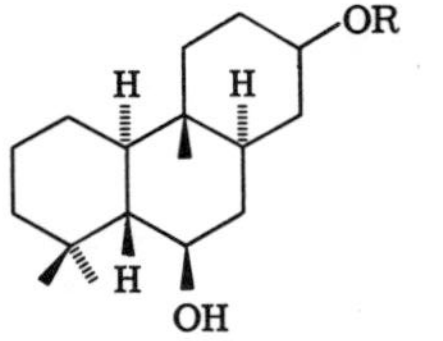	1. LTA/$CaCO_3$ cyclohexane 2. $I_2/h\nu$	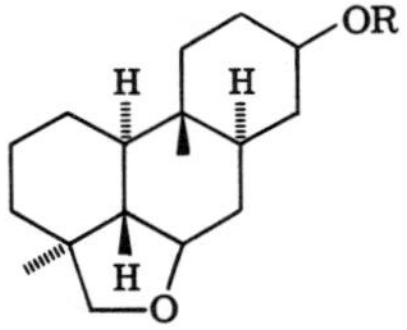	30	61, 62
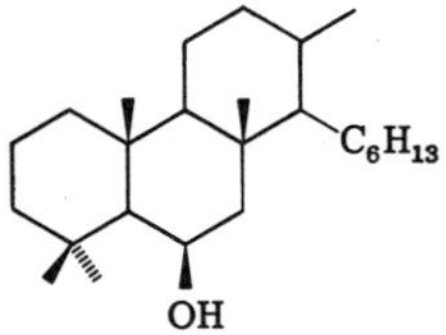	LTA/$I_2/h\nu$	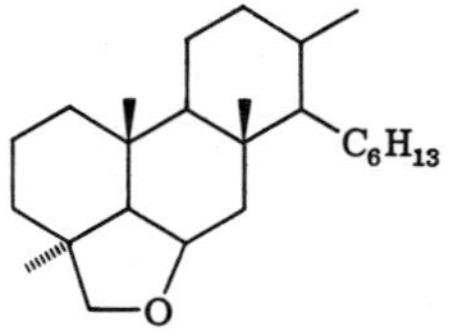	100	63
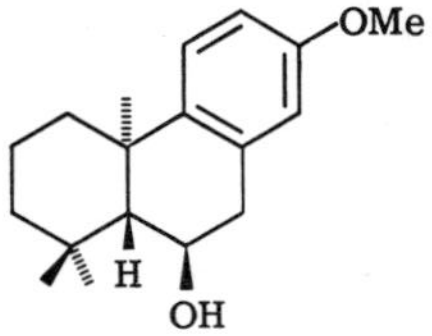	LTA/$CaCO_3/I_2$ cyclohexane	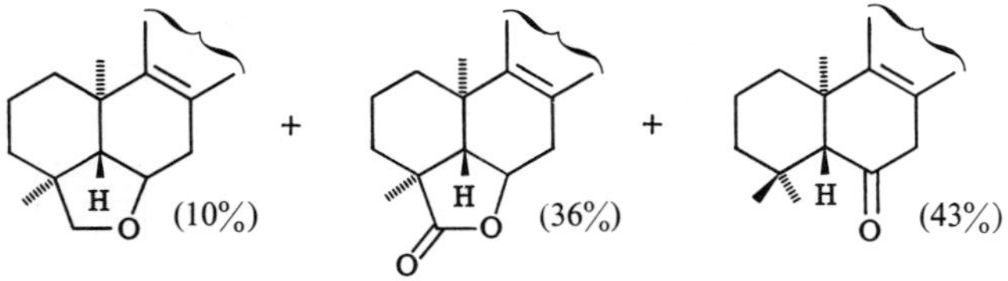		63a

(*cont.*)

TABLE II (*cont.*)

Substrate	Conditions	Product	% Yield	Reference
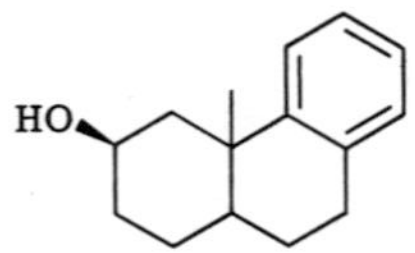	LTA/I_2/*hν* benzene	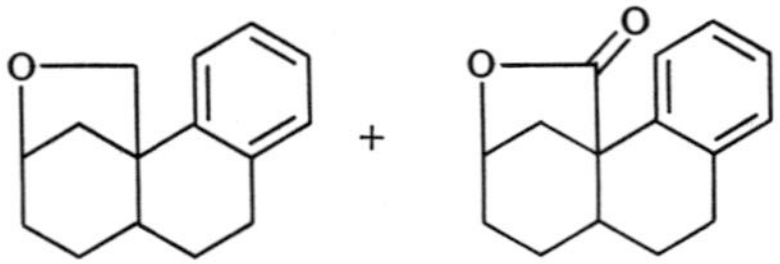	68, mixture	63b
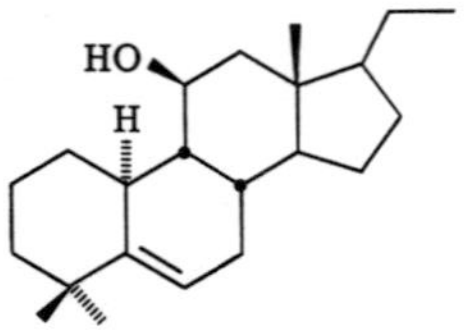	LTA/I_2 benzene	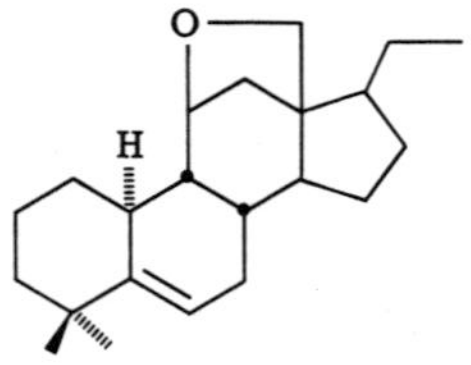	10	64
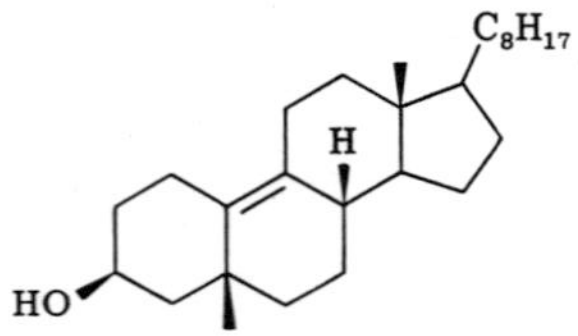	LTA/I_2 benzene	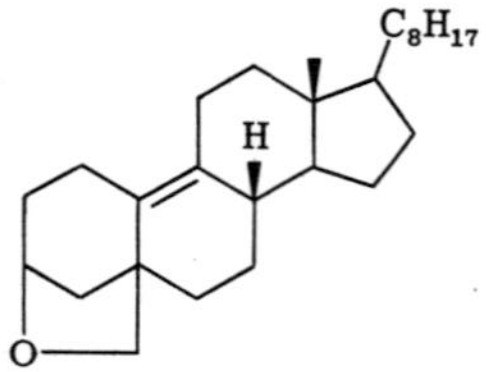	—	65

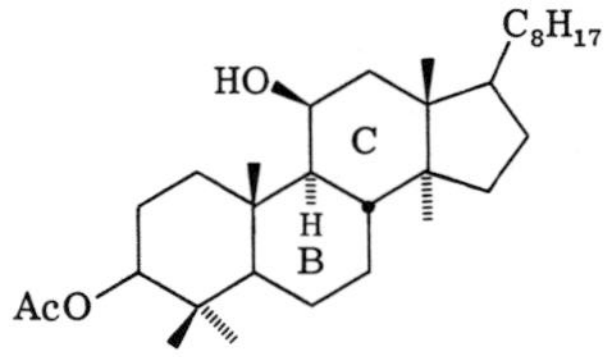	LTA/I_2/$h\nu$/HOAc cyclohexane	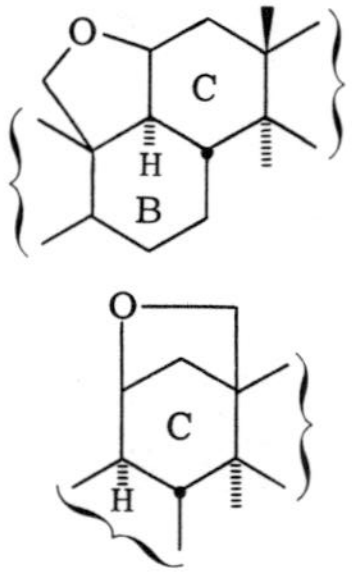	— —	66
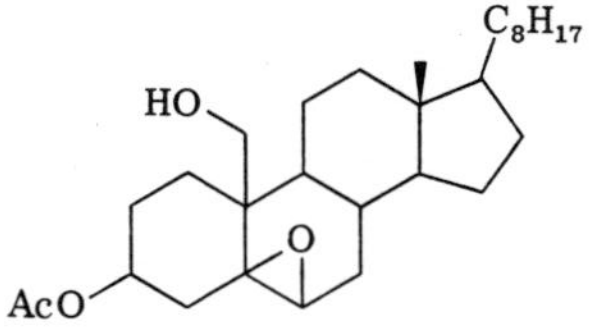	1. LTA/$CaCO_3$ cyclohexane 2. I_2/$h\nu$	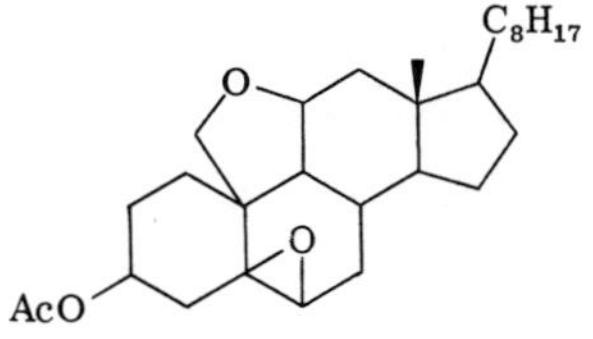	43	67
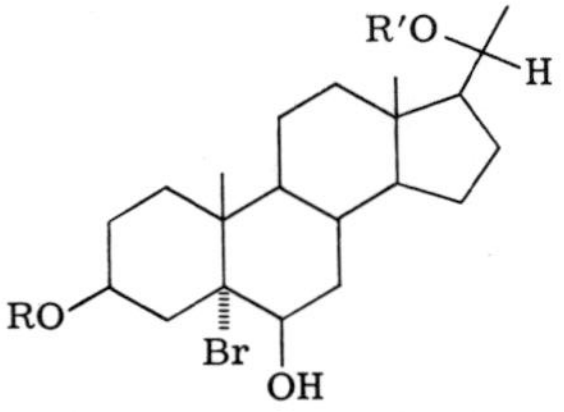	LTA/$CaCO_3$/I_2 benzene	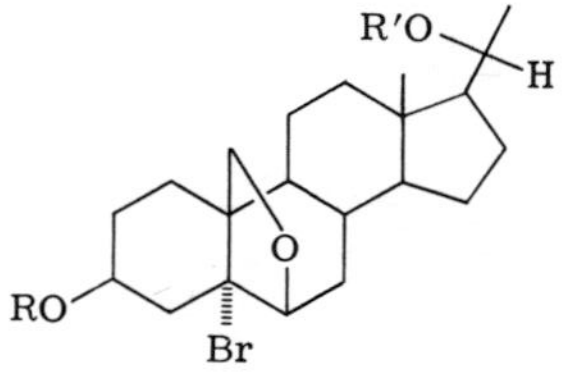	R R′ (%) Ac Bz 43 Bz Bz 80 Bz Ac 25	67a

(*cont.*)

TABLE II (*cont.*)

Substrate	Conditions	Product	% Yield	Reference
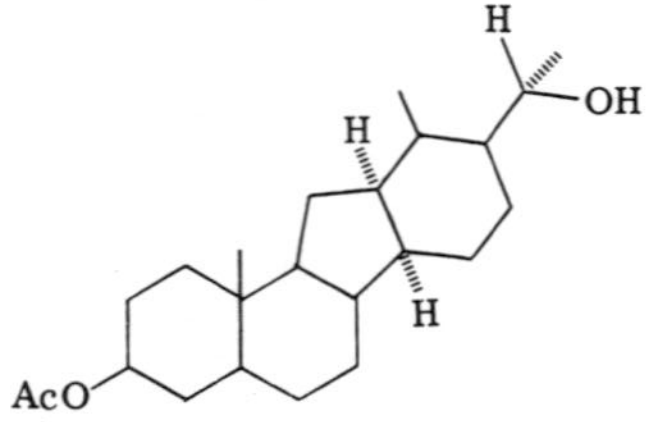	LTA/$CaCO_3$/I_2 *hv*/cyclohexane	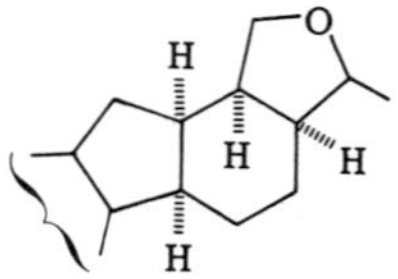	27	67b
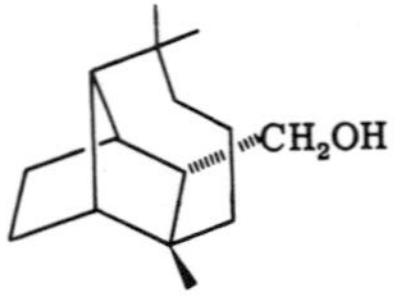	LTA/I_2/*hv* cyclohexane	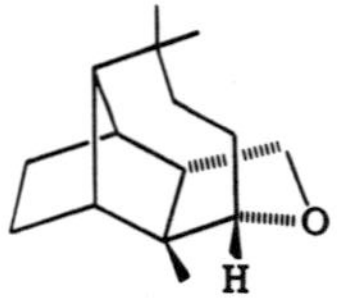 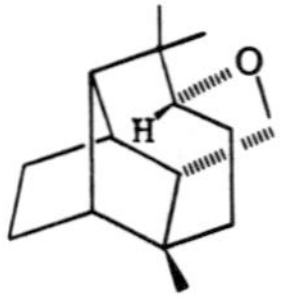	~80 as 1/1 mixture	67c
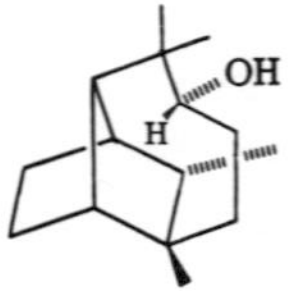	LTA/I_2/*hv* cyclohexane	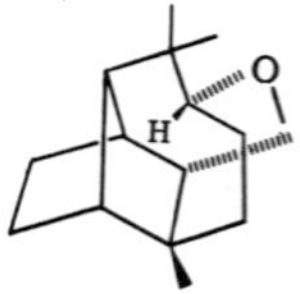	73	67c

The generally accepted mechanism of the reaction[60] involves initial transformation of the alcohol to a hypoiodite **24** which then undergoes homolysis to afford **25**. Hydrogen abstraction followed by iodohydrin formation **26** and ring closure via an S_N2 process produces **27**. In the presence of a second

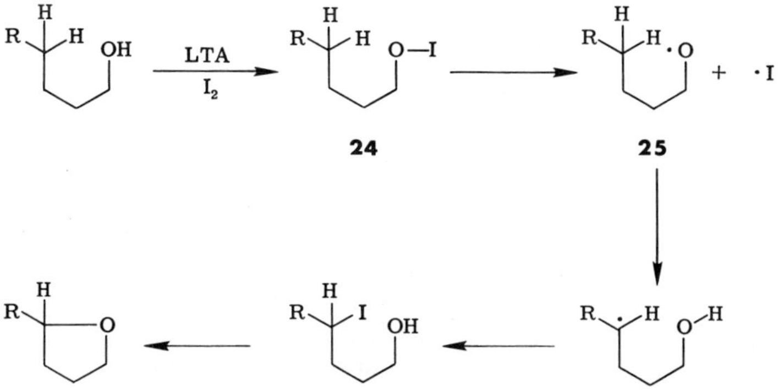

24 **25** **27** **26**

equivalent of LTA, the iodohydrin **26** can be diverted to **28** and thus to **29**. This reaction has been put to practical use as a lactone synthesis.[29] When

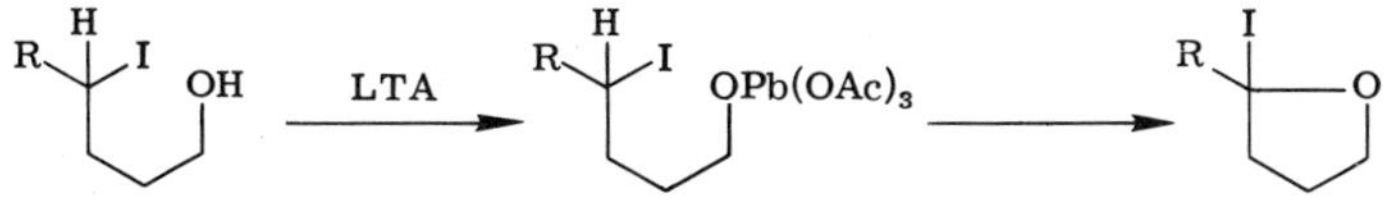

26 **28** **29**

access of iodine radical (or iodine) to the radical **30** is sterically blocked,

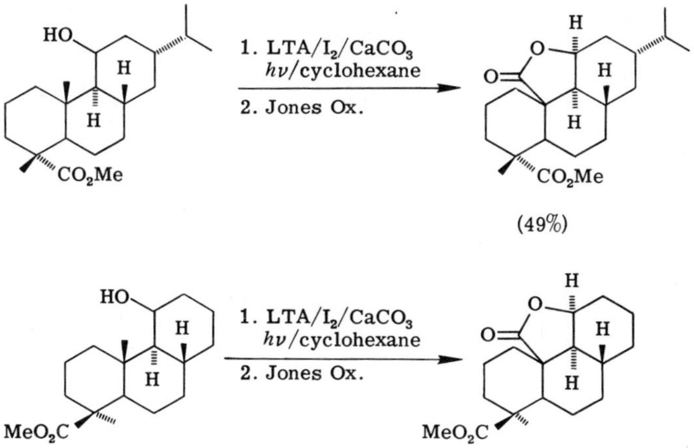

(49%)

(ref. 29)

(40%)

or S_N2 displacement from **31** is sterically or geometrically unfavored, the formation of **32** can predominate.[29,68] Subsequent transformations of the

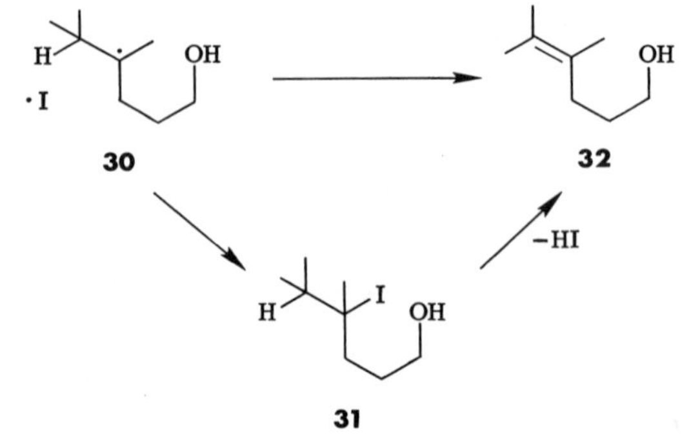

alkene **32** with the LTA/iodine reagent led to the products isolated from the reaction. The product spectrum obtained from the LTA/iodine treatment of 16-*ent*-17-kauranol is illustrative.[68]

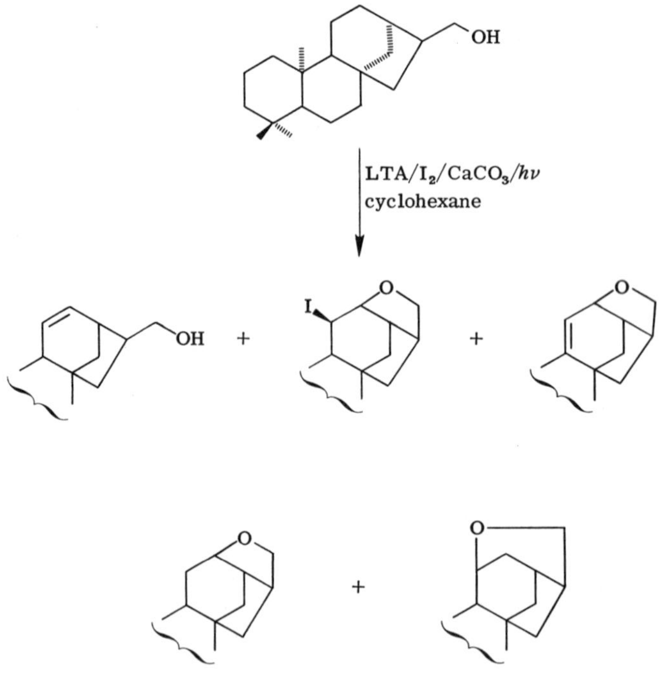

(ref. 68)

[68] A. J. McAlees and R. McCrindle, *Can. J. Chem.* **51**, 4103 (1973).
[69] C. Singh, J. Singh, and S. Dev, *Tetrahedron* **33**, 1759 (1977).
[70] W. H. W. Lunn, *J. Chem. Soc. C*, 2124 (1970).

Cyclic ether formation from straight-chain aliphatic alcohols has also been observed with the LTA/iodine reagent.[43,69] With **33** cyclization is

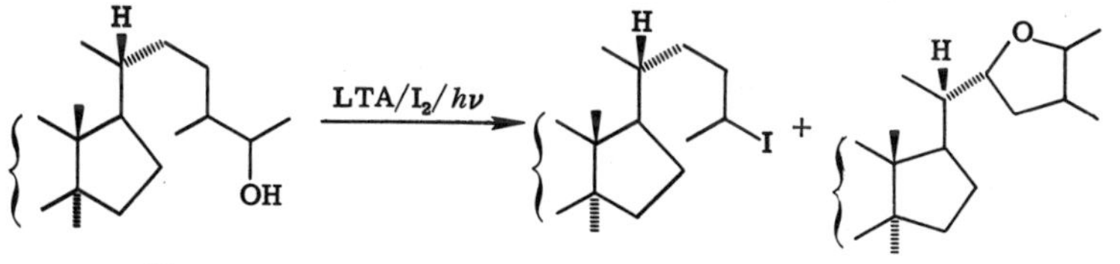

33 (40%) (ref. 69)

accompanied by fragmentation.[69] This latter process becomes the major pathway when cyclization is impossible. This reaction has been put to excellent synthetic use in the fragmentation of a number of adamantanoid systems.[70–72] Fragmentation was also noted with both the enmein-type

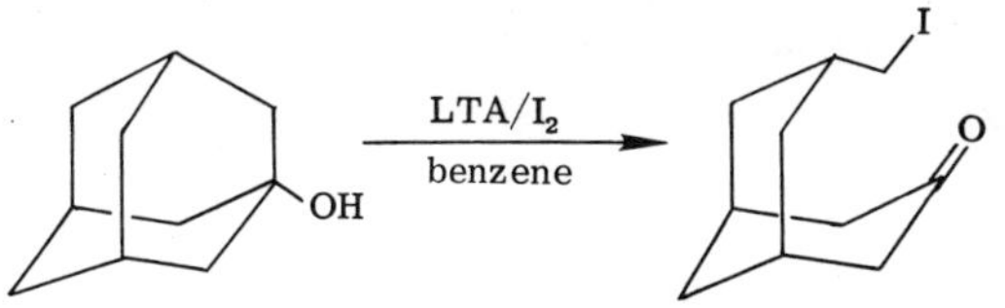

(45%) (ref. 70, 70a)

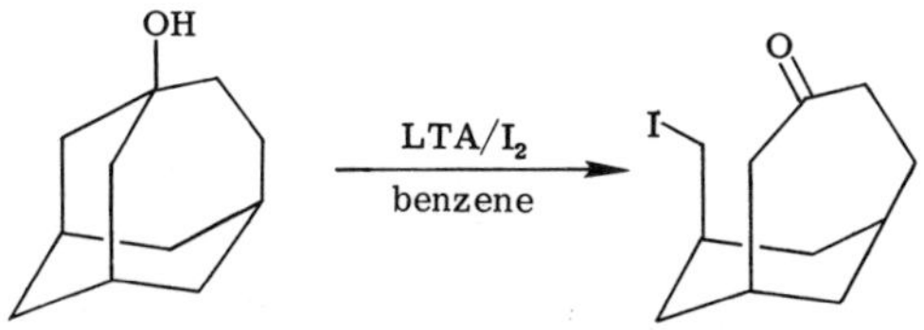

(?) (ref. 71)

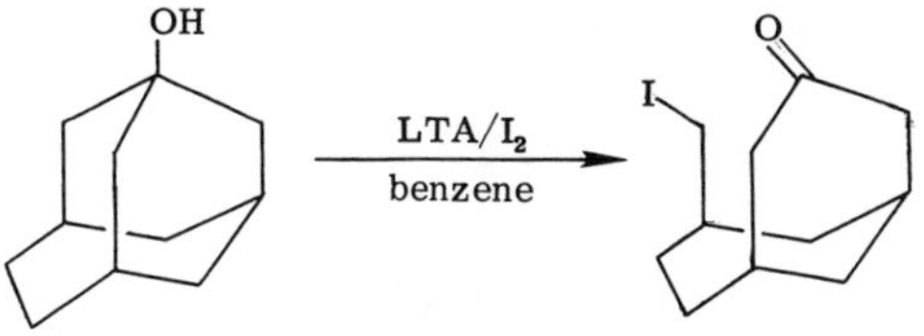

(?) (refs. 71a, 71b)

[70a] Z. Majerski and Z. Hameršak, *Org. Synth.* **59**, 147 (1979).
[71] Z. Hameršak, D. Škare, and Z. Majerski, *J. Chem. Soc. Chem. Commun.*, 478 (1977).
[71a] Z. Majerski, Z. Hameršak, and D. Škare, *Tetrahedron Lett.*, 3943 (1977).
[71b] Z. Majerski, S. Djigaš, and V. Vinković, *J. Org. Chem.* **44**, 4064 (1979).
[71c] Z. Majerski and J. Janjatovic, *Tetrahedron Lett.*, 3977 (1979).
[72] R. M. Black and G. B. Gill, *J. Chem. Soc. Chem. Commun.*, 172 (1971).

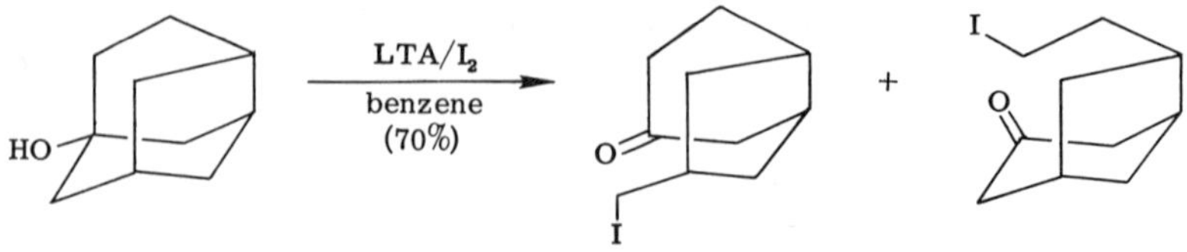

3 : 2

(ref. 71c)

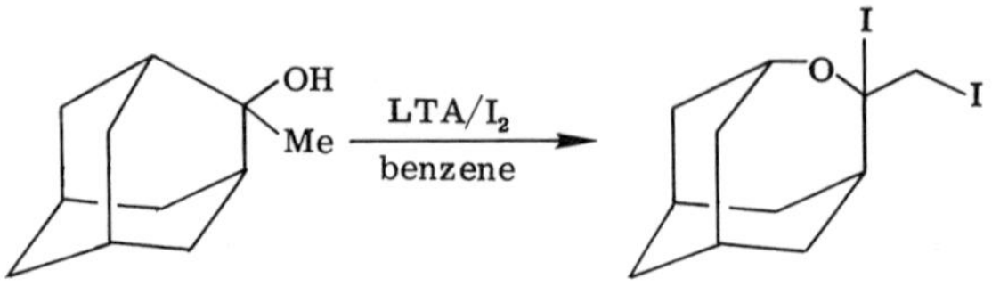

(32–52%)

(ref. 72)

compounds **34**[73] and the cage molecule **35**.[73a] In each instance, the respective iodo formates **36** and **37** were obtained. Finally, a combination of

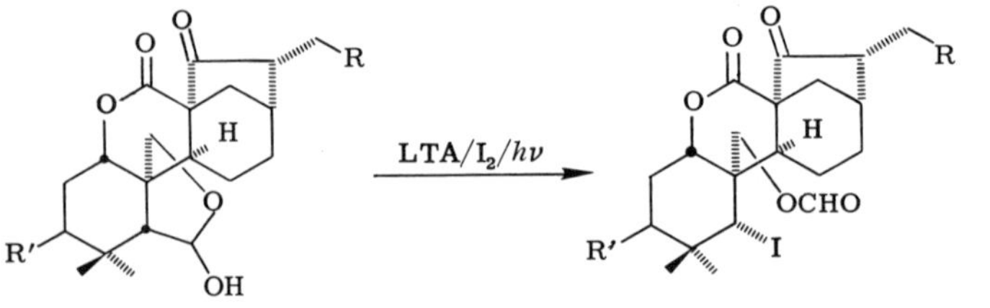

34 **36**

(ref. 73)

R	R′
H	H
OMe	OAc
H	OAc

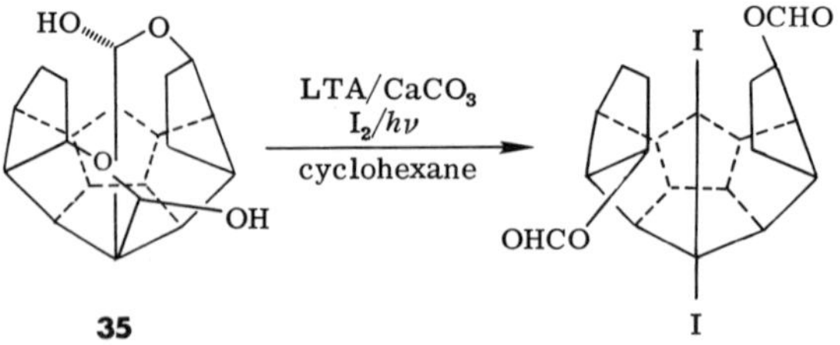

35 **37** (81%)

(ref. 73a)

[73] E. Fujita, I. Uchida, and T. Fujita, *Chem. Pharm. Bull.* **22**, 1656 (1974).

[73a] L. A. Paquette, W. J. Begley, D. Balogh, M. J. Wyvratt, and D. Bremner, *J. Org. Chem.* **44**, 3630 (1979).

fragmentation, oxidation, and cyclization was observed with **38**.[29]

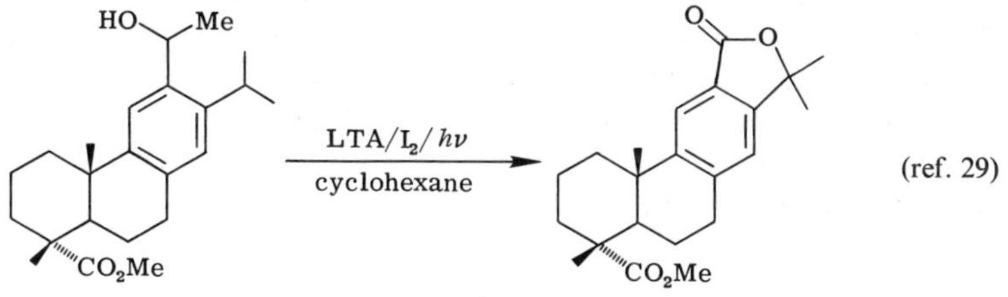

38 (55%)

B. 1,2-DIOLS AND RELATED SYSTEMS

The cleavage of 1,2-diols to afford carbonyl compounds is one of the most commonly used LTA reactions. The ability of LTA to cleave *trans*-diols and the compatibility of the reagent with aprotic as well as protic solvents recommend LTA when comparison is made with other standard 1,2-diol cleaving agents such as periodate. The mechanistic aspects of the reaction have been well studied and the results reviewed,[9,10a,74,74a] and the picture shown in Scheme 2 emerges. In cases where geometry is favorable, oxidation

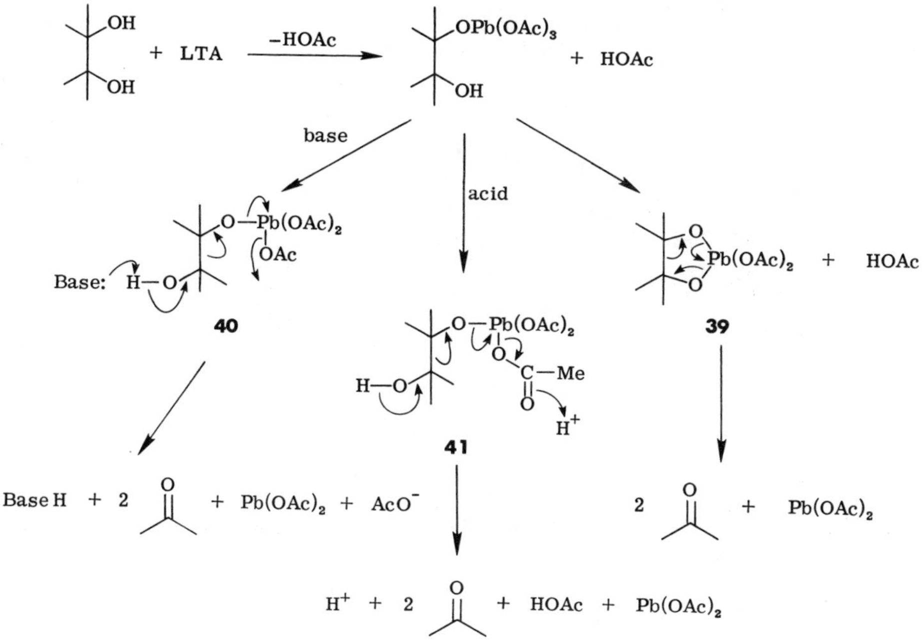

SCHEME 2

via **39** occurs in a two-electron process. With *trans*-diols this pathway is not feasible and trans-periplanar fragmentation becomes important. The role of both base and acid in enhancing the cleavage of *trans*-diols has been rationalized by involving transition states such as **40** and **41**.[10a]

A large body of work, beyond the scope of this chapter, exists concerning the use of glycol cleavage in structural determination and in the degradation of sugars. A number of reviews on the latter have appeared.[75-77] The following summary will outline some of the specific synthetic uses of LTA diol cleavage. For instance, the sequential oxidation of 1,2-diols followed by treatment of the ensuing dicarbonyl compound with base offers a general high yield method for the preparation of cyclic systems,[78-80] whereas the

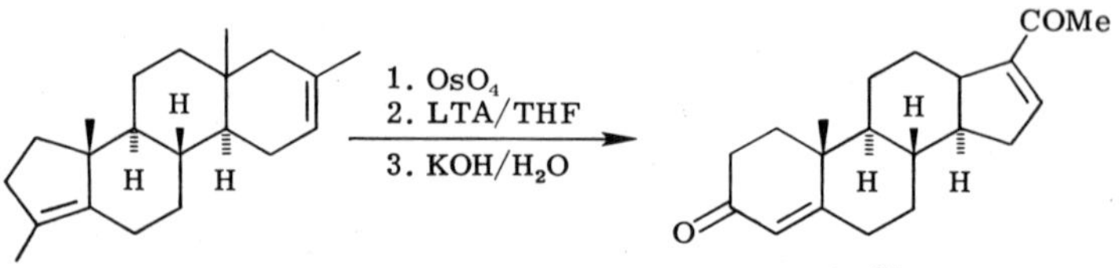

(29%) (ref. 78)

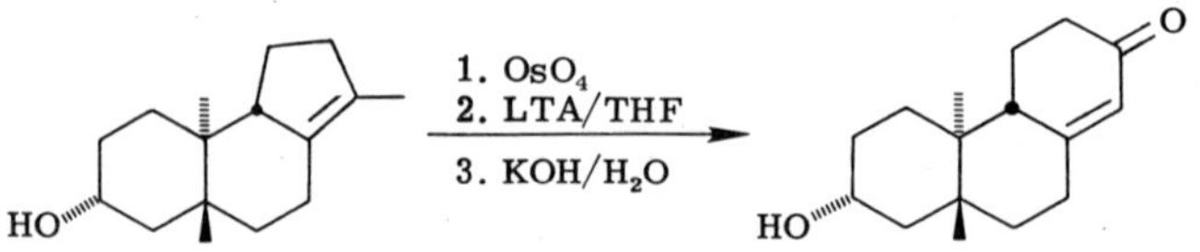

(63%) (ref. 79)

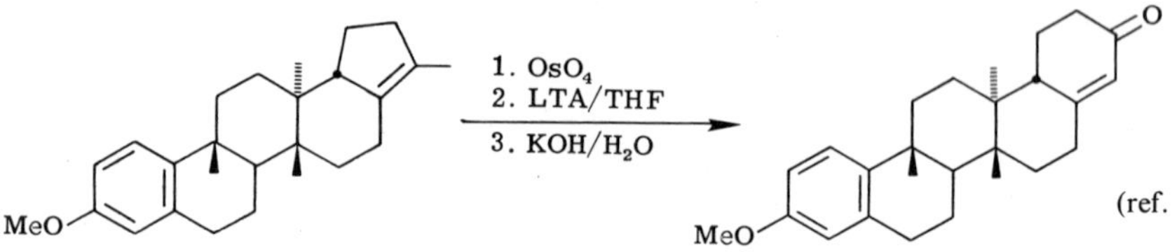

(58%) (ref. 79)

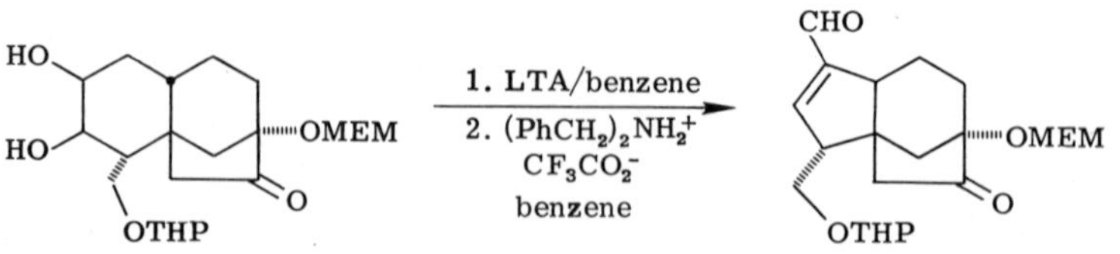

(64%) (ref. 80)

cleavage of bridged diols allows entry into medium or large rings as indicated by the following examples.[81–85]

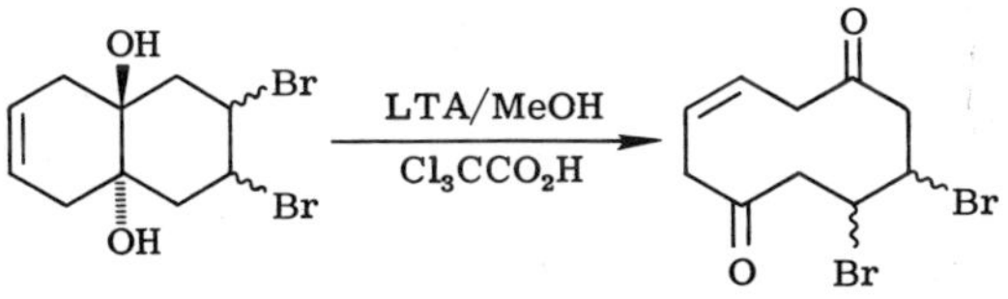

(55%)

(ref. 81)

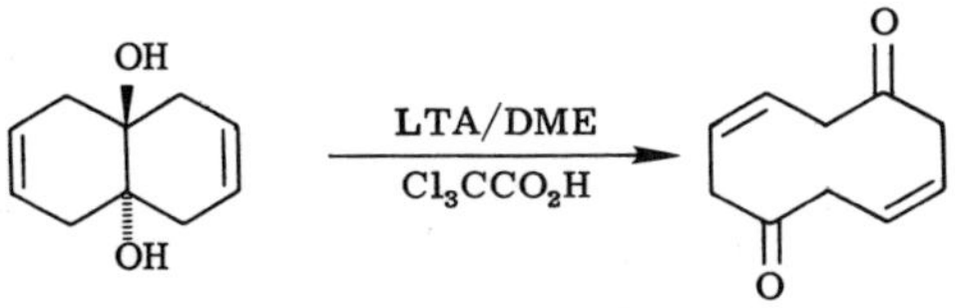

(60%)

(ref. 82)

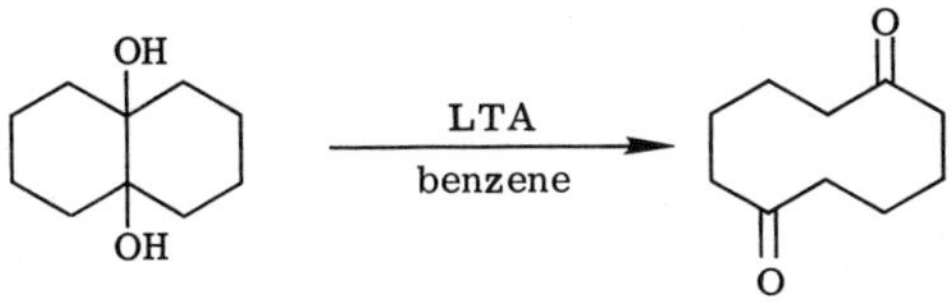

(59%)

(ref. 83)

---

[74] C. A. Bunton, *in* "Oxidation in Organic Chemistry" (K. B. Wiberg, ed.), Part A, p. 367. Academic Press, New York, 1965.

[74a] W. S. Trahanovsky, J. R. Gilmore, and P. C. Heaton, *J. Org. Chem.* **38**, 760 (1973).

[75] A. S. Perlin, *Adv. Carbohydr. Chem.* **14**, 9 (1959).

[76] C. T. Bishop, *Methods Carbohydr. Chem.* **6**, 350 (1972).

[77] P. S. O'Colla, *Methods Carbohydr. Chem.* **5**, 382 (1965).

[78] W. S. Johnson, M. F. Semmelhack, M. U. S. Sultanbawa, and L. A. Dolak, *J. Am. Chem. Soc.* **90**, 2994 (1968).

[79] R. E. Ireland, P. Bey, K. F. Cheng, R. J. Czarny, J. F. Moser, and R. I. Trust, *J. Org. Chem.* **40**, 1000 (1975).

[80] E. J. Corey, R. L. Danheiser, S. Chandrasekaran, P. Siret, G. E. Keck, and J.-L. Gras, *J. Am. Chem. Soc.* **100**, 8031 (1978).

[81] P. J. Mulligan and F. Sondheimer, *J. Am. Chem. Soc.* **89**, 7118 (1967).

[82] B. W. Roberts, J. J. Vollmer, and K. L. Servis, *J. Am. Chem. Soc.* **90**, 5264 (1968).

[83] I. J. Borowitz, A. Liberles, K. Megerle, and R. D. Rapp., *Tetrahedron* **30**, 4209 (1974).

[84] A. Tahara and T. Ohsawa, *Tetrahedron Lett.*, 2469 (1969).

[85] D. Termont, P. De Clereq, D. De Keukeleire, and M. Vandewalle, *Synthesis*, 46 (1977).

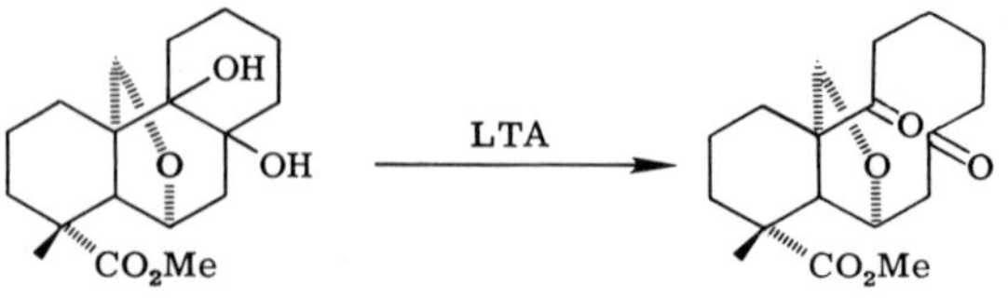

(?)

(ref. 84)

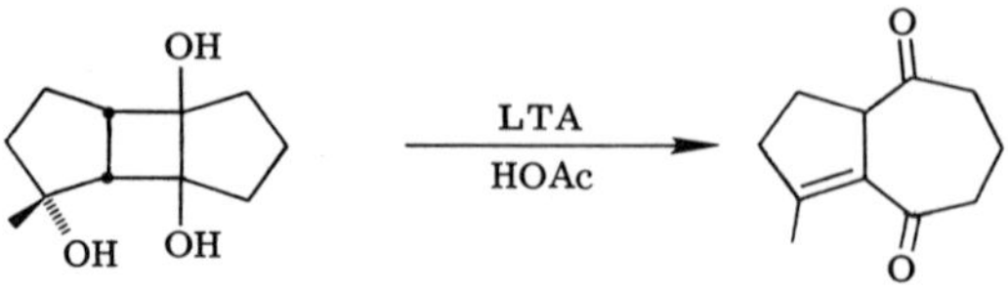

cis-syn-cis and
cis-anti-cis mixture

(74%)

(ref. 85)

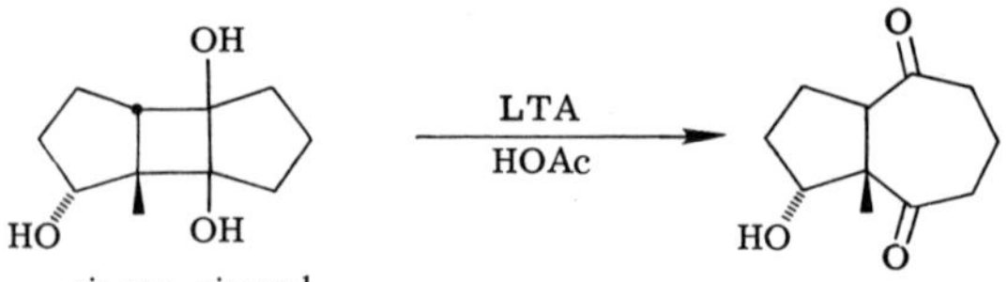

cis-syn-cis and
cis-anti-cis mixture

(83%)

(ref. 85)

The strategic placement of the 1,2-diol system in cyclic precursors also allows controlled placement of remote functional groups in the resulting cleavage products.[86–89] The examples presented show not only the feasibility of this approach but also indicate the wide range of substitution that

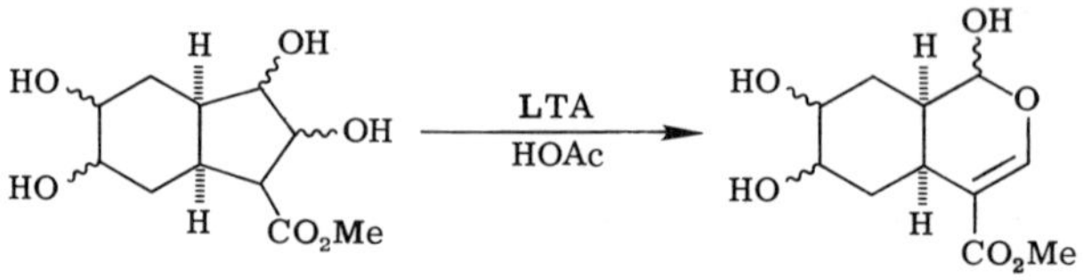

(45%, 3 isomers)

(ref. 86)

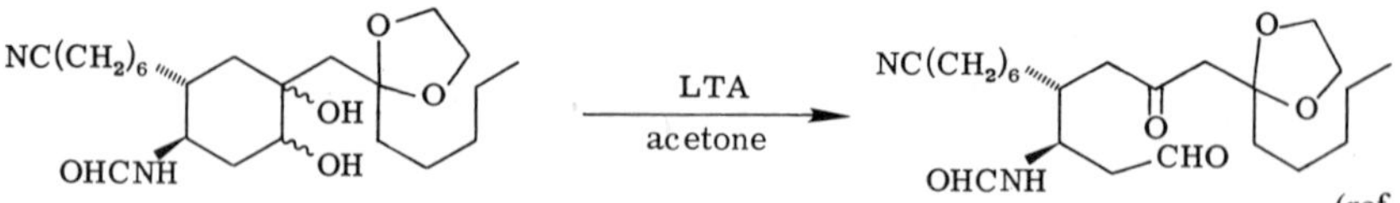

(?)

(ref. 87)

[86] G. Büchi, B. Gubler, R. S. Schneider, and J. Wild, *J. Am. Chem. Soc.* **89**, 2776 (1967).

[87] E. J. Corey, N. H. Andersen, R. M. Carlson, J. Paust, E. Vedejs, I. Vlattas, and R. E. K. Winter, *J. Am. Chem. Soc.* **90**, 3245 (1968).

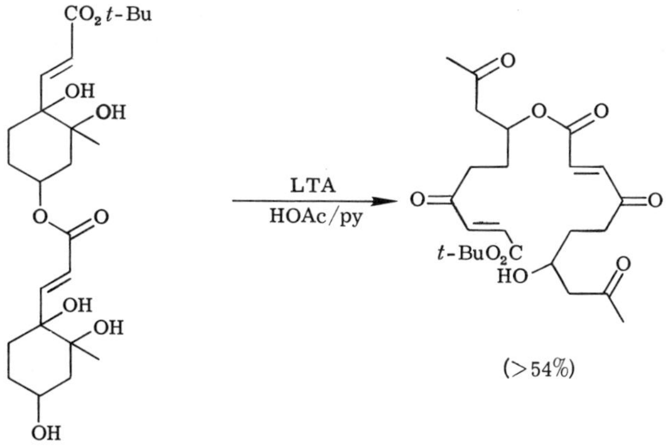

(ref. 88)

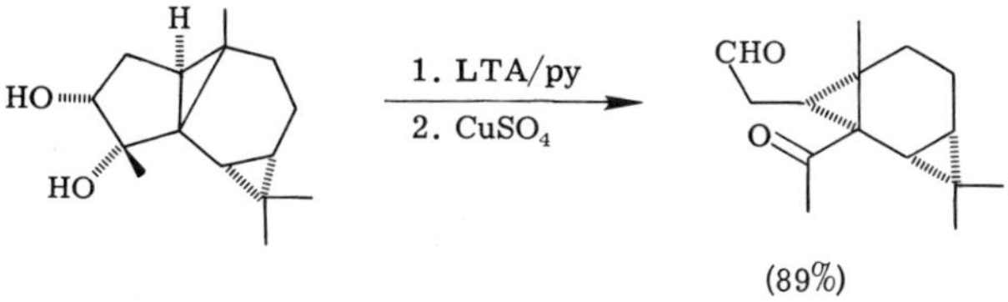

(ref. 89)

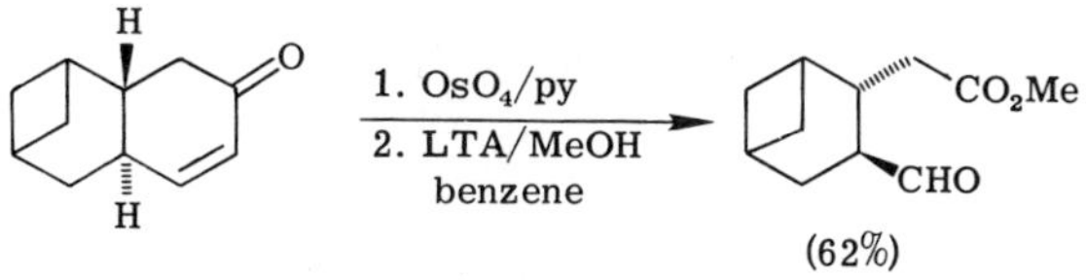

(ref. 92a)

is compatible with the cleavage reaction. This is especially noteworthy in cases where the molecule contains either a cyclopropyl[89-92] or cyclobutyl[92a] ring.

Oxidation of α-hydroxy carbonyl compounds is also an efficient process, and although reports exist wherein cleavage is affected in aprotic solvent or the solvent is not noted,[93-95] the general conditions for the transforma-

---

[88] Y. Fukuyama, C. L. Kirkemo, and J. D. White, *J. Am. Chem. Soc.* **99**, 646 (1977).

[89] N. Andersen, Y. Ohta, A. Moore, and C. W. Tseng, *Tetrahedron* **34**, 41 (1978).

[90] H. E. Zimmerman and P. S. Mariano, *J. Am. Chem. Soc.* **91**, 1718 (1969).

[91] H. E. Zimmerman, P. Hackett, D. F. Juers, J. M. McCall, and B. Schröder, *J. Am. Chem. Soc.* **93**, 3653 (1971).

[92] A. S. Narula and S. Dev, *Tetrahedron* **29**, 569 (1973).

[92a] S. Ohuchida, N. Hamanaka, and M. Hayashi, *Tetrahedron Lett.*, 3661 (1979).

[93] D. Helmlinger, P. de Mayo, M. Nye, L. Westfelt and R. B. Yeats, *Tetrahedron Lett.*, 349 (1970).

[94] J. Cadet and R. Teoule, *Tetrahedron Lett.*, 3229 (1972).

[95] F. M. Beringer, P. Ganis, G. Avitabile, and H. Jaffe, *J. Org. Chem.* **37**, 879 (1972).

tion include the use of protic media such as water or alcohol. These solvents apparently encourage hydrate or hemiketal formation and thus increase

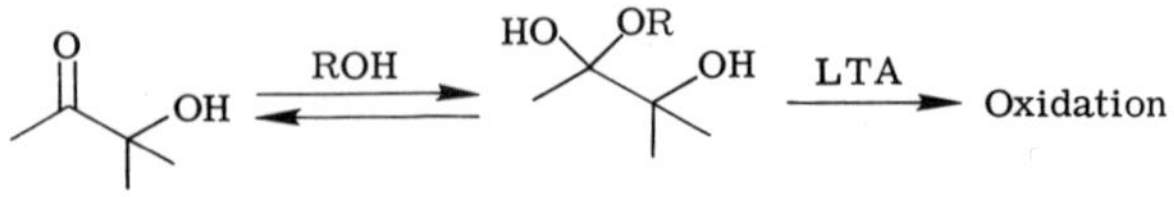

the rate of oxidation.[74] Table III gives some typical examples.[96–100] When

TABLE III

LTA CLEAVAGE OF α-HYDROXY KETONES

Substrate	Solvent	Product	% Yield	Reference
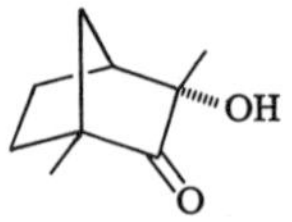	90% HOAc	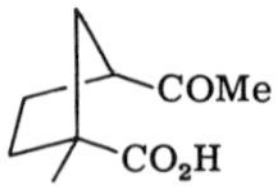	—	96
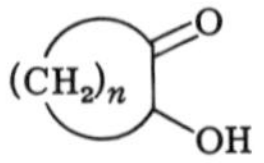	1. MeOH/LTA 2. MeOH/H^+	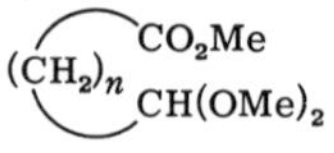	*n* 2 42 3 58	97
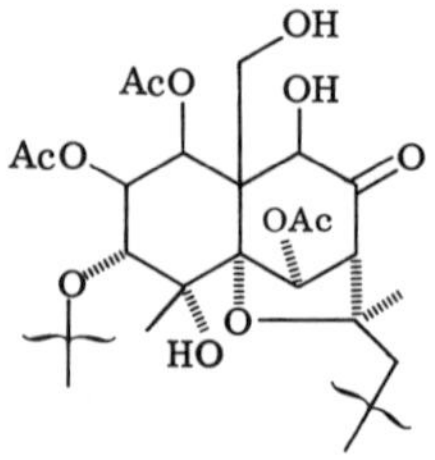	MeOH/HOAc	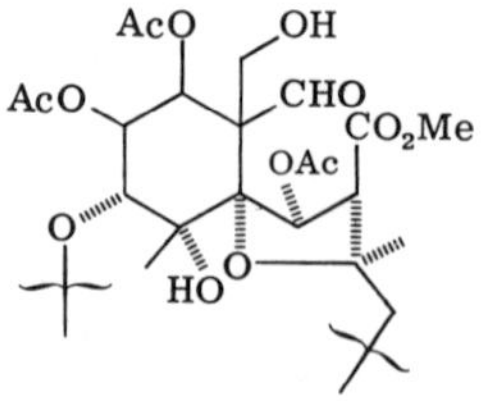	49	98
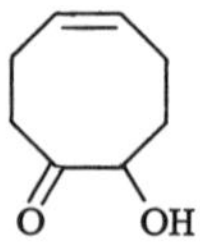	EtOH/benzene	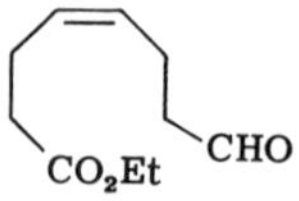	91	99
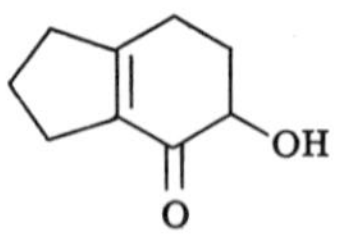	1. MeOH/LTA 2. MeOH/H^+	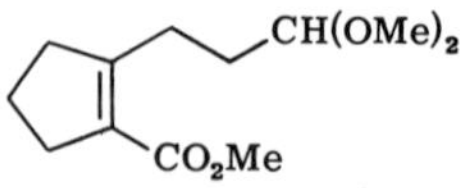	86	100

[96] R. L. Cargill, D. M. Pond, and S. O. LeGrand, *J. Org. Chem.* **35**, 359 (1970).
[97] J. R. Hazen, *J. Org. Chem.* **35**, 973 (1970).
[98] Y. Shizuri, H. Wada, K. Sugiura, K. Yamada, and Y. Hirata, *Tetrahedron* **29**, 1773 (1973).
[99] R. J. Anderson and C. A. Henrick, *J. Am. Chem. Soc.* **97**, 4327 (1975).
[100] A. Barco, S. Benetti, G. P. Pollini, P. G. Baraldi, M. Guarneri, and C. B. Vincentini, *Synth. Commun.* **7**, 13 (1977).

pyridine, benzene, or acetic acid is used as solvent, α-oxygenated carbonyl compounds can sometimes be isolated.[101–103] It is also possible to prepare

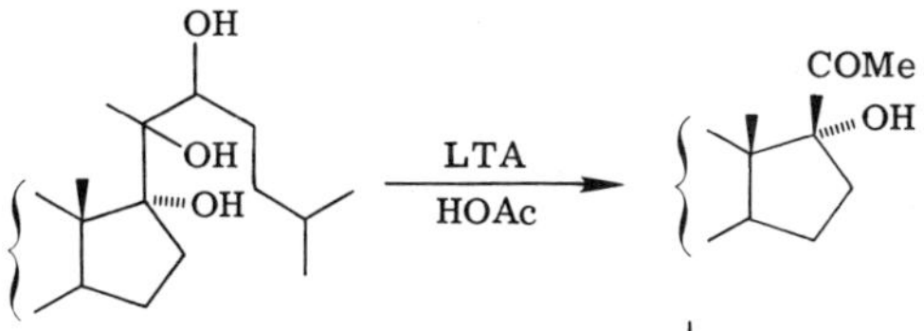

(ref. 101)

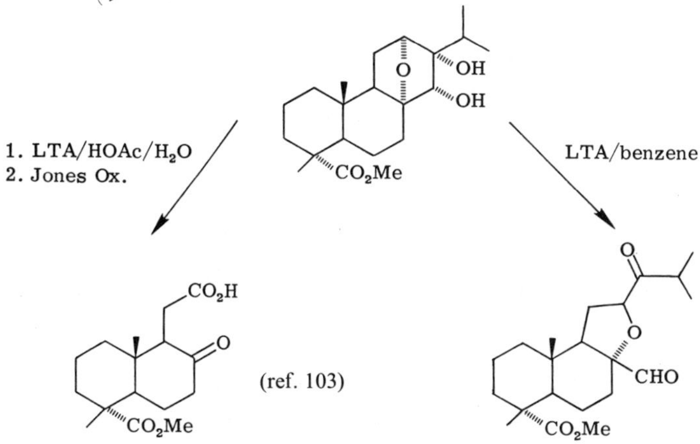

(ref. 103)

(ref. 102)

1,2-diones **42** by this modification.[104–106] The use of pyridine is crucial to the success of this particular reaction. For instance, when Z = S, oxidation

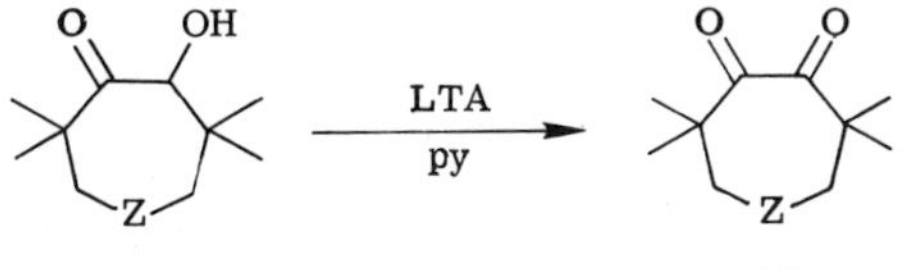

42

Z	% yield **42**	Reference
O	65	104
t-BuN	68	105
S	80	106

at sulfur occurs with benzene as solvent.[106] This reaction also has been used as a synthetic procedure.[107]

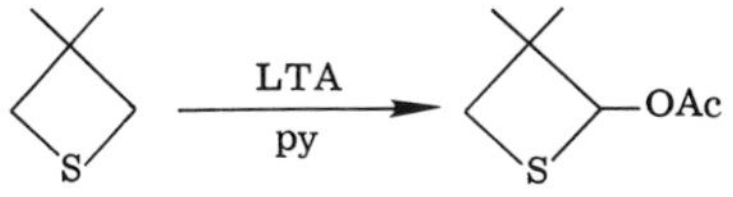

(30%)

(ref. 107)

[101] R. C. Nickolson and M. Gut, *J. Org. Chem.* **37**, 2119 (1972).
[102] W. Herz and R. C. Ligon, *J. Org. Chem.* **37**, 1400 (1972).
[103] S. W. Pelletier, K. N. Iyer, and C. W. J. Chang, *J. Org. Chem.* **35**, 3535 (1970).
[104] P. Y. Johnson, J. Zitsman, and C. E. Hatch, *J. Org. Chem.* **38**, 4087 (1973).
[105] P. Y. Johnson, I. Jacobs, and D. J. Kerkman, *J. Org. Chem.* **40**, 2710 (1975).

Extension of the oxidation process to α,β-dihydroxy ketones affords excellent yields of the corresponding bis-cleavage products.[108–111]

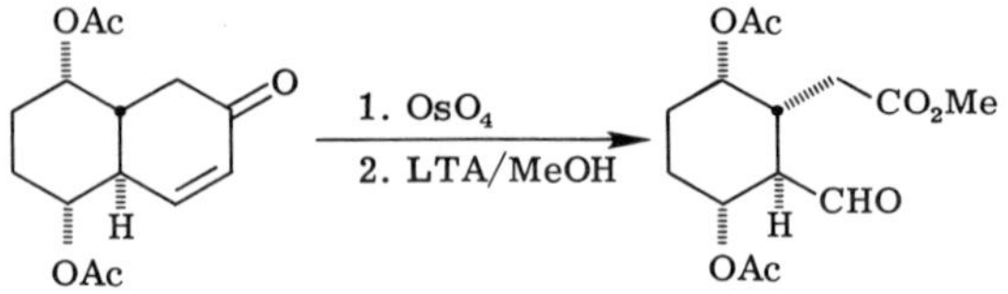

(ref. 108)

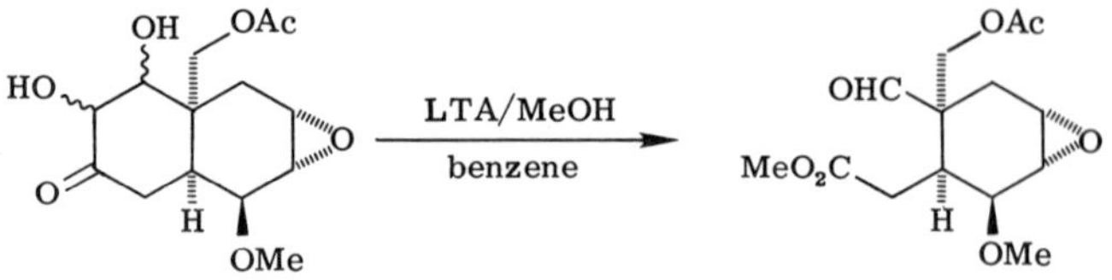

(ref. 109)

(100%)

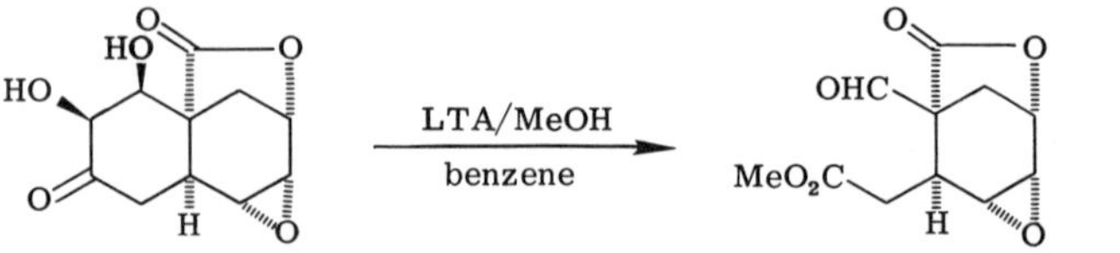

(ref. 110)

(86%)

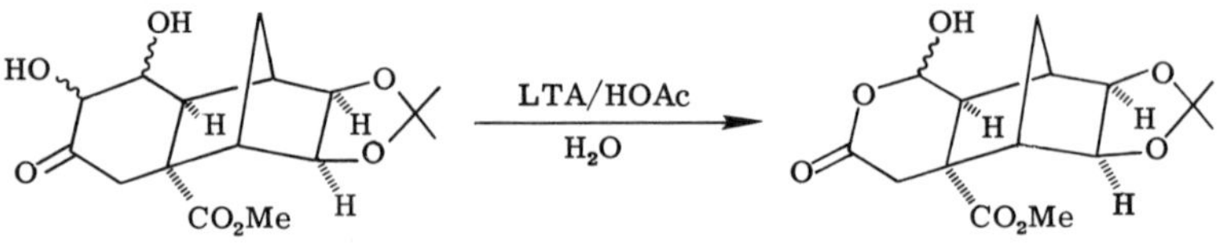

(ref. 111)

(100%)

---

[106] A. Krebs and H. Kimling, *Liebigs Ann. Chem.* **740**, 126 (1970).

[107] P. Y. Johnson and M. Berman, *J. Org. Chem.* **40**, 3046 (1975).

[108] N. S. Crossley, *Tetrahedron Lett.*, 3327 (1971).

[109] S. Danishefsky, P. Schuda, and K. Kato, *J. Org. Chem.* **41**, 1081 (1976).

[110] S. Danishefsky, P. F. Schuda, T. Kitahara, and S. J. Etheredge, *J. Am. Chem. Soc.* **99**, 6066 (1977).

[111] S. Danishefsky, M. Hirama, K. Gombatz, T. Harayama, E. Berman, and P. Schuda, *J. Am. Chem. Soc.* **100**, 6536 (1978).

The cleavage of 1,2-amino alcohols with LTA proceeds in a manner analogous to the 1,2-diol cleavage. Thus, both **43** and **44** afford moderate yields of the corresponding cleavage products.[112,113] The reaction of **45**

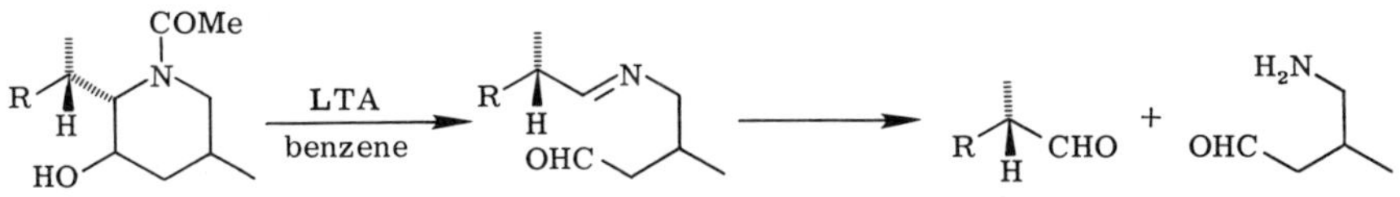

43

(ref. 112)

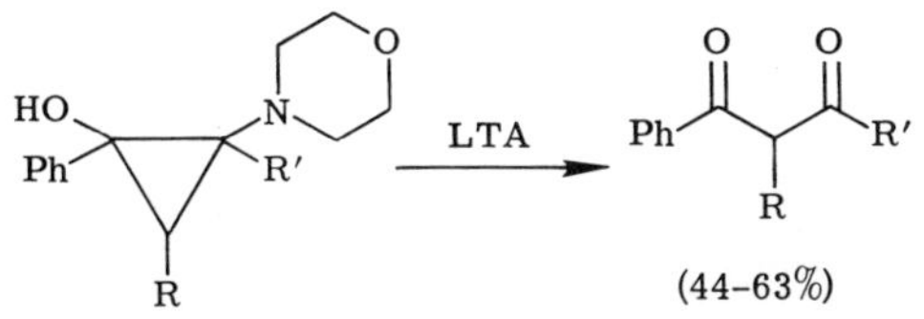

(ref. 113)

44

with LTA results in a 68% yield of **46**.[114] In this case, it is postulated that the initially formed immonium salt **47** undergoes further oxidation and re-

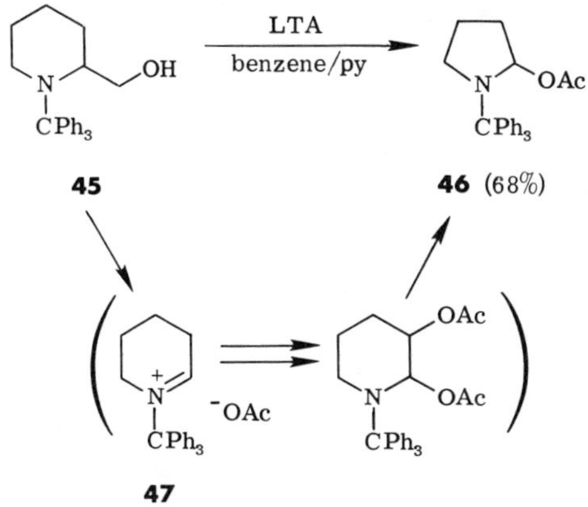

(ref. 114)

arrangement leading to **46**. A similar pathway could account for the appearance of **48**.[115]

[112] H. Suginome, H. Umeda, and T. Masamune, *Tetrahedron Lett.*, 4571 (1970).

[113] H. J. Roth, T. Schrauth, and M. H. El Raie, *Arch. Pharm.* (*Weinheim*) 307, 482 (1974).

[114] H. De Koning, A. Springer-Fidder, M. J. Moolenaar, and H. O. Huisman, *Recl. Trav. Chim. Pays-Bas* **92**, 237 (1973).

[115] M. Narisada, F. Watanabe, and W. Nagata, *Tetrahedron Lett.*, 3681 (1971).

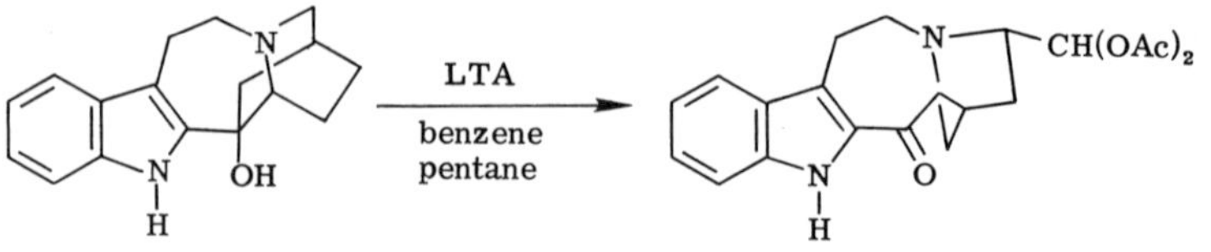

48

(ref. 115)

The oxidation of α-substituted α-amino ketones results in cleavage between the carbonyl and carbinamine functions. In the presence of an alcohol, an ester is formed from the acyl group and a nitrile from the carbinamine portion of the original substrate.[116]

$$\mathrm{ArCOCH(NH_2)R} \xrightarrow[\mathrm{CH_2Cl_2}]{\mathrm{LTA/R'OH}} \mathrm{ArCO_2R'} + \mathrm{RCN}$$

(ref. 116)

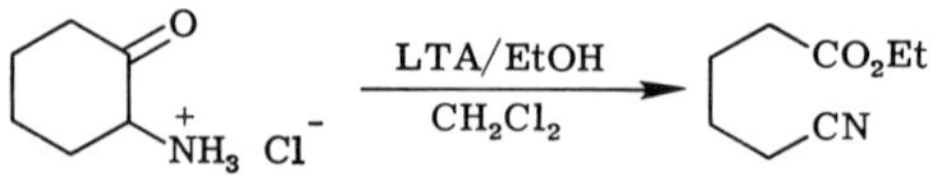

(25%)

A series of recent investigations indicates that the cleavage of β-hydroxy sulfides[117,118,118a] and β-hydroxydithianes[119,119a,120] has great synthetic

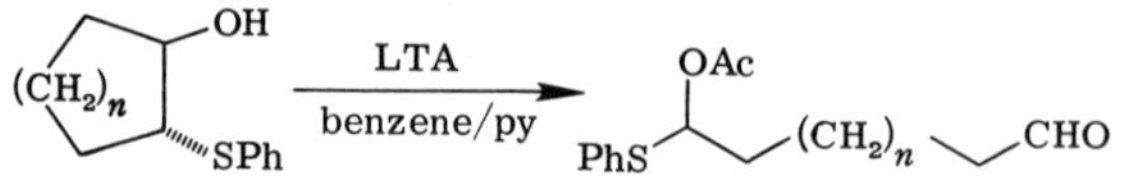

(ref. 118)

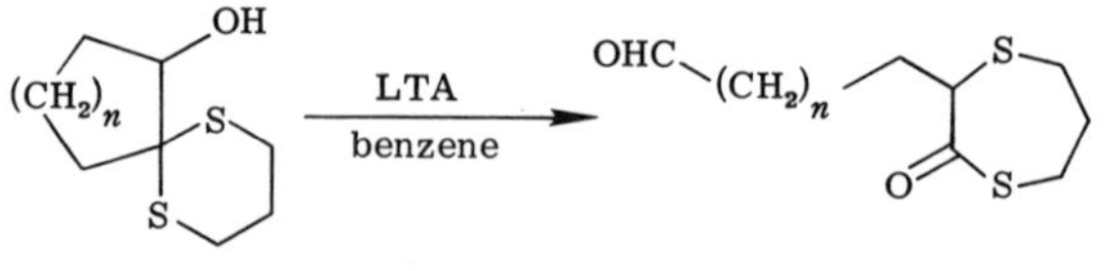

(ref. 119a)

potential. The mechanistic aspects of the former reaction have yet to be clearly defined, but it has been noted that a *trans* diaxial disposition of the hydroxy and thiophenoxy groupings is required.[118] The mechanism of the

[116] H. E. Baumgarten, D. F. McLaen, and H. W. Taylor, Jr., *J. Org. Chem.* **36**, 3668 (1971).
[117] B. M. Trost and K. Hiroi, *J. Am. Chem. Soc.* **97**, 6911 (1975).
[118] B. M. Trost, M. Ochiai, and P. G. McDougal, *J. Am. Chem. Soc.* **100**, 7103 (1978).
[118a] D. Caine and A. S. Frobese, *Tetrahedron Lett.*, 3107 (1977).

latter reaction is also complex.[119] Evidence has been obtained implicating **49** as a key intermediate in the ring expansion process.[119]

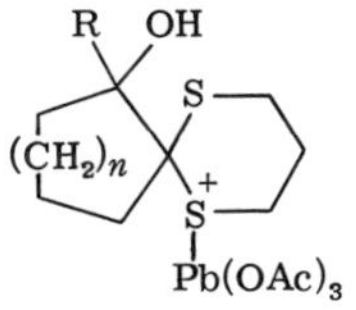

49

Treatment of the β-amidosulfide **50** with LTA yields a mixture containing two products resulting from the oxidation of sulfur and **51** a unique product in which migration of the thiomethyl group occurred.[121]

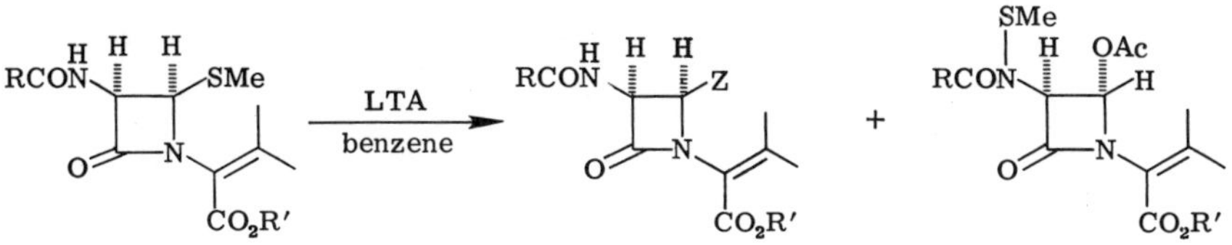

50 **51** (40%)

Z	% yield
SOMe	15
SCH_2OAc	35

(ref. 121)

C. ENOLS AND RELATED SYSTEMS

The reaction of enolizable carbonyl compounds with LTA has emerged as a standard method for α-acyloxylation. An extensive review of synthetic applications through 1971 exists[122] and discussions concerning the mechanisms of the transformation are also available.[10a,12,52,122] Evidence points to rate-determining enolization,[12,123] with a fast second step involving either radical or ionic pathways,[10a,12,124] most workers favoring the latter

---

[119] B. M. Trost, K. Hiroi, and L. N. Jungheim, *J. Org. Chem.* **45**, 1839 (1980).

[119a] B. M. Trost and K. Hiroi, *J. Am. Chem. Soc.* **98**, 4313 (1976).

[120] W. Lottenbach and W. Graf, *Helv. Chim. Acta* **61**, 3087 (1978).

[121] E. G. Brain, A. J. Eglington, J. H. C. Nayler, M. J. Pearson, and R. Southgate, *J. Chem. Soc. Chem. Commun.*, 229 (1972).

[122] D. J. Rawlinson and G. Sosnovsky, *Synthesis*, 567 (1973).

[123] P. S. Radhakrishnamurti and S. H. Pati, *Indian J. Chem. Sect. A* **16**, 319 (1978).

[124] R. O. C. Norman and D. R. Harvey, *J. Chem. Soc.*, 4860 (1964).

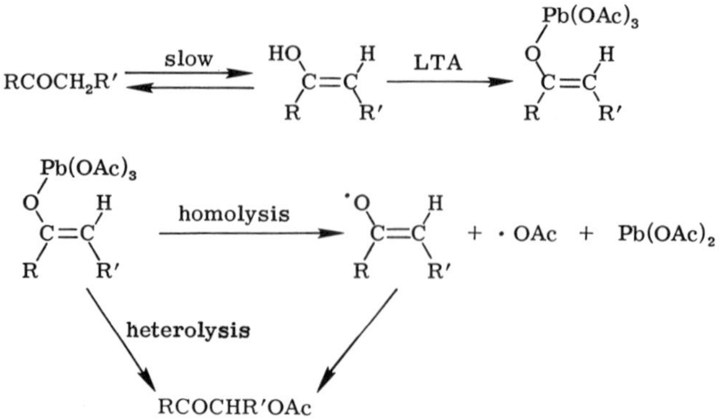

route. Intramolecular acetoxy transfer has also been suggested as a synchronous mechanism.[10a] Boron trifluoride has been used successfully to

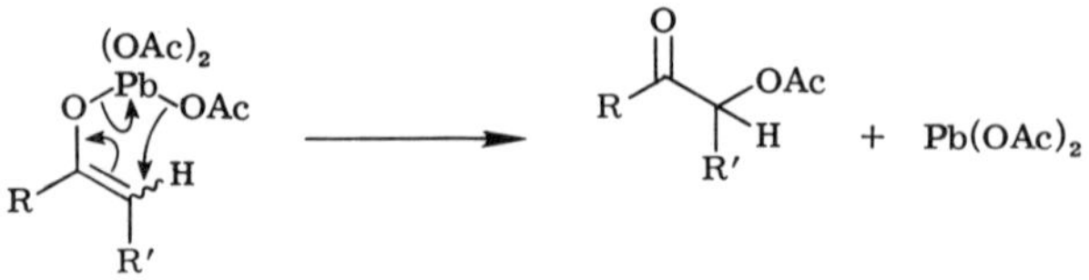

facilitate the reaction, causing more rapid enol formation.[12,122] Enolates have also been found to undergo acetoxylation with LTA[125] but this process may not be general.[126]

An indication of the synthetic use of α-acetoxylation with LTA is provided in Table IV.[127–139a] Reference 122 may also be consulted for a comprehensive listing. The sequential ring B, and ring A aromatization of **52** represents

[125] J. W. Ellis, *J. Chem. Soc. Chem. Commun.*, 406 (1970).

[126] R. K. Boeckman, Jr. and M. Ramaiah, *J. Org. Chem.* **42**, 1581 (1977).

[127] D. Gorenstein and F. H. Westheimer, *J. Am. Chem. Soc.* **92**, 634 (1970).

[128] T. A. Spencer, R. A. Ariel, D. S. Rouse, and W. P. Dunlap, Jr., *J. Org. Chem.* **37**, 2349 (1972).

[129] J. B. Press and H. Shechter, *J. Org. Chem.* **40**, 2446 (1975).

[130] M. Lj. Mahailović, J. Foršek, and Lj. Lorenc, *Tetrahedron* **33**, 235 (1977).

[130a] V. S. Kamat, G. K. Trivedi, and S. C. Bhattacharyya, *Indian J. Chem. Sect. B* **16**, 184 (1978).

[131] Y. Fukuyama, T. Tokoroyama, and T. Kubota, *Tetrahedron Lett.*, 4869 (1973).

[132] G. A. Russell, R. L. Blankespoor, K. D. Trahanovsky, C. S. C. Chung, P. R. Wittle, J. Mattox, C. L. Myers, R. Penny, T. Ku, Y. Kosugi, and R. S. Givens, *J. Am. Chem. Soc.* **97**, 1906 (1975).

[133] D. Caine, A. A. Boucugnani, S. T. Chao, J. B. Dawson, and P. F. Ingwalson, *J. Org. Chem.* **41**, 1539 (1976).

[133a] K. Shimada, T. Nambara, I. Uchida, and S. M. Kupchan, *Heterocycles* **12**, 1445 (1979).

[134] G. R. Pettit, C. L. Herald, and J. P. Yardley, *J. Org. Chem.* **35**, 1389 (1970).

[135] T. Ibuka, K. Tanaka, and Y. Inubushi, *Tetrahedron Lett.*, 4811 (1970).

[136] D. N. Kirk and M. S. Rajagopalan, *J. Chem. Soc. Perkin Trans. 1*, 1860 (1975).

TABLE IV

LTA ACETOXYLATION OF ENOLIZABLE CARBONYL COQPOUNDS

Substrate	Solvent	Product	% Yield	Reference
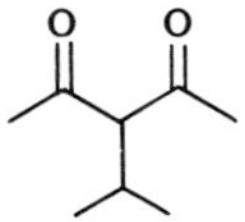	Benzene	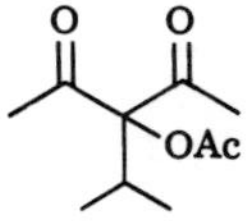	60 30	127 128
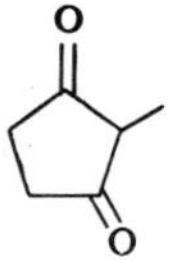	Benzene/HOAc Ac_2O	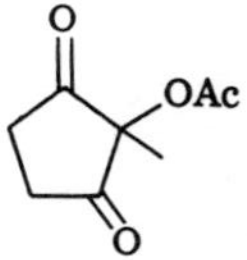	3	128
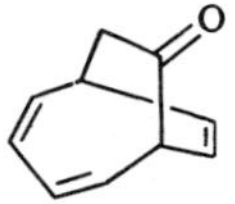	HOAc	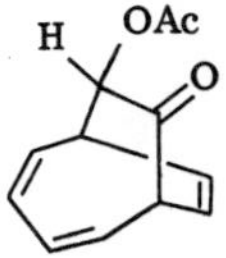	100	129
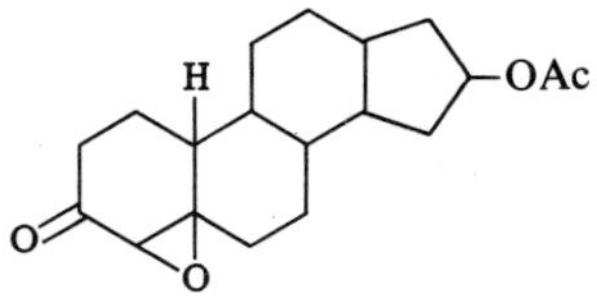	HOAc/Ac_2O	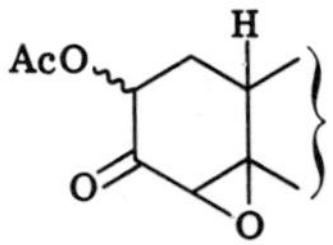	100	130
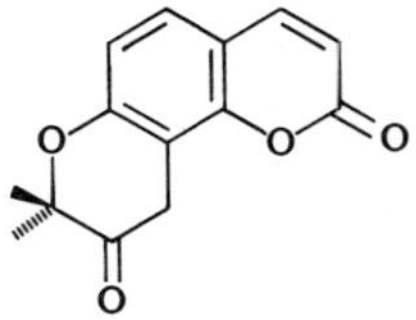	HOAc	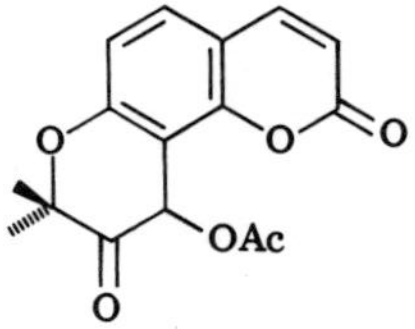	65	130a

(*cont.*)

TABLE IV (*cont.*)

Substrate	Solvent	Product	% Yield	Reference
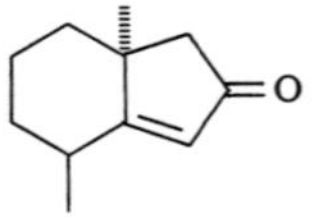	Benzene	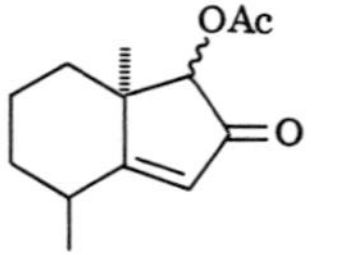	—	131
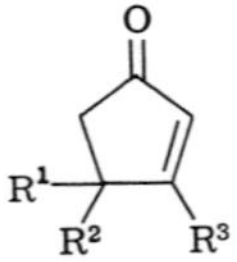	Benzene	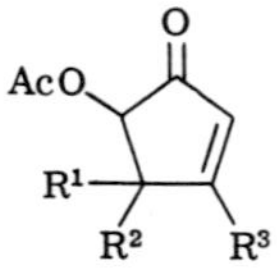	—	132
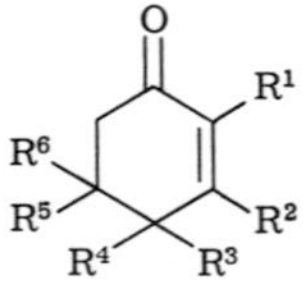	Benzene	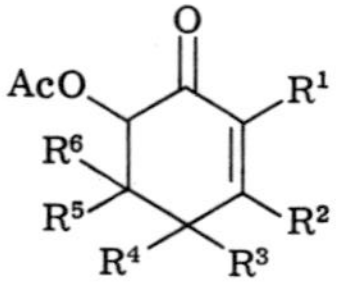	—	132
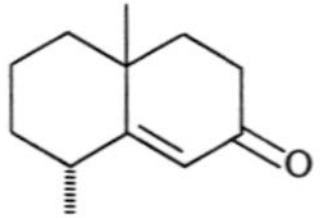	HOAc/Ac_2O	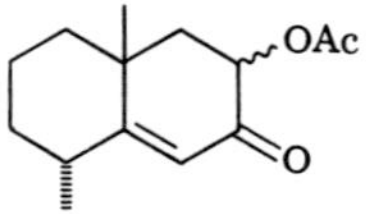	44	133
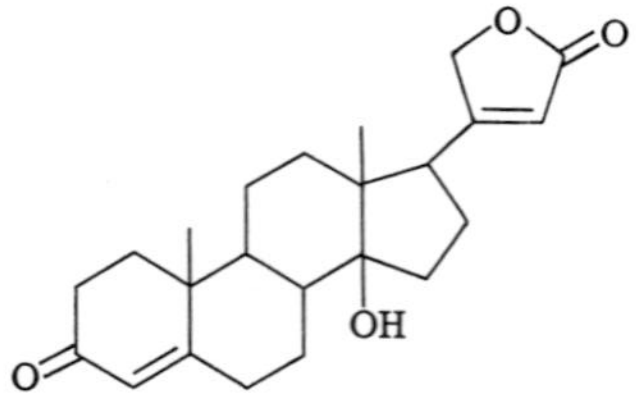	—	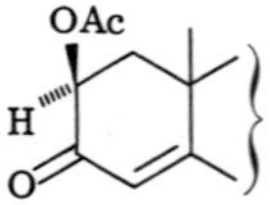	—	133a

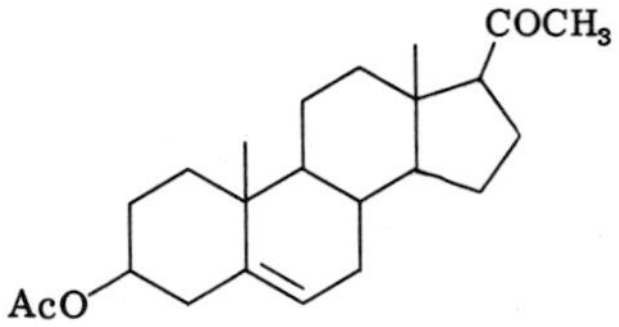	Benzene/MeOH $BF_3 \cdot Et_2O$	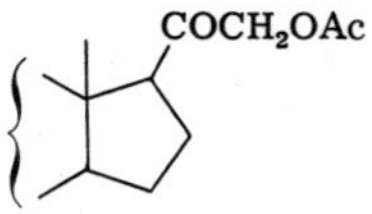	69	134
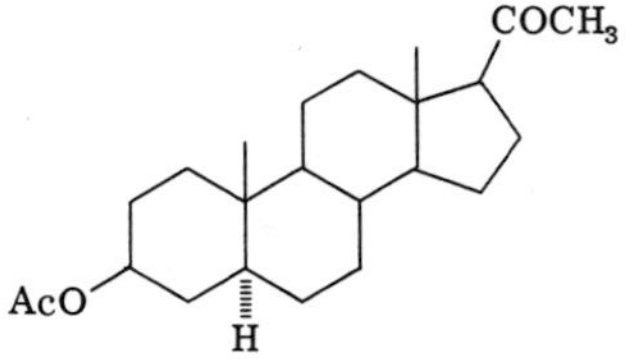	Benzene/MeOH $BF_3 \cdot Et_2O$	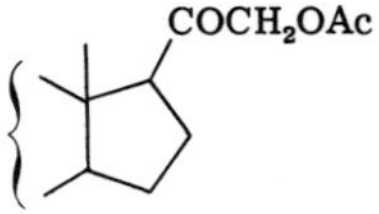	71	134
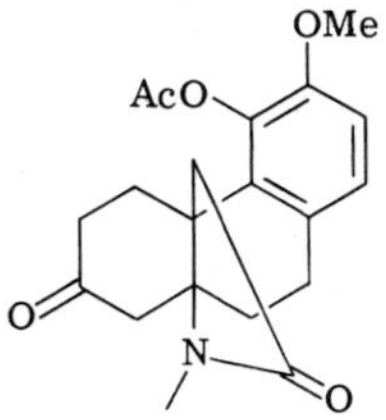	Benzene $BF_3 \cdot Et_2O$	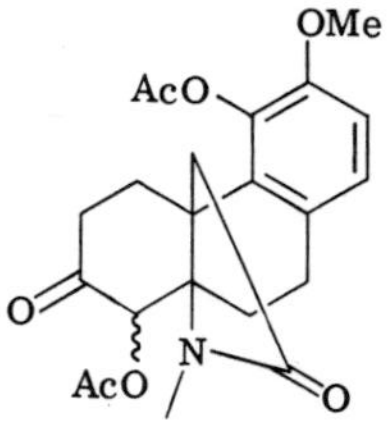	65	135
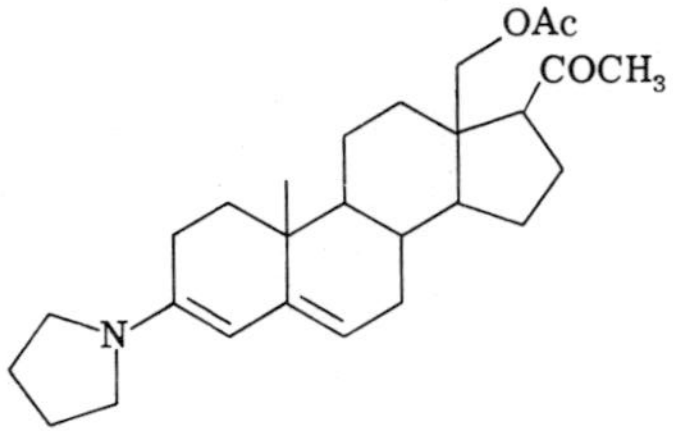	1. Benzene/MeOH $BF_3 \cdot Et_2O$ 2. $NaHCO_3$ H_2O/EtOH	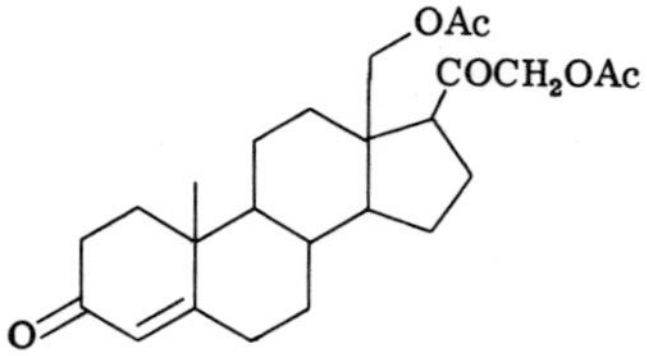	—	136

(*cont.*)

TABLE IV (*cont.*)

Substrate	Solvent	Product	% Yield	Reference
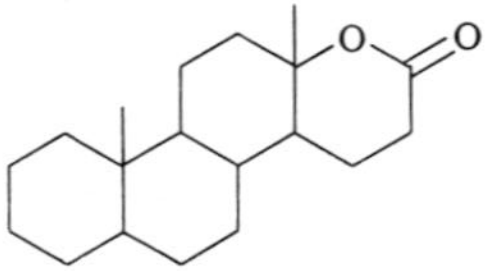	—	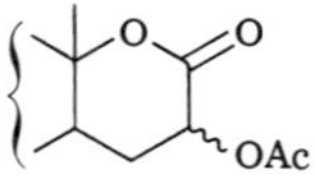	α-45–50 β-10–15	137
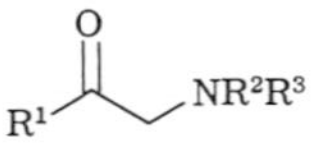 R^1 = Me, Ph $R^2 = R^3$ = Et or $NR^2 = R^3$ = *N*-piperidinyl, *N*-morpholino	—	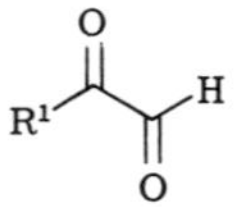	—	138
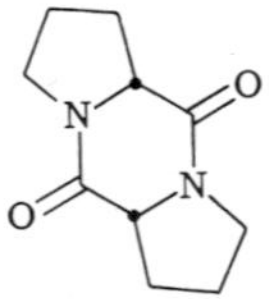	Benzene	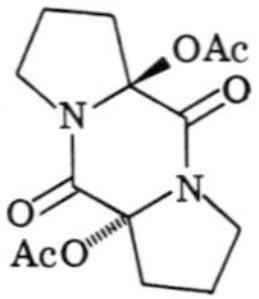	32	139
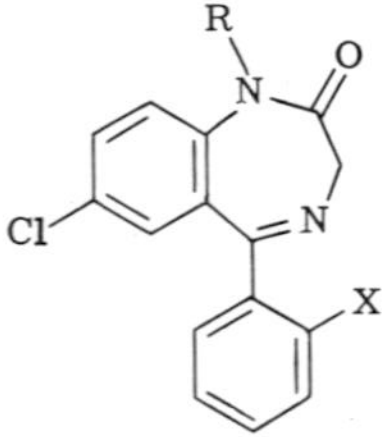	HOAc (I_2 also present)	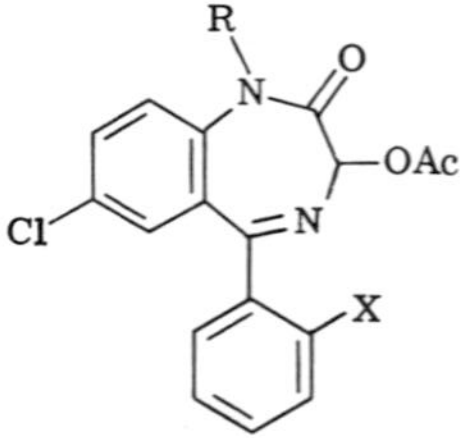	X R (%) H H >85 H Me >78 Cl H >85	139a

a new approach to this type of transformation and a novel use for LTA acetoxylation.[140]

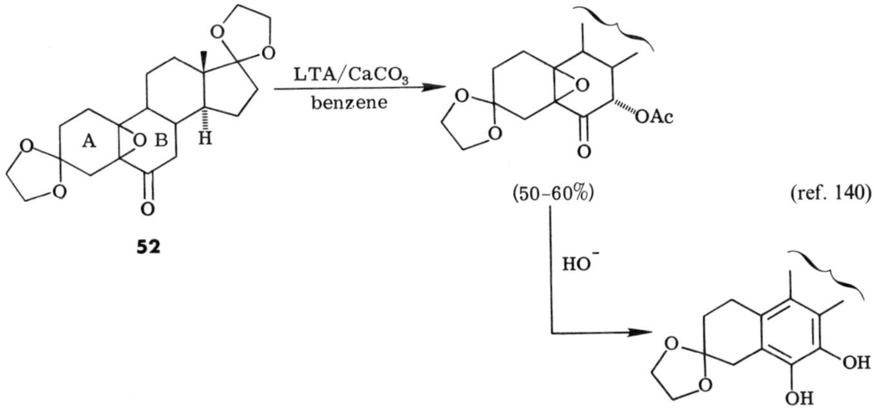

52

(50–60%) (ref. 140)

(80%)

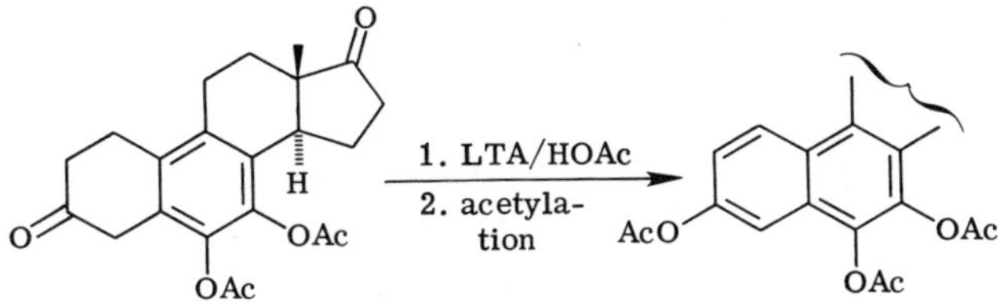

(50%)

Several drawbacks to the acetoxylation procedure have been reported. For instance, in isolated cases, rearrangement occurs during the reaction.[141,141a,142] A more common problem involves the fact that enolization under the reaction conditions employed is not regiospecific and hence,

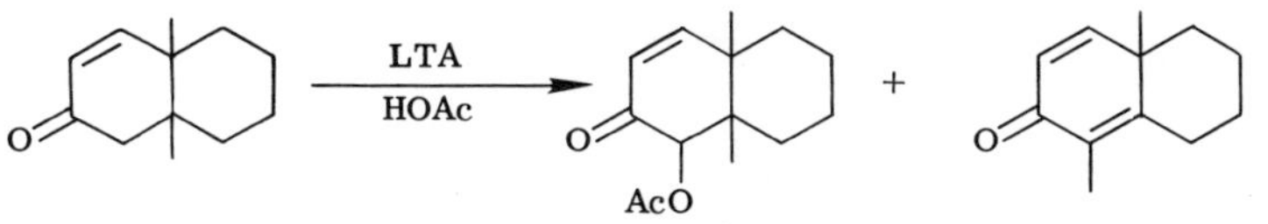

(ref. 141)

(?) (30%)

[137] M. Stefonic, Z. Djarmati, and M. Gasic, *Glas. Hem. Drus., Beograd.* **37**, 373 (1972).
[138] H. Moehrle, W. Haug, and E. Federolf, *Arch. Pharm.* (*Weinheim*) **306**, 44 (1973).
[139] E. Öhler, F. Tataruch, and U. Schmidt, *Chem. Ber.* **106**, 396 (1973).
[139a] T. Kovač, M. Oklobdžija, V. Šunjić, and F. Kajfež, *J. Heterocycl. Chem.* **16**, 1449 (1979).
[140] M. Lj. Mihailović, J. Foršek, and Lj. Lorenc, *J. Chem. Soc. Chem. Commun.*, 916 (1978).
[141] J. A. Marshall and G. L. Bundy, *J. Chem. Soc. Chem. Commun.*, 500 (1966).
[141a] H. Miura, Y. Fujimoto, and T. Tatsuno, *Synthesis*, 898 (1979).
[142] M. H. A. Elgamal, B. A. H. El-Tawil, and M. B. E. Fayez, *Tetrahedron* **30**, 4083 (1974).

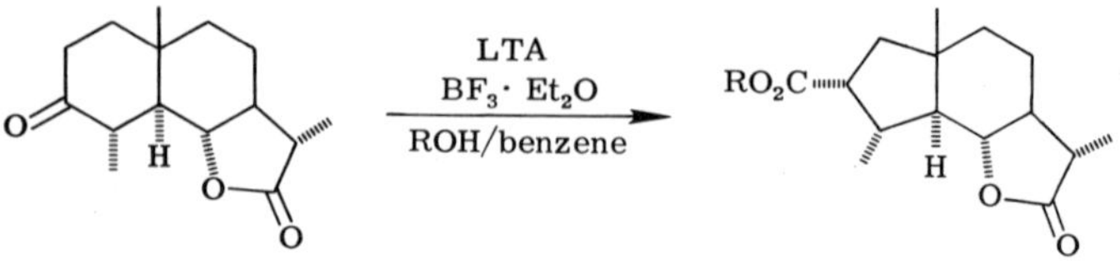

(ref. 141a)

(54–80%)

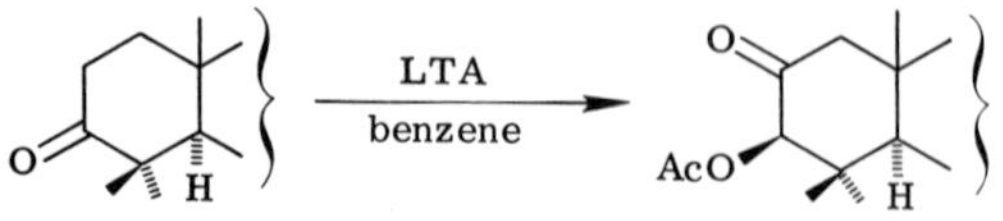

(ref. 142)

(20%)

regioisomers result[143,144] as well as the production of diacetates.[144] It was also discovered that rates of enolization do not correspond to product ratios in acyclic cases.[144] Several efforts have been made toward solving

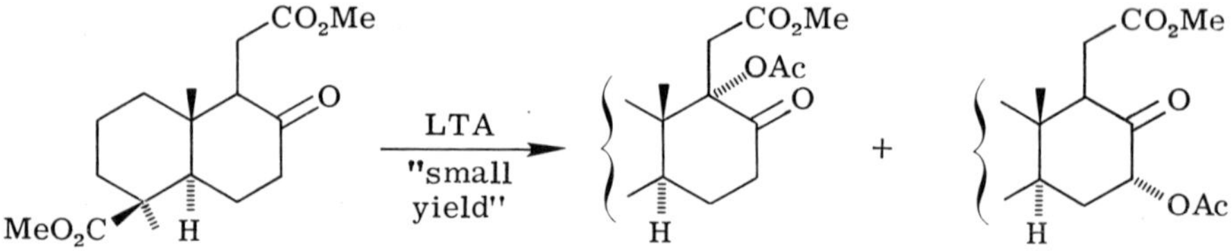

1 : 1 (ref. 143)

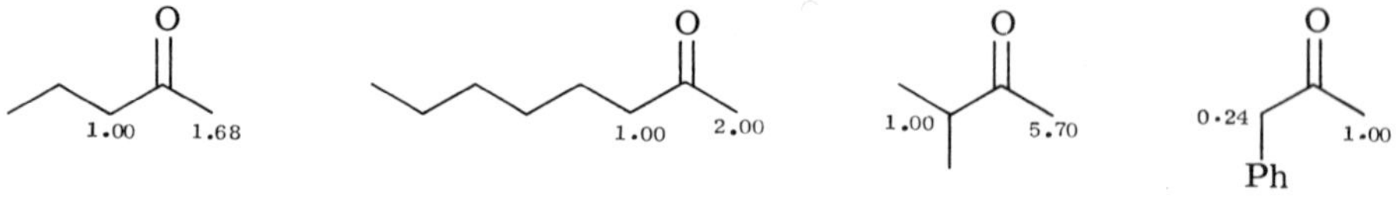

(ref. 144)

this problem. The use of enamines,[145–152] enol acetates,[144,153] and enol silyl ethers[154–156] appear to offer the best alternatives.

[143] G. A. Ellestad, R. H. Evans, Jr., and M. P. Kunstmann, *Tetrahedron Lett.*, 497 (1971).
[144] S. Moon and H. Bohm, *J. Org. Chem.* **37**, 4338 (1972).
[145] F. Corbani, B. Rindone, and C. Scolastico, *Tetrahedron* **31**, 455 (1975).
[146] F. Corbani, B. Rindone, and C. Scolastico, *Tetrahedron* **29**, 3253 (1973).
[147] F. Corbain, B. Rindone, and C. Scolastico, *Tetrahedron Lett.*, 2597 (1972).
[148] S. K. Khetan, *J. Chem. Soc. Chem. Commun.*, 917 (1972).
[149] P. L. Majumder, S. Joardar, and T. K. Chanda, *Tetrahedron* **34**, 3341 (1978).
[150] M. Ohno, T. F. Spande, and B. Witkop, *J. Am. Chem. Soc.* **92**, 343 (1970).
[150a] H. H. Eckhardt and H. Perst, *Tetrahedron Lett.*, 2125 (1979).

The oxidation of enamines with LTA results in products consistent with diacetate intermediates and, in systems capable of imine–enamine tautomerization, oxidation of the latter form predominates.[145–147] Although the wide

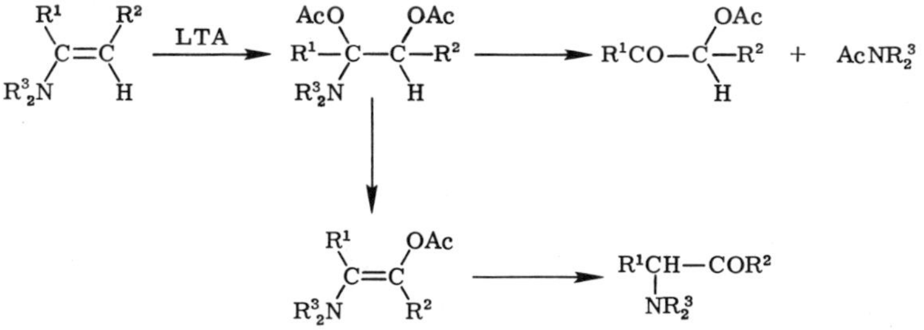

range of products obtained generally precludes the use of the LTA oxidation of enamines for synthetic purposes, useful applications have been reported for **53**,[148] **54**,[149] and for a series of enamides.[150–152]

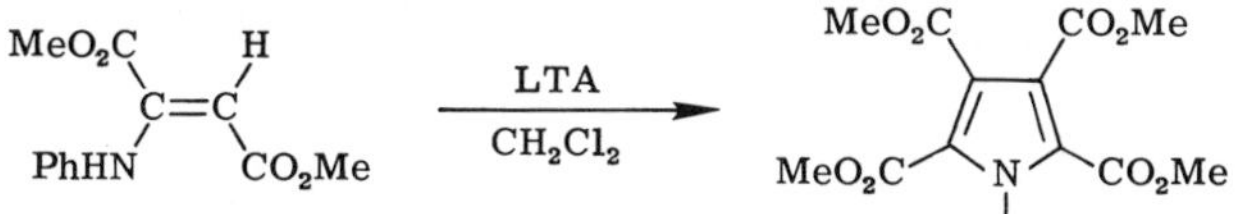

53

(ref. 148)

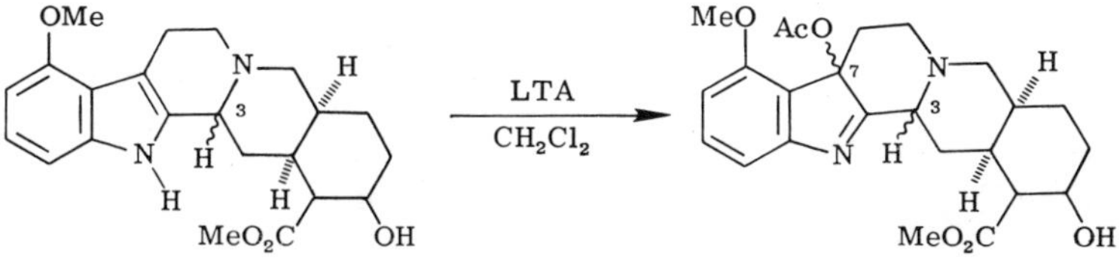

54

(ref. 149)

C-3 Hβ ⟶ C-3 Hβ, C-7 OAc β (24%)
C-3 Hα ⟶ C-3 Hα, C-7 OAc α (22%)

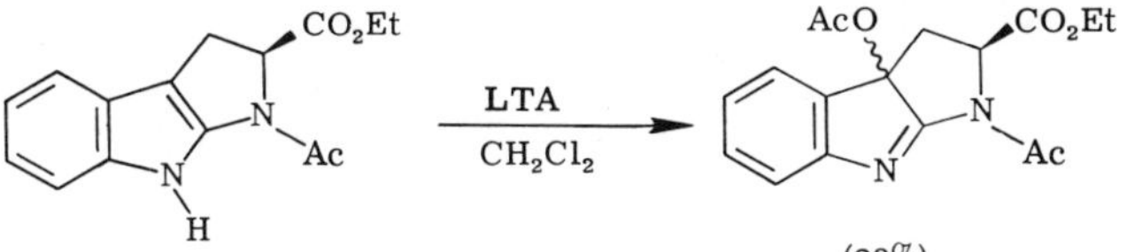

(23%)

(ref. 150)

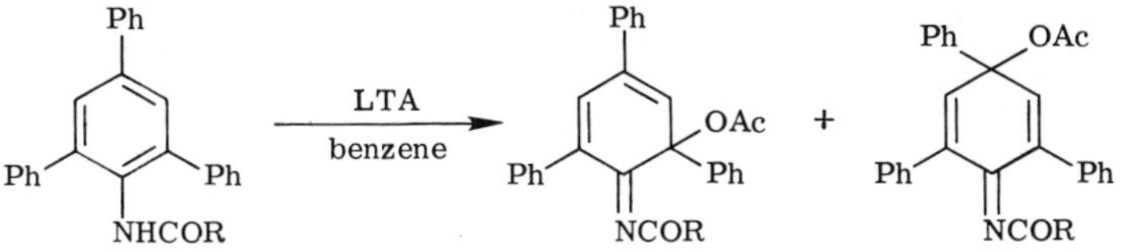

(57–68%) (10–15%)

(ref. 150a)

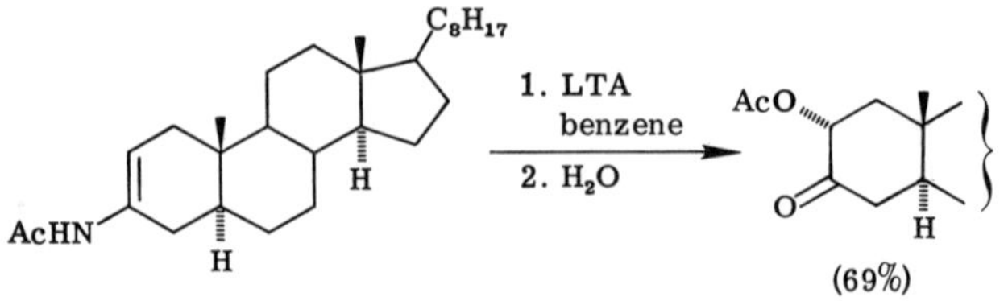

(69%)

(ref. 151)

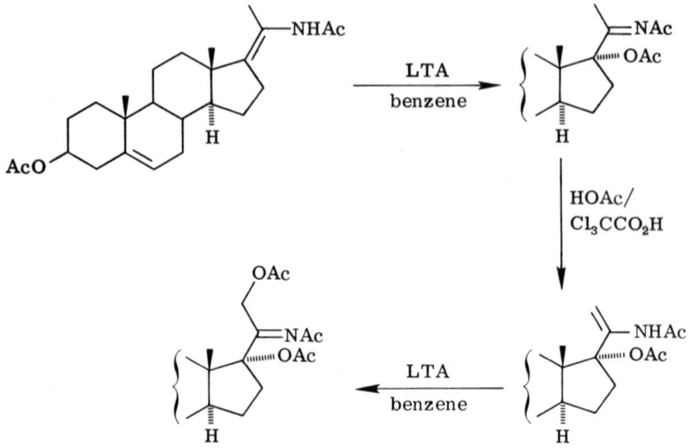

(ref. 152)

(84%) (100%)

Oxidation of enol acetates[144,153] and enol silyl ethers[154] gives high yields of the corresponding α-acetoxy derivatives. The several examples presented

(46%)

(ref. 144)

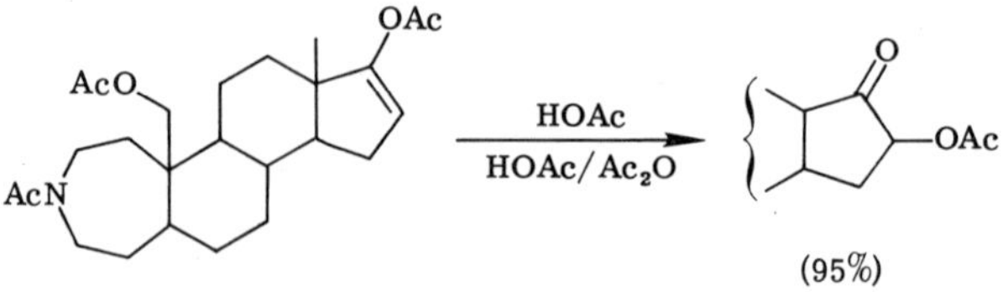

(95%)

(ref. 153)

---

[151] R. B. Boar, J. F. McGhie, M. Robinson, D. H. R. Barton, D. C. Horwell, and R. V. Stick, *J. Chem. Soc. Perkin Trans. 1*, 1237 (1975).

[152] R. B. Boar, J. F. McGhie, M. Robinson, and D. H. R. Barton, *J. Chem. Soc. Perkin Trans. 1*, 1242 (1975).

[153] K. Oka and S. Hara, *J. Am. Chem. Soc.* **99**, 3859 (1977).

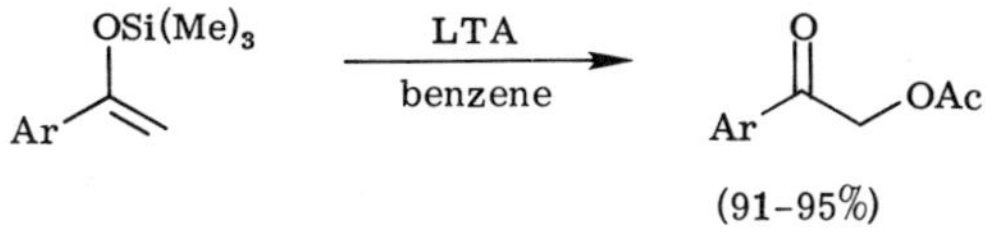

(ref. 154)

are illustrative. With enol silyl ethers the use of lead(IV) benzoate (LTB) is to be recommended.[155,156] In general, the reaction of both enol acetates and enol silyl ethers can be carried out at ambient temperatures, another

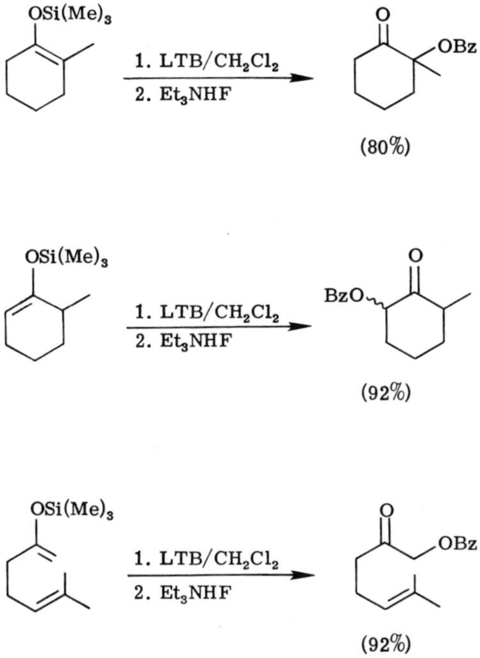

(ref. 155)

advantage of the method. Isolation of intermediates such as **55**[154] and **56**[35] point to a normal alkene addition process in the LTA reaction with enol

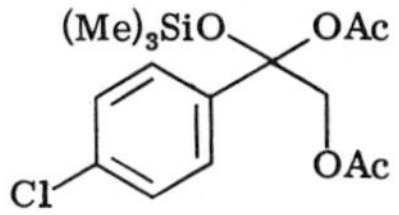

55

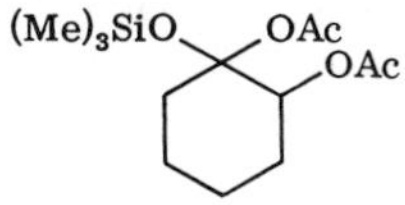

56

---

[154] G. M. Rubottom, J. M. Gruber, and K. Kincaid, *Synth. Commun.* **6**, 59 (1976).
[155] G. M. Rubottom, J. M. Gruber, and G. M. Mong, *J. Org. Chem.* **41**, 1673 (1976).
[156] G. M. Rubottom and J. M. Gruber, *J. Org. Chem.* **42**, 1051 (1977).

silyl ethers. The treatment of trimethylsiloxy-1,3-dienes with LTB is also a

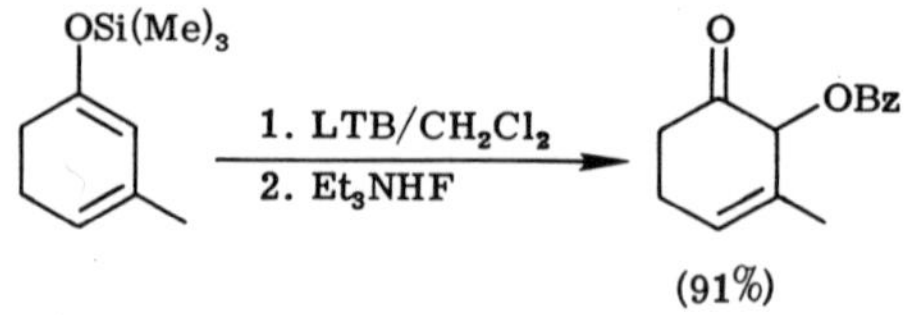

(ref. 156)

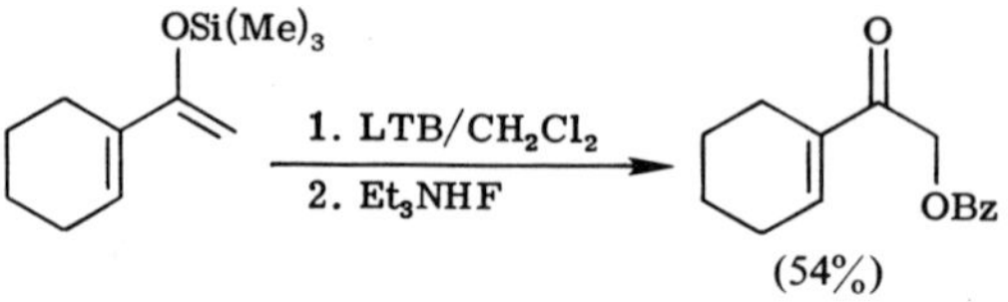

high yield process; however, certain cyclic cases afford products rationalized by 1,4-addition.[156]

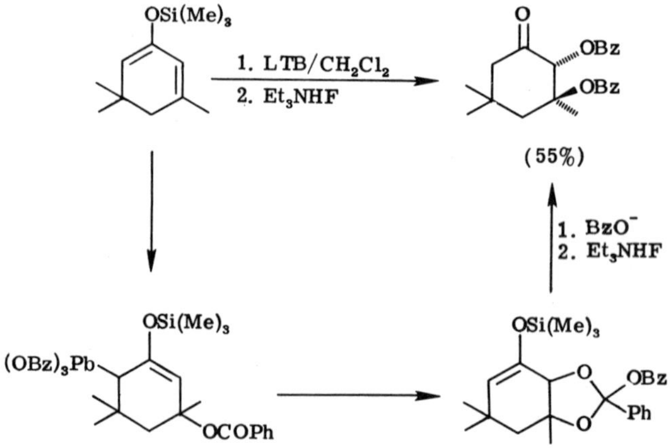

(ref. 156)

Vinyl ethers undergo LTA oxidation in a manner similar to that of enol silyl ethers. Such a system has been implicated[136] in the conversion of **57** to **58**.[136,157,157a] The analogous systems **59** and **60** also afford high yields of oxidation product.[158]

[157] W. Slaunwhite and A. Solo, *J. Pharm. Sci.* **64**, 168 (1975).
[157a] M. Hossain and D. N. Kirk, *Steroids* **34**, 677 (1979).
[158] G. R. Lenz, *J. Chem. Soc. Chem. Commun.*, 241 (1978).

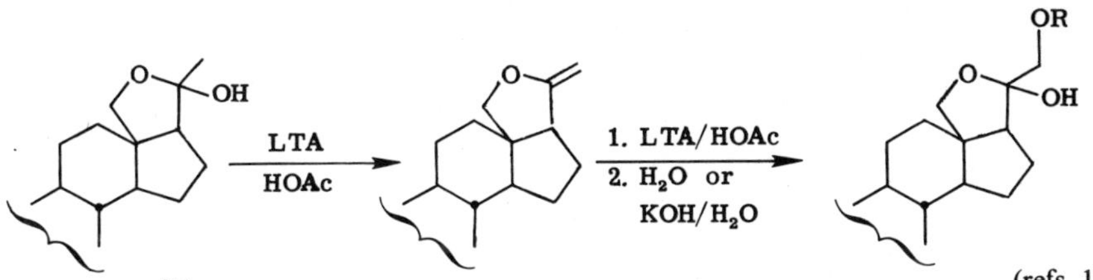

57

58

R= Ac or H

(refs. 136, 157, 157a)

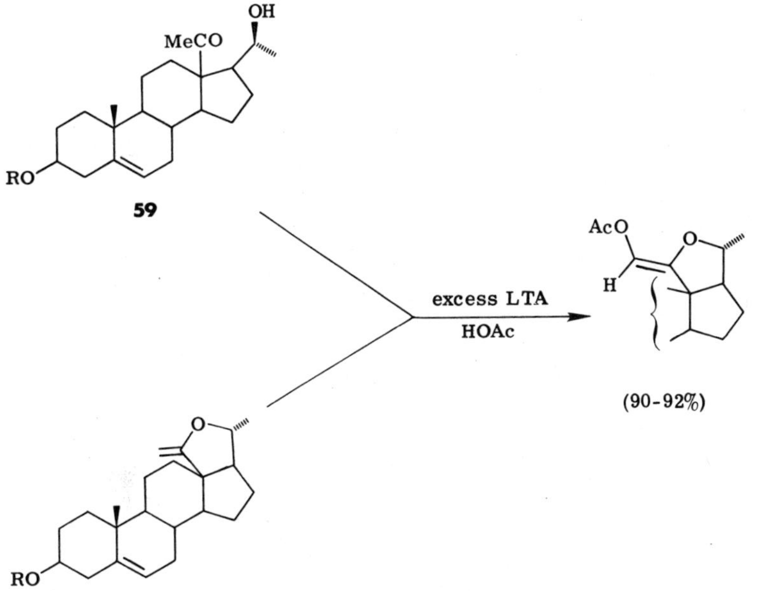

59

60

(90-92%)

(ref. 158)

Systems in which enol formation is directed are also useful in solving the problem of regioselectivity. The treatment of β-keto sulfides with LTA in hot benzene produces α-acetoxy-α-phenylthioketones in excellent yield.[159,160,160a] The latter represent a versatile class of synthetic intermediates.

[159] B. M. Trost and A. J. Bridges, *J. Am. Chem. Soc.* **98**, 5017 (1976).
[160] B. M. Trost and G. S. Massiot, *J. Am. Chem. Soc.* **99**, 4405 (1977).
[160a] B. M. Trost, W. C. Vladuchick, and A. J. Bridges, *J. Am. Chem. Soc.* **102**, 3554 (1980).

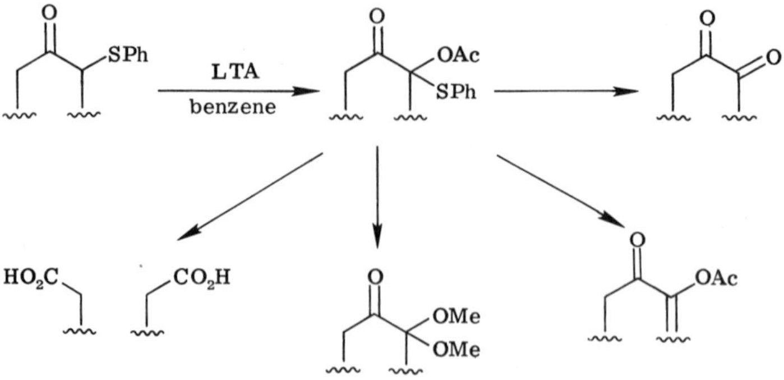

(ref. 160)

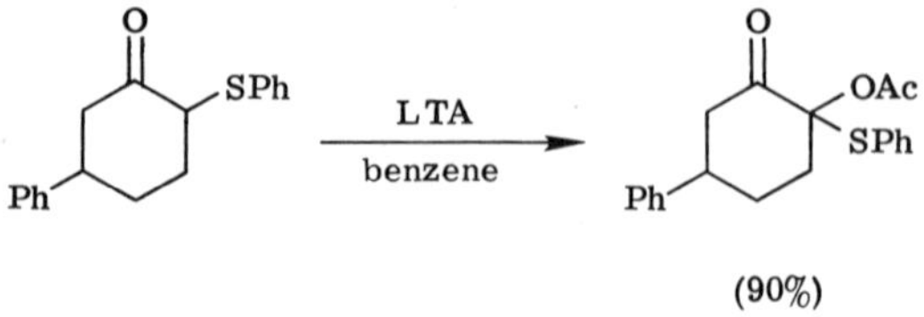

Enol thioethers react with LTA to give thionium ions which then partition to produce allylic acetates and/or bis-acetoxylation products.[161] The formation of the allylic acetates is favored in THF whereas methylene chloride

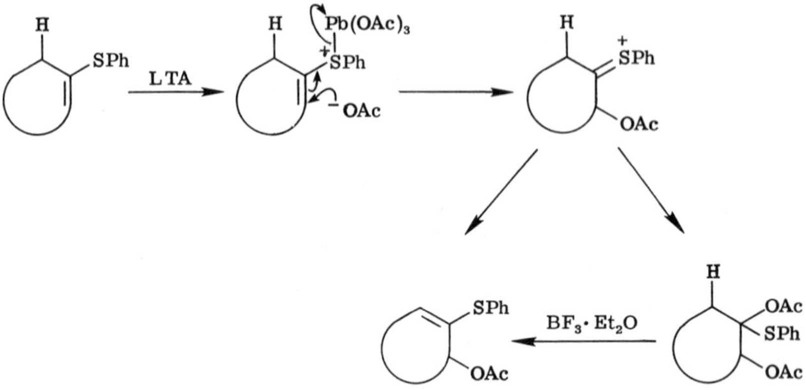

favors diacetate formation. Also, treatment of the latter with boron trifluoride etherate or 5 *N* KOH leads to excellent yields of the corresponding allylic acetates. The reaction of **61** to **62** indicates that isomerization of the

[161] B. M. Trost and Y. Tanigawa, *J. Am. Chem. Soc.* **101**, 4413 (1979).

kinetically formed **63** to the thermodynamically more stable **62** is occurring.[161] Alkyl-substituted ketene thioacetals also react with LTA.[161a,161b]

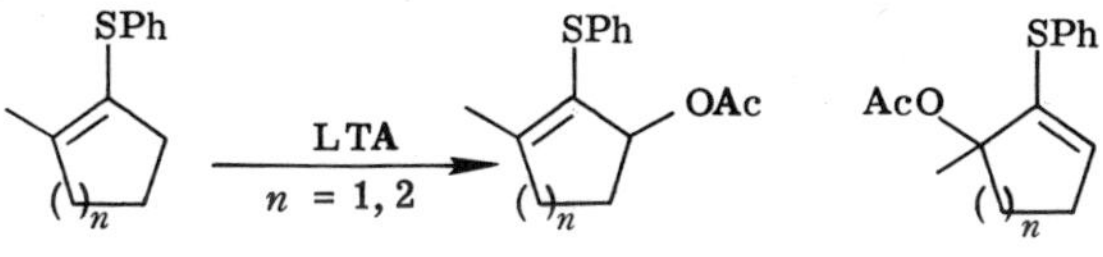

61 **62** **63**

In this case oxidation affords the corresponding 3-alkyl-1,4-dithiepan-2-ones in moderate yield.

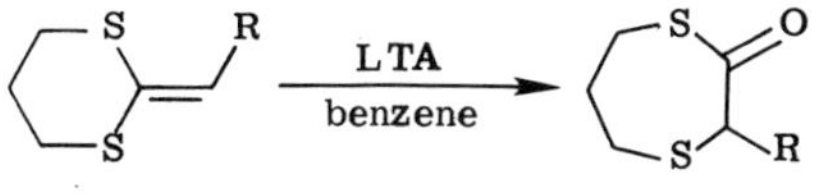

(ref. 161a)

(34-74%)

The oxidation of 1,3-diones leads not only to the production of α-acetoxy derivatives, but to 1,2-diones as well.[162,163] Use of ^{14}C-labeling studies shows that to a large extent the carbon flanked by the two carbonyls is the one expelled and intermediates such as **64** have been implicated.[162] The

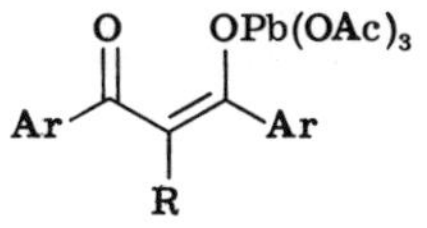

64

R = H,OAc

picture is not complete since **65** and **66** do not give similar results and loss

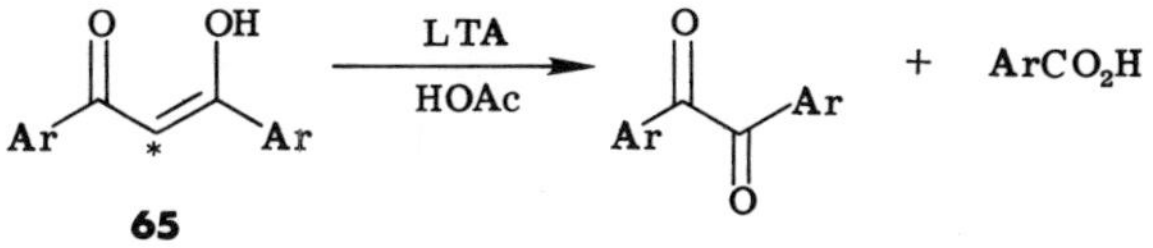

(ref. 162)

65

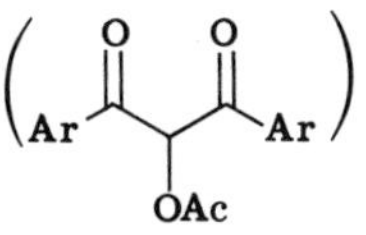

66

[161a] K. Hiroi and S. Sato, *Chem. Lett.*, 923 (1979).
[161b] K. Hiroi, S. Sato, and K. Matsuo, *Heterocycles* **12**, 213 (1979).
[162] K. Kurosawa and A. Moriyama, *Bull. Chem. Soc. Jpn.* **47**, 2717 (1974).
[163] K. Ezoe and K. Kurosawa, *Bull. Chem. Soc. Jpn.* **50**, 443 (1977).

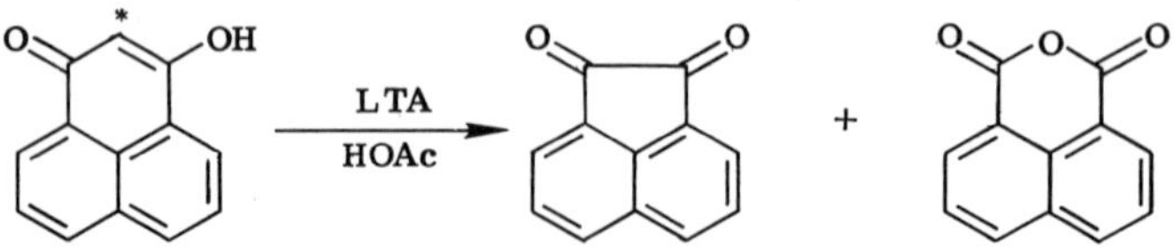

(ref. 163)

of ^{14}C is not total. An intermediate analogous to **64** may also be involved in the oxidation of 1-phenylheptane-1,3,4,6-tetrone to give **67** and **68**.[164]

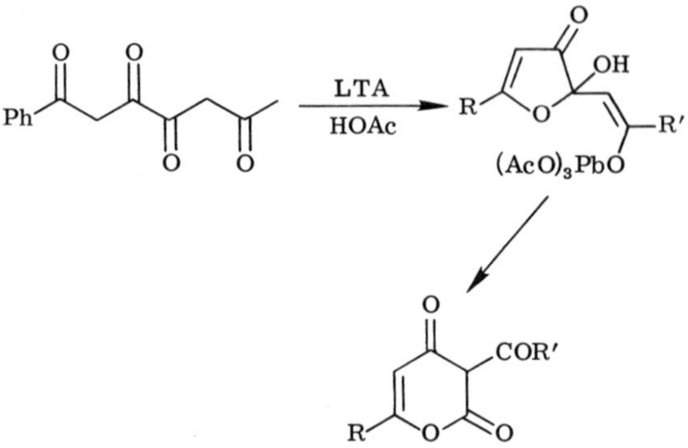

(ref. 164)

67 R = Ph, R′ = Me (43%)
68 R = Me, R′ = Ph (14%)

Another example of an interesting LTA reaction with a 1,3-dione involves the oxidation of humulone to give TCD.[165] Although the TCD structure shown is favored, **69** is also possible, whereas **70**, a previously postulated structure, is ruled out by spectral data. The mechanism of the transformation is not clear at present.

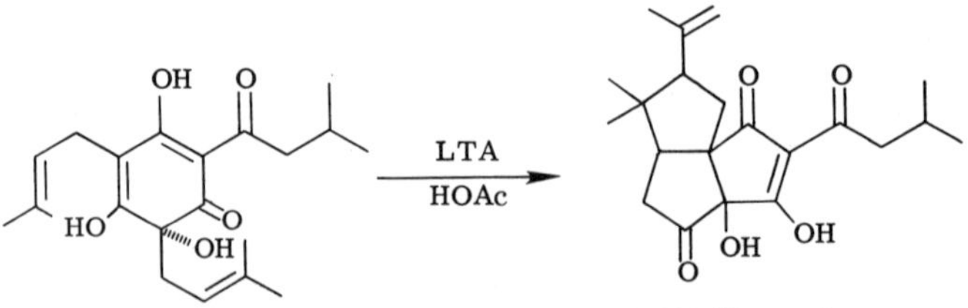

(ref. 165)

TCD (Tricyclo-dehydroisohumulone)

[164] M. Poje, B. Gašpert, and K. Balenovic, *Recl. Trav. Chim. Pays-Bas* **97**, 242 (1978).

[165] L. De Taeye, D. De Keukeleire, A. De Bruyn, and M. Verzele, *Bull. Soc. Chim. Belg.* **87**, 471 (1978).

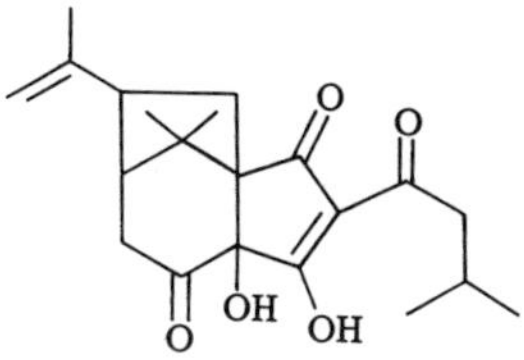

69

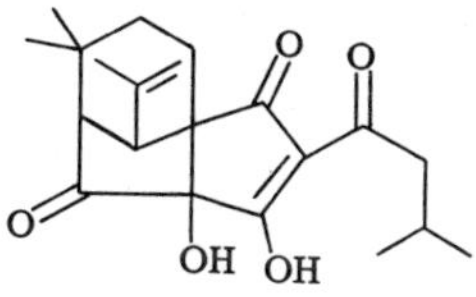

70

Both acyclic and cyclic 1,3-diones react with aryllead(IV) tricarboxylates to produce the corresponding arylated derivatives.[165a] The examples cited are representative of the method.

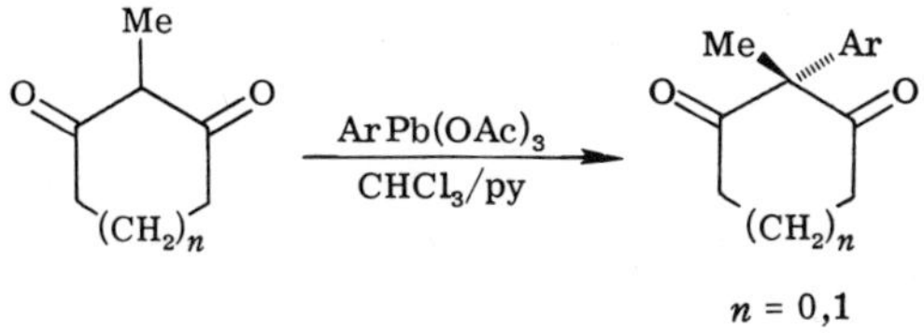

(ref. 165a)

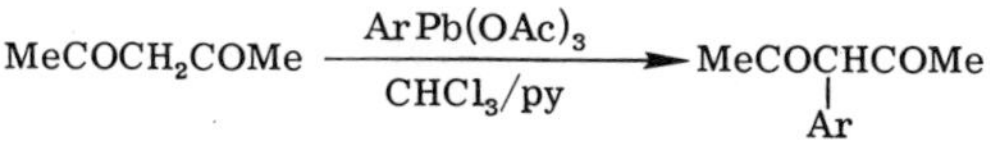

Nonenolizable 1,2-diones are cleaved by LTA in benzene. In rigid systems, anhydrides are obtained in high yield, and when a solvent mixture of 1:1 benzene:methanol is employed, diesters are formed. The latter oxidation is postulated to occur from the corresponding hemiketal.[166] The oxidation of

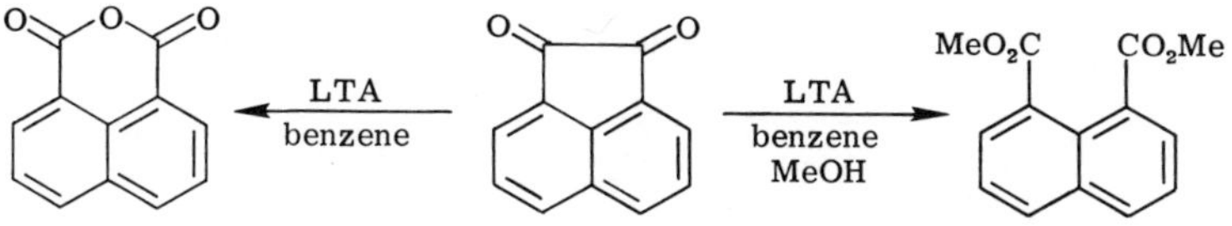

(91%) (85%)

(ref. 166)

the nonenolizable ketone **71** gives a 55% yield of **72** and the reaction is rationalized by involving carbonium ion intermediates.[167]

[165a] J. T. Pinhey and B. A. Rowe, *Aust. J. Chem.* **32**, 1561 (1979).
[166] L. Canonica, B. Danieli, P. Manitto, and G. Russo, *Gazz. Chim. Ital.* **100**, 1026 (1970).
[167] H. Miura, K. Hirao, and O. Yonemitsu, *Tetrahedron* **34**, 1805 (1978).

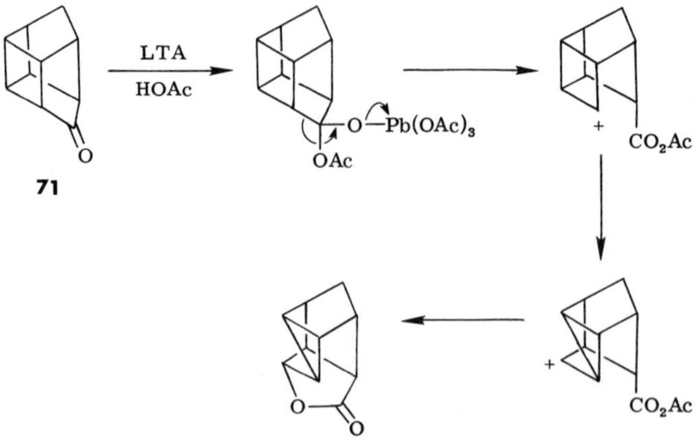

71

(ref. 167)

72 (55%)

Nonenolizable thiocarbonyl derivatives also react with LTA and generally yield the corresponding carbonyl compound.[168–170] With sugar derivatives,

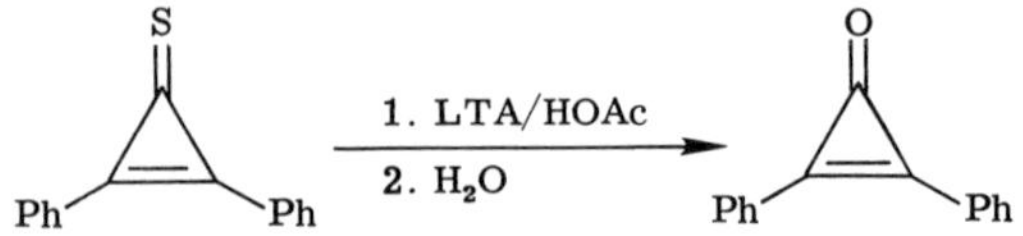

(ref. 168)

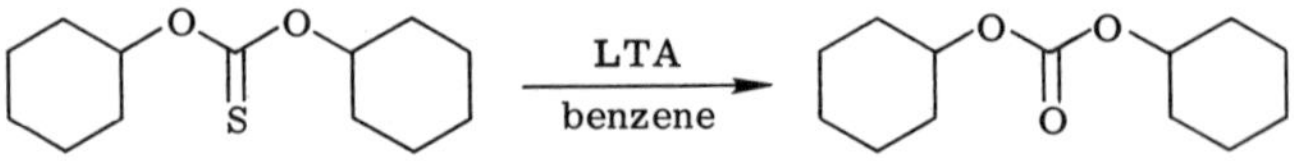

(ref. 169)

(62%)

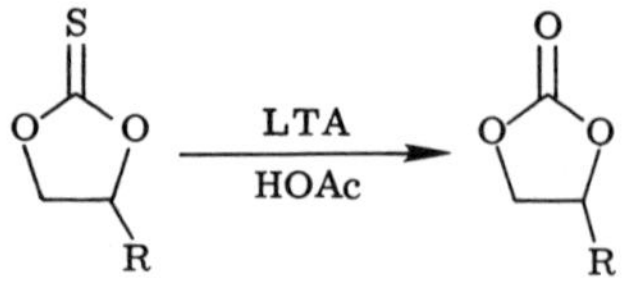

(ref. 170)

production of disulfides can subvert the reaction and trithiocarbonates lead to **73**.[170] Finally, the selenocarbonyl compound **74** is reported to give a high

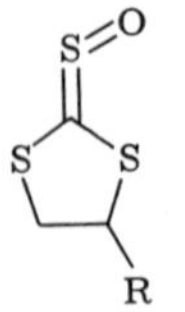

73

yield of **75** with **76** postulated as an intermediate.[171]

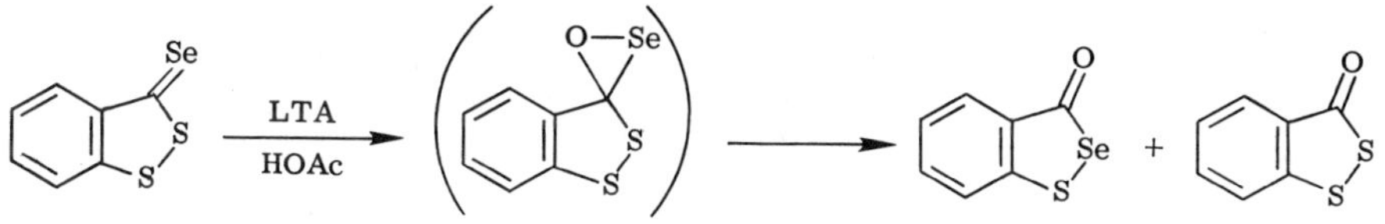

74 **76** **75** (87%) (4%)

(ref. 171)

D. PHENOLS

Phenols react readily with LTA, usually in acetic acid, to give excellent yields of the corresponding acetoxy cyclohexadienones.[12] Although homo-

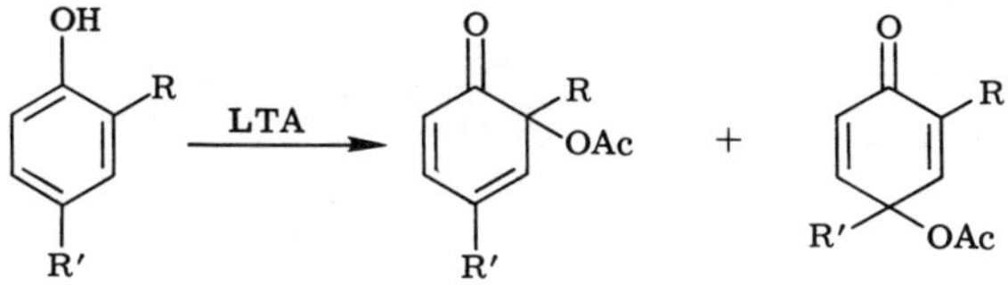

lytic pathways were originally postulated for the reaction,[12] more recent work favors either heterolysis of **77**[172] or electrophilic attack by the phenol

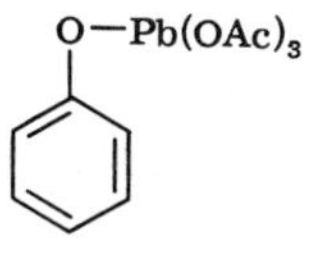

77

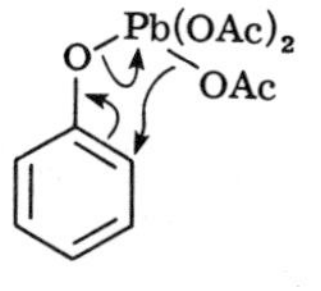

78

on LTA.[173] Concerted transfer of acetate from lead to carbon (**78**) can account for the generally encountered predominance of α-acetoxylation and is analogous to the mechanism proposed for enol oxidation.[174] The systems **79**–**81** are some typical examples.[175–177]

---

[168] J. W. Lown and T. W. Maloney, *J. Org. Chem.* **35**, 1716 (1970).
[169] M. Mori and T. Taguchi, *Synthesis*, 469 (1975).
[170] W. M. Doane, B. S. Shasha, C. R. Russell, and C. E. Rist, *J. Org. Chem.* **30**, 3071 (1965).
[171] S. Tamagaki, S. Sakaki, and S. Oae, *Heterocycles* **2**, 45 (1974).
[172] M. J. Harrison and R. O. C. Norman, *J. Chem. Soc. C*, 728 (1970).
[173] W. A. Bubb and S. Sternhell, *Tetrahedron Lett.*, 4499 (1970).
[174] A. Lethbridge, R. O. C. Norman, and C. B. Thomas, *J. Chem. Soc. Perkin Trans. 1*, 2465 (1975).
[175] J. R. van der Veck, H. Steinberg, and Th. J. de Boer, *Tetrahedron Lett.*, 2157 (1970).
[176] R. J. Andersen and D. J. Faulkner, *J. Am. Chem. Soc.* **97**, 936 (1975).
[177] J. B. Henderson, R. W. Alder, D. R. Dalton, and D. G. Hey, *J. Org. Chem.* **34**, 2667 (1969).

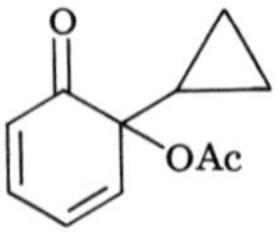

79 (50%)

(ref. 175)

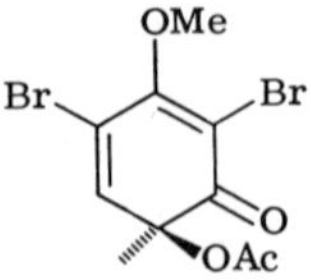

80 (75%)

(ref. 176)

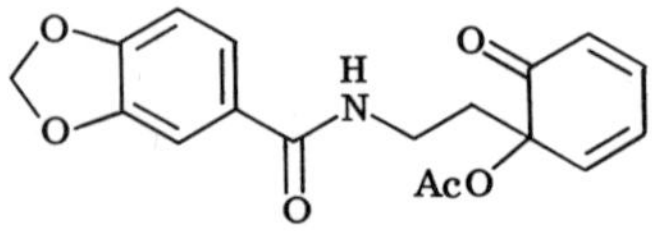

81 (62%)

(ref. 177)

The application of the reaction to the synthesis of a wide range of isoquinoline alkaloids has been reviewed.[178] In general, tetrahydroisoquinolinols of type **82** give *p*-quinol acetates, while type **83** gave 4-acetoxytetra-

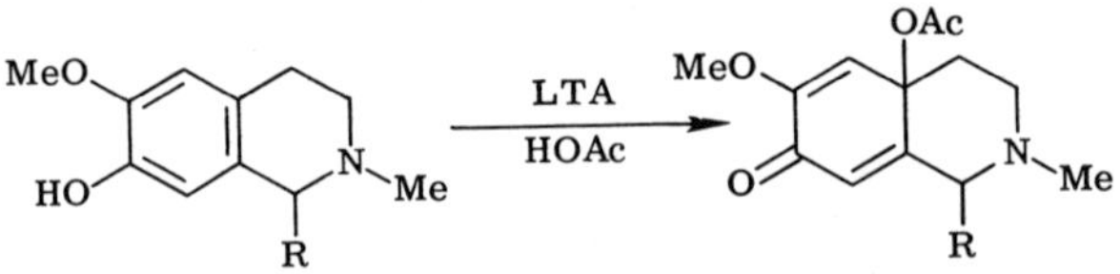

82 (~20%)

(ref. 178)

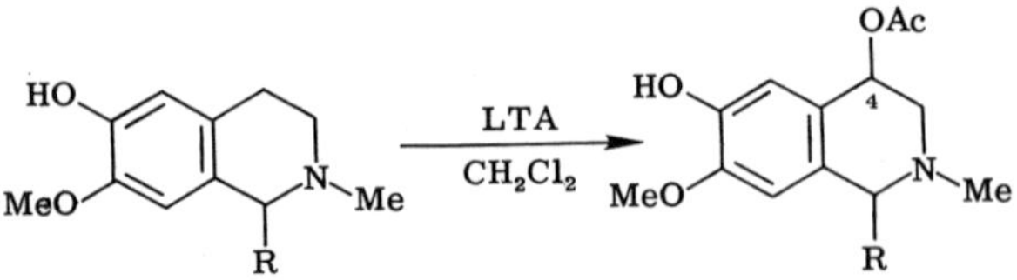

83 (~50%)

hydroisoquinolinols. Intermediate **84** has been implicated in the latter

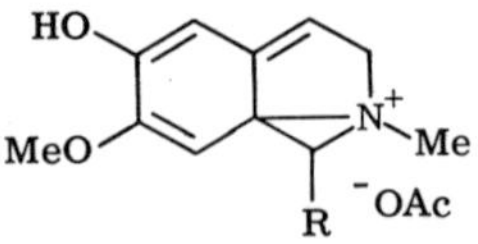

84

case.[178] A third mode of behavior has recently been discovered, demonstrated by the **85** to **86** transformation.[179] A mechanism has been proposed

[178] B. Umezawa and O. Hoshino, *Heterocycles* **3**, 1005 (1975).

[179] H. Hara, M. Hosaka, O. Hoshino, and B. Umezawa, *Tetrahedron Lett.*, 3809 (1978).

involving cleavage of a *p*-quinol acetate to produce **87** which then recyclizes to give **86**.

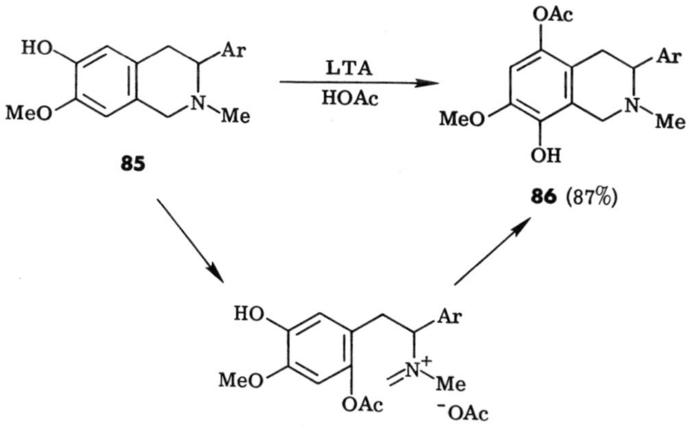

85 **86** (87%) **87**

(ref. 179)

LTA oxidation of phenols carried out in the presence of alcohols gives rise to the production of alkoxydienones.[172,174] Whereas this reaction has been used as evidence against a homolytic mechanism for the oxidation, it is not clear whether the actual oxidizing agent is LTA or a mixed Pb(IV) species resulting from alcohol–LTA ligand exchange.[175] The oxidation of pentafluorophenol with LTA gives a 75% yield of **88** and **89** in a reaction where no acetoxy products are reported.[180] When a phenol is treated with LTA using

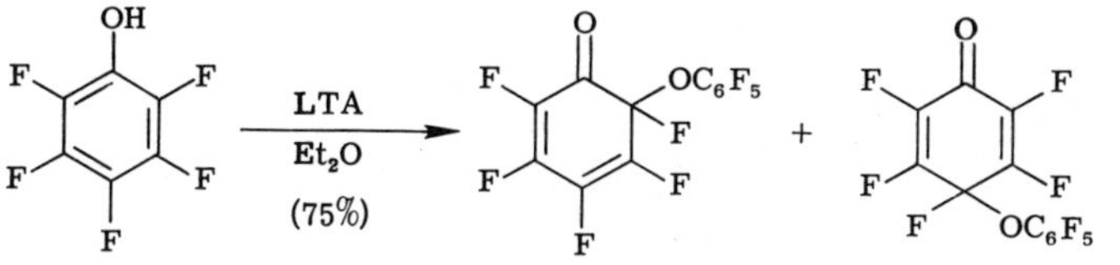

88 2 : 1 **89**

(ref. 180)

a carboxylic acid other than acetic acid as solvent, the external nucleophile can be effectively incorporated into the oxidation product. Two interesting examples involve the reactions of mesitol and **90**, respectively, with LTA in conjunction with acrylic acid as solvent.[181]

[180] L. S. Kobrina, V. N. Kovtonyuk, and G. G. Yakobson, *Zh. Org. Khim.* **13**, 1447 (1977).
[181] D. J. Bichan and P. Yates, *Can. J. Chem.* **53**, 2054 (1975).

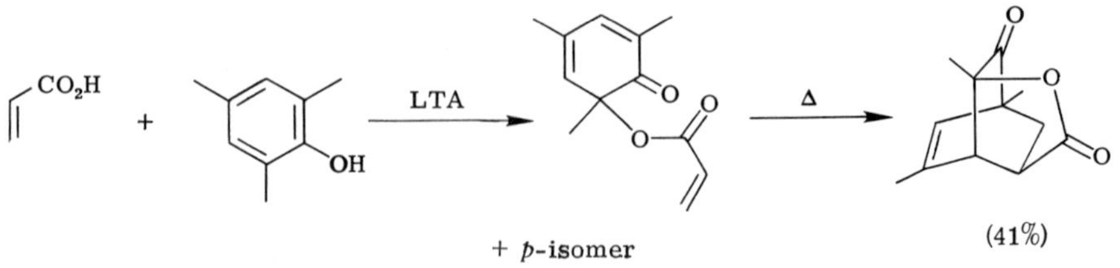

(ref. 181)

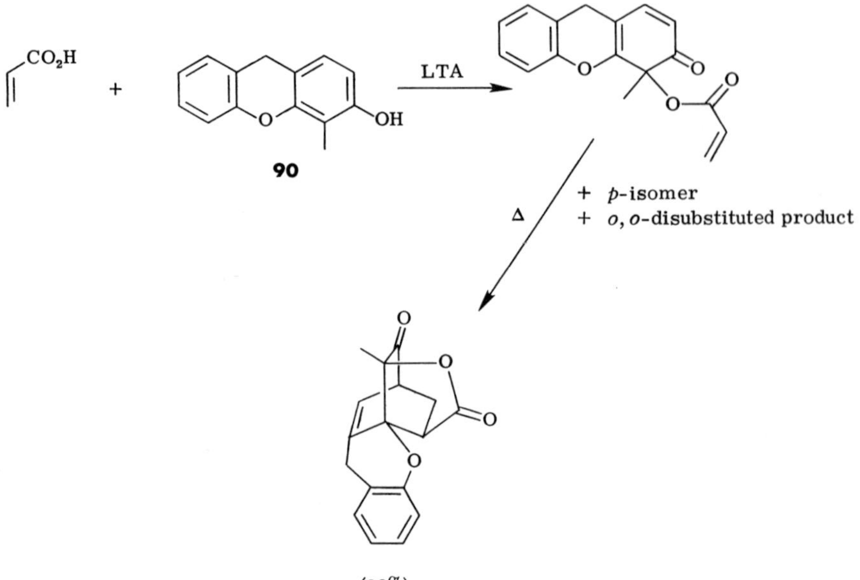

90

(23%)

The oxidation of **91** with LTA in acetic acid gives **92** as the major product[174,182] with no apparent interaction with the remote double bond.

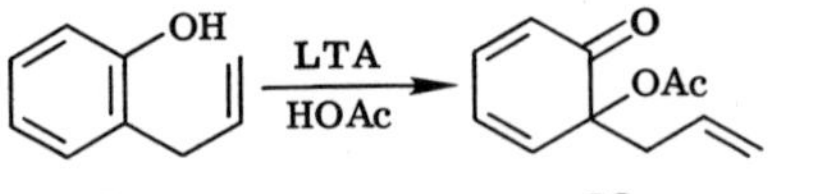

(refs. 174, 182)

91 **92**

However, with chalcones,[183] 2-hydroxybenzophenones,[184] 2-hydroxystilbenes,[185] and 3-(2-hydroxphenyl)coumarins[186] treatment with LTA leads

[182] E. Zibral, W. Wessely, and J. Jörg, *Monatsh. Chem.* **92**, 654 (1961).
[183] K. Kurosawa and J. Higuchi, *Bull. Chem. Soc. Jpn.* **45**, 1132 (1972).
[184] K. Kurosawa, Y. Sasaki, and M. Ikeda, *Bull. Chem. Soc. Jpn.* **46**, 1498 (1973).

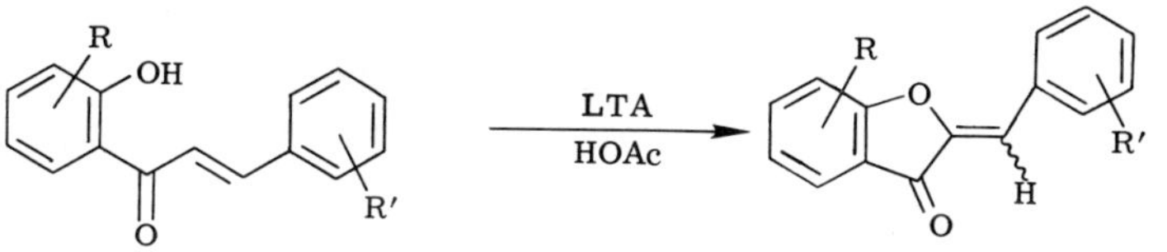

(ref. 183)

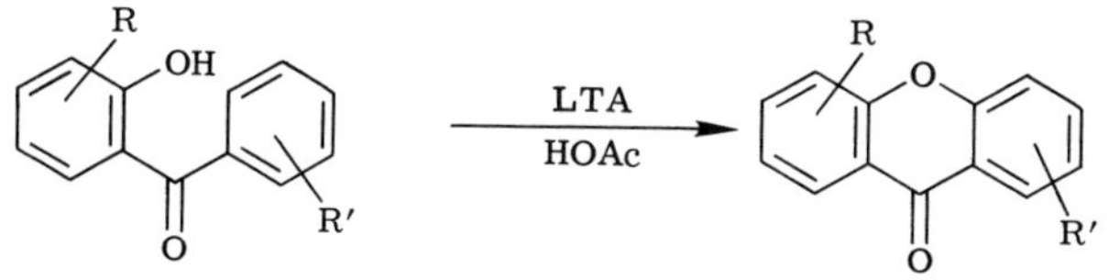

(ref. 184)

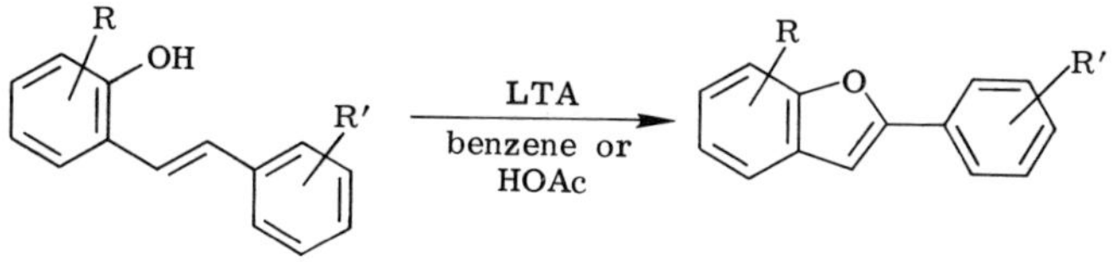

(ref. 185)

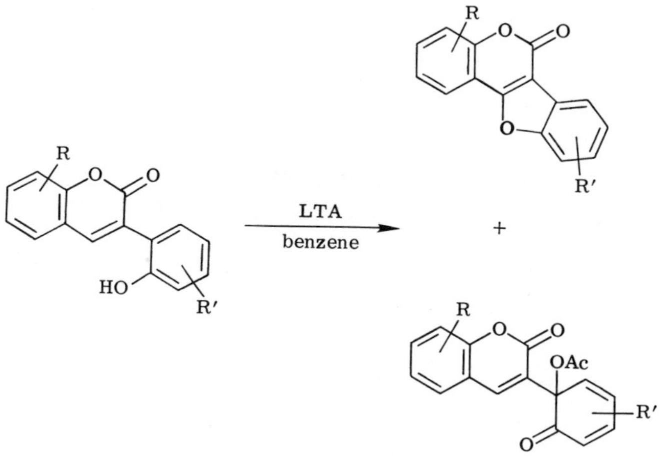

(ref. 186)

[185] K. Nogami and K. Kurosawa, *Bull. Chem. Soc. Jpn.* **47**, 505 (1974).
[186] K. Kurosawa and K. Nogami, *Bull. Chem. Soc. Jpn.* **49**, 1955 (1976).

to oxidative cyclization. The presence of methoxy groups on the phenolic substrate can lead to predominant formation of quinones.[187] The use of

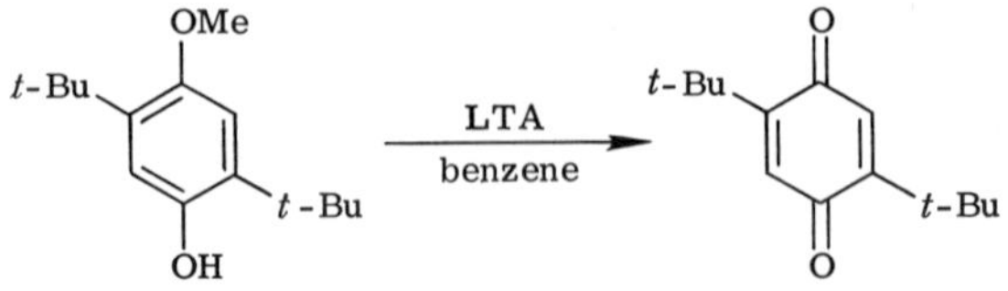

(31%)

(ref. 187)

acetic acid enhances the process whereas benzene can lead to other reaction pathways.[188,189] Treatment of **93** with 2 equivalents of LTA in benzene

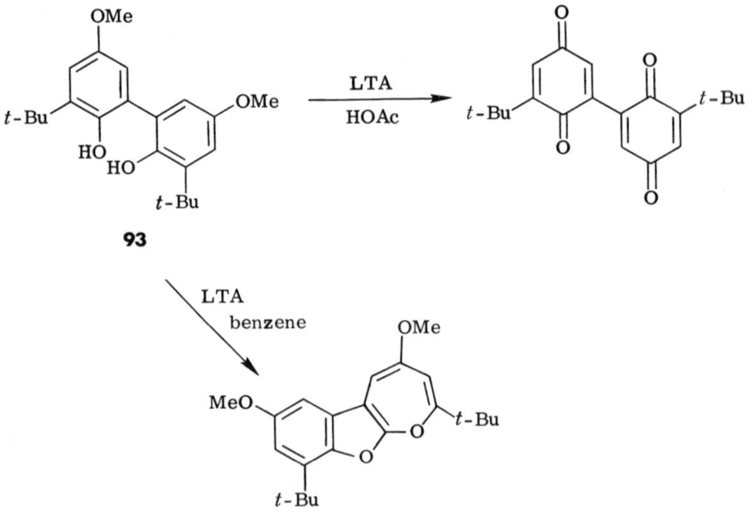

93

94

(refs. 188, 189)

gives further oxidation (via **94**) to afford **95**, **96**, and **97**.[188–190] The presence of water in the benzene accounts for the production of **96** and **97**, and thus, a coherent mechanism is presented.[188,190] Finally, the reaction of phenols with aryl lead(IV) triacetates in chloroform/pyridine produces moderate yields of *o*- or *p*-dienones with no acetoxylation occurring.[191,191a] With

[187] F. R. Hewgill and S. L. Lee, *J. Chem. Soc. C*, 1556 (1968).

[188] F. R. Hewgill, D. G. Hewitt, G. B. Howie, C. L. Raston, R. J. Webb, and A. H. White, *J. Chem. Soc. Perkin Trans. 1*, 290 (1979).

[189] F. R. Hewgill and D. G. Hewitt, *J. Chem. Soc. C*, 726 (1967).

[190] F. R. Hewgill and D. G. Hewitt, G. B. Howie, and W. L. Spencer, *Aust. J. Chem.* **30**, 1971 (1977).

[191] H. C. Bell, G. L. May, J. T. Pinhey, and S. Sternhell, *Tetrahedron Lett.*, 4303 (1976).

[191a] H. C. Bell, J. T. Pinhey, and S. Sternhell, *Aust. J. Chem.* **32**, 1551 (1979).

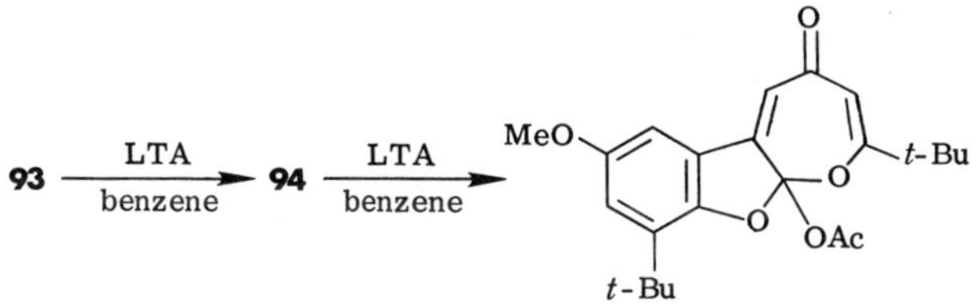

95

(refs. 188, 189)

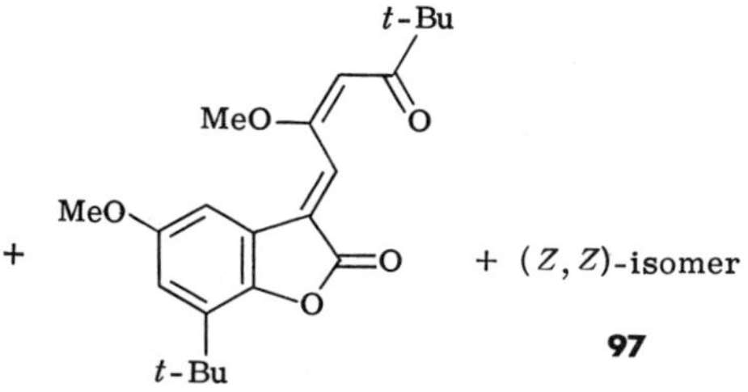

96 97

(refs. 188, 190)

methylated phenols, the yields are highest when both ortho positions are substituted and the reaction fails when the phenol in question bears only electron-withdrawing groups.[191a]

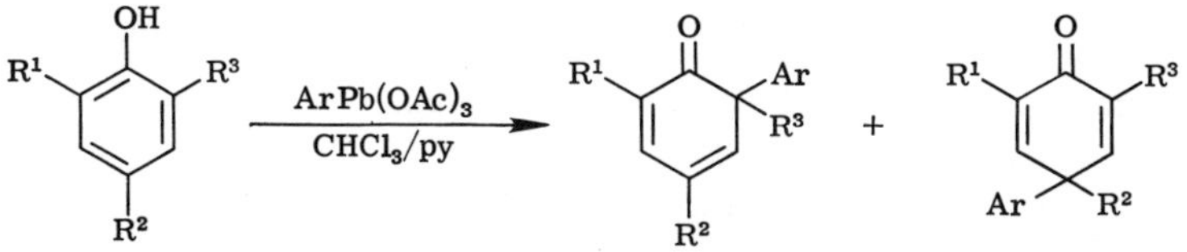

(refs. 191, 191a)

E. MONOCARBOXYLIC ACIDS

The reaction of monocarboxylic acids with LTA has been studied extensively and an excellent review of the topic is available.[192] A general consensus as to mechanism exists as outlined below in Scheme 3.[192]

Metathesis:	$RCO_2H + LTA \longrightarrow Pb(IV)(OAc)_{4-n}(OCOR)_n + HOAc$
Initiation:	$Pb(IV)OCOR \xrightarrow{\Delta \text{ or } h\nu} Pb(III) + R\cdot + CO_2$
Propagation:	$R\cdot + Pb(IV)(OCOR) \longrightarrow R^+ + Pb(III)(OCOR)$
	$Pb(III)(OCOR) \longrightarrow Pb(II) + R\cdot + CO_2$
Termination:	$R\cdot + Pb(III)(OCOR) \longrightarrow R^+ + Pb(II)(OCOR)$
	$R\cdot \longrightarrow$ hydrogen abstraction; dimerization; etc.

SCHEME 3

[192] R. A. Sheldon and J. K. Kochi, *Org. React.* **19**, 279 (1972).

Product formation can thus occur from either R· or R^+ leading to the formation of alkanes, alkenes, and acetates. The structure of the carboxylic acid is crucial to the outcome of the reaction since oxidation rates of primary, secondary, and tertiary R· vary considerably with tertiary R· being oxidized most rapidly to R^+. Addition of $Cu(X)_2$ (X is normally OAc) with both primary and secondary carboxylic acids, gives enhanced rates of R· oxidation with concomitant proton loss and thus, high yields of alkenes. Actual rates of oxidation of primary radicals with Cu(II) approach diffusion

$$R\cdot\,(1^\circ \text{ or } 2^\circ) + Cu(II) \rightarrow R(-H) + H^+ + Cu(I)$$
$$Cu(I) + Pb(IV) \rightarrow Cu(II) + Pb(III)$$

control whereas secondary radicals are oxidized at least 100 times faster by Cu(II) than by Pb(IV).[192] As noted, the Cu(II) can be used in catalytic amounts since efficient Cu(I) to Cu(II) oxidation is accomplished by Pb(IV). Because tertiary R· is oxidized by Pb(IV), addition of $Cu(OAc)_2$ to reactions involving tertiary acids generally has little effect.

Solvents such as HMPA, THF, and DMF in conjunction with pyridine are employed most commonly and are, in general, to be preferred over benzene, acetonitrile, or ethyl acetate.[192]

1. Primary Acids

In the presence of pyridine the Pb(IV)/Cu(II) reagent has been used to degrade steroidal acids to give excellent yields of the corresponding alkenes,[193,194] and a similar procedure gave **98**, an intermediate in the

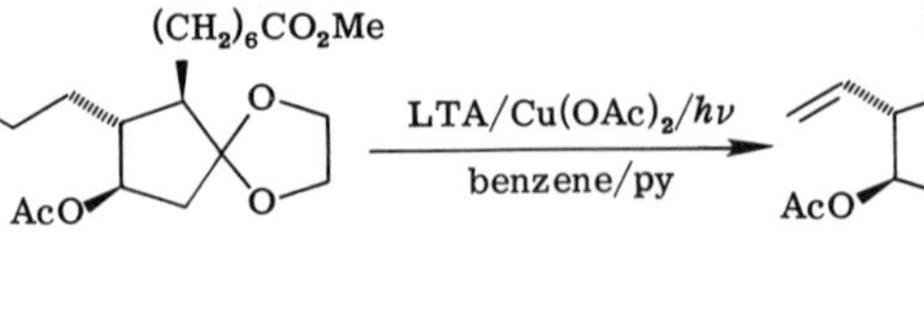

(refs. 193, 194)

synthesis of prostaglandin E_1.[195] Extending the LTA/$Cu(OAc)_2$ method to

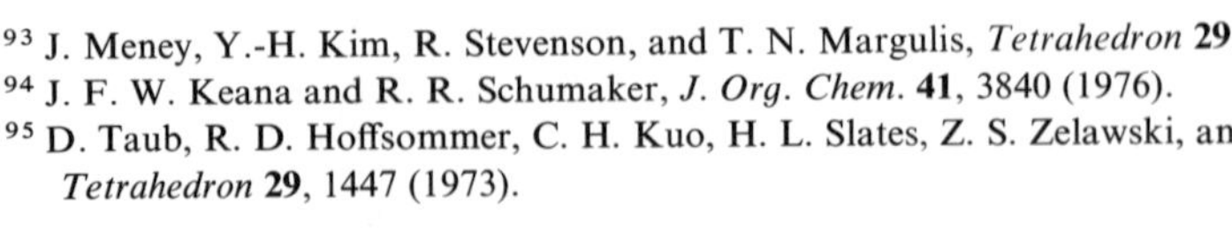

(ref. 195)

98 (37%)

[193] J. Meney, Y.-H. Kim, R. Stevenson, and T. N. Margulis, *Tetrahedron* **29**, 21 (1973).

[194] J. F. W. Keana and R. R. Schumaker, *J. Org. Chem.* **41**, 3840 (1976).

[195] D. Taub, R. D. Hoffsommer, C. H. Kuo, H. L. Slates, Z. S. Zelawski, and N. L. Wendler, *Tetrahedron* **29**, 1447 (1973).

include β-carboxy ketones gives a useful synthesis of α,β-unsaturated ketones.[196,197] Model studies indicate that the procedure has general syn-

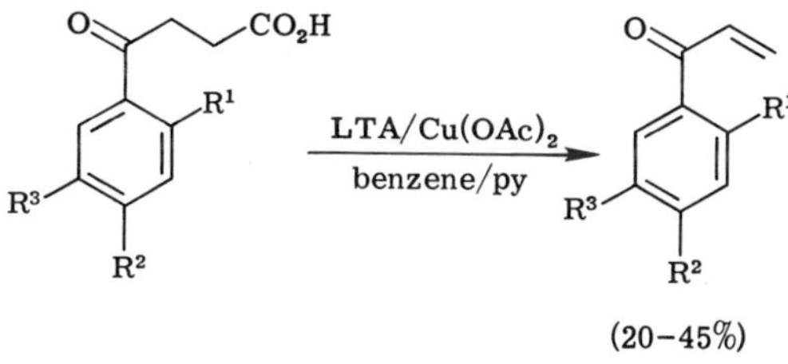

(ref. 196)

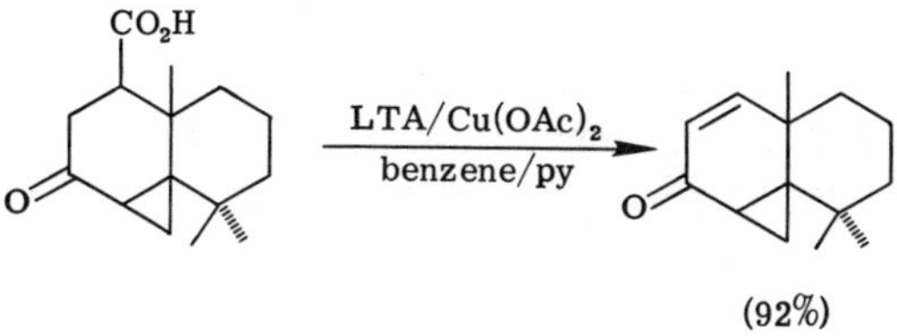

(ref. 197)

thetic potential. The variety of compounds available from **99** is illustrative.[197]

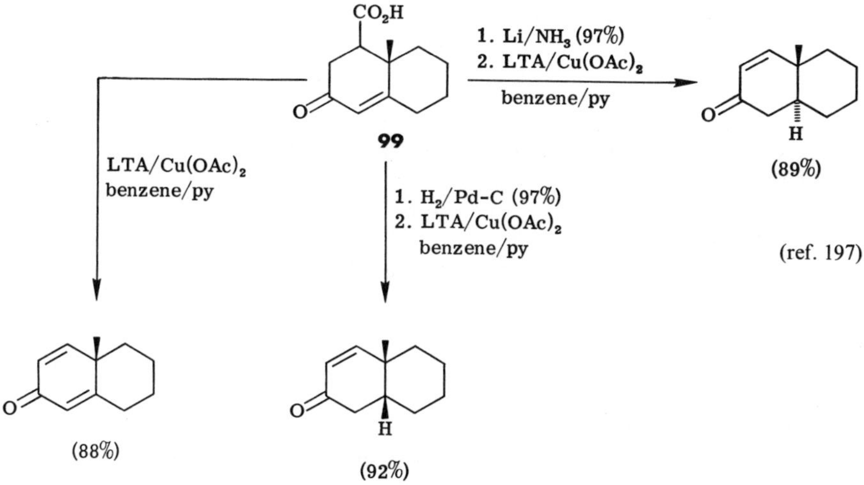

(ref. 197)

The oxidation of β-carboxy lactones with LTA/$Cu(OAc)_2$ gives rise to the corresponding α-methylene lactones.[198]

[196] P. P. Sane, K. J. Divakar, and A. S. Rao, *Synthesis*, 541 (1973).
[197] J. E. Mc Murry and L. C. Blaszczak, *J. Org. Chem.* **39**, 2217 (1974).
[198] K. J. Divakar, P. P. Sane, and A. S. Rao, *Tetrahedron Lett.*, 399 (1974).

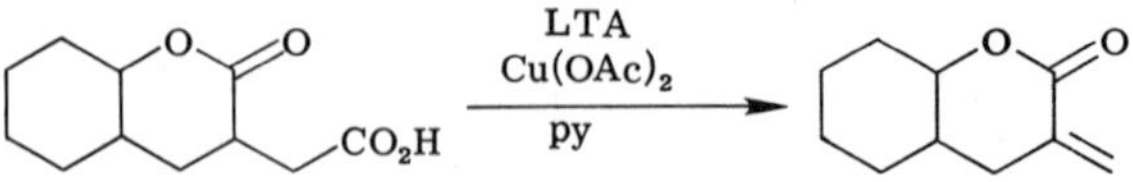

(30%)

(ref. 198)

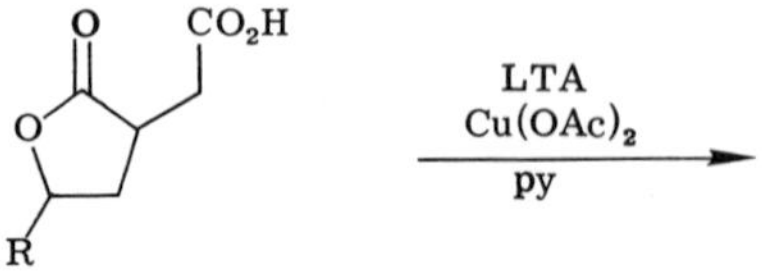

(~50%)

Decarboxylation of 5-(1′-naphthyl)-pentanoic acid with LTA in benzene gives a mixture of tetrahydrophenanthrene and **100** in overall yields of 58%.[199,200] Deuteration studies indicate that the primary radical formed in

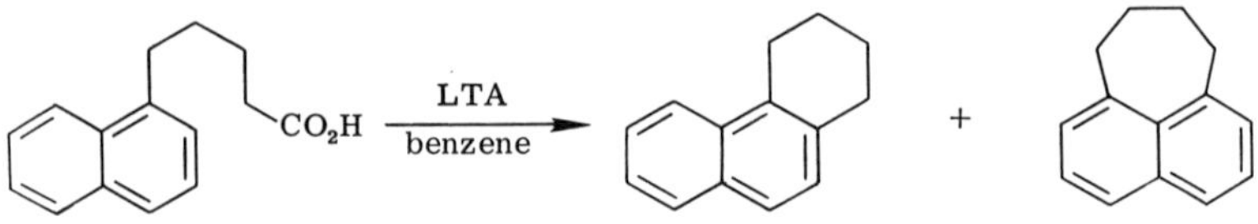

(53.5%) **100** (4.5%)

(refs. 199, 200)

the reaction gives tetrahydrophenanthrene via competitive formation of both **101** and **102**.

101

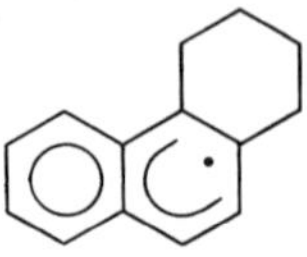

102

In a related case, **103** affords 20% of the cyclic product **104**,[201] and 4-pentonic acid has been found to yield substituted γ-lactones when treated with either LTA or LTA/LiCl mixtures.[37]

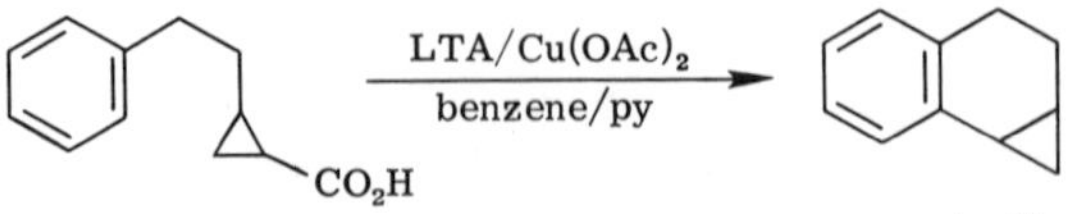

(ref. 201)

103 **104** (20%)

2. SECONDARY ACIDS

The LTA/$Cu(OAc)_2$ decarboxylation of **105** in benzene/pyridine is optimized by removal of acetic acid from the LTA prior to the reaction. In this manner, a 35% yield of **106** was realized.[202] Decarboxylation of **107** is

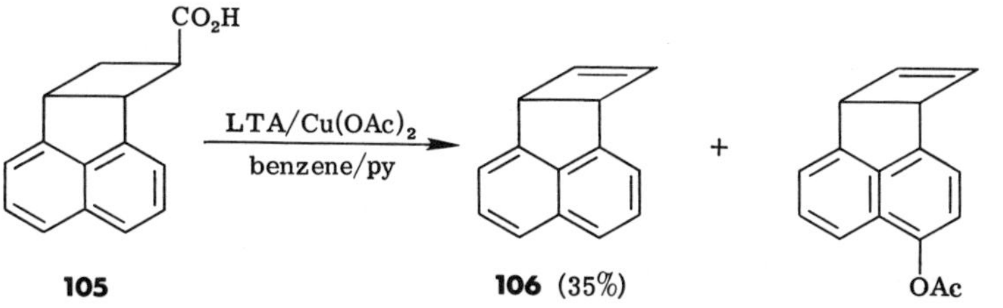

(ref. 202)

105 **106** (35%) (minor product isolated as phenol)

accomplished in high yield in the presence of HOAc/KOAc.[202a]

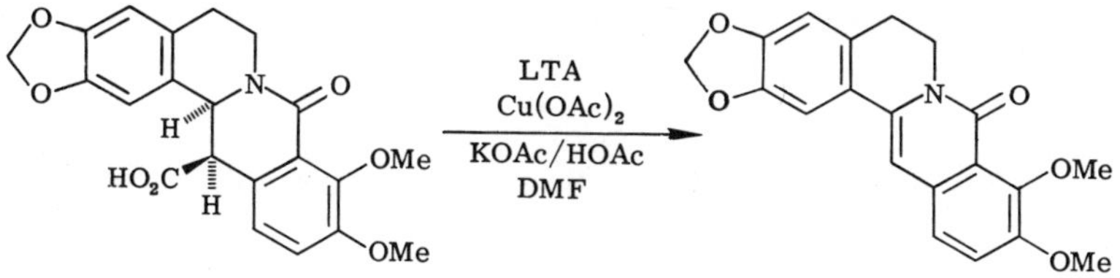

(ref. 202a)

107 (65%)

LTA oxidation of **108** in benzene/pyridine produces a number of products including alkenoic acids, acetates, and lactones from selective loss of the secondary carboxy group over the primary.[203,204] Addition of $Cu(OAc)_2$

$$R{-}CH_2{-}\underset{|}{\overset{CO_2H}{C}}H(CH_2)_nCO_2H$$

108

---

[199] J. C. Chottard and M. Julia, *Tetrahedron Lett.*, 2561 (1971).

[200] J. C. Chottard and M. Julia, *Tetrahedron* **28**, 5615 (1972).

[201] C. Descoins, M. Julia, and H. V. Sang, *Bull. Soc. Chim. Fr.*, 2037 (1972).

[202] R. L. Petty, M. Ikeda, G. E. Samuelson, C. J. Boriack, K. D. Onan, A. T. McPhail, and J. Meinwald, *J. Am. Chem. Soc.* **100**, 2464 (1978).

[202a] M. Cushman and F. W. Dekow, *J. Org. Chem.* **44**, 407 (1979).

[203] Yu. N. Ogibin, M. I. Katsin, and G. I. Nikishin, *Izv. Akad. Nauk SSSR, Ser. Khim.*, 1345 (1975).

[204] Yu. N. Ogibin, M. I. Katsin, and G. I. Nikishin, *Izv. Akad. Nauk SSSR, Ser. Khim.*, 1225 (1972).

subverts acetate and lactone formation thereby giving alkenoic acids as the major products. There is a preference for alkene formation occurring away from the primary carboxy group and varying the nature of X in CuX_2 has little effect upon product ratios.[204] Decarboxylation of **109**, regardless of the CuX_2 used, gave mixtures of alkenes, again favoring the structure with the double bond placement farthest from the OAc group.[205]

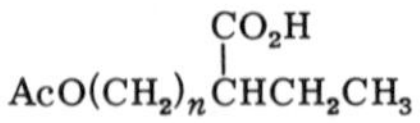

109

Selectivity between carboxy groups was also noted in the oxidation of **110**.[206] The stereochemical relationship of the carboxy groups is crucial

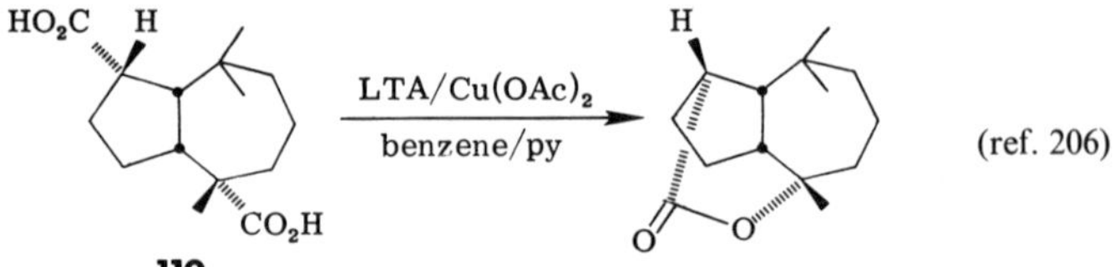

110 (70%)

since **111** gives **112** in 30% yield along with several other products. The latter

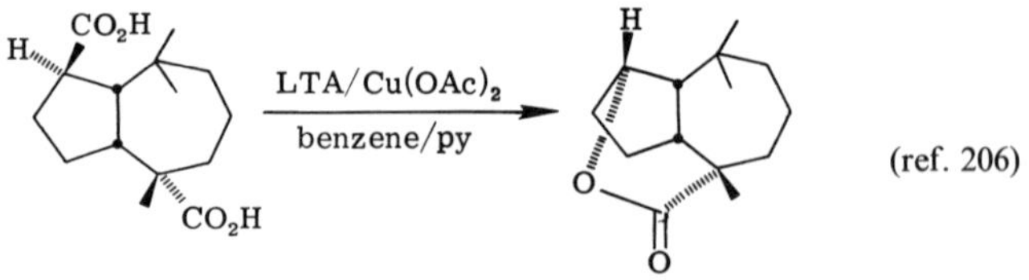

111 **112** (30%)

case is a unique example of a secondary carboxyl decarboxylation in preference to a tertiary. It is not clear what role steric crowding might play in the ability of the two diacids to undergo a ligand exchange reaction with LTA, and thus affect decarboxylation rates.

LTA decarboxylation of secondary acids in the absence of $Cu(OAc)_2$ can lead to the production of moderate to good yields of the corresponding acetates. Table V summarizes some typical examples.[207–212]

[205] M. I. Katsin, Yu. N. Ogibin, and G. I. Nikishin, *Izv. Akad. Nauk SSSR, Ser. Khim*, 139 (1974).

[206] P. K. Jadhav, V. S. Dalavoy, A. S. C. Prakasa Rao, and U. R. Nayak, *Indian J. Chem. Sect. B* **15**, 589 (1977).

[207] M. Fetizon, F. J. Kakis, and V. Ignatiadou-Ragoussis, *Tetrahedron* **30**, 3981 (1974).

[208] J. B. Lambert, F. R. Koeng, and J. W. Hamersma, *J. Org. Chem.* **36**, 2941 (1971).

TABLE V

LTA DECARBOXYLATION OF SECONDARY CARBOXYLIC ACIDS

Substrate	Solvent	Product	% Yield	Reference
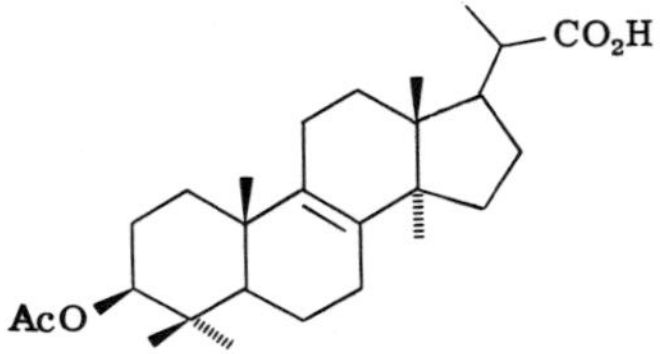	Benzene/py	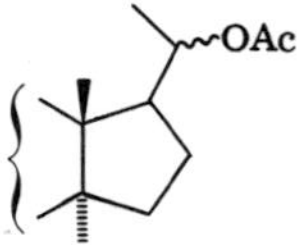	45	207
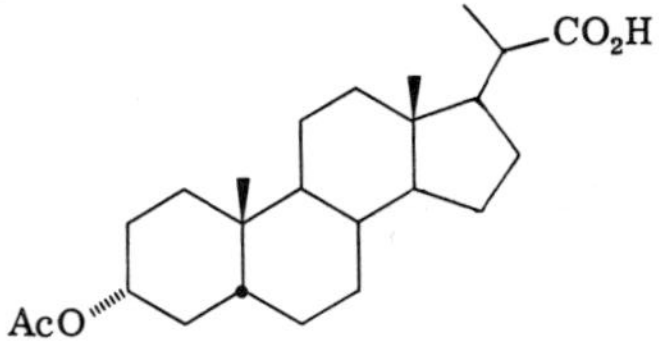	Benzene/py	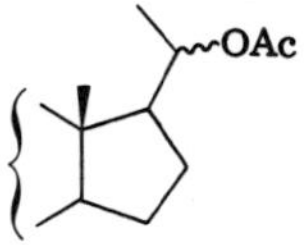	66	207
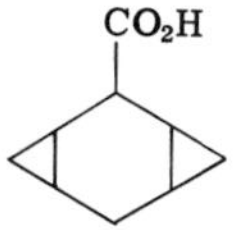	Benzene/py	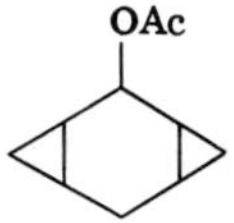	73	208
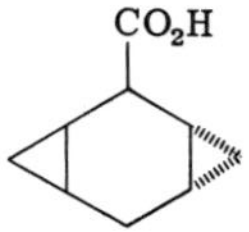	Benzene/py	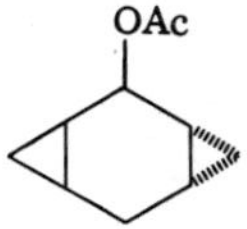	—	208
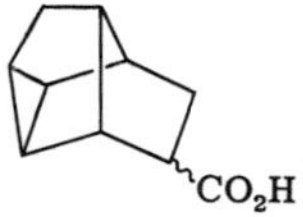	Benzene/py	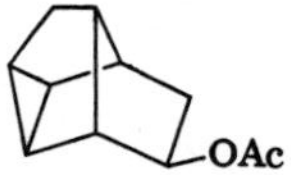 (sole product)	—	209

(*cont.*)

TABLE V (*cont.*)

Substrate	Solvent	Product	% Yield	Reference
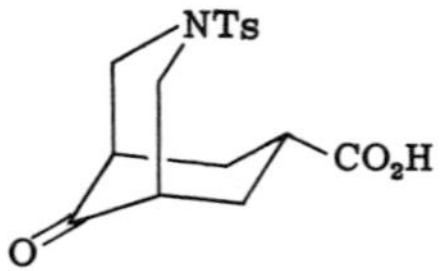	Py	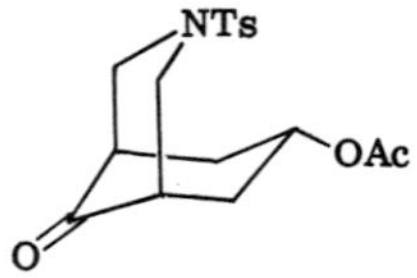	18 (+14% alkene)	210
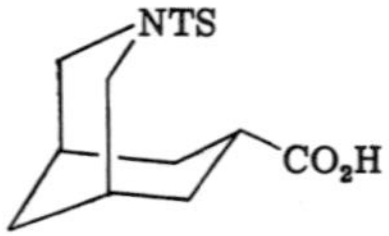	Py	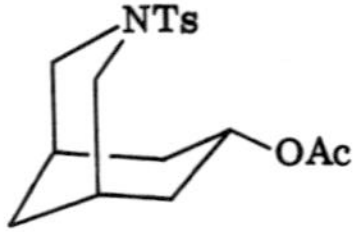	>47 (+16% alkene)	210
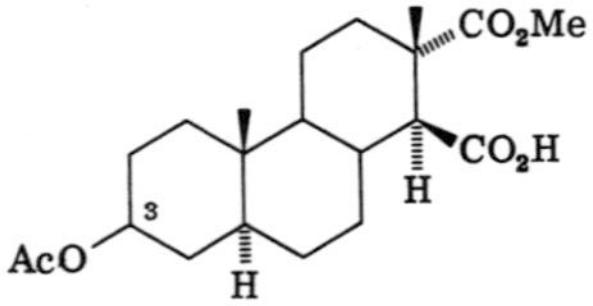	Benzene/py	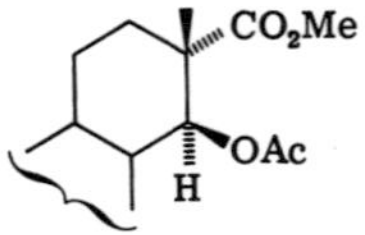 + 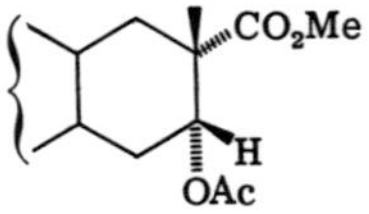 (60%) + (—) (isolated as 3β-alcohol 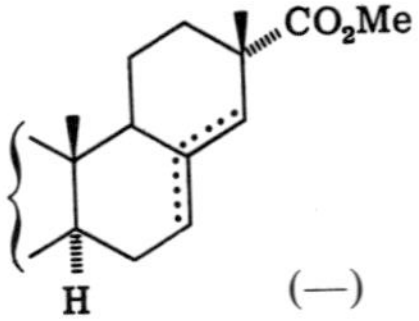 (—)		211
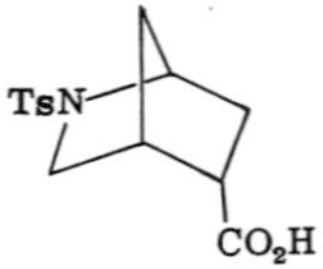	Benzene/py	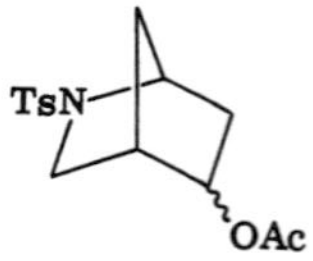	35 90:10, exo:endo	212

Perusal of Table V indicates that the decarboxylation of secondary acids is not a stereospecific process. A number of examples are accommodated by the intervention of carbonium ion intermediates, i.e., the conversions of **113**[213] and **114**.[213a] The reaction of system **115** is not so clear-cut since, in

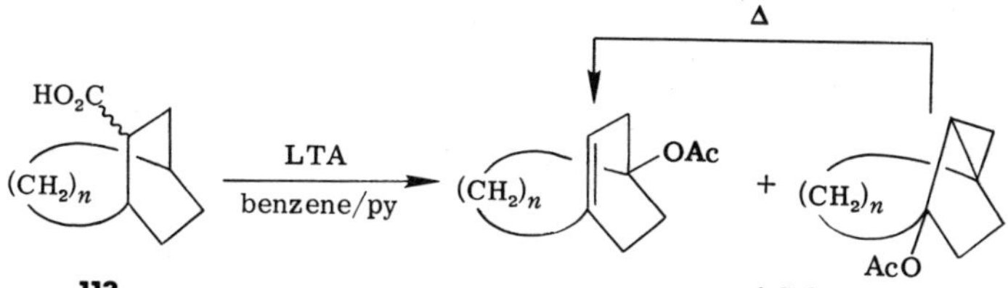

113 n = 4,5,6 (ref. 213)

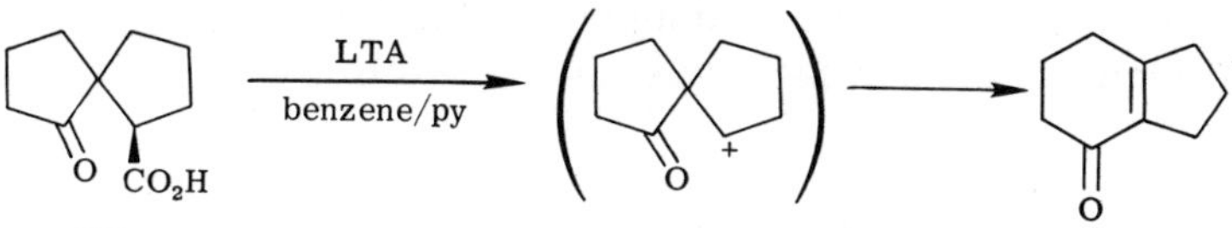

114 Sole product (ref. 213a)

the absence of a carbonyl group, mixtures of both rearranged and unrearranged products result with the latter predominating.[214] The fact that

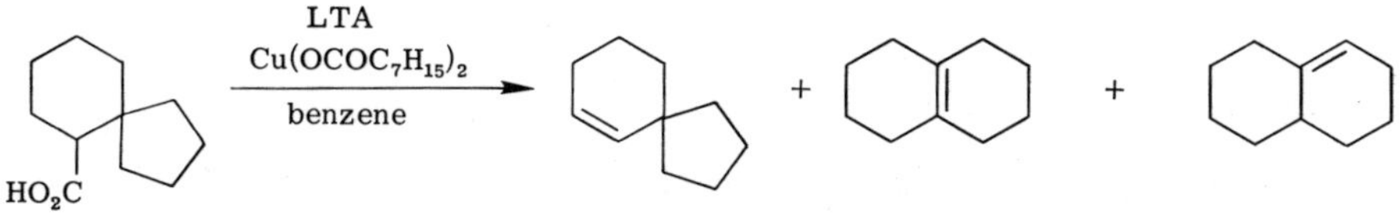

115 (81%) (4.9%) (4.9%)

(ref. 214)

both epimers **116** and **117** give identical mixtures of axial and equatorial acetates and that *erythro*- and *threo*-**118** give equivalent amounts of *erythro*- and *threo*-**119** has been cited as evidence for cationic intermediates as

---

[209] P. K. Freeman, D. M. Balls, and J. N. Blazevich, *J. Am. Chem. Soc.* **92**, 2051 (1970).
[210] Th. Reints Bok and W. N. Speckamp, *Tetrahedron* **35**, 267 (1979).
[211] P. Rosen and G. Oliva, *J. Org. Chem.* **38**, 3040 (1973).
[212] R. D. Gleim and L. A. Spurlock, *J. Org. Chem.* **41**, 1313 (1976).
[213] Y. Sakai, S. Toyotani, Y. Tobe, and Y. Odaira, *Tetrahedron Lett.*, 3855 (1979).
[213a] R. L. Cargill and A. M. Foster, *J. Org. Chem.* **35**, 1971 (1970).
[214] A. L. J. Beckwith, G. E. Gream, and D. L. Struble, *Aust. J. Chem.* **25**, 1081 (1972)

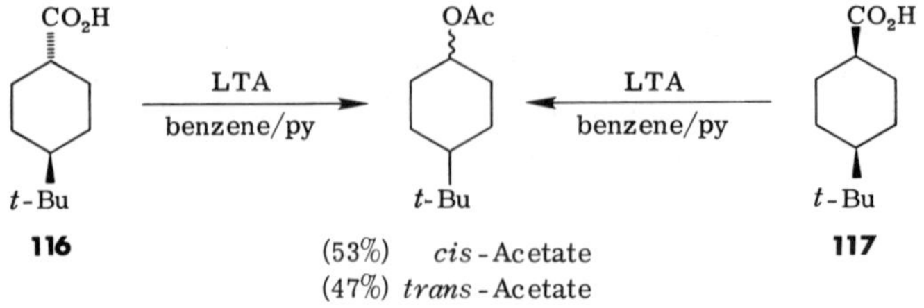

(refs. 36, 215)

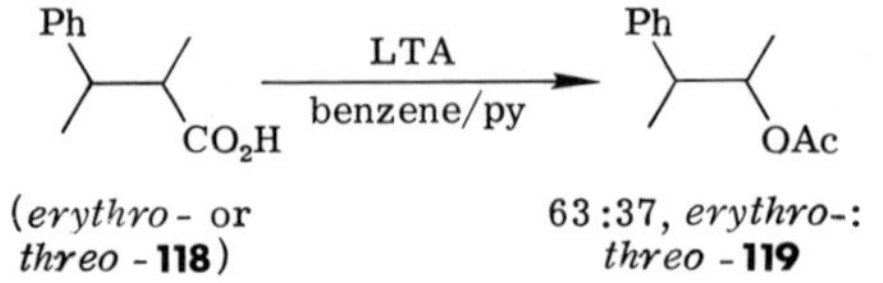

(ref. 216)

well.[36,215,216] Decarboxylation of both **120** and **121** in the presence of $Cu(OAc)_2$ in pyridine also yields similar amounts of *cis*- and *trans*-alkenes, among other products. Therefore, a common, nonstereospecifically generated carbonium ion intermediate is implicated.[217] The foregoing conclusions

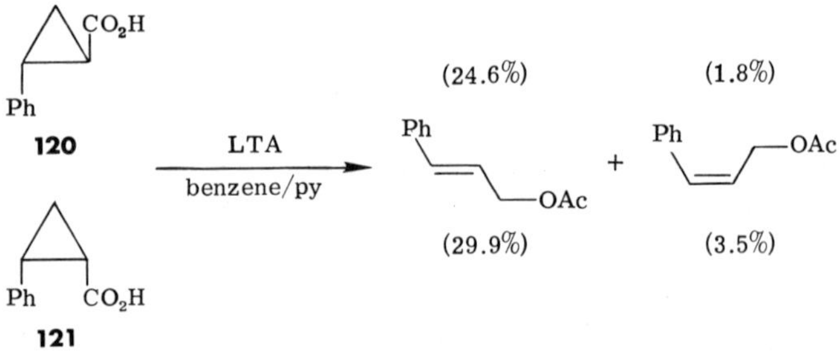

cannot be extrapolated to a number of examples of LTA decarboxylation involving bicyclic systems. Extensive studies of the LTA reactions of a series of bicyclic acids lead to a more complex view of the oxidation process.[218–220a] Initial decarboxylation of Pb(IV) carboxylates generates secondary radicals that can participate in typical radical reactions. Secondary radicals can also couple with Pb(III) or Pb(IV) carboxylates to give Pb(IV) alkyls. Heterolysis leads to carbonium ions, products of cis elimination, or the S_Ni production

[215] S. D. Elakovich and J. G. Traynham, *J. Org. Chem.* **38**, 873 (1973).
[216] A. Theine and J. G. Traynham, *J. Org. Chem.* **39**, 153 (1974).
[217] T. Shono, I. Nishiguchi, and R. Oda, *Tetrahedron Lett.*, 373 (1970).
[218] G. E. Gream, C. F. Pincombe, and D. Wege, *Aust. J. Chem.* **27**, 603 (1974).
[219] J. B. Stothers and K. C. Teo, *Can. J. Chem.* **54**, 1222 (1976).
[220] B. C. C. Cantello, J. M. Mellor, and G. Scholes, *J. Chem. Soc. Perkin Trans. 2*, 348 (1974).
[220a] P. K. Jadhav and U. R. Nayak, *Indian J. Chem. Sect, B* **16**, 947 (1978).

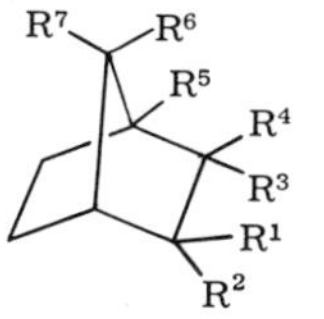

R^1	R^2	R^3	R^4	R^5	R^6	R^7	Reference
CO_2H	H	H	H	H	H	H	218
H	CO_2H	H	H	H	H	H	218
CO_2H	Me	Me	Me	H	H	H	218
Me	CO_2H	Me	Me	H	H	H	218
H	H	H	CO_2H	Me	Me	Me	218
H	H	CO_2H	H	Me	Me	Me	218
Me	H	H	CO_2H	H	H	H	219
Me	H	CO_2H	H	H	H	H	219
H	Me	H	CO_2H	H	H	H	219, 220
H	Me	CO_2H	H	H	H	H	219, 220
Me	Me	CO_2H	H	H	H	H	220a
Me	Me	H	CO_2H	H	H	H	220a

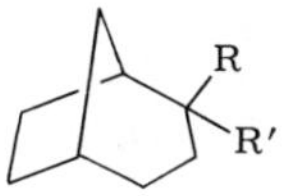

(ref. 220)

R	R′
CO_2H	H
H	CO_2H

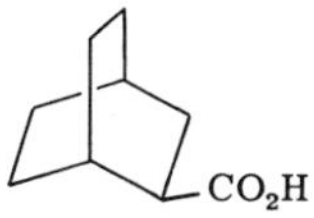

(ref. 220)

of esters. Initially formed radicals, in the presence of Cu(II), give Cu(II) derivatives that, in turn, lead to Cu(II) alkyls. The Cu(II) alkyls then give products by heterolysis, elimination, and substitution via S_Ni or ligand exchange pathways.[218] The high yield production of **122** from either longifolic acid, **123**, or its epimer, indicates that ring formation can be a consequence of LTA-promoted decarboxylation as well.[220a] Even in acyclic cases the nature of the reaction can be complex. For instance, the LTA

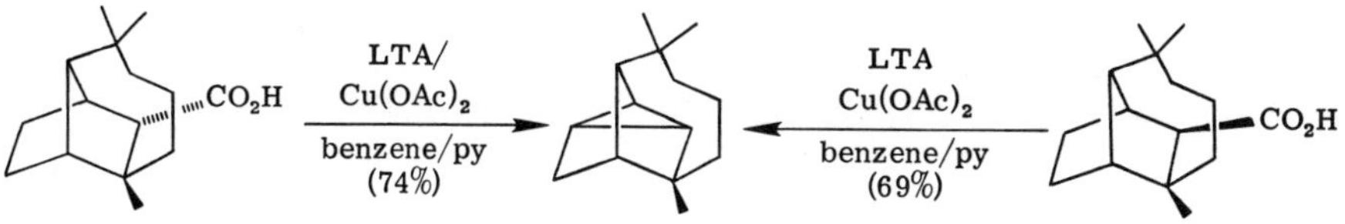

123 **122**

(ref. 220a)

oxidation of 2,3,3-trimethylbutanoic acid, in the presence or absence of CuX_2, gives inordinate amounts of both substitution and elimination products in which the carbon skeleton has not rearranged, thus militating against free cationic intermediates.[221]

It should be noted that treatment of (*S*)-(+)-2-methylbutanoic acid with LTA in the presence of cinnamic acid gives (*E*)-3-methyl-1-phenyl-1-pentene via free radical intermediates.[222] The product is racemic and substitution of

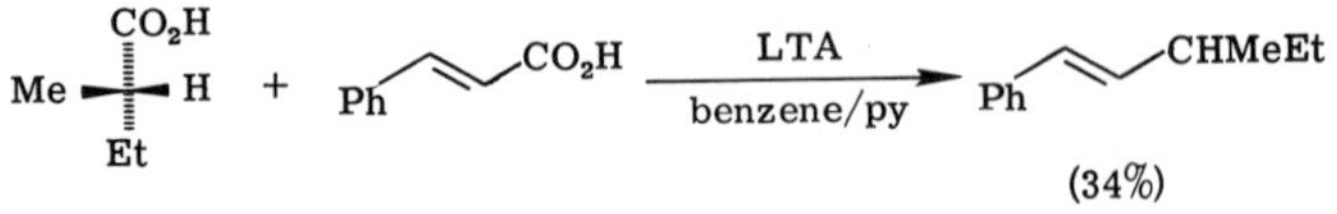

(ref. 222)

cinnamic by *p*-methoxycinnamic acid diverts the reaction and *β*-acetoxy-*p*-methoxy styrene is formed as an *E*/*Z*-mixture.[223]

3. Tertiary Acids

The mechanistic parameters discussed above are also applicable to the reaction of LTA with tertiary carboxylic acids. As an example, the decarboxylations of both *cis*- and *trans*-decalin-9-carboxylic acids afford similar amounts of $\Delta^{1,9}$- and $\Delta^{9,10}$-octalin. Differences in product ratios engendered by CuX_2 addition are lessened with tertiary acids due to the comparable oxidation rates of R· by Cu(II) and Pb(IV).[224]

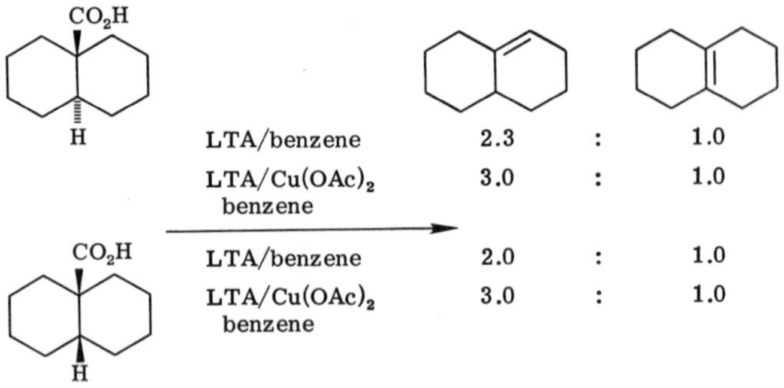

(ref. 224)

The oxidation reaction generally leads to good yields of alkenes, but regioselectivity is not particularly high. Three examples of olefin formation

[221] A. L. J, Beckwith, R. T. Cross, and G. E. Gream, *Aust. J. Chem.* **27**, 1693 (1974).
[222] B. Danieli, P. Manitto, F. Ronchetti, and G. Russo, *Chem. Ind.* (*London*), 430 (1973).
[223] B. Danieli, F. Ronchetti, and G. Russo, *Chem. Ind.* (*London*), 1067 (1973).
[224] A. L. J. Beckwith, R. T. Cross, and G. E. Gream, *Aust. J. Chem.* **27**, 1673 (1974).

with terpenoid acid derivatives are given below.[225,226,226a] Tertiary acids

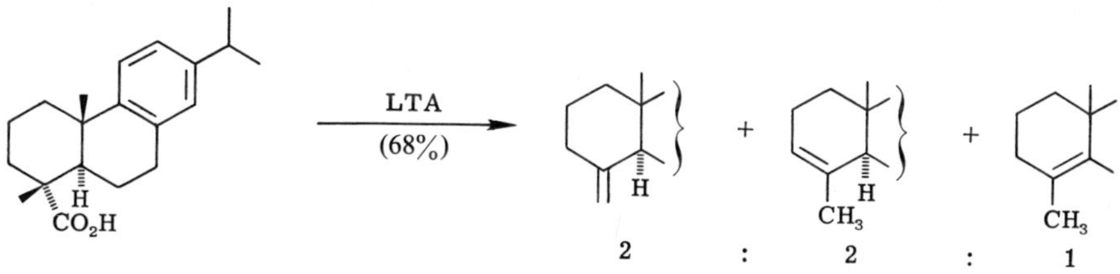

(ref. 225)

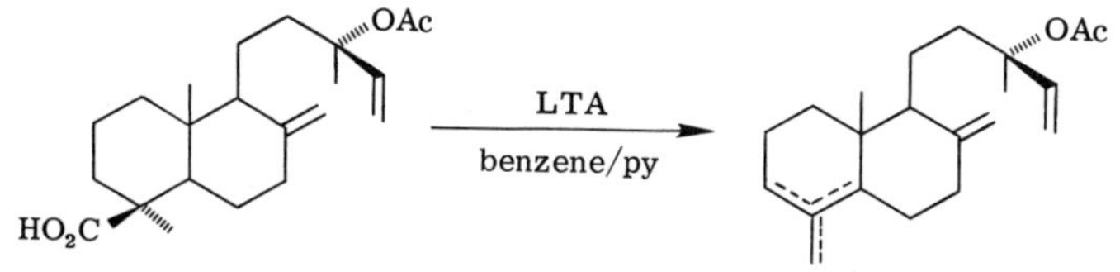

(ref. 226)

(61%)

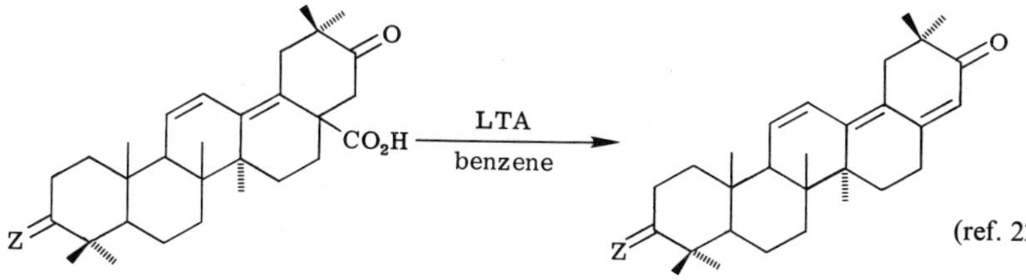

(ref. 226a)

Z = O

Z = OH/H

Z = O (67%)

Z = OH/H (90%)

such as **124** that are structurally not able to give alkenes can give substitution products in excellent yield.[224] The acid **125** underwent decarboxylation

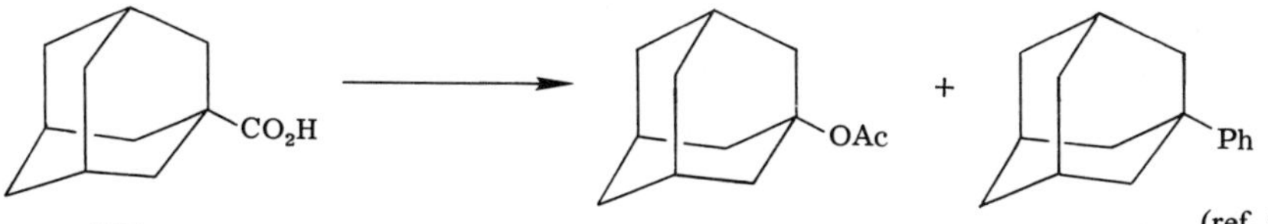

124

(ref. 224)

LTA/benzene	(89%)	(2%)
LTA/$Cu(OAc)_2$ benzene	(93%)	(1%)

[225] J. W. Huffman, *J. Org. Chem.* **35**, 478 (1970).

[226] R. Caputo, L. Mangoni, L. Previtera, and R. Iaccarino, *Tetrahedron* **29**, 2047 (1973).

[226a] S. Amagaya, T. Takeda, Y. Ogihara, and K. Yamasaki, *J. Chem. Soc. Perkin Trans. 1*, 2044 (1979).

followed by cyclization during experiments conducted to obtain structural proof for **126**.[181] Cyclization prior to decarboxylation was noted for LTA

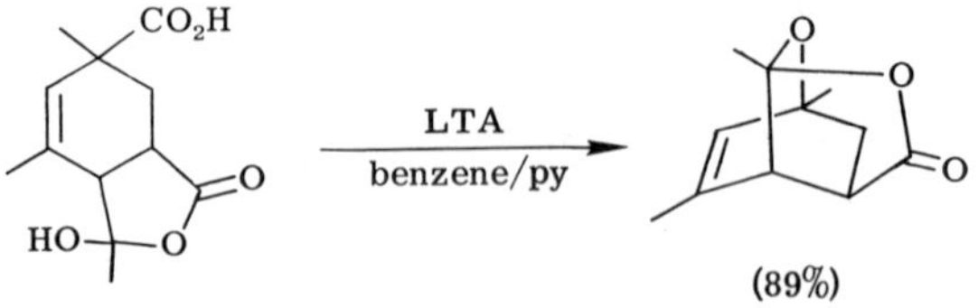

125 (89%) (ref. 181)

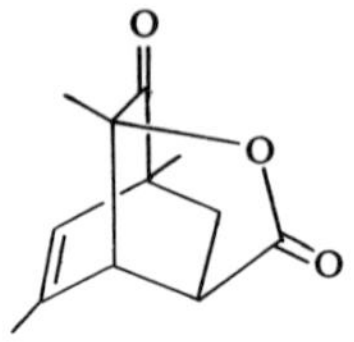

126

treatment of acid **127**.[227]

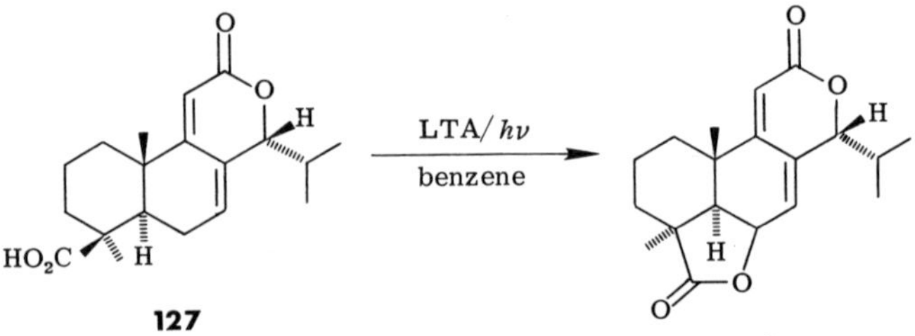

127 (50%) (ref. 227)

An interesting decarboxylation involving **128** gave a mixture containing **129**, **130**, and **131**.[227a] These results are best rationalized by cyclization of an intermediate radical followed by oxidation leading to the products observed.

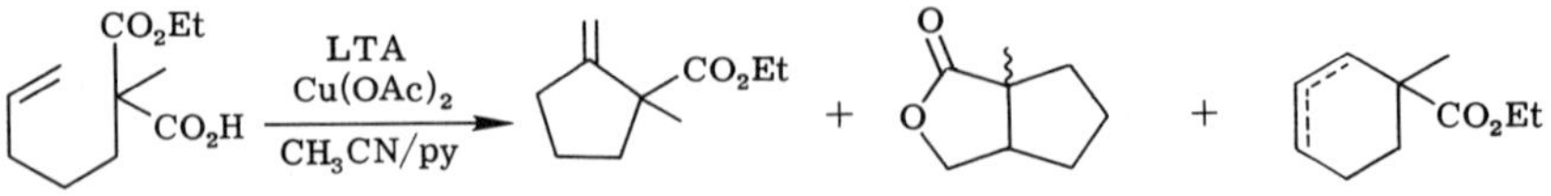

128 **129** (22%) **130** (11%) **131** (4%)

(ref. 227a)

[227] Y. Hayashi, T. Matsumoto, T. Hyono, N. Nishikawa, M. Uemura, M. Nishizawa, M. Togami, and T. Sakan, *Tetrahedron Lett.*, 3311 (1979).

[227a] M. Julia, J. M. Salard, and J. C. Chottard, *Bull. Soc. Chim. Fr.*, 2478 (1973).

4. α-SUBSTITUTED ACIDS

Oxidation of α-hydroxy carboxylic acids is a high yield method for the production of ketones. One example of the usefulness of the reaction in the presence of multiple oxidizable functionality involves **132**.[228] The mechanism of the decarboxylation has been investigated[229–232] and intermediates such as **133** postulated.[230,231] The catalysis of the reaction by bases, such as amines, water, and acetate ion, has been rationalized in terms of oxidant modification rather than by proton removal.[232]

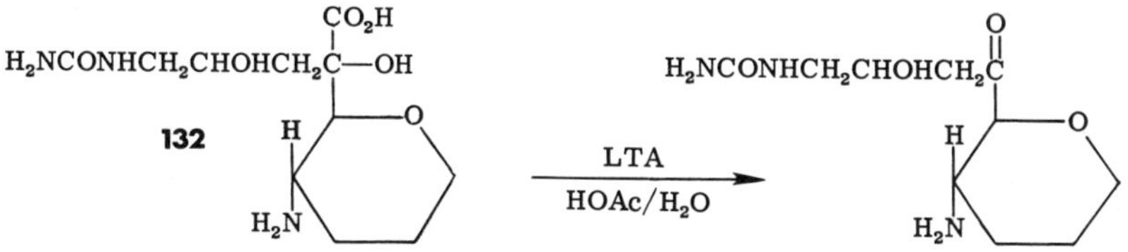

(ref. 228)

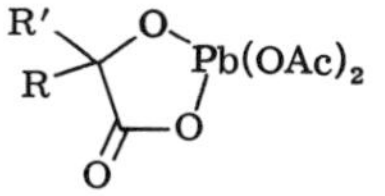

133

α-Alkoxy acids also decarboxylate readily and the degradation of oligoglycosides represents a synthetically useful case.[233] Treatment of sodium

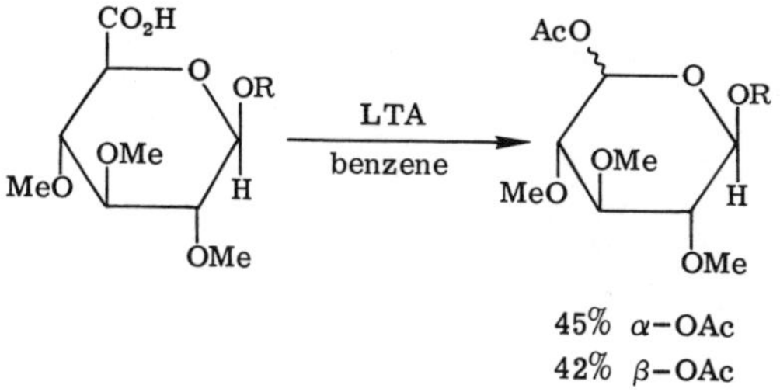

(ref. 233)

[228] S. Harada, E. Mizuta, and T. Kishi, *J. Am. Chem. Soc.* **100**, 4895 (1978).
[229] R. Shanker, S. K. Banerjee, and O. P. Sachdeo, *Z. Naturforsch. B. Anorg. Chem., Org. Chem.* **28**, 375 (1973).
[230] Y. Pocker and B. C. Davis, *J. Am. Chem. Soc.* **95**, 6216 (1973).
[231] K. Swaminathan, S. Sundaram, and N. Venkatasubramanian, *Indian J. Chem.* **13**, 1163 (1975).
[232] Y. Pocker and B. C. Davis, *J. Org. Chem.* **40**, 1625 (1975).
[233] I. Kitagawa, M. Yoshikawa, Y. Ikenishi, K. S. Im, and I. Yosioka, *Tetrahedron Lett.*, 549 (1976).

glycidates with LTA gives moderate to high yields of the corresponding α-acetoxy aldehydes[234] or ketones.[234a] Thermolysis of the latter leads to the

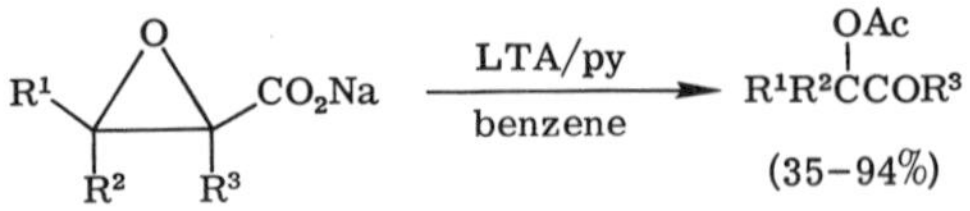

(refs. 234, 234a)

production of α,β-unsaturated ketones, again in high yield.[234a] Decarboxylation of 2-methyl-2-*tert*-butylperoxypropanoic acid has also been investigated.[235] The decomposition of **134** by LTA/$Cu(OAc)_2$ is a convenient method for alcohol regeneration in a sequence designed to obtain optically pure alcohol.[236]

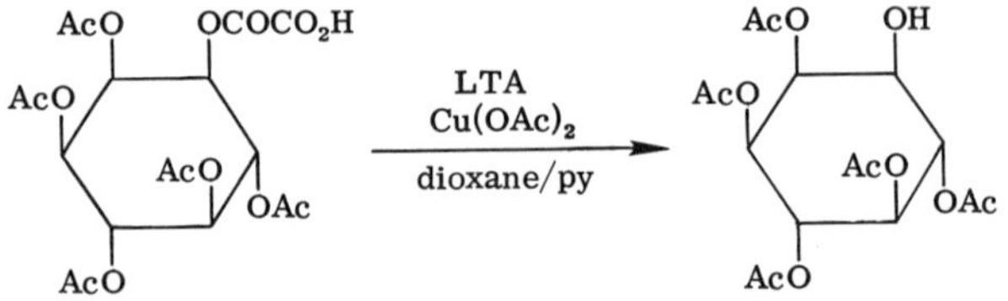

(ref. 236)

134

Oxidation of α-amino acids such as *N*,*N*-dimethylglycine has been used as a degradative procedure for documenting biosynthetic pathways.[237,238] Some question as to the reliability of the method was raised,[238] but control experiments seem to justify the approach.[237] The method is also successful

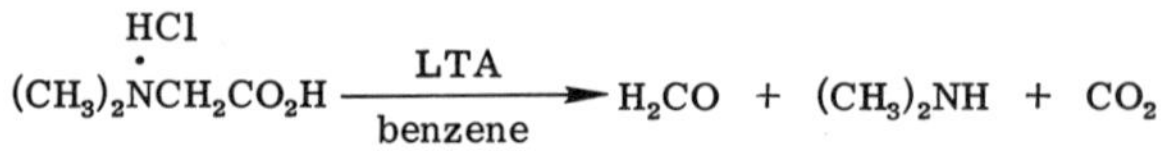

(refs. 237, 238)

for the degradation of hippuric acid.[239] Sequential treatment of oxazolin-5-ones with hydroxide or acid and then LTA provides a general synthesis for both aldehydes and ketones.[240] The preparation of **135** and **136** is illustrative.

[234] B. D. Kulkarni and A. S. Rao., *Synthesis*, 454 (1975).

[234a] V. Reutrakul, S. Nimgirawath, S. Panichanun, and Y. Srikirin, *Tetrahedron Lett.*, 1321 (1979).

[235] W. H. Richardson and W. C. Koskinen, *J. Org. Chem.* **41**, 3188 (1976).

[236] J. G. Molotkovsky and L. D. Bergelson, *Tetrahedron Lett.*, 4791 (1971).

[237] A. A. Liebman, B. P. Mundy, M. L. Rueppel, and H. Rapoport, *J. Chem. Soc. Chem. Commun.*, 1022 (1972).

[238] E. Leete, *J. Chem. Soc. Chem. Commun.*, 1524 (1971).

[239] E. Leete and S. A. Slattery, *J. Am. Chem. Soc.* **98**, 6326 (1976).

[240] R. Lohmar and W. Steglich, *Angew. Chem. Int. Ed. Engl.* **17**, 450 (1978).

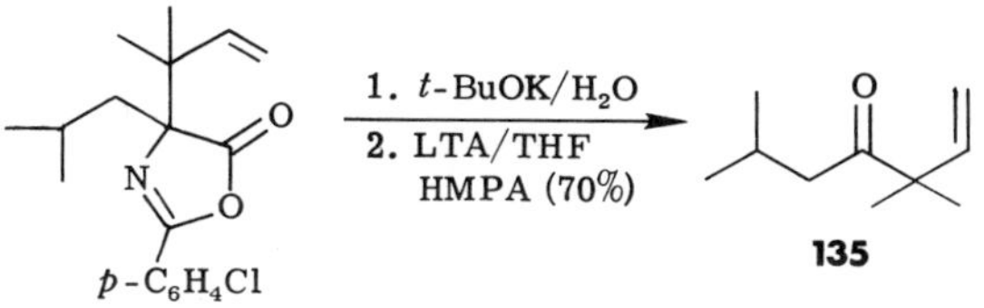

135

(ref. 240)

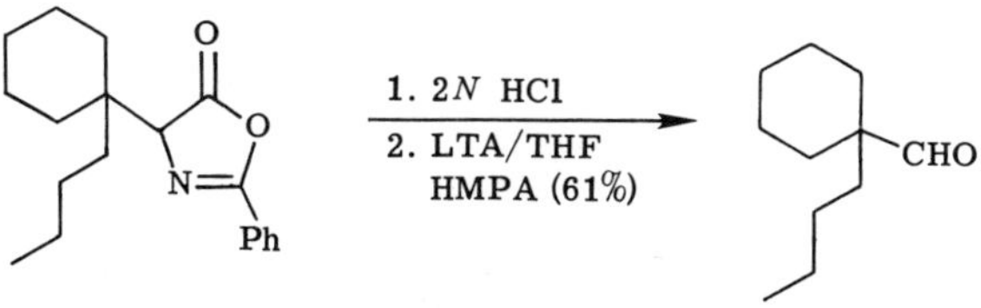

136

Two additional examples of α-amino acid oxidation involve the β-lactam **137**[241] and the preparation of both cyclic and acyclic α-acetoxy nitrosamines from the corresponding nitrosoamino acids.[242,243] *N*-Nitrosopipecolic acid

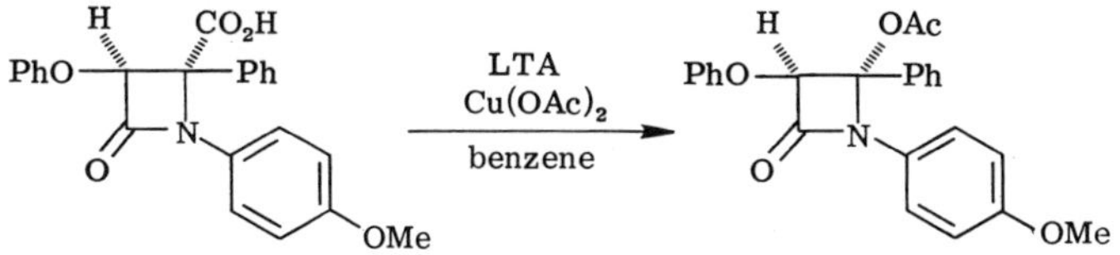

137 (68%) (ref. 241)

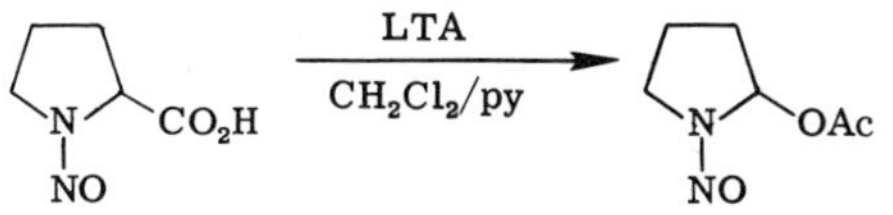

(35–44%)

(refs. 242, 243)

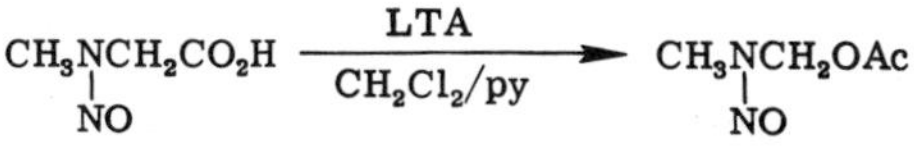

(37%)

[241] A. K. Bose, M. Tsai, J. C. Kapur, and M. S. Manhas, *Tetrahedron* **29**, 2355 (1973).
[242] J. E. Saavedra, *J. Org. Chem.* **44**, 4511 (1979).
[243] J. E. Saavedra, *Tetrahedron Lett.*, 1923 (1978).

affords a complex mixture of products in which the α-acetoxynitrosamine is accompanied by compounds derived from alkene formation. With the nitrosamine derivatives, the oxidation is materially enhanced by pyridine and inhibited by oxygen.[242]

5. Halodecarboxylation

Perhaps the most useful modification of the LTA reaction with carboxylic acids involves carrying out the normal procedure in the presence of halide ion. In this manner excellent yields of the corresponding alkyl halide are produced. Table VI gives some representative examples[209,210,212,244–250]

TABLE VI

LTA-Halide Halodecarboxylation of Carboxylic Acids

Substrate	Conditions	Product	% Yield	Reference
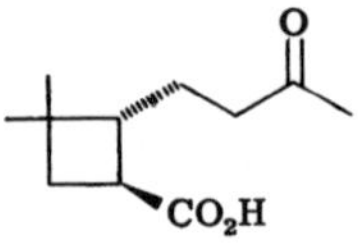	LTA/LiCl	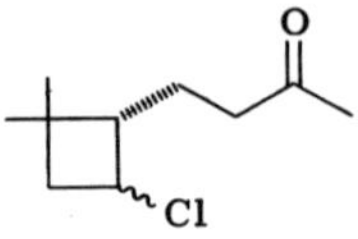	—	244
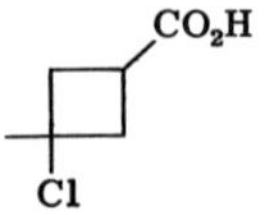	LTA/LiCl/benzene	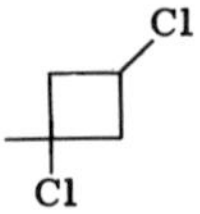	43	245
	LTA/LiCl	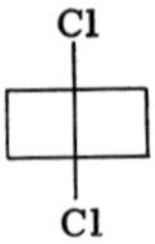	—	246
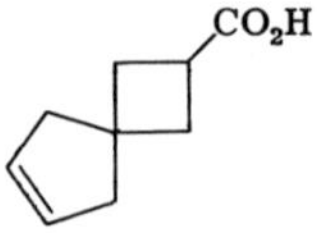	LTA/LiCl benzene/ether	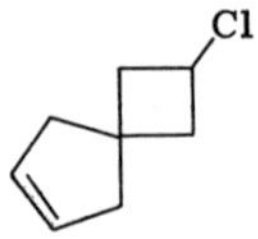	65	247
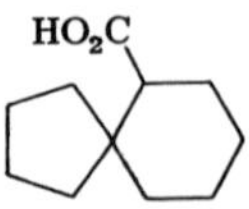	LTA/LiCl benzene	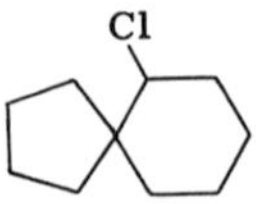	46	248

(*cont.*)

[244] A. F. Thomas and M. Ozainne, *J. Chem. Soc. Chem. Commun.*, 746 (1973).
[245] H. K. Hall, Jr., C. D. Smith, E. P. Blanchard, Jr., S. C. Cherkofsky, and J. B. Sieja, *J. Am. Chem. Soc.* **93**, 121 (1970).

TABLE VI (*cont.*)

Substrate	Conditions	Product	% Yield	Reference
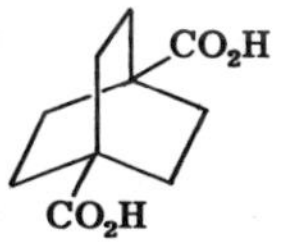	—		—	249
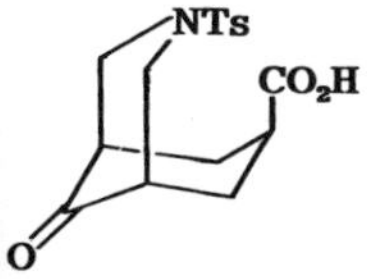	LTA/LiCl/benzene	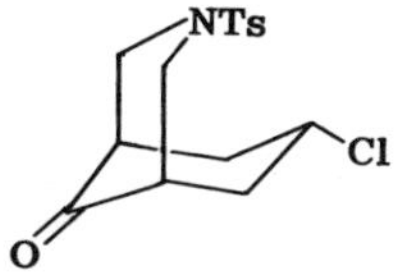	60	210
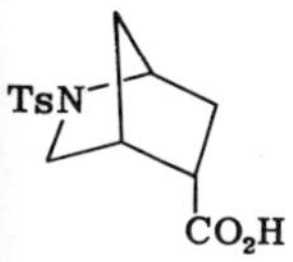	1. LTA/benzene/py 2. 40% HCl		39	212
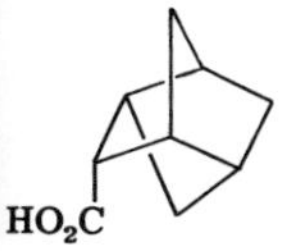	LTA/LiX/benzene	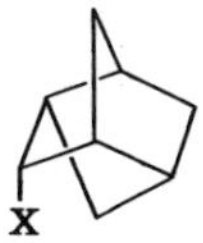	X Br 29 Cl 70 +5 other products 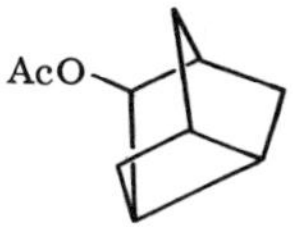	250
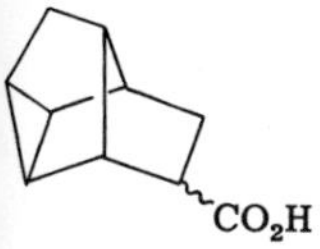	LTA/LiCl/benzene	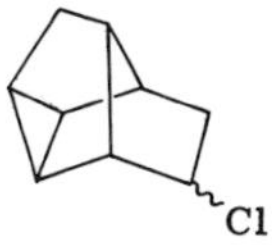	23 exo:endo 70–83:30–17 variable	209
		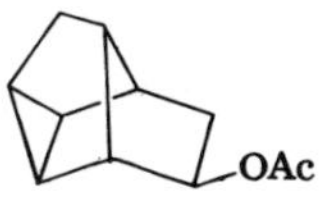	7	

[246] E. N. Cain, *Tetrahedron Lett.*, 1865 (1971).
[247] R. D. Miller and M. Schneider, *Tetrahedron Lett.*, 1557 (1975).
[248] G. E. Gream, *Aust. J. Chem.* **25**, 1051 (1972).
[249] K. B. Wiberg, G. A. Epling, and M. Jason, *J. Am. Chem. Soc.* **96**, 912 (1974).
[250] R. R. Sauers, K. W. Kelly, and B. R. Sickles, *J. Org. Chem.* **37**, 537 (1972).

and reference 192 includes many more. Another modification uses treatment of the acid with LTA followed by iodine and light.[35,251,252] The mechanism

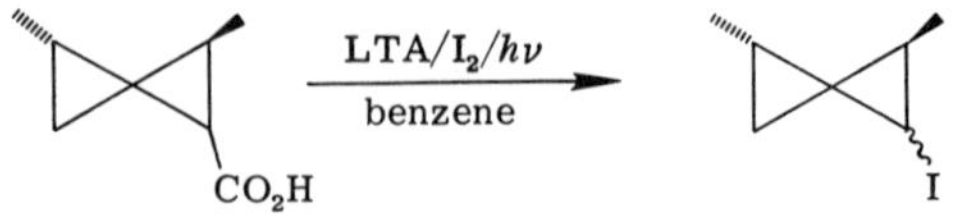

(ref. 251)

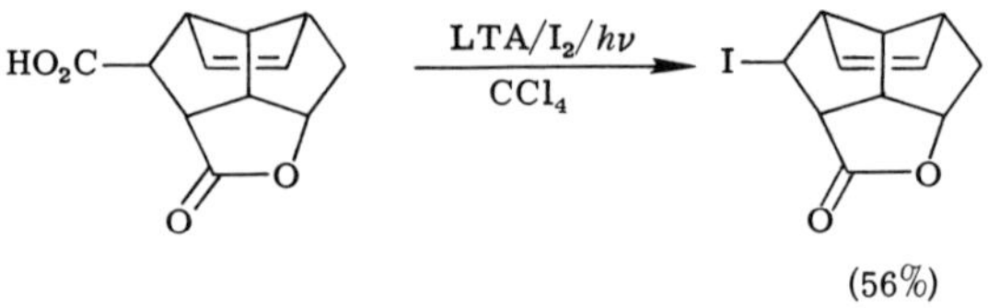

(ref. 252)

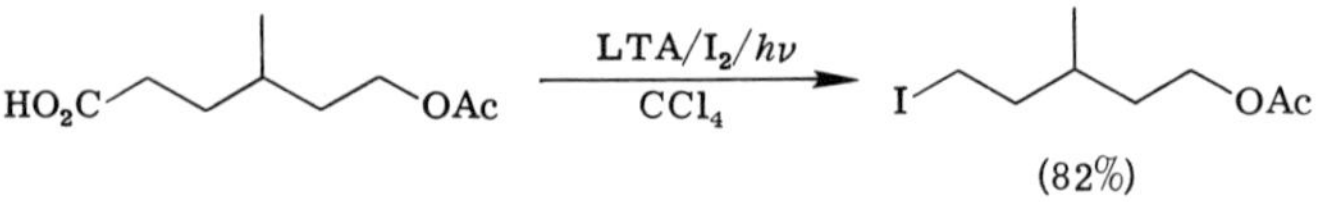

(ref. 35)

of the halodecarboxylation process is rationalized in terms of rapid ligand-transfer oxidation of alkyl radicals by halolead(IV)[192]. Studies on the

$$R\cdot + Pb(IV)X \longrightarrow RX + Pb(III)$$

stereochemistry of the reaction reveal that the reaction is nonstereospecific.[36,215,216,253,254] Remote functionality can alter epimer ratios by polar

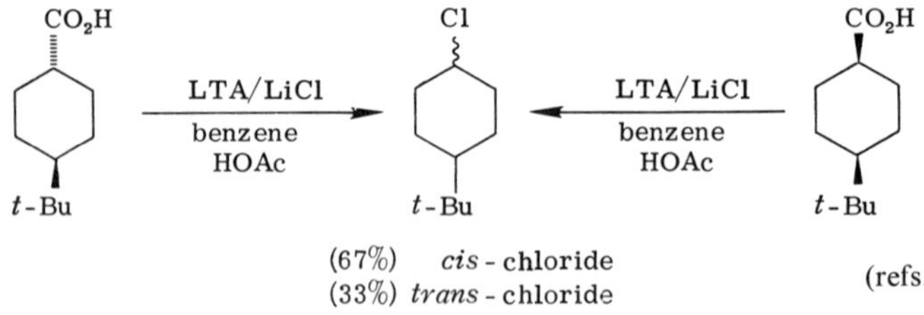

(refs. 36, 215, 253)

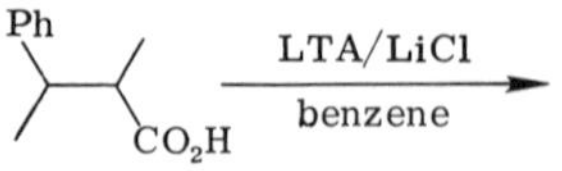

(ref. 216)

erythro	1.43 : 1.00 *threo*: *erythro*
threo	1.37 : 1.00 *threo*: *erythro*

effects as noted by comparing the products formed from **138** versus **139**.[255]

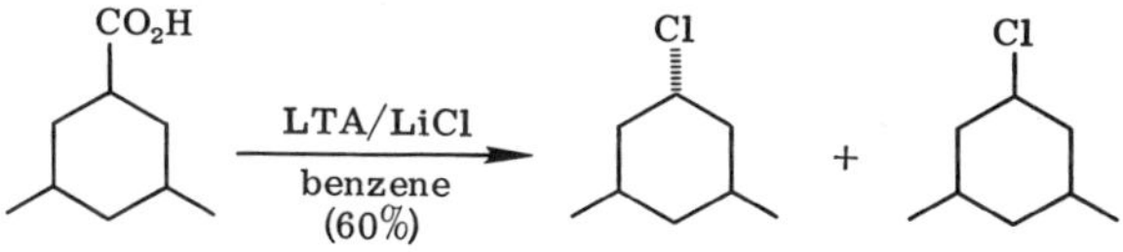

138

64:35

(ref. 255)

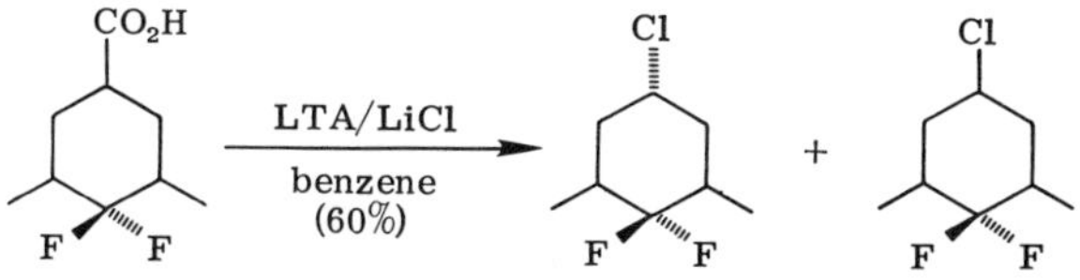

139

79:21

F. DICARBOXYLIC ACIDS

The bis-decarboxylation of 1,2-dicarboxylic acids by LTA, generally carried out in pyridine solvent, has proven to be an extremely useful method for the introduction of an alkene linkage, especially in cyclic systems. One example involves the preparation of 1,4-cyclohexadienes, compounds which are inaccessible by other standard methodology such as the Birch reduction of aromatics.[256] Yields are generally low, but the certainty of placement of

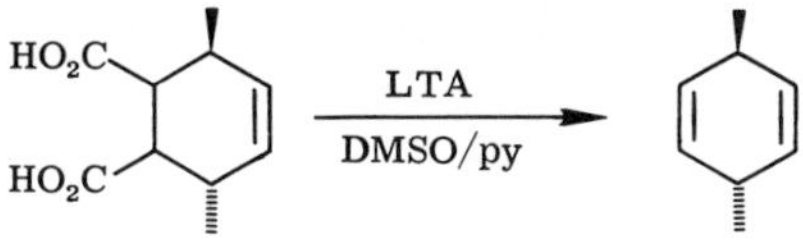

(ref. 256)

(35%)

the double bond makes the reaction attractive. Some additional cases, including examples of the synthesis of bridged polycyclic alkenes, are noted in Table VII.[256–264] References 12 and 192 contain many further examples.

---

[251] J. J. Gajewski and L. T. Burka, *J. Am. Chem. Soc.* **94**, 8860 (1972).

[252] L. A. Paquette, W. B. Farnham, and S. V. Ley., *J. Am. Chem. Soc.* **97**, 7273 (1975).

[253] R. D. Stolow and T. W. Giants, *Tetrahedron Lett.*, 695 (1971).

[254] See also reference 37.

[255] R. D. Stolow and T. W. Giants, *J. Am. Chem. Soc.* **93**, 3536 (1971).

[256] F. Stoos and J. Roček, *J. Am. Chem. Soc.* **94**, 2719 (1972).

TABLE VII

LTA BIS-DECARBOXYLATION OF 1,2-DICARBOXYLIC ACIDS

Substrate	Solvent	Product	% Yield	Reference
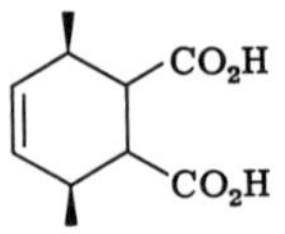	DMSO/py		11	256
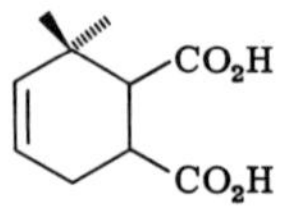	DMSO/py	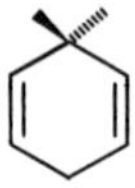	30	256
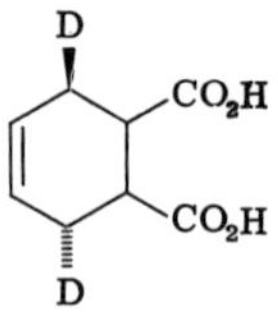	Py		—	257
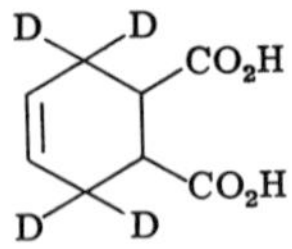	Benzene/py	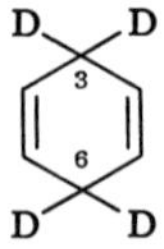	32 12% H at C-3 and C-6	257a
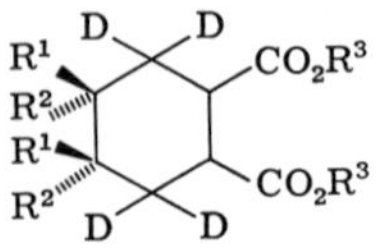	Benzene/py	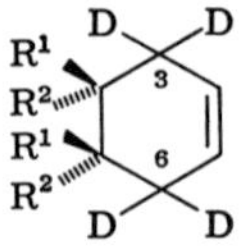		257a
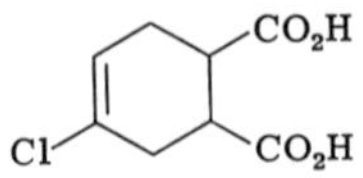	DMSO/py	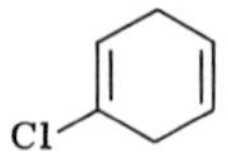	—	258
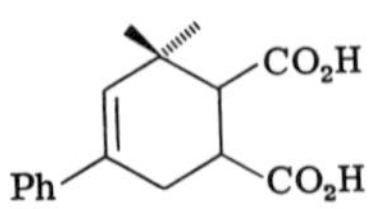	Dioxane/py	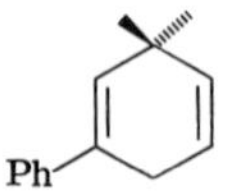	10	259

R^1	R^2	R^3	R^1	R^2	% yield	% H at C-3 and C-6
H	H	H	H	H	35–42	10
H	H	D	H	H	41	8
H	H	Na	H	H	14	9
D	H	H	D	H	38	9

TABLE VII (*cont.*)

Substrate	Solvent	Product	% Yield	Reference
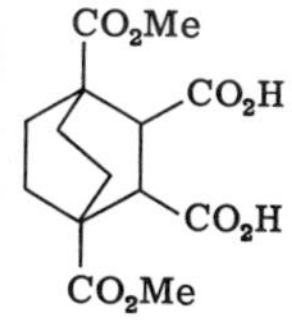	DMSO/py or dioxane	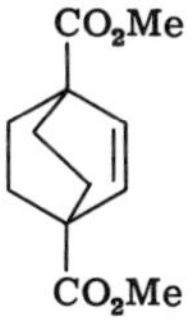	45	260
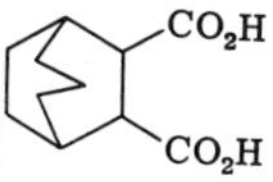	Py		—	261
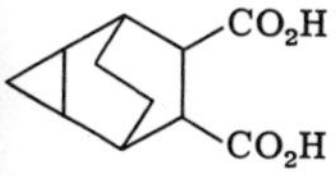	Benzene/py	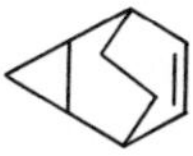	36	262
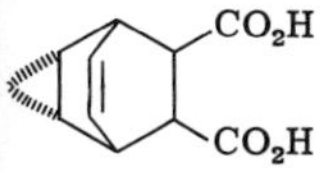	—	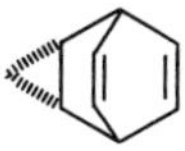	30–40	263
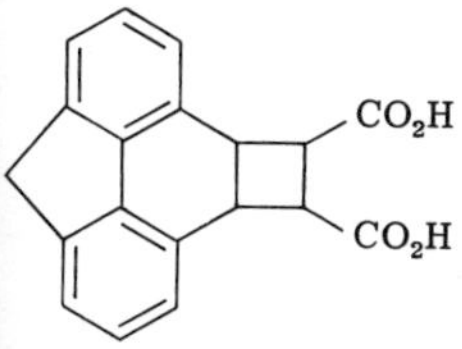	Benzene/py	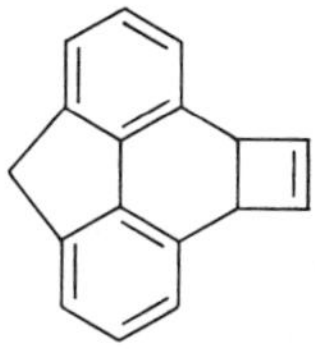	21	263a
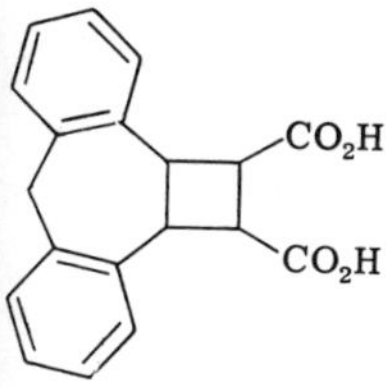	Benzene/py/N_2	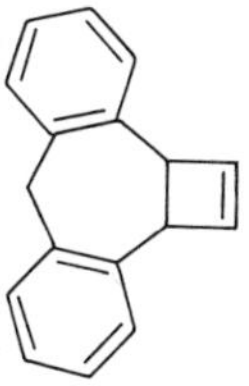	25	263b
	Py/O_2		31	
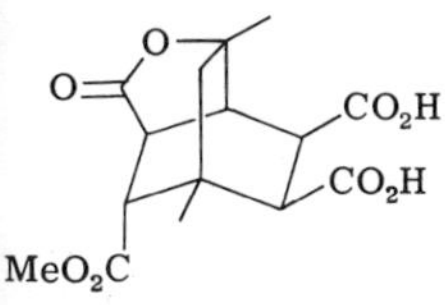	Py/O_2	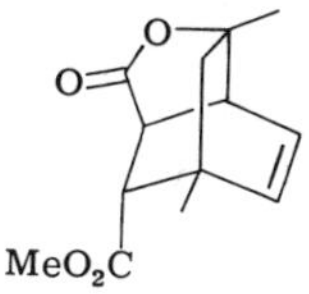	27	264

The last entry in Table VII is illustrative of the fact that the presence of molecular oxygen often enhances the yields from the reaction. This same oxygen effect was also noted in the preparation of **140** from the corresponding dicarboxylic acids.[265]

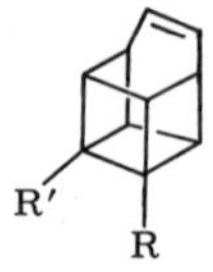

140

R	R′
H	H
H	CH_3
CH_3	CH_3

(ref. 265)

Use of excess LTA, or the sequential treatment of the appropriate dicarboxylic acid with LTA and a dehydrogenation agent, affords a method for the preparation of substituted aromatic systems **141**–**144**.[266–268a] Several reports have appeared concerning the preparation of acenaphthylene

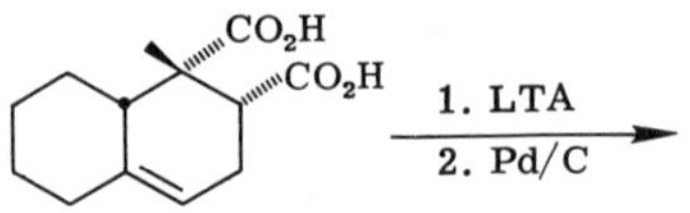

(ref. 266)

141

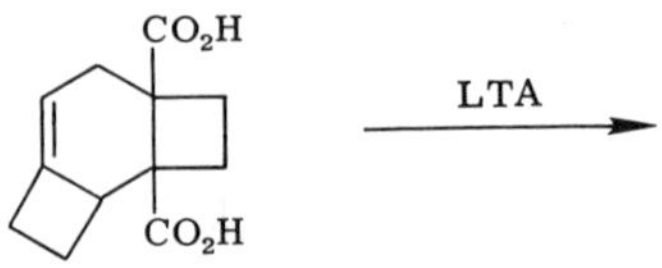

(ref. 267)

142 (~5%)

(ref. 268)

143 (12%)

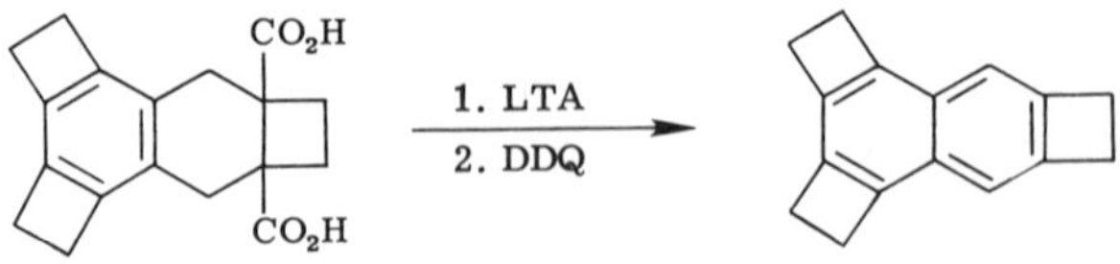

(ref. 268a)

144 (10%)

derivatives **145**.[202,269–271] Reported yields are variable and reproducibility of experimental conditions has been questioned.[202] The use of anhydrides

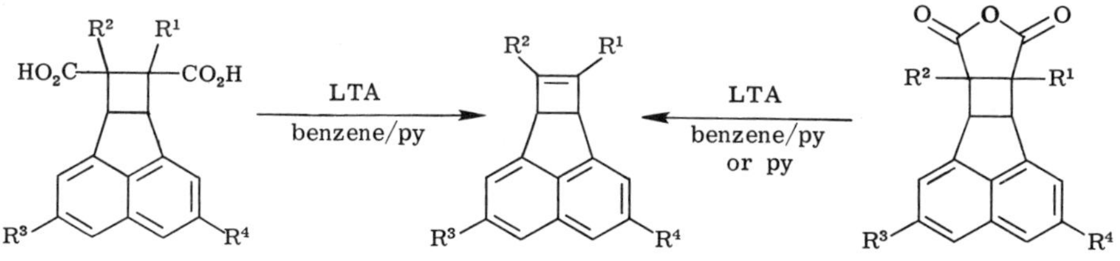

145 **146**

such as **146** for producing alkenes[269] has a special advantage in that many anhydrides are available via [2 + 2]- or [4 + 2]cycloadditions of alkenes or dienes, respectively, with maleic anydride. The synthesis of Dewar benzene (**147**) represents another pertinent example.[272] In certain cases anhydrides

147 (5–20%) (ref. 272)

---

[257] I. Fleming and E. Wildsmith, *J. Chem. Soc. Chem. Commun.*, 223 (1970).
[257a] S. Wolfe and J. R. Campbell, *Synthesis*, 117 (1979).
[258] H. G. Selander, D. M. Jerina, D. E. Piccolo, and G. A. Berchtold, *J. Am. Chem. Soc.* **97**, 4428 (1975).
[259] H. E. Zimmerman and L. M. Tolbert, *J. Am. Chem. Soc.* **97**, 5497 (1975).
[260] N. B. Chapman, S. Sotheeswaran, and K. J. Toyne, *J. Org. Chem.* **35**, 917 (1970).
[261] G. A. Russell, R. G. Keske, G. Holland, J. Mattox, R. S. Givens, and K. Stanley, *J. Am. Chem. Soc.* **97**, 1892 (1975).
[262] P. E. Schueler and Y. E. Rhodes, *J. Org. Chem.* **39**, 2063 (1974).
[263] Y. E. Rhodes, P. E. Schueler, and V. G. DiFate, *Tetrahedron Lett.*, 2073 (1970).
[263a] I. Willner and M. Rabinovitz, *J. Org. Chem.* **45**, 1628 (1980).
[263b] I. Willner and M. Rabinovitz, *Tetrahedron* **35**, 2359 (1979).
[264] J. Wolinsky and R. B. Login, *J. Org. Chem.* **37**, 121 (1972).
[265] L. A. Paquette, R. S. Beckley, and W. B. Farnham, *J. Am. Chem. Soc.* **97**, 1089 (1975).
[266] T. Tokoroyama, K. Matsuo, and T. Kubota, *Tetrahedron* **34**, 1907 (1978).
[267] R. P. Thummel, *J. Am. Chem. Soc.* **98**, 628 (1976).
[268] R. P. Thummel and W. Nutakul, *J. Am. Chem. Soc.* **100**, 6171 (1978).
[268a] K. G. Bilyard, P. J. Garratt, A. J. Underwood, and R. Zahler, *Tetrahedron Lett.*, 1815 (1979).
[269] J. E. Shields, D. Gavrilovic, J. Kopecký, W. Hartmann, and H.-G. Heine, *J. Org. Chem.* **39**, 515 (1974).
[270] S. F. Nelsen and J. P. Gillespie, *J. Am. Chem. Soc.* **95**, 1874 (1973).
[271] J. Meinwald, G. E. Samuelson, and M. Ikeda, *J. Am. Chem. Soc*, **92**, 7604 (1970).
[272] E. E. van Tamelen, S. P. Pappas, and K. L. Kirk, *J. Am. Chem. Soc.* **93**, 6092 (1971).

are formed as reaction side products which, in turn, do not afford alkenes under the conditions employed.[264,273] In the latter example, a 15–20% yield of diarylallenes could be realized without concomitant anhydride formation.

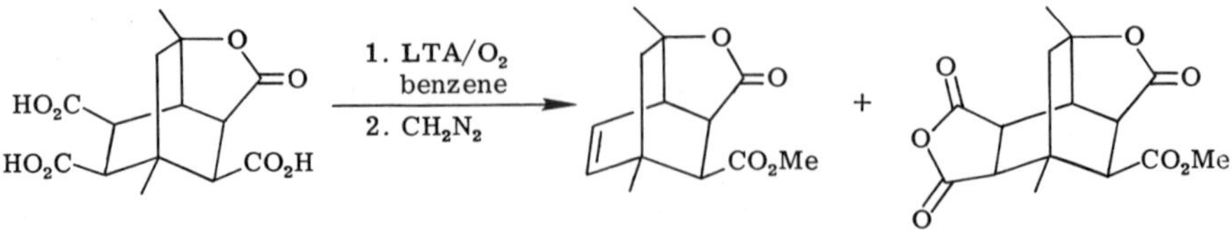

(ref. 264)

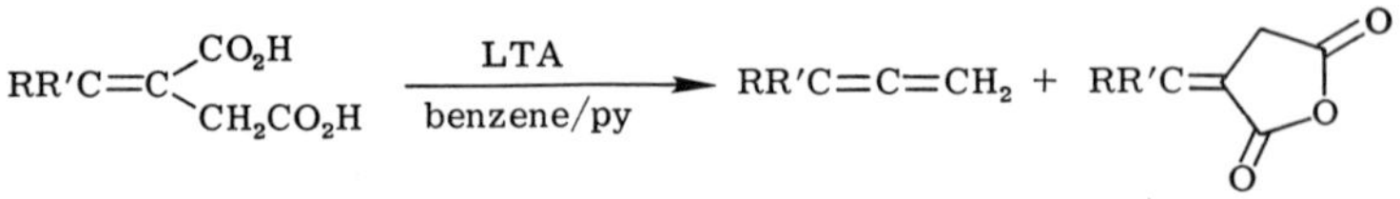

(ref. 273)

The possibility of concerted bis-decarboxylation seems to be ruled out by orbital symmetry arguments.[274] A transition state such as **148** requires a

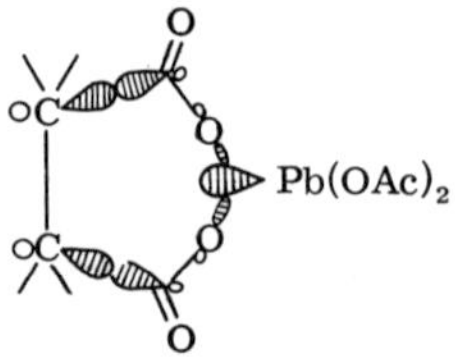

148

Hückel array of eight electrons and is therefore forbidden. The probable mechanism is stepwise and involves carbonium ion intermediates.[192] Sup-

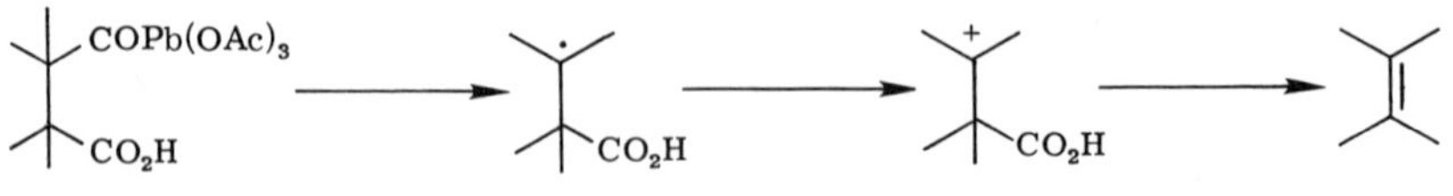

port for this picture arises from a number of observations. First, *trans*-dicarboxylic acids can be successfully bis-decarboxylated.[257a,275,276] The results noted in reference 257a are especially interesting in that bis-decarboxylation of the corresponding *cis*-1,2-dicarboxylic acids leads to significant

[273] J. K. Crandall, R. A. Colyer, and D. C. Hampton, *Synth. Commun.* **3**, 13 (1973).
[274] J. S. Littler, *Tetrahedron* **27**, 81 (1971).
[275] J. D. Slee and E. LeGoff, *J. Org. Chem.* **35**, 3897 (1970).
[276] W. B. Dauben, C. H. Schallhorn, and D. L. Whalen, *J. Am. Chem. Soc.* **93**, 1446 (1971).

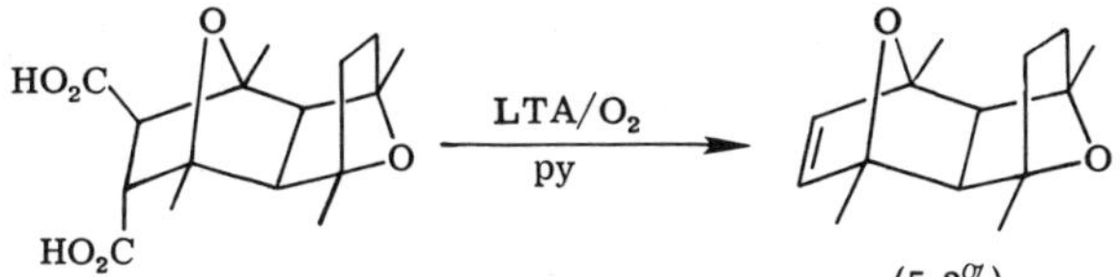

(5.2%) (ref. 275)

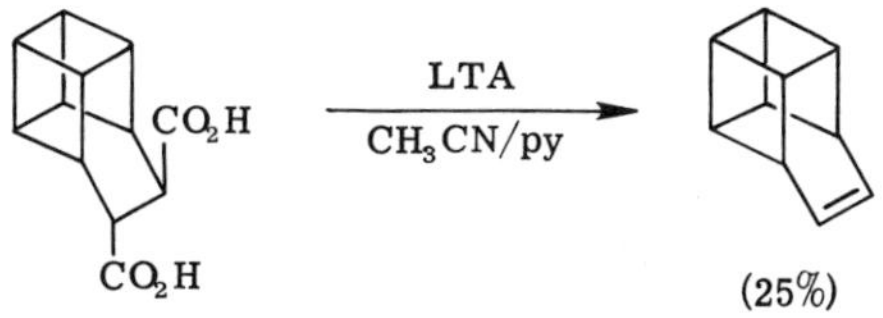

(25%) (ref. 276)

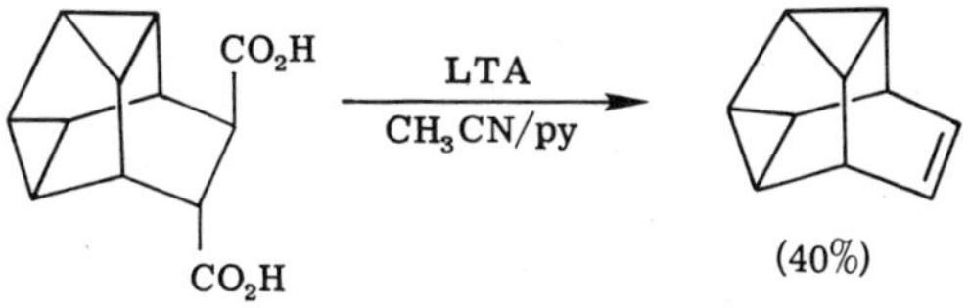

(40%) (ref. 276)

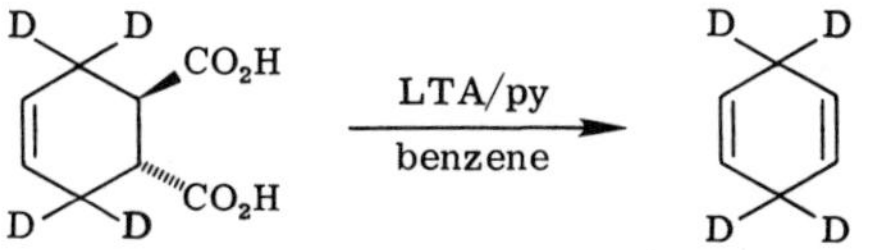

(40%)
(no scrambling) (ref. 257a)

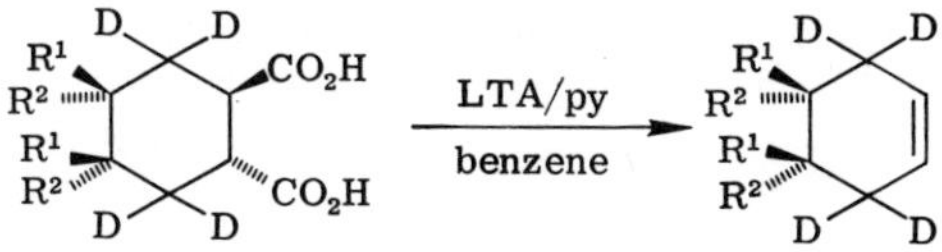

(ref. 257a)

R^1	R^2
D	H
H	H
C_6H_5	H

R^1	R^2	% yield
D	H	43
H	H	46
C_6H_5	H	69

(no scrambling noted)

deuterium scrambling (see Table VII). Second, products arising from carbonium ion (or radical) intermediates have been isolated.[202,277,277a]

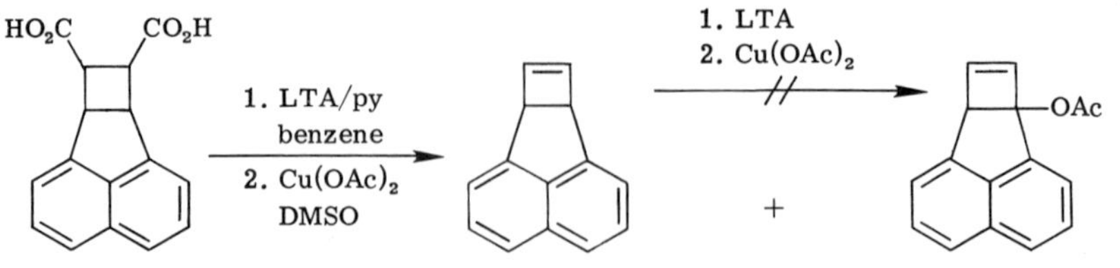

(7.2%) (6.1%)

(ref. 202)

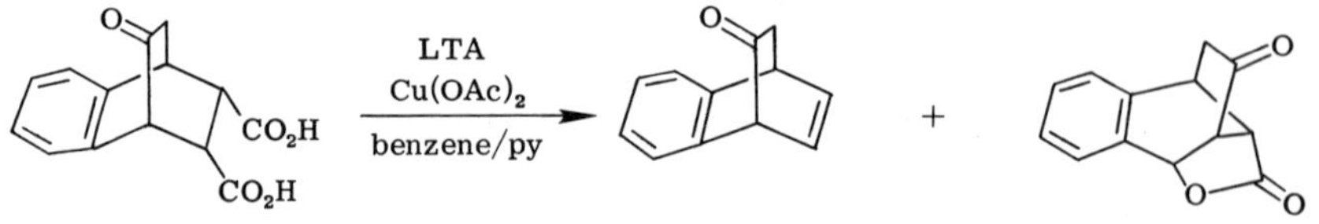

(13.3%) (7%)

(ref. 277)

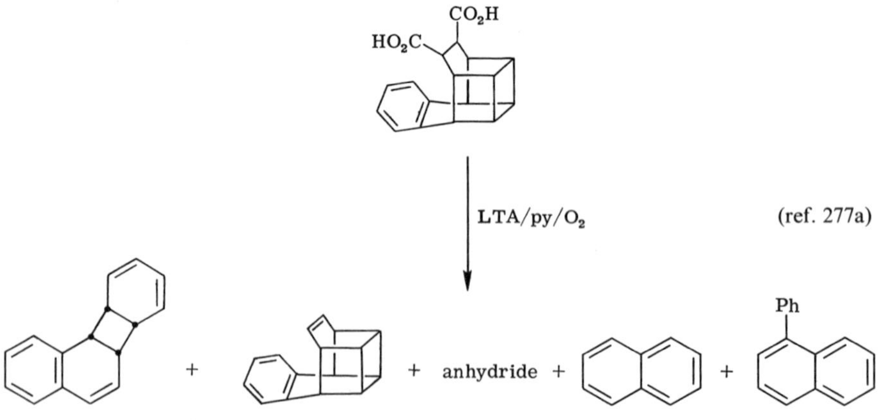

(ref. 277a)

(40–50%) (10–15%) (20–25%) (20–25%)

On the other hand, the reactions of **149**[264] and **150**[277b] with LTA might be concerted since a Hückel array involving six electrons may be involved.[274]

[277] R. S. Givens and W. F. Oettle, *J. Am. Chem. Soc.* **93**, 3963 (1971).

[277a] K. T. Mak, K. Srinivasachar, and N.-C. C. Yang, *J. Chem. Soc. Chem. Commun.*, 1038 (1979).

[277b] E. J. Corey and H. L. Pearce, *J. Am. Chem. Soc.* **101**, 5841 (1979).

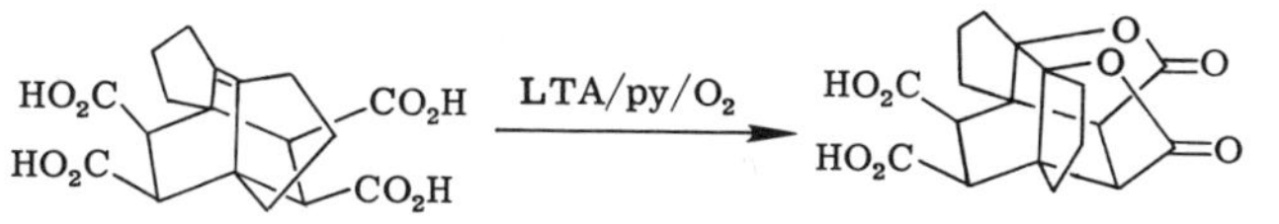

149

(ref. 264)

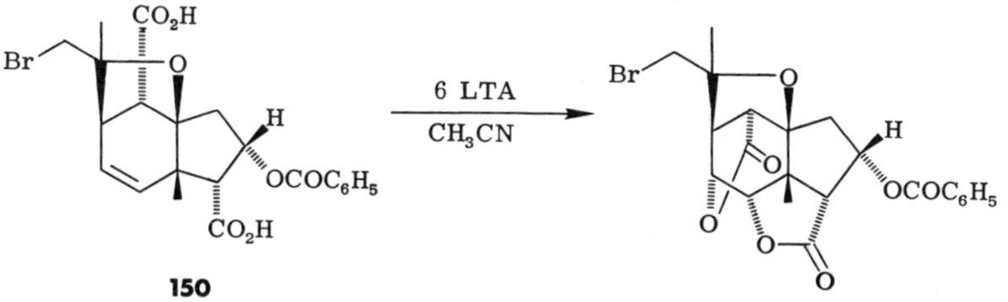

150

(99%)

(ref. 277b)

It is also of interest to note that **151** is reported to give a high yield of the *trans*-alkene **152**.[278]

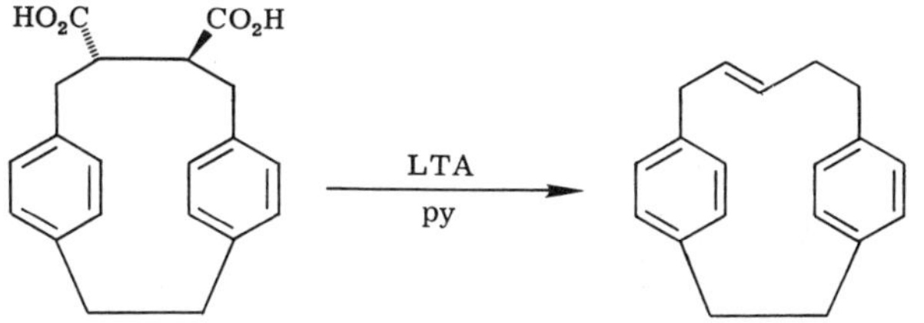

151

152 (70%)

(ref. 278)

III. LTA Reactions with Nitrogen-Containing Compounds

A. Amines and Related Compounds

1. Alkylamines

The reaction of primary alkylamines with LTA to produce nitriles and/or aldimines as well as the reaction of primary aryl amines to form azo compounds has been reviewed.[12,279] Recent studies have been mainly concerned with the effects of neighboring groups during amine oxidation. One extremely useful example involves the LTA treatment of primary amines

[278] D. J. Cram, R. B. Hornsby, E. A. Truesdale, H. J. Reich, M. H. Delton, and J. M. Cram, *Tetrahedron* **30**, 1757 (1974).

[279] J. B. Aylward, *Quart. Rev.* (*London*) **25**, 407 (1971).

containing a δ–ϵ double bond. With these systems, aziridines are formed in high yield; representative examples are given.[280–283] Oxidation of **153** re-

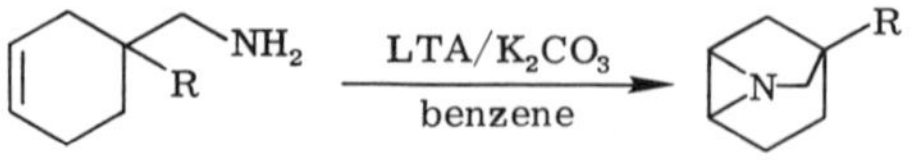

(ref. 280)

(55–60%)
R = H, Me

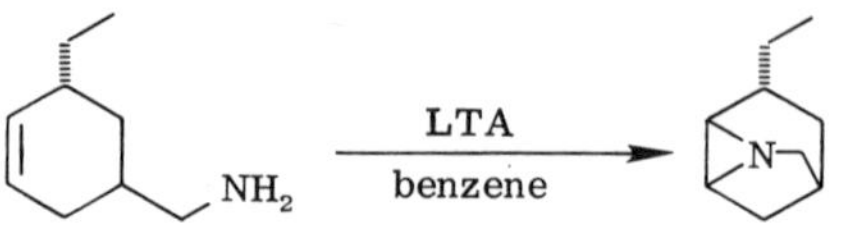

(ref. 281)

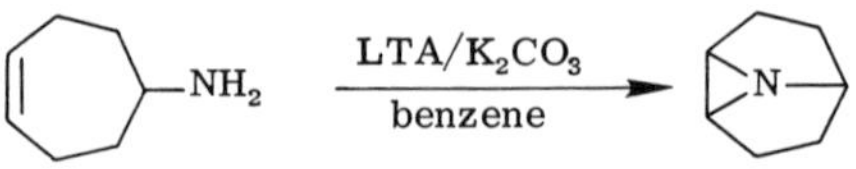

(ref. 282)

(90%)

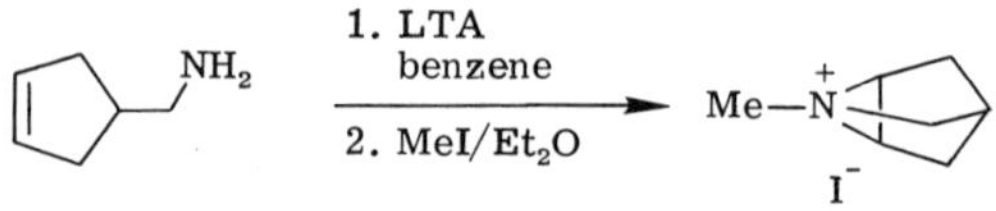

(ref. 283)

(52%)

sults in a complex mixture with the nitrile **154** the only identifiable product, thus indicating that δ–ϵ double bond placement in the amine is crucial.[282]

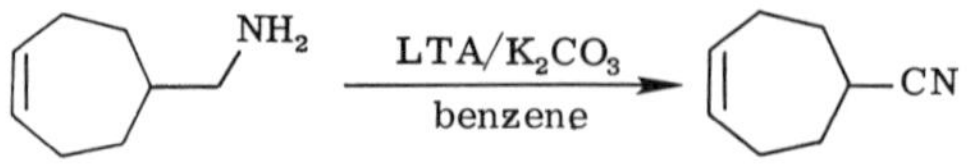

(ref. 282)

153 **154**

With **155**, the aziridine **156** was obtained in 78% yield and no trace of the 1,4-addition product **157** was detected.[284] Aziridenes constructed by the LTA method have been used as synthetic intermediates for the preparation

[280] W. Nagata, S. Hirai, K. Kawata, and T. Aoki, *J. Am. Chem. Soc.* **89**, 5045 (1967).
[281] W. Nagata, S. Hirai, T. Okumura, and K. Kawata, *J. Am. Chem. Soc.* **90**, 1650 (1968).
[282] W. Nagata, T. Wakabayashi, and N. Haga, *Synth. Commun.* **2**, 11 (1972).
[283] P. S. Portoghese and D. T. Sepp, *Tetrahedron* **29**, 2253 (1973).
[284] M. Narisada, F. Watanabe, and W. Nagata, *Synth. Commun.* **2**, 21 (1972).

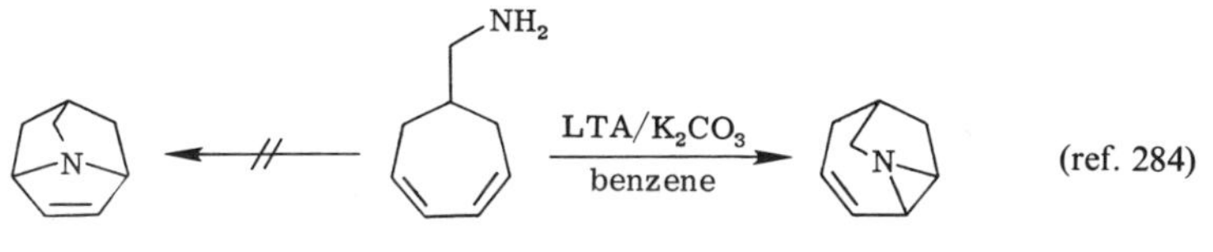

157 **155** **156** (78%)

of diverse structural types.[281–283] The mechanistic aspects of the aziridine-forming reactions are not clear. However, several instances where the behavior of primary amines with LTA has been directly compared to the behavior of nitrenes generated from azides of analogous structure indicate that nitrenes are not involved in the former cases.[285–287] For instance, whereas pyrolysis of **158** affords **159**, the treatment of **160** with LTA gives an 85% yield of **161**, probably via a concerted rearrangement.[285] Either concerted fragmentations or reactions involving nitrenium ions have been

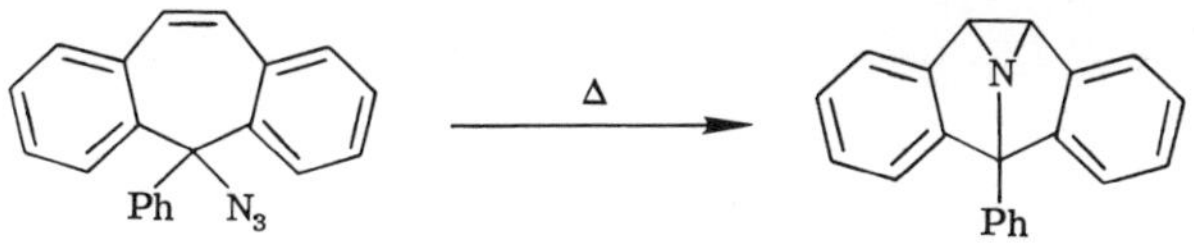

158 **159**

(ref. 285)

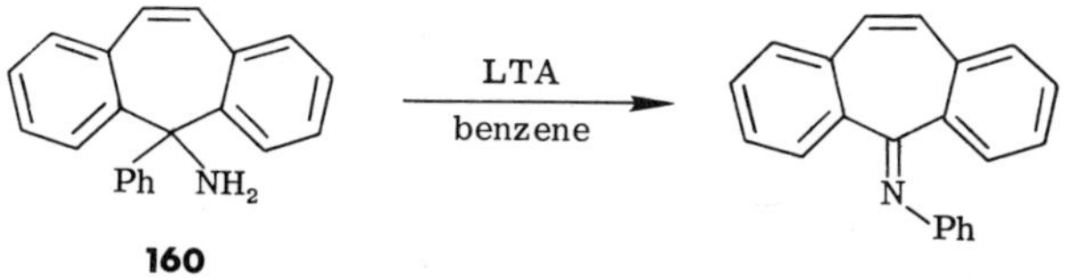

160

161 (85%)

postulated to explain the formation of *trans*-cinnamaldehyde from a 2-phenycyclopropylamine and the production of benzonitrile from LTA treatment of 1-phenylcyclopropylamine.[286,287]

Alkoxyamines react with LTA in the presence of alkenes to yield *N*-alkoxyaziridines.[288–292] Initially, a stepwise addition of nitrenium ion to alkene

[285] A. J. Sisti and S. R. Milstein, *J. Org. Chem.* **39**, 3932 (1974).
[286] T. Hiyama, H. Koide, and H. Nozaki, *Tetrahedron Lett.*, 2143 (1973).
[287] T. Hiyama, H. Koide, and H. Nozaki, *Bull. Chem. Soc. Jpn.* **48**, 2918 (1975).
[288] S. J. Brois, *J. Am. Chem. Soc.* **92**, 1079 (1970).
[289] B. V. Ioffe and E. V. Koroleva, *Khim. Geterotsikl. Soedin*, 1514 (1972).
[290] B. V. Ioffe and E. V. Koroleva, *Tetrahedron Lett.*, 619 (1973).
[291] F. A. Carey and L. J. Hayes, *J. Org. Chem.* **38**, 3107 (1973).
[292] B. V. Ioffe, Yu. P. Artsybasheva, and I. G. Zenkovich, *Dokl. Akad. Nauk SSSR* **231**, 1130 (1976).

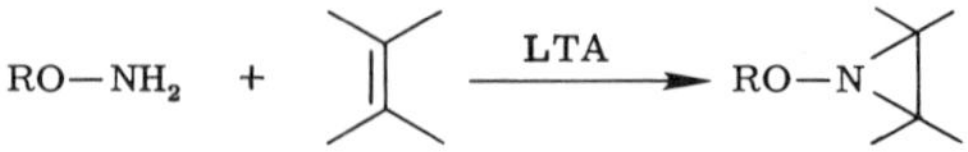

was postulated since it was believed that the reaction of *n*-butoxyamine with *cis*- and *trans*-2-butene was nonstereospecific.[291,293] Recent work, however, has shown that this reaction is in fact stereospecific so that a singlet nitrene may well be involved.[292] At least a two-step addition process involving a

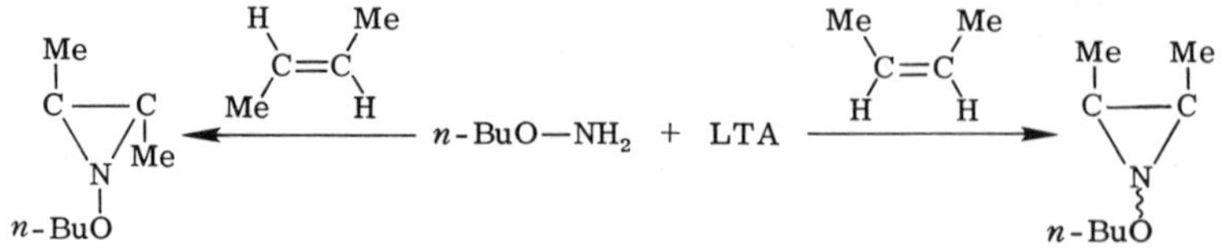

(ref. 292)

nitrenium ion can be excluded. When alkene is omitted, a major pathway for alkoxyamine oxidation is hyponitrite ester **162** formation with subsequent decomposition to alkoxy radicals and finally, production of alcohols.[291,294–296]

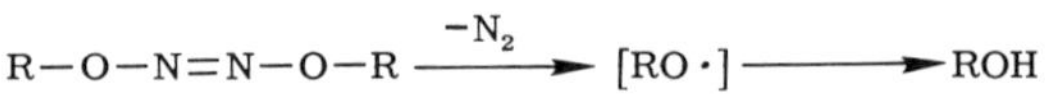

162

The cyclic system **163** is oxidized to the corresponding oxazine by LTA.[296a]

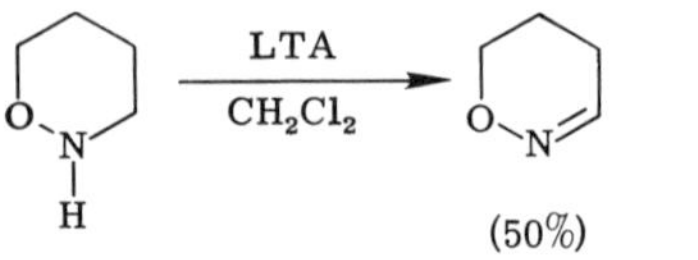

(ref. 296a)

163

Oxidation of **164** in the presence of electron-rich alkenes leads to the formation of aziridines.[296b] The reaction has been shown to be nonstereospecific with (*Z*)-1-phenylpropene and an equilibrating singlet and triplet nitrene has been implicated to explain the stereochemical results.[296c] When treatment of **164** with LTA is carried out in the presence of allyl aryl sufides,

[293] L. J. Hayes, F. P. Billingsley II, and Carl Trindle, *J. Org. Chem.* **37**, 3924 (1972).

[294] R. Partch, B. Stokes, D. Bergman, and M. Budnik, *J. Chem. Soc. Chem. Commun.*, 1504 (1971).

[295] A. J. Sisti and S. Milstein, *J. Org. Chem.* **38**, 2408 (1973).

[296] R. O. C. Norman, R. Purchase, and C. B. Thomas, *J. Chem. Soc. Perkin Trans. 1*, 1701 (1972).

[296a] V. J. Lee and R. B. Woodward, *J. Org. Chem.* **44**, 2487 (1979).

[296b] R. S. Atkinson and B. D. Judkins, *J. Chem. Soc. Chem. Commun.*, 832 (1979).

[296c] R. S. Atkinson and B. D. Judkins, *J. Chem. Soc. Chem. Commun.*, 833 (1979).

N-arylthiosulfenamides are produced. This reaction may be specific for the singlet nitrene,[296c] and the mechanism of product formation is discussed in more detail in Section III, C with regard to diazene trapping.

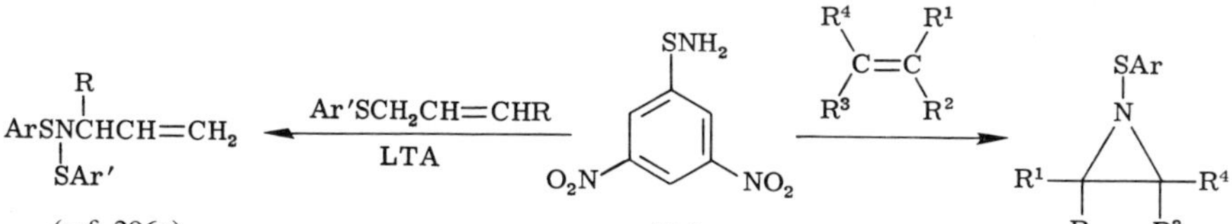

(ref. 296c) **164**

(refs. 296b, 296c)

Secondary and tertiary alkylamines react with LTA to form imine derivatives when α-hydrogen is available; thus **165** is oxidized to a mixture of

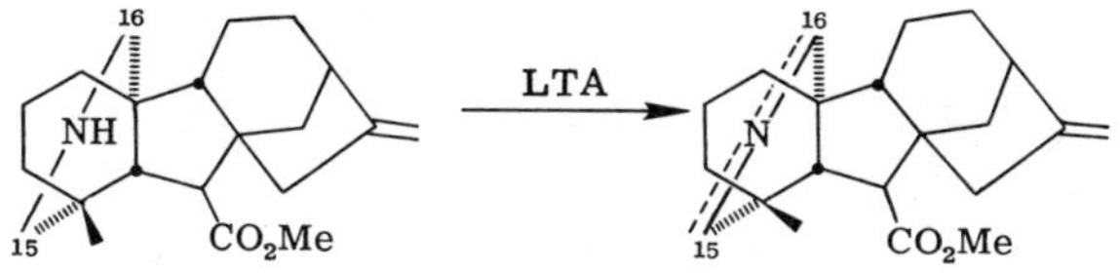

165 **166**

$\Delta^{15(N)}$- and $\Delta^{16(N)}$-azomethine isomers **166**.[297] The oxidation of alkaloid **167** is accompanied by fragmentation to yield **168**,[298] whereas a series of bicyclic

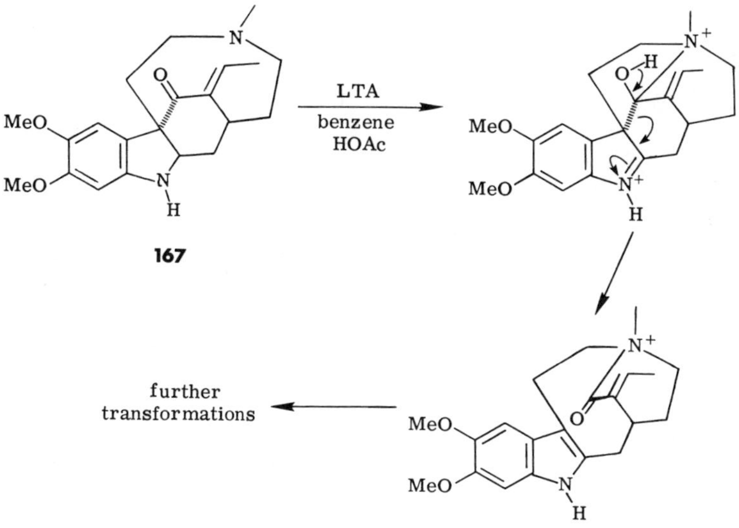

167

(ref. 298)

168

aziridines **169** fragment to produce the useful ω-cyanocarbonyl derivatives

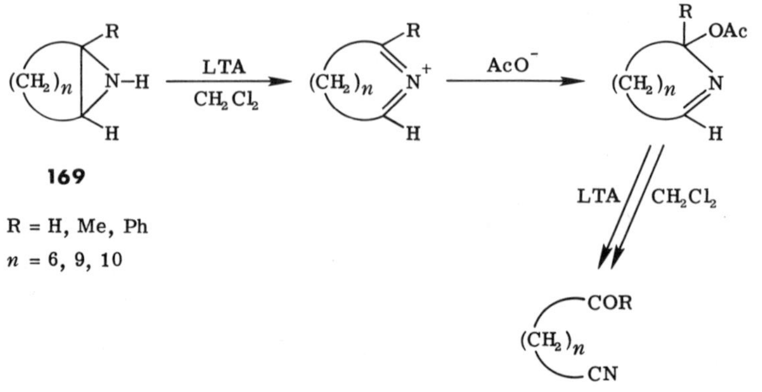

169

R = H, Me, Ph
n = 6, 9, 10

170 (29–85%)

(ref. 287)

170.[287] Treatment of either **171** or **172** with LTA affords the immonium salt **173**, and subsequent $NaBH_4$ reduction of **173** generates **172**.[299,300] Further, it has been shown that with base **173** can be transformed into the corresponding enamine **174** enabling R-group epimerization to occur.[300] Enamine

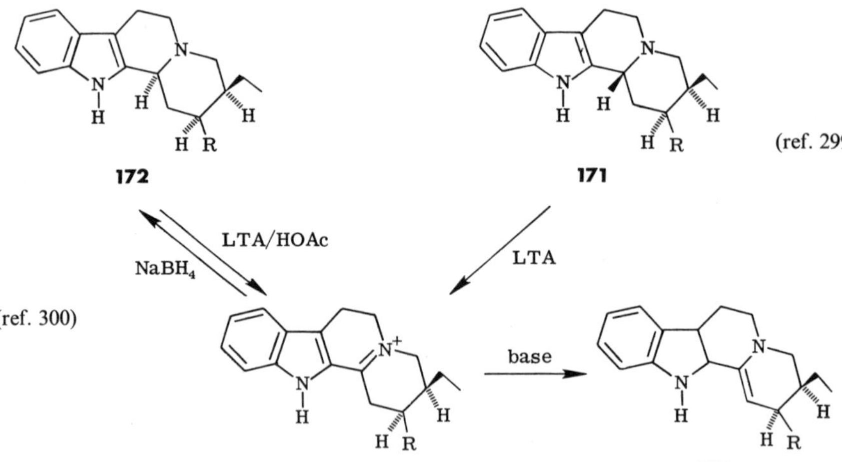

172 **171** (ref. 299)

(ref. 300)

173 **174** (ref. 300)

---

[297] W. Nagata, T. Wakabayashi, Y. Hayase, M. Narisada, and S. Kamata, *J. Am. Chem. Soc.* **92**, 3202 (1970).

[298] R. E. Moore and H. Rapoport, *J. Org. Chem.* **38**, 215 (1973).

[299] R. T. Brown, C. L. Chapple, and A. A. Charalambides, *J. Chem. Soc. Chem. Commun.*, 756 (1974).

[300] M. Bárczai-Beke, G. Dörnyei, M. Kajtár, and Cs. Szántay, *Tetrahedron* **32**, 1019 (1976).

formation is also noted in the oxidation of **175**,[301] **176**,[302] **177**,[303] and

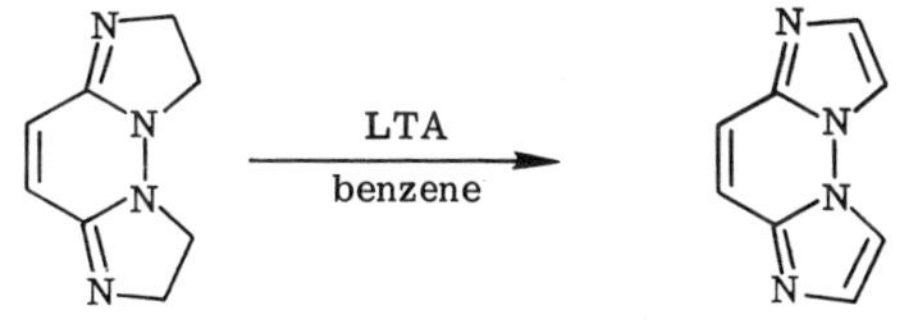

175 (46%) (ref. 301)

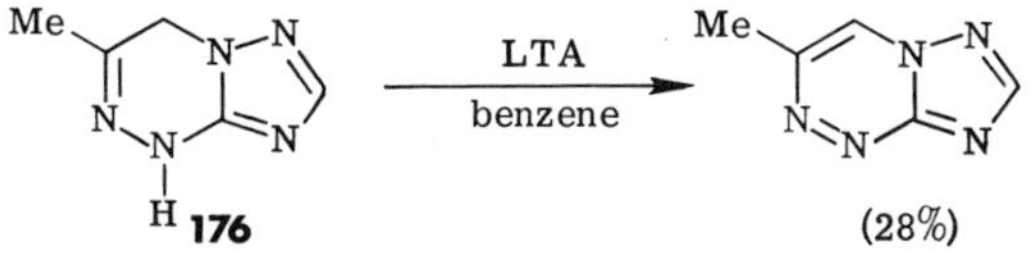

176 (28%) (ref. 302)

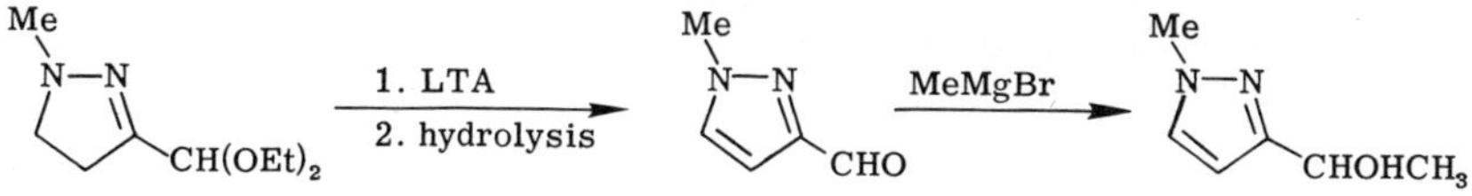

177 (68%)

(ref. 303)

several alkaloid systems,[304–306] and demethylation has been noted as well.[305,306] In one case, the use of pyridine enhanced the demethylation

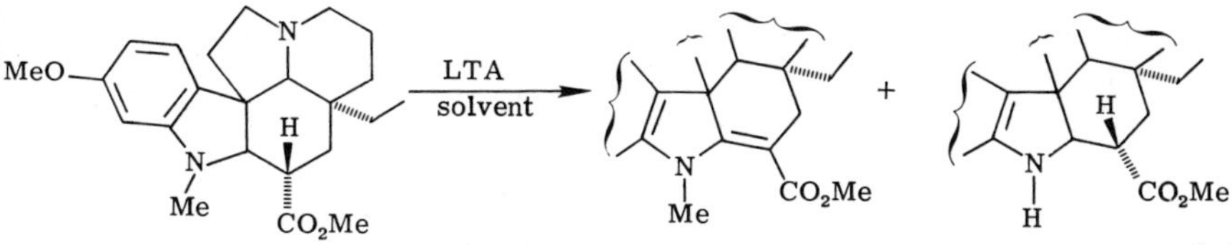

(ref. 305)

solvent		
benzene	20%	17%
pyridine	10%	41%

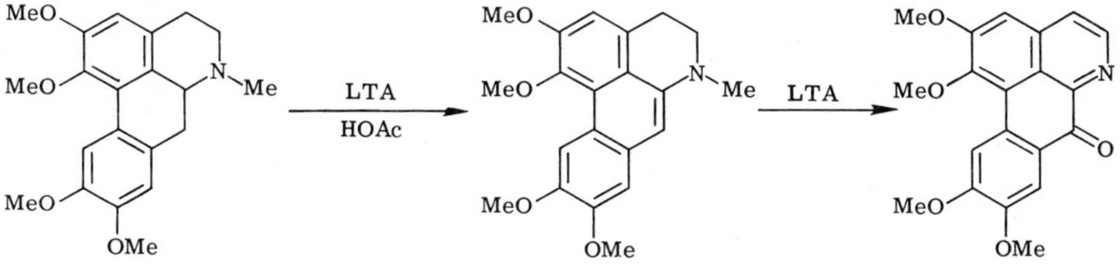

(7–10%) (69%)

(ref. 306)

process,[305] and when substituted *N*-alkyl-*N*-methylanilines are treated with LTA/Ac_2O/$CHCl_3$, high yields of the corresponding acetamides result, presumably via **178**.[307–309]

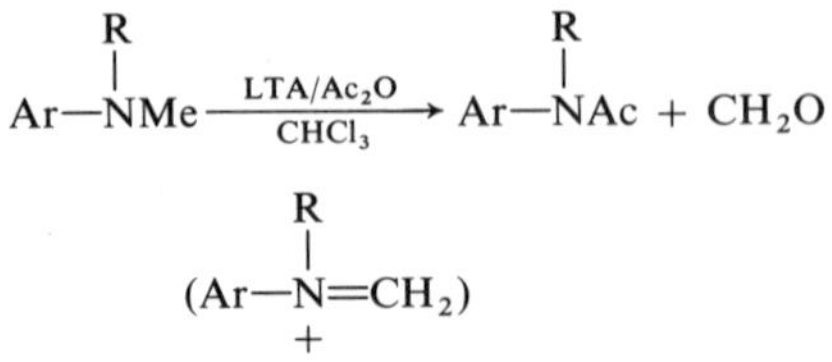

(**178**)

(ref. 307–309)

2. Aromatic Amines

Neighboring groups also effect the course of the LTA oxidation of aromatic amines.[279] For example, **179**[310] **180**[311] and **181**[312] give cyclic derivatives, often in high yield. The reaction has been rationalized in terms of both cation and radical intermediates.[312a] Oxidative cyclization can also be

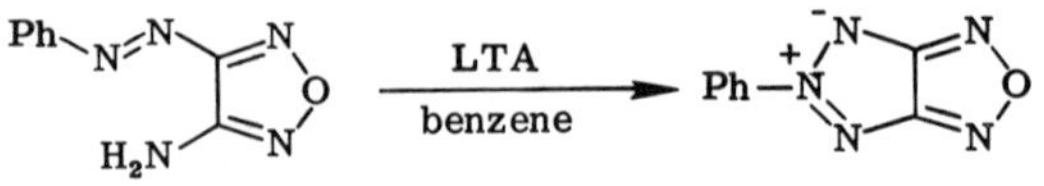

179 (28%) (ref. 310)

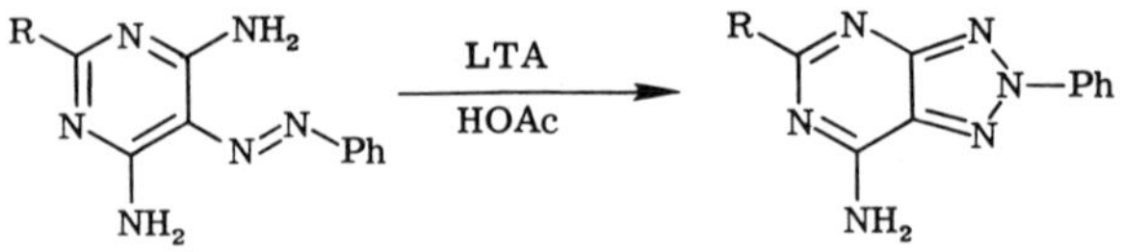

180

(ref. 311)

R	% yield
Ph	90
NH_2	75
H	86

(ref. 312)

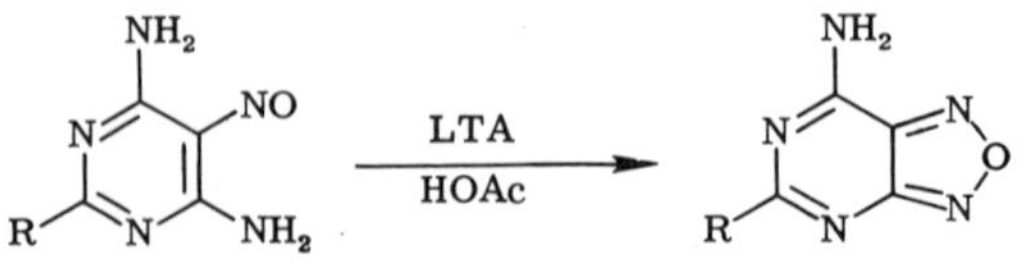

181

R	% yield	R	% yield
H	82	NMe_2	75
Me	86	SMe	72
NH_2	87	Ph	82

used with quinoidal systems, again with excellent results.[311,313] However, in one instance, the *N*-oxide **182** was obtained as major product.[314] As yet the mechanism for the formation of **182** is unclear.

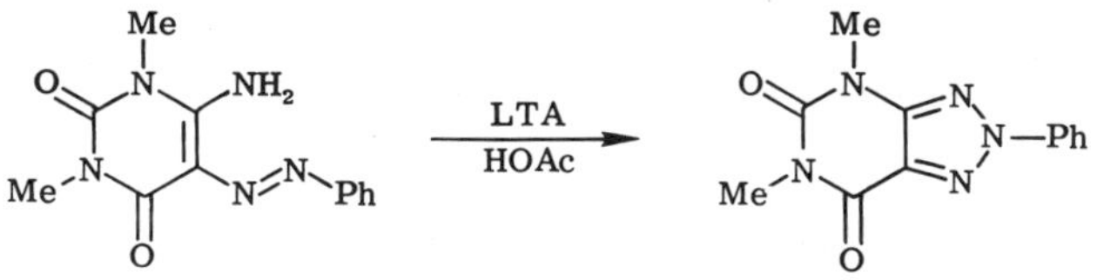

(25%)

(ref. 311)

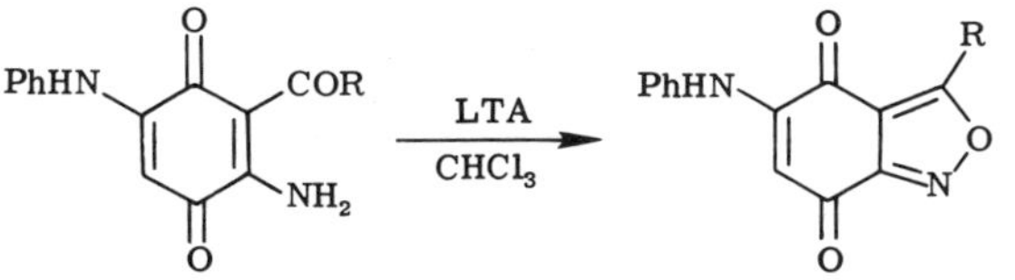

R	% yield
Me	75
OMe	72
NMe_2	65

(ref. 313)

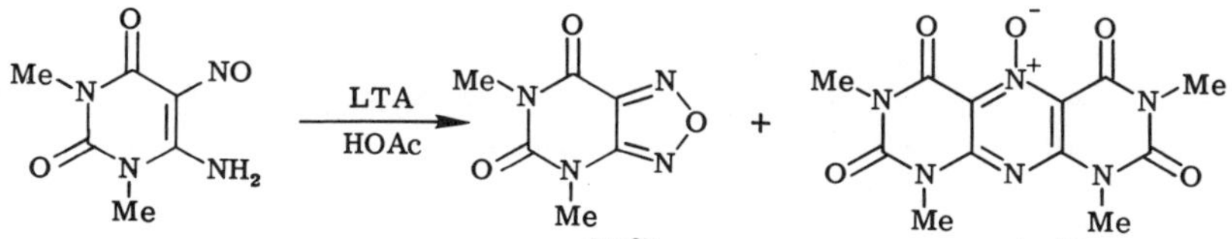

(19%) **182** (67%)

(ref. 314)

Pyrrolopyrimidines when treated with LTA undergo ring expansion to the pyrimido–pyrimidine system.[315,316] This high-yield transformation has

---

[301] M. Watanabe, S. Kajigaeshi, and S. Kanemasa, *Synthesis*, 761 (1977).
[302] J. Daunis, H. Lopez, and G. Maury, *J. Org. Chem.* **42**, 1018 (1977).
[303] D. S. Noyce and B. B. Sandel, *J. Org. Chem.* **41**, 3640 (1976).
[304] G. Stork and A. G. Schultz, *J. Am. Chem. Soc.* **93**, 4074 (1971).
[305] J. P. Kutney, U. Bunzli-Trepp, K. K. Chan, J. P. de Souza, Y. Fujise, T. Honda, J . Katsube, F. K. Klein, A. Leutwiler, S. Morehead, M. Rohr, and B. R. Worth, *J. Am. Chem. Soc.* **100,** 4220 (1978).
[306] L. Castedo, R. Suau, and A. Mourino, *Heterocycles* **3**, 449 (1975).
[307] G. Galliani, B. Rindone, and P. L. Beltrame, *J. Chem. Soc. Perkin Trans. 2*, 1803 (1976).
[308] G. Galliani, B. Rindone, and C. Scolastico, *Tetrahedron Lett.*, 1285 (1975).
[309] B. Rindone and C. Scolastico, *Tetrahedron Lett.*, 3379 (1974).
[310] A. Matsumoto, M. Yoshida, and O. Simamura, *Bull. Chem. Soc. Jpn.* **47**, 1493 (1974).
[311] Y. Maki and E. C. Taylor, *Chem. Pharm. Bull.* **20**, 605 (1972).
[312] E. C. Taylor, G. P. Beardsley, and Y. Maki, *J. Org. Chem.* **36**, 3211 (1971).
[312a] L. K. Dyall, *Aust. J. Chem.* **32**, 643 (1979).
[313] W. Schäfer, H. W. Moore, and A. Aquado, *Synthesis*, 30 (1974).
[314] E. C. Taylor, Y. Maki, and A. McKillop, *J. Org. Chem.* **37**, 1601 (1972).
[315] F. Yoneda and M. Higuchi, *J. Chem. Soc. Chem. Commun.*, 402 (1972).
[316] F. Yoneda and M. Higuchi, *Bull. Chem. Soc. Jpn.* **46**, 3849 (1973).

been shown to involve **183** as an intermediate, and an attempted trapping experiment with cyclohexene failed to provide evidence for a free nitrene.

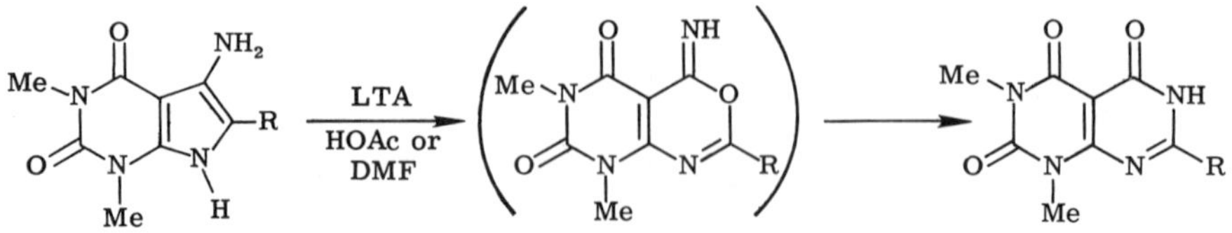

183

(refs. 315, 316)

R	% yield
Ph	90
Cl	86
Br	72

B. AMIDES

The reaction of primary amides with LTA in the presence of an alcohol is a very useful alternative to classical methods for the oxidative rearrangement of amides into isocyanates and, hence, carbamates.[317–319] The latter step is catalyzed by addition of triethylamine.[317] When the oxidation is performed in DMF, and the resulting isocyanate treated with *tert*-butylamine, excellent yields of *unsym*-ureas result.[317] The *tert*-butyl carbamates

$$\text{RNHCONH—}t\text{-Bu} \xleftarrow[\text{2. } t\text{-BuNH}_2]{\text{1. LTA/DMF}} \text{RCONH}_2 \xrightarrow[\text{2. Et}_3\text{N}]{\text{1. LTA/}t\text{-BuOH}} \text{RNHCO}_2t\text{-Bu} \qquad \text{(ref. 317)}$$

formed by the procedure also have the advantage of being rapidly cleaved to the corresponding amine hydrochlorides by treatment with anhydrous HCl in dry ethanol. Discrete nitrene intermediates could not be trapped from the oxidation and rearrangement from an intermediate **184** has

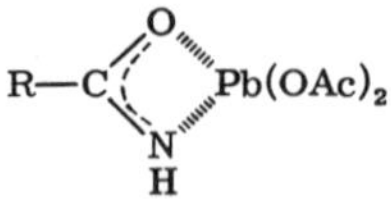

184

been proposed.[317] Rearrangement of **185** into **186** indicates that the reaction occurs with retention of configuration.[317] This point has also been proven with the LTA/pyridine rearrangement of β-hydroxy primary

[317] H. E. Baumgarten, H. L. Smith, and A. Staklis, *J. Org. Chem.* **40**, 3554 (1975).
[318] B. Acott, A. L. J. Beckwith, and A. Hassanali, *Aust. J. Chem.* **21**, 185 (1968).
[319] B. Acott, A. L. J. Beckwith, and A. Hassanali, *Aust. J. Chem.* **21**, 197 (1968).

amides which afford extremely high yields (73–100%) of the corresponding 2-oxazolidinones.[320,321]

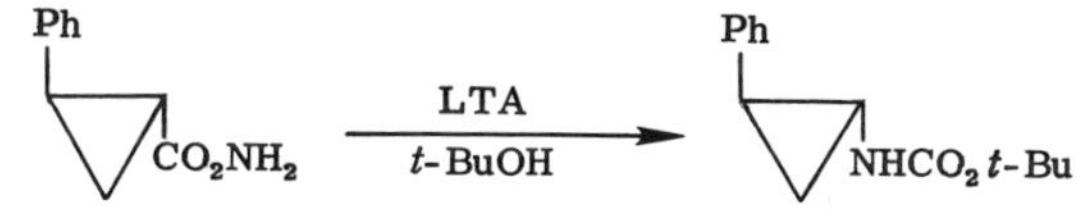

185 **186** (13%) (ref. 317)

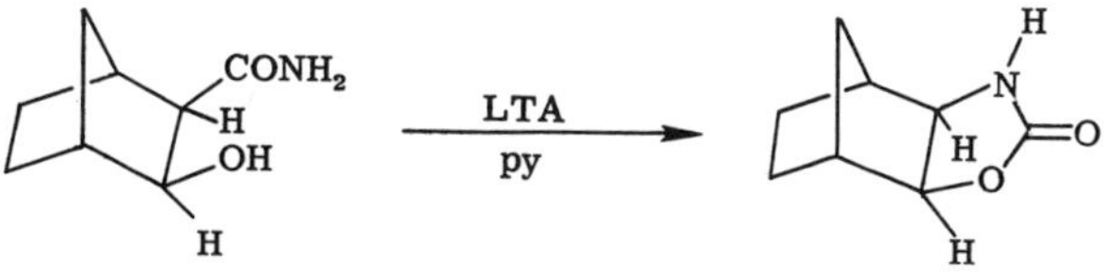

(95%) (refs. 320, 321)

When the above reaction is carried out in the presence of iodine and light, *N*-iodoamides are formed which subsequently homolyze and cyclize. After treatment with dilute alkali γ- and δ-lactones are produced.[279] A recent example involves the conversion of **187** to **188** (68% yield).[322] Similar experiments with lupanoic acid amides **189** gives mainly the corresponding

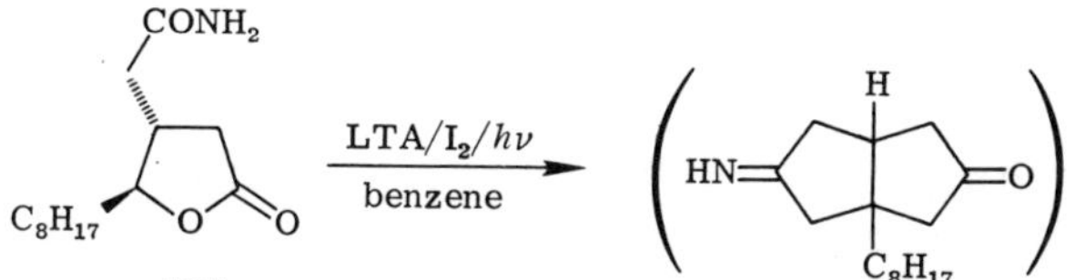

187

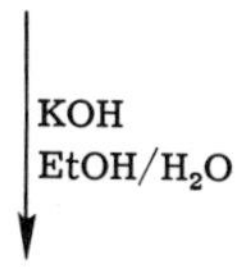

$C_8H_{17}COCH(CH_2CO_2H)_2$

188 (68%) (ref. 322)

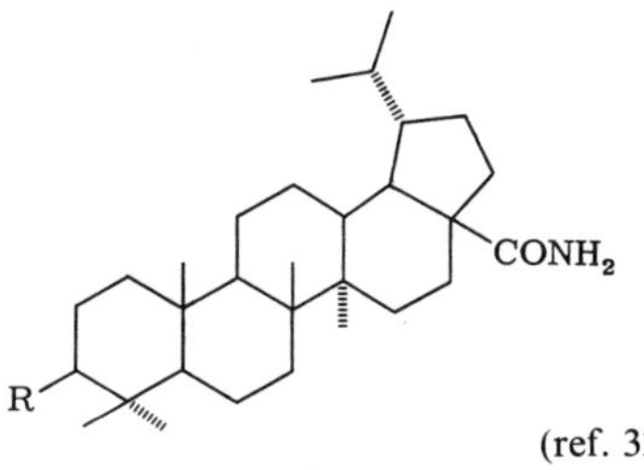

(ref. 323)

189

R = OAc, H

[320] S. S. Simons, Jr., *J. Org. Chem.* **38**, 414 (1973).
[321] S. S. Simons, Jr., *J. Am. Chem. Soc.* **96**, 6492 (1974).
[322] W. L. Parker and F. Johnson, *J. Org. Chem.* **38**, 2489 (1973).

isocyanates.[323] Cyclization was also observed in the LTA oxidation of **190**.[324,324a] In this case electron-withdrawing substituents inhibited ring formation and increased the amount of dimeric products **191** produced.[324]

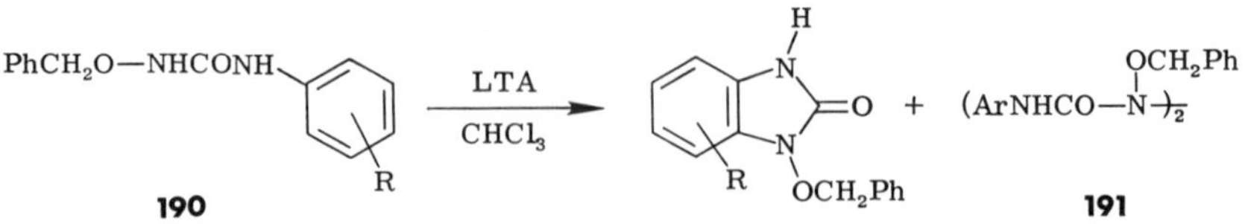

190 **191**

(ref. 324)

A novel ring contraction occurs when **192** is treated with LTA.[325] Product **193** may well arise from an intramolecular acetyl transfer involving **194**.

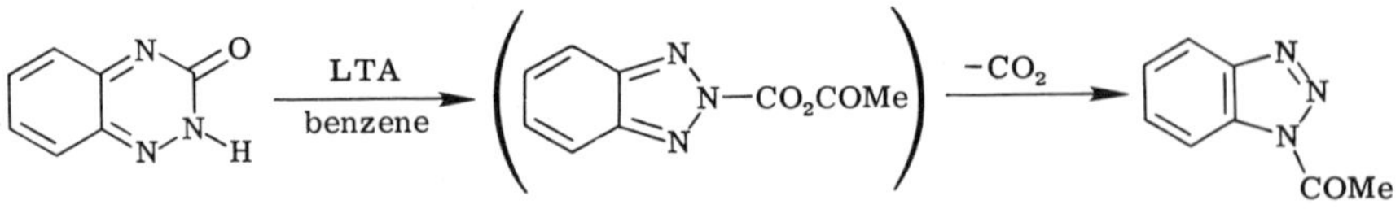

192 **194** **193** (86%)

(ref. 325)

Finally, oxidation of a series of bicyclic secondary amides **195** gives products consistent with the formation of the highly strained **196**.[326]

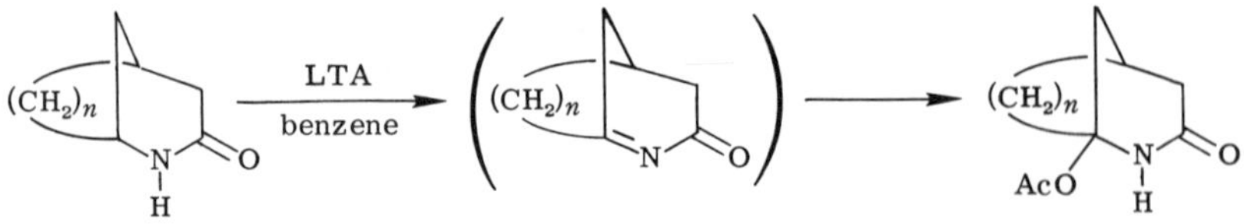

195 **196**

(ref. 326)

n	% yield
3	30
2	2

Sulfonamides can also undergo nitrenoid reactions when treated with LTA. The examples presented below are typical.[327,328,328a] Failure of

[323] J. Protiva and A. Vystrčil, *Collect. Czech. Chem. Commun.* **41**, 1200 (1976).
[324] J. H. Cooley and P. T. Jacobs, *J. Org. Chem.* **40**, 552 (1975).
[324a] A. R. Forrester, E. M. Johansson, and R. H. Thomson, *J. Chem. Soc. Perkin Trans. 1*, 1112 (1979).
[325] C. W. Rees and A. A. Sale, *J. Chem. Soc. Perkin Trans. 1*, 545 (1973).
[326] M. Toda, H. Niwa, K. Ienaga, Y. Hirata, and S. Yamamura, *Tetrahedron Lett.*, 335 (1972).
[327] M. Okahara, K. Matsunaga, and S. Komori, *Synthesis*, 203 (1972).
[328] T. Ohashi, K. Matsunaga, M. Okahara, and S. Komori, *Synthesis*, 96 (1971).
[328a] T. A. Chaudri, *Pak. J. Sci. Ind. Res.* **18**, 1 (1976).

$$\mathrm{RR'N{-}SO_2NH_2 + SMe_2 \xrightarrow{LTA} RR'N{-}SO_2{-}N{=}S(Me)_2} \qquad \text{(ref. 327)}$$

$$\mathrm{ArSO_2N{=}SRR' \xleftarrow{RSR'} ArSO_2NH_2 + LTA \xrightarrow{DMSO} ArSO_2N{=}\overset{\overset{\displaystyle O}{\|}}{S}(Me)_2} \qquad \text{(refs. 328,328a)}$$

alkenes to trap intermediates is taken as negative evidence for free nitrene formation.[328a] The same conclusion was reached from the observation that the arsinimines **197** produced by LTA oxidation of sulfonamides in the presence of triphenylarsine actually arise from nitrogen attack on intermediate **198**.[329,330] The same considerations are thought to apply to the

$$\mathrm{RSO_2NH_2 + Ph_3As \xrightarrow{LTA} \underset{\textbf{(197)}}{Ph_3As{=}NSO_2R}} \qquad \text{(refs. 329,330)}$$

$$\mathrm{(Ph_3As(OAc)_2)}$$

198

reactions of sulfonamides with LTA and sulfides noted above. In this case **199** is proposed as a likely intermediate.[329] Analogous oxidations have also been reported to give the tellurium derivatives **200**.[331] The bis-sulfonamide

$$\underset{\textbf{199}}{\mathrm{R_2S(OAc)_2}} \quad \text{(ref. 329)} \qquad \underset{\textbf{200}}{\mathrm{R_2Te(OAc)_2}} \quad \text{(ref. 331)}$$

201 gives a 93% yield of (*E*/*Z*)-**202** when allowed to react with LTA in acetic acid.[332]

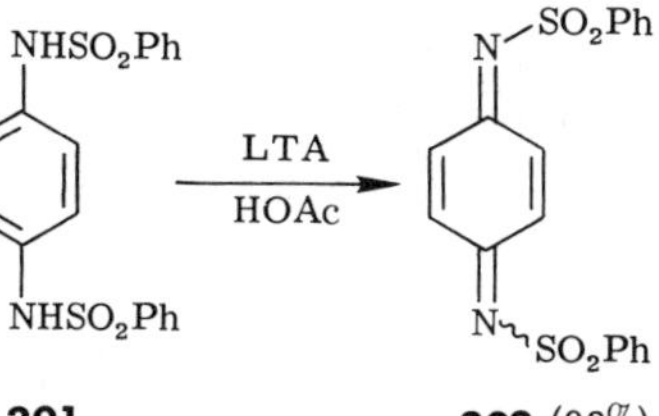

(ref. 332)

201 **202** (93%)

C. HYDRAZINES AND RELATED COMPOUNDS

The reaction of hydrazines with LTA generally produces azo compounds or diazines, depending upon substitution pattern and this behavior has been reviewed.[279,333] With hydrazine, treatment with two equivalents of LTA

[329] J. I. G. Cadogan and I. Gosney, *J. Chem. Soc. Perkin Trans. 1*, 466 (1974).
[330] J. I. G. Cadogan and I. Gosney, *J. Chem. Soc. Chem. Commun.*, 586 (1973).
[331] B. C. Pant, *Tetrahedron Lett.*, 4779 (1972).
[332] A. G. Pinkus and J. Tsuji *J. Org. Chem.* **39**, 497 (1974).
[333] D. M. Lemal, *in* "Nitrenes" (W. Lwowski, ed.), p. 345. Wiley (Interscience), New York, 1970.

gives a quantitative yield of N_2 in a reaction that has been used to detect small amounts of hydrazine using coulometric techniques.[334,335] Oxidation of thiadiazolidines produces excellent yields of the corresponding thiadiazolines as part of a high-yield method for the synthesis of symmetrically substituted alkenes. Systems **203**[336] and **204**[337] are typical.

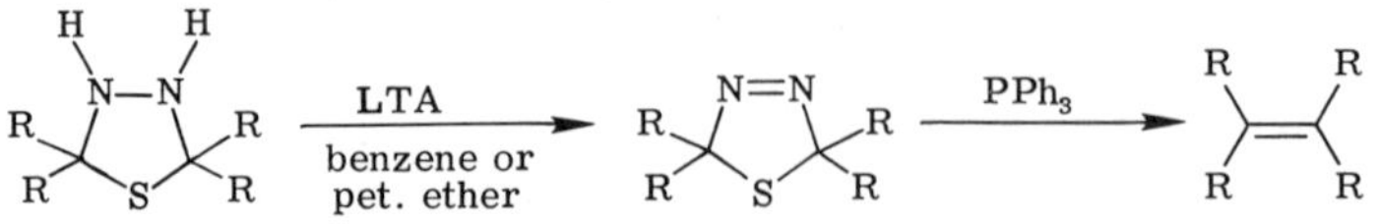

(refs. 336, 337)

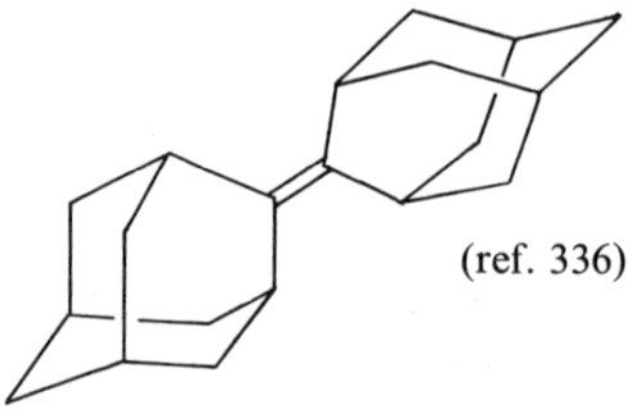

(ref. 336)

203 (69%)

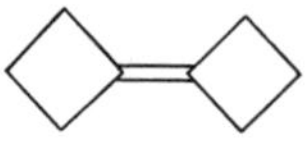

204 (85%)

(ref. 337)

The reaction of benzylhydrazine with LTA is reported to afford benzyldiimide which then reacts further to give benzyl acetate with a second molar equivalent of LTA. Since the production of benzyl acetate predominates in the presence of excess methanol, it is proposed that this compound arises via the intramolecular reaction of **205**.[338] The diimide initially formed from

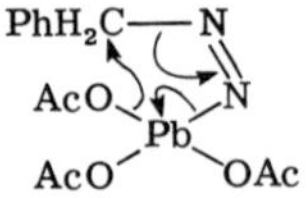

205

206 cyclizes to give **207** in low yield.[339]

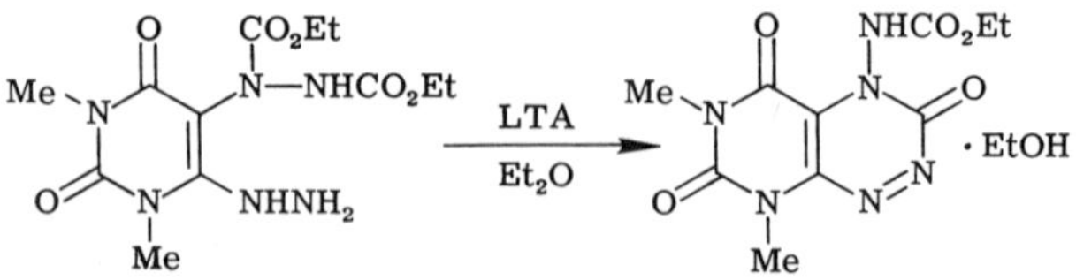

206 **207** (19%)

(ref. 339)

Diacylhydrazides are readily oxidized to the corresponding diacyldiimides by LTA,[338,340–342] and in the presence of 1,3-dienes, high yields of [4 + 2]-cycloaddition products are obtained.[340–343]

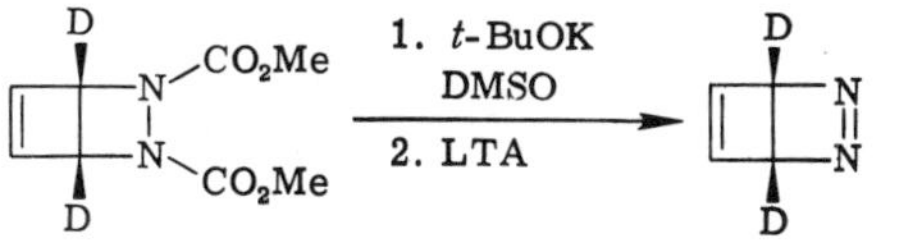

(ref. 340)

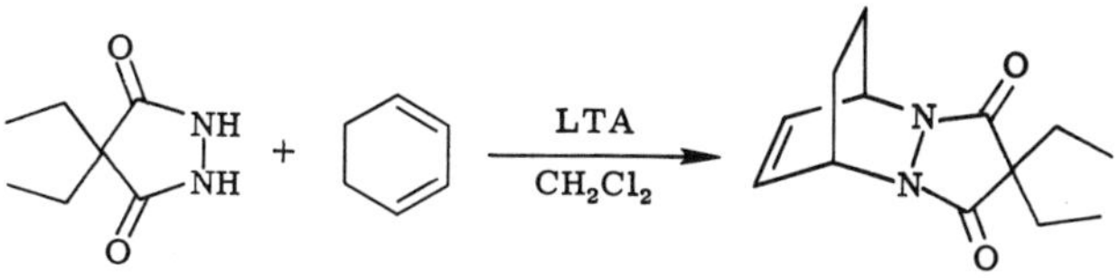

(refs. 340a, 340b)

(76%)

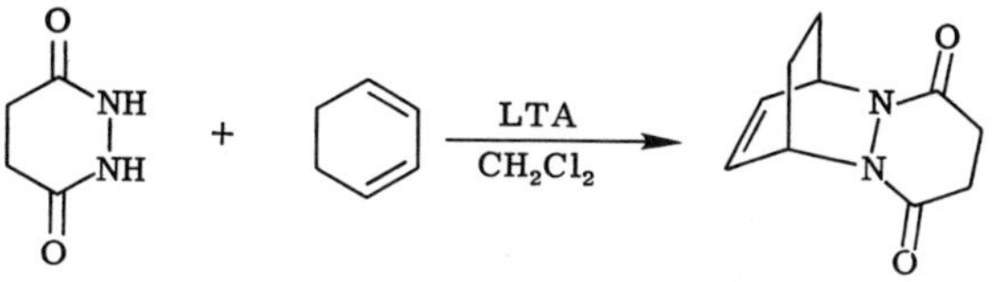

(ref. 340a)

(35%)

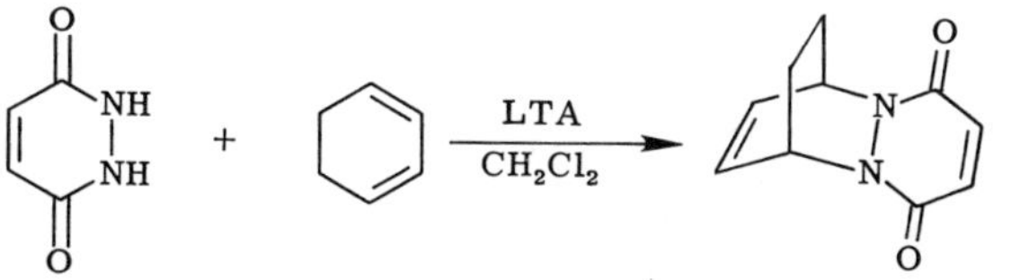

(ref. 340a)

(48%)

---

[334] T. J. Pastor, V. J. Vajgand, V. V. Vojka, and V. Antonijević, *Mikrochim. Acta*, 131 (1978).

[335] M. R. Mahmoud, M. S. El-Meligy, and I. M. Issa, *Indian J. Chem.* **54**, 872 (1977).

[336] A. P. Schaap and G. R. Faler, *J. Org. Chem.* **38**, 3061 (1973).

[337] L. K. Bee, J. Beeby, J. W. Everett, and P. J. Garratt, *J. Org. Chem.* **40**, 2212 (1975).

[338] R. O. C. Norman, R. Purchase, C. B. Thomas, and J. B. Aylward, *J. Chem. Soc. Perkin Trans. 1*, 1692 (1972).

[339] E. C. Taylor and F. Sowinski, *J. Org. Chem.* **40**, 2321 (1975).

[340] D. W. Whitman and B. K. Carpenter, *J. Am. Chem. Soc.* **102**, 4272 (1980).

[340a] S. F. Nelsen, W. C. Hollinsed, L. A. Grezzo, and W. P. Parmelee, *J. Am. Chem. Soc.* **101**, 7347 (1979).

[340b] B. T. Gillis and R. A. Izydore, *J. Org. Chem.* **34**, 3181 (1969).

[341] M. E. Jung and J. A. Lowe, *J. Org. Chem.* **42**, 2371 (1977).

[342] R. J. Cremlyn, M. J. Frearson, and D. R. Milnes, *J. Chem. Soc. Chem. Commun.*, 319 (1974).

[343] R. J. Weinkam and B. T. Gillis, *J. Org. Chem.* **37**, 1696 (1970).

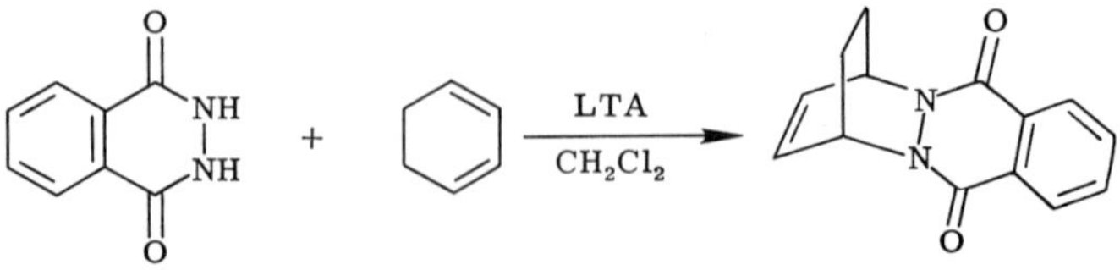

(80%) (ref. 340a)

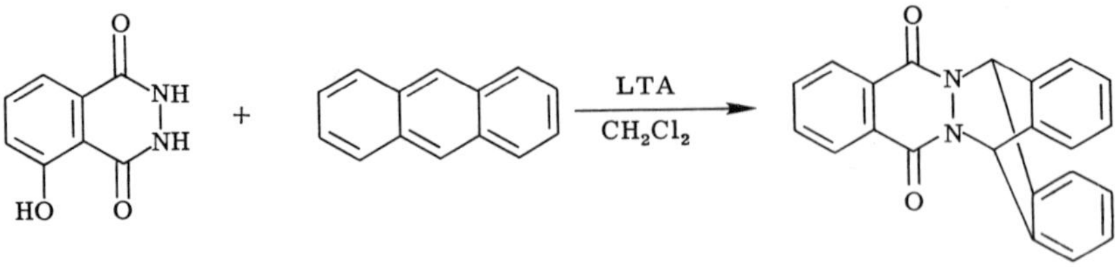

(72%) (ref. 341)

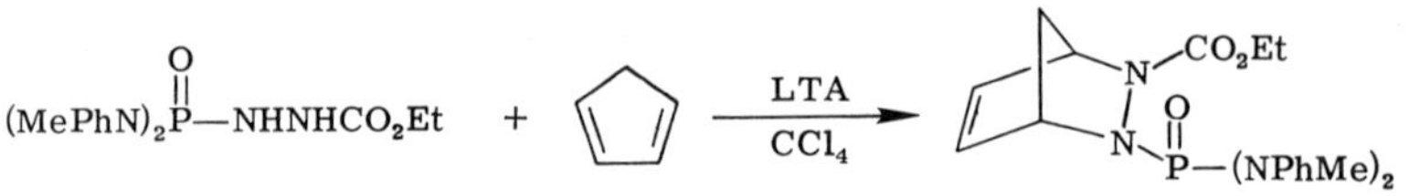

(?) (ref. 342)

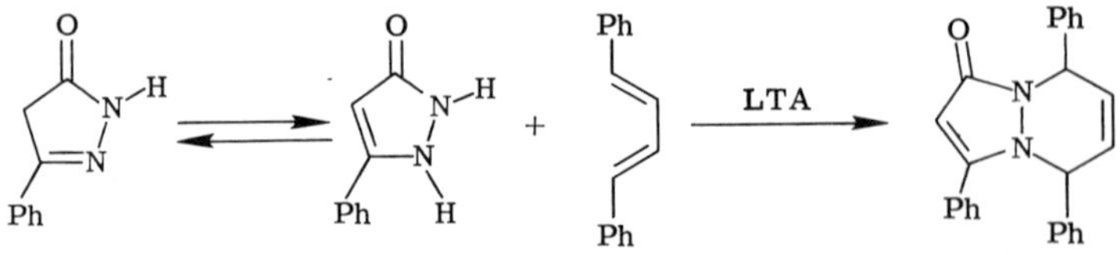

(ref. 343)

(88%)

The formation of azo compounds has also been noted from the deacylation of certain hydrazine derivatives,[338,344] and in this regard *N*,*N*-diisopropyl-

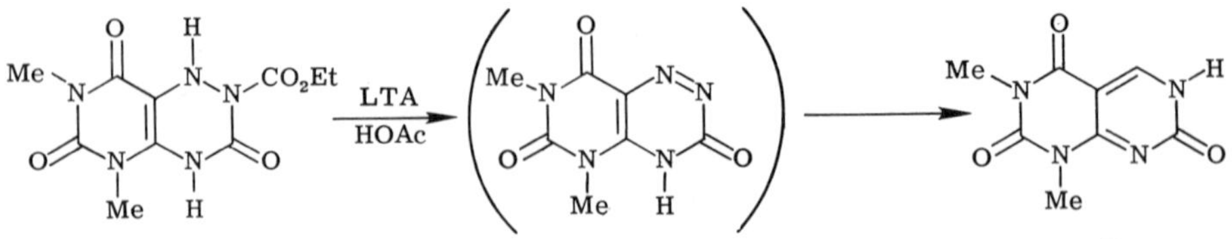

(66%)

(ref. 344)

[344] E. C. Taylor and F. Sowinski, *J. Org. Chem.* **40**, 2329 (1975).

hydrazine has been used as an easily removed protecting group for carboxylic acids.[345] *N,N*-Dialkylhydrazides[338,346] and *N,N*-dialkylhydrazines[338] can

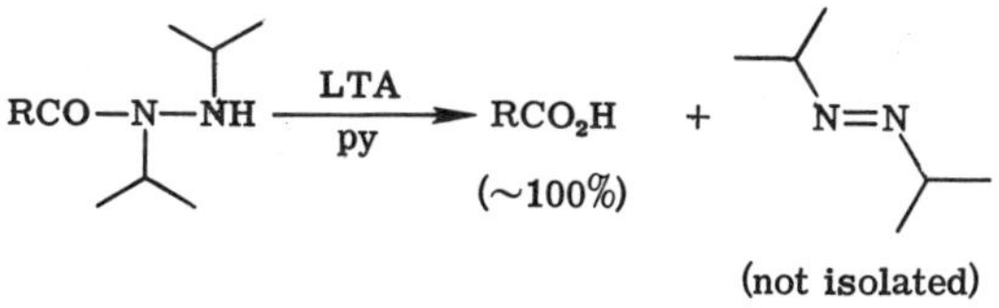

(ref. 345)

also generate dipoles such as **208** when treated with LTA. A series of di-

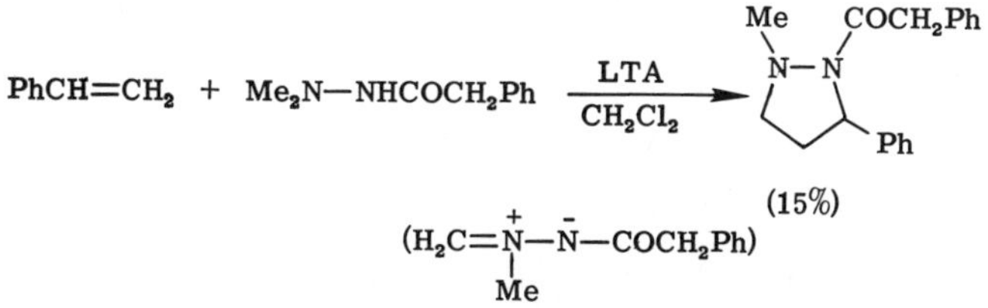

(ref. 346)

208

aziridines **209** was cleaved with LTA to give **210** and **211**.[347]

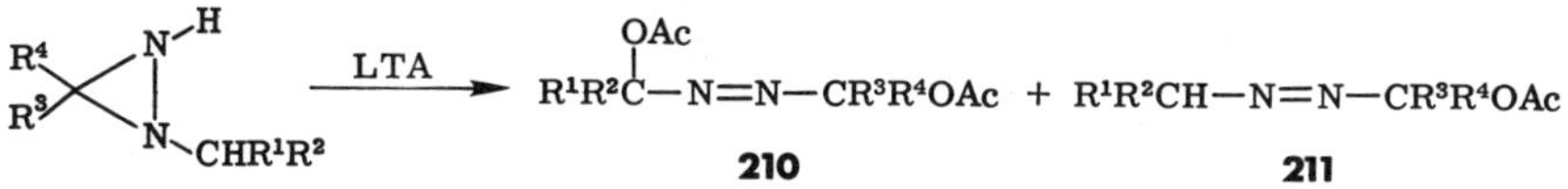

209 **210** **211**

(ref. 347)

When the 1,1-disubstituted hydrazine is incorporated into a ring system, a number of products arise which probably come from an intermediate diazine (nitrene). For instance, the formation of a tetrazene and the appropriate amine derivative can occur from the oxidation or fragmentation of the initially formed **212**.[348,349] Several other examples of the latter process

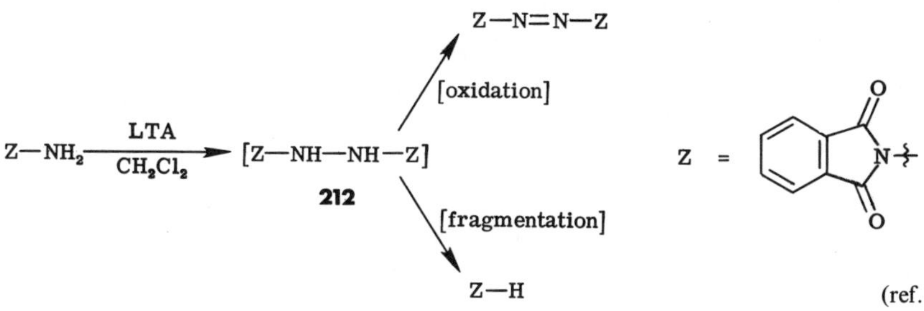

212

(ref. 348)

[345] D. H. R. Barton, M. Girijavallabhan, and P. G. Sammes, *J. Chem. Soc. Perkin Trans. 1*, 929 (1972).

[346] W. Oppolzer and H. P. Weber, *Tetrahedron Lett.*, 1711 (1972).

are available.[350,351] In systems where nitrogen extrusion can take place, fragmentation occurs in high yield.[350,352–355] Similar products are also observed from the LTA oxidation of substituted 1,2-diaminobenzenes.[356]

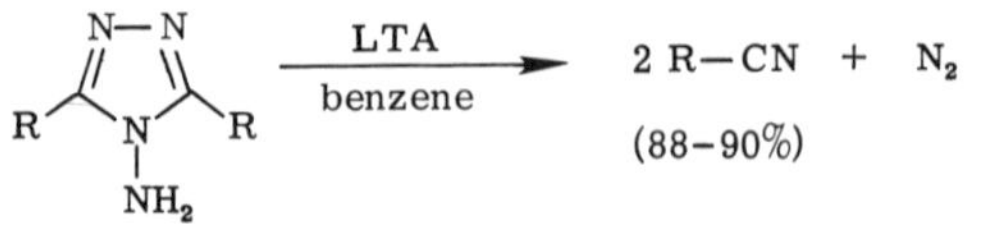

(refs. 350, 354, 355)

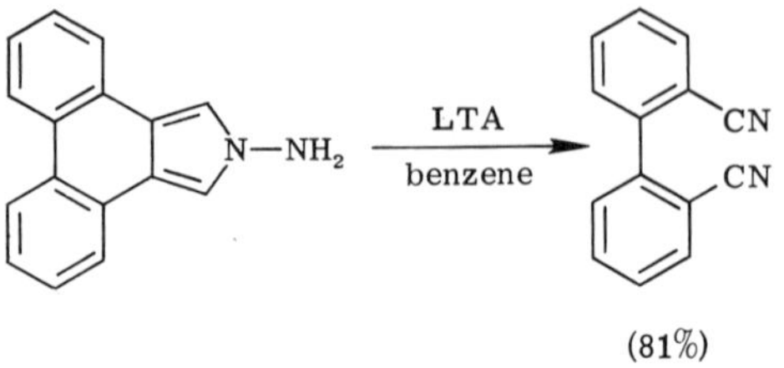

(ref. 352)

Ring expansion has also been observed and used as a high-yield method for the synthesis of both 1,2,3-benzotriazines[357] and 1,2,4-benzotriazines.[358,359]

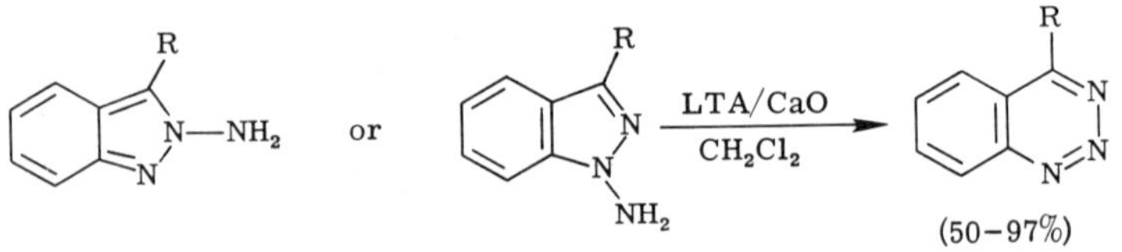

(ref. 357)

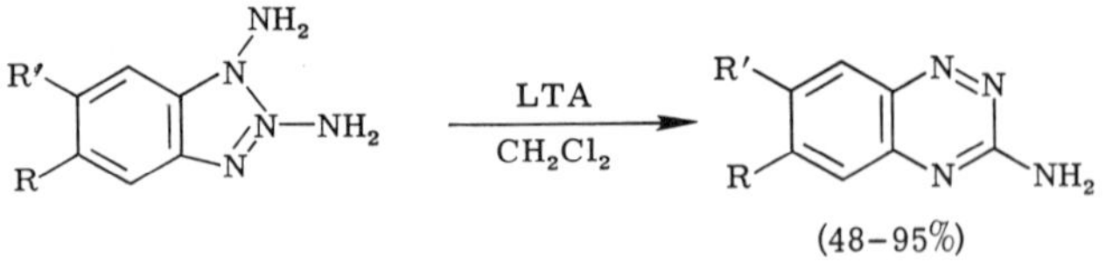

(refs. 358, 359)

---

347 V. N. Yandovskii, P. M. Adrov, and L. B. Koroleva, *Zh. Org. Khim.* **11**, 156 (1975).
348 L. Hoesch and A. S. Dreiding, *Helv. Chim. Acta* **58**, 980 (1975).
349 D. J. Anderson, T. L. Gilchrist, and C. W. Rees, *J. Chem. Soc. Chem. Commun.*, 800 (1971).
350 K. Sakai and J.-P. Anselme, *Tetrahedron Lett.*, 3851 (1970).
351 K. Sakai and J.-P. Anselme, *J. Org. Chem.* **37**, 2351 (1972).
352 J. W. Barton and A. R. Grinham, *J. Chem. Soc. Perkin Trans. 1*, 634 (1972).
353 T. L. Gilchrist, G. E. Gymer, and C. W. Rees, *J. Chem. Soc. Perkin Trans. 1*, 1747 (1975).
354 K. K. Mayer, F. Schröppel, and J. Sauer, *Tetrahedron Lett.*, 2899 (1972).
355 F. Schröppel and J. Sauer, *Tetrahedron Lett.*, 2945 (1974).
356 L. S. Kobrina, N. V. Akulenko, and G. G. Yakobson, *Zh. Org. Khim.* **8**, 2375 (1972).
357 B. M. Adger, S. Bradbury, M. Keating, C. W. Rees, R. C. Storr, and M. T. Williams, *J. Chem. Soc. Perkin Trans. 1*, 31 (1975).
358 A. V. Zeiger and M. M. Joullié, *J. Org. Chem.* **42**, 542 (1977).
359 A. V. Zeiger and M. M. Joullié, *Synth. Commun.* **6**, 457 (1976).

When 1-aminopyridinium,[360] 1-aminoquinolinium,[360] 1-aminoisoquinolinium,[360] or certain 2-aminoimidazo[1,5-*a*]pyridinium bromides[360a] are treated with LTA, the corresponding acylated cyclic hydrazides result in moderate to excellent yield. The sequence shown portrays the postulated mechanism for the pyridine derivative.[360]

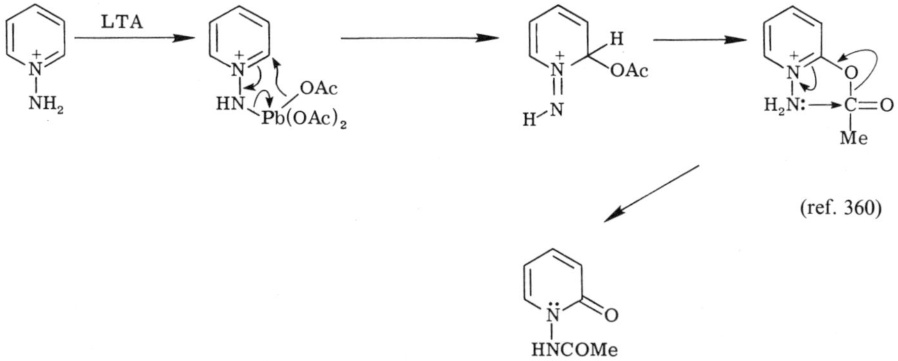

(ref. 360)

Treatment of 1-aminobenzotriazole with LTA affords a mild, high-yield method for the generation of benzyne.[361] The procedure, carried out in the presence of suitable 1,3-dienes gives products derived from [4 + 2]cycloaddition.[362–364] Benzyne produced by the above method also reacts with CS_2

(ref. 361)

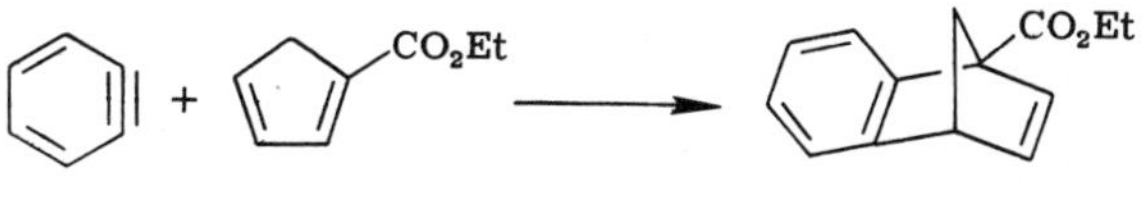

(88%)

(ref. 362)

[360] J. T. Boyers and E. E. Glover, *J. Chem. Soc. Perkin Trans. 1*, 1960 (1977).
[360a] E. E. Glover, L. W. Peck, and D. G. Doughty, *J. Chem. Soc. Perkin Trans. 1*, 1833 (1979).
[361] C. D. Campbell and C. W. Rees, *J. Chem. Soc. C*, 742 (1969).
[362] T. Irie and H. Tanida, *J. Org. Chem.* **43**, 3274 (1978).
[363] H. Kato, S. Nakazawa, T. Kiyosawa, and K. Hirakawa, *J. Chem. Soc. Perkin Trans. 1*, 672 (1976).
[364] T. J. Barton, A. J. Nelson, and J. Clardy, *J. Org. Chem.* **37**, 895 (1972).

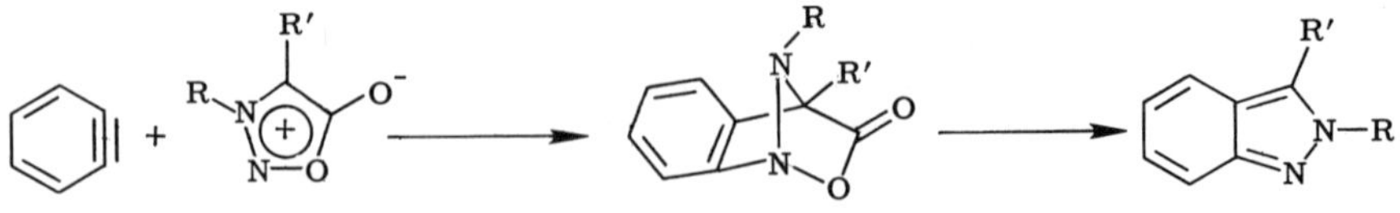

(64–73%)

(ref. 363)

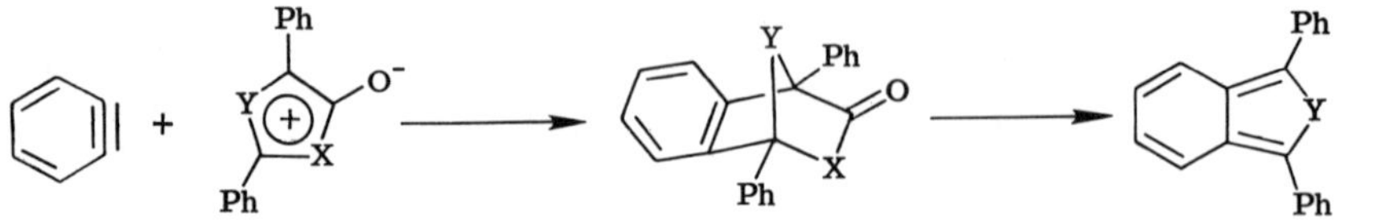

(ref. 363)

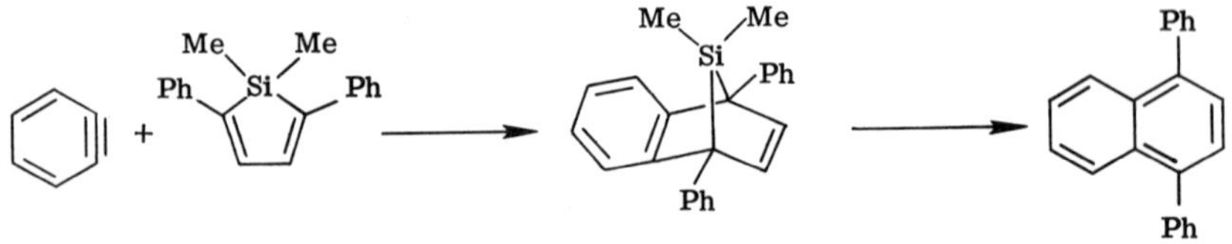

(37%)

(ref. 364)

to give good yields of **213** or **214** depending upon the solvent employed.[365]

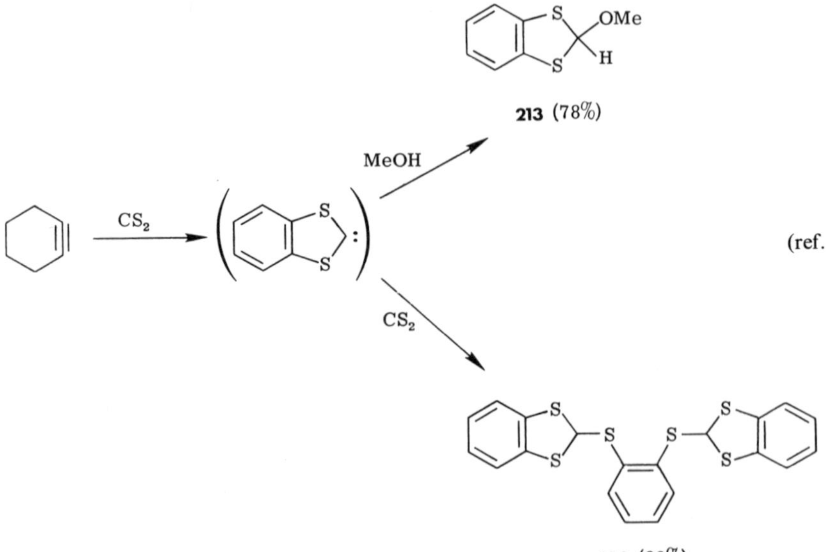

213 (78%)

(ref. 365)

214 (86%)

The lifetime of an aryne can be lengthened dramatically by immobilizing the system on carboxylated polystyrene resin. The persistence of **215** for 70 seconds has been reported.[366]

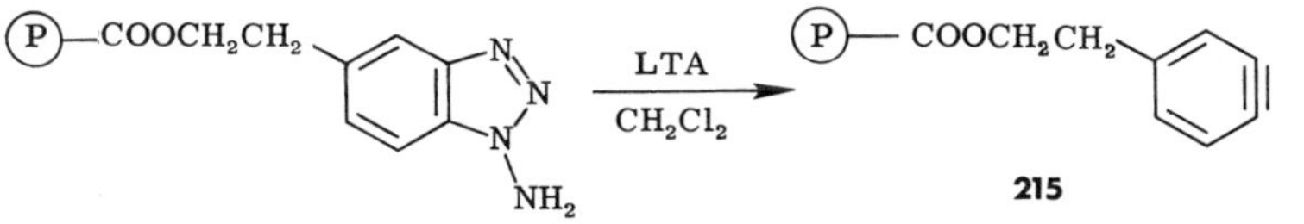

215 (ref. 366)

In the absence of an "aryne trap," the formation of biphenylenes becomes important,[367] and in several instances crossed aryne couplings have been demonstrated.[352] The typical [4 + 2] adducts obtained from LTA treatment

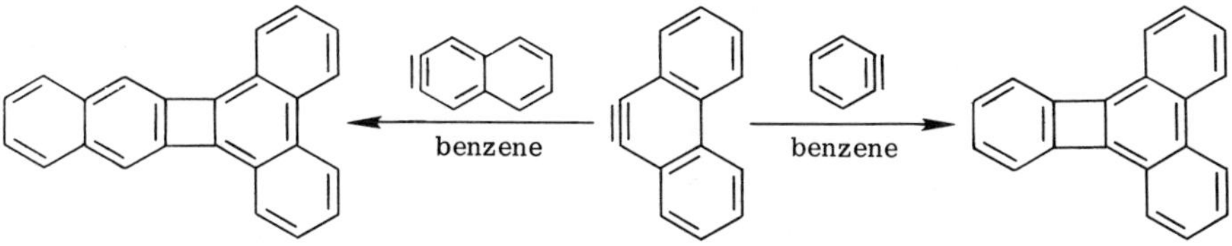

(31%) (15%)

(ref. 352)

of **216**[353] or **217**[368] in the presence of dienes are indicative of the fact that nonbenzenoid arynes are accessible by the method too.

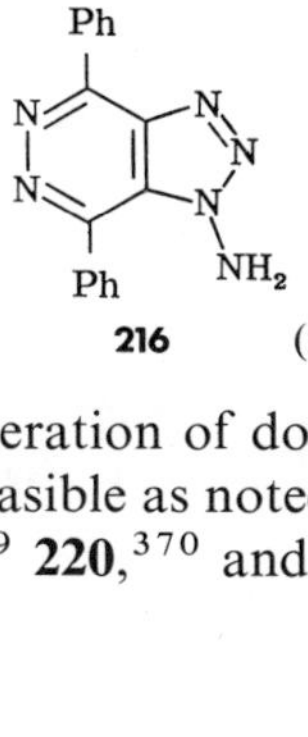

216 (ref. 353)

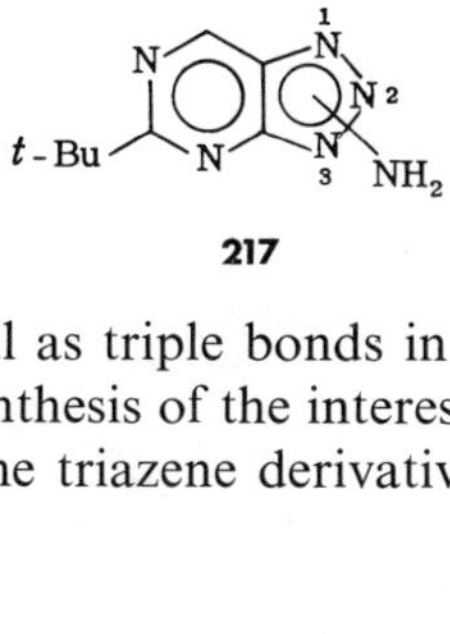

N_1 or N_3-isomer

217 (ref. 368)

Generation of double as well as triple bonds in nonaromatic systems is also feasible as noted by the synthesis of the interesting compounds **218**,[369] **219**,[369] **220**,[370] and **221**.[371] The triazene derivative **222** loses nitrogen on

---

[365] J. Nakayama, *J. Chem. Soc. Perkin Trans. 1*, 525 (1975).

[366] P. Jayalekshmy and S. Mazur, *J. Am. Chem. Soc.* **98**, 6710 (1976).

[367] C. F. Wilcox, Jr., J. P. Uetrecht, G. D. Grantham, and K. G. Grohmann, *J. Am. Chem. Soc.* **97**, 1914 (1975).

[368] D. Christophe, R. Promel, and M. Maeck, *Tetrahedron Lett.*, 4435 (1978).

[369] R. L. Viavattene, F. D. Greene, L. D. Chueng, R. Majeste, and L. M. Trefonas, *J. Am. Chem. Soc.* **96**, 4342 (1974).

[370] T. Nakazawa and I. Murata, *Angew. Chem., Int. Ed. Engl.* **14**, 711 (1975).

[371] C. W. Rees and D. E. West, *J. Chem. Soc. C*, 583 (1970).

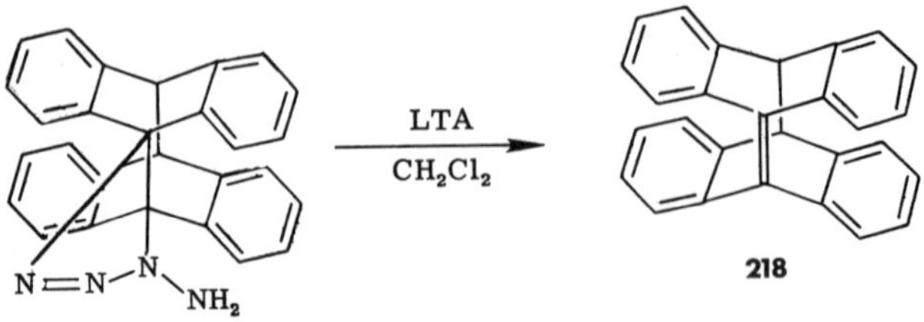

218

(ref. 369)

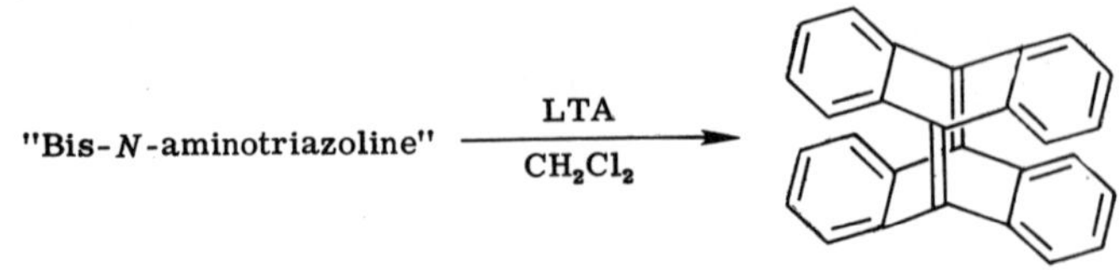

219 (20%)

(ref. 369)

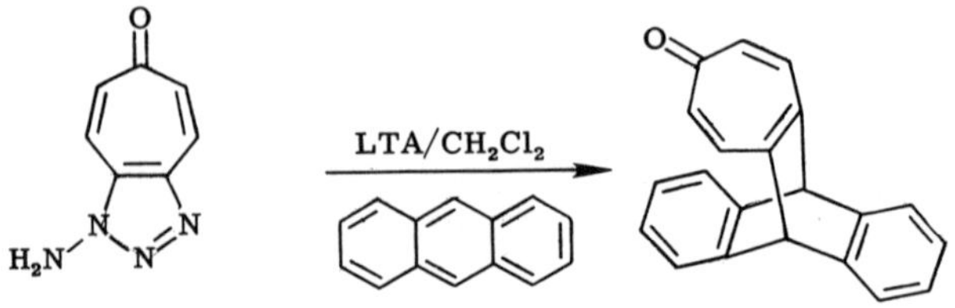

220 (12%)

(ref. 370)

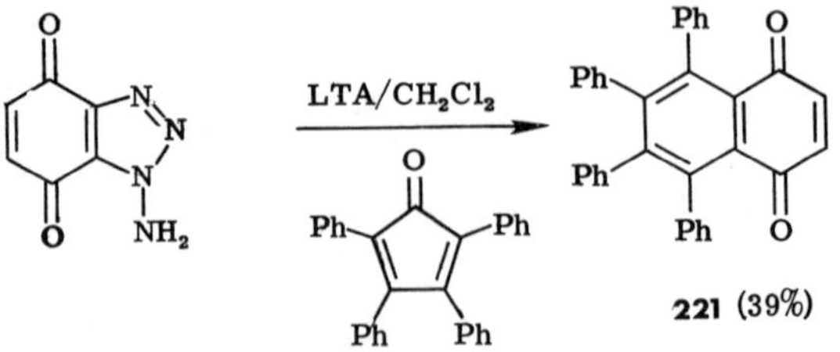

221 (39%)

(ref. 371)

treatment with LTA to afford **223**, or the equivalent, which gives 1,3- as well as 1,2- and 1,4-adducts when trapped with a number of 1,3-dienes.[372,373]

[372] J. Meinwald and G. W. Gruber, *J. Am. Chem. Soc.* **93**, 3802 (1971).
[373] J. Meinwald, L. V. Dunkerton, and G. W. Gruber, *J. Am. Chem. Soc.* **97**, 681 (1975).

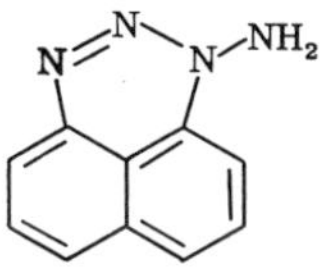

222

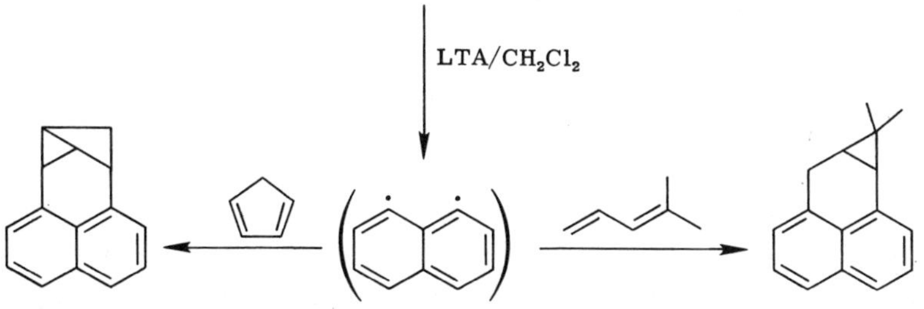

(44%)
(ref. 372)

223

(72%)
(ref. 373)

Diazenes which do not undergo the fragmentation reactions noted above afford high yields of aziridines by cycloaddition to alkenes. Stereochemical studies[355,374–376] indicate stereospecific addition which implicates singlet

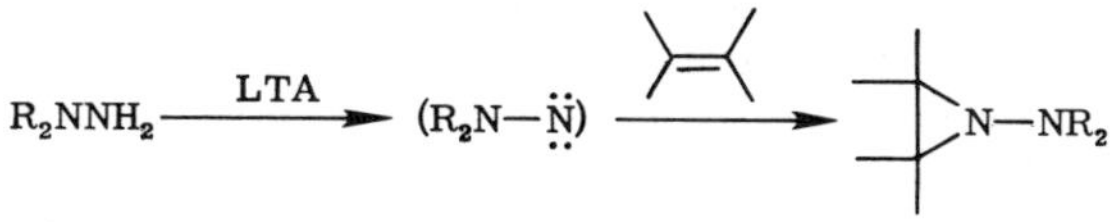

nitrene as the reactive intermediate, and INDO calculations even favor a preferred singlet ground-state.[293] Systems **224**–**230** have all been used successfully and even **230**, which fragments readily in the absence of alkene, gives aziridines when alkenes are present.[354,355] When the diazene generated from **230** was trapped with a series of aryl substituted styrenes it was found that electron-donating groups in the styrene promote aziridine formation

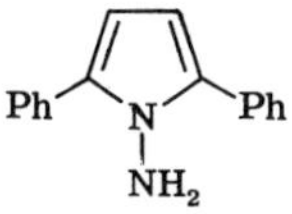

224 (ref. 354)

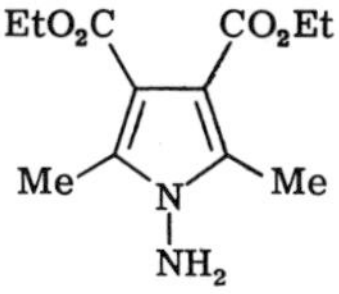

225 (ref. 354)

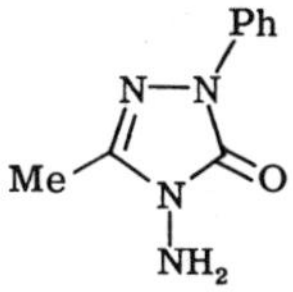

226 (ref. 354)

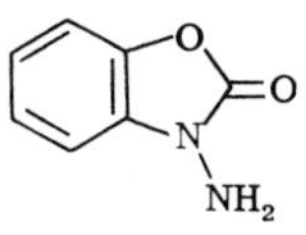

227 (ref. 374)

[374] D. J. Anderson, T. L. Gilchrist, D. C. Horwell, and C. W. Rees, *J. Chem. Soc. C*, 576 (1970).
[375] H. Person, C. Fayat, F. Tonnard, and A. Foucaud, *Bull. Soc. Chim. Fr.*, 635 (1974).
[376] H. Person, F. Tonnard, A. Foucaud, and C. Fayat, *Tetrahedron Lett.*, 2495 (1973).

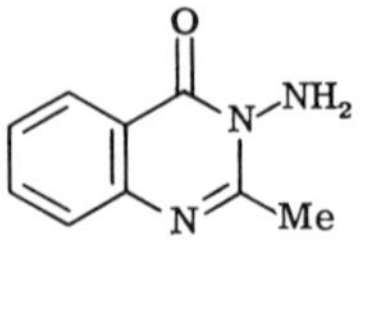

228 (ref. 374)

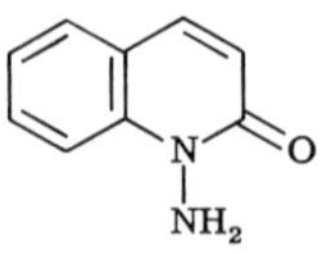

229 (ref. 374)

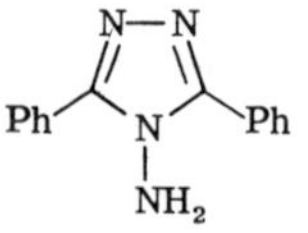

230 (refs. 354, 355)

at the expense of fragmentation.[355] The fact that diazenes might have some nucleophilic character due to tervalent nitrogen lone pair delocalization should enhance the reactivity of the nitrene toward electrophilic alkenes.

$$R_2\ddot{N}-\underset{..}{\ddot{N}} \leftrightarrow R_2N^+=\underset{..}{\ddot{N}}^-$$

This behavior has been observed.[374–376] The effect of substituents on both the nitrene and alkene have also been discussed in terms of frontier orbital theory.[375,376]

A large number of aziridines have been synthesized using *N*-aminophthalimide as the nitrene precursor. The examples given in Table VIII are illustrative.[374–383]

TABLE VIII

LTA-Promoted Aziridine Formation with *N*-Aminophthalimide in Methylene Chloride

Substrate		Product		% Yield	Reference
	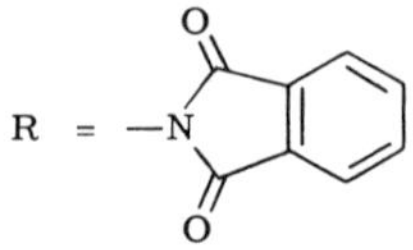				
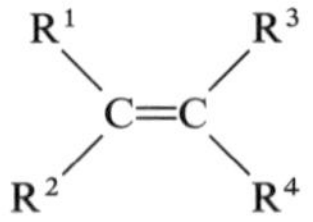		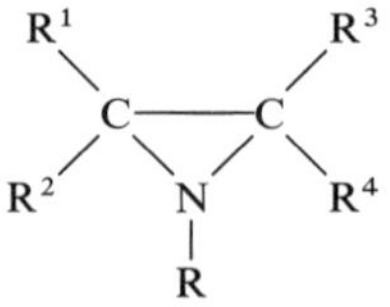			
R^1	R^2	R^3	R^4		
C_6H_5	H	H	H	42	374
t-Bu	H	Me	Me	61	374
Me	H	H	Me	36	374
Me	H	Me	H	19	374
H	Me	Me	Me	57	375
Cl	H	H	Cl	60	374
Cl	Cl	H	Cl	90	374

TABLE VIII (*cont.*)

Substrate	Product			% Yield	Reference

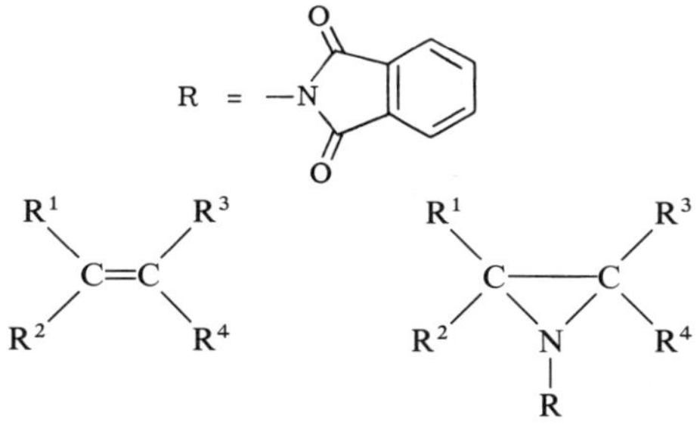

R^1	R^2	R^3	R^4	% Yield	Reference
CO_2Me	H	H	H	73	374
CO_2Et	H	H	CO_2Et	20	374
CO_2Me	Me	H	H	100	374
CO_2Me	H	H	Me	90	374
COMe	H	Me	Me	88	374
C_6H_5	H	CN	CO_2Me	73	375, 376
p-ClC_6H_4	H	CN	CO_2Me	83	375, 376
p-$NO_2C_6H_4$	H	CN	CO_2Me	95	375, 376
m-$MeOC_6H_4$	H	CN	$CONH_2$	74	375, 376
C_6H_5	H	CN	$CONH_2$	81	375, 376
p-ClC_6H_4	H	CN	$CONH_2$	90	375, 376
p-$NO_2C_6H_4$	H	CN	$CONH_2$	83	375, 376
p-$MeOC_6H_4$	H	CO_2Et	CO_2Et	40	375, 376
m-$MeOC_6H_4$	H	CO_2Et	CO_2Et	40	375
C_6H_5	H	CO_2Et	CO_2Et	48	375
p-ClC_6H_4	H	CO_2Et	CO_2Et	51	375
p-$NO_2C_6H_4$	H	CO_2Et	CO_2Et	96	375, 376
p-$MeOC_6H_4$	H	Et	NO_2	50	375, 376
C_6H_5	H	Et	NO_2	50	375, 376
p-ClC_6H_4	H	Et	NO_2	62	375, 376
p-$NO_2C_6H_4$	H	Et	NO_2	70	375, 376
p-$MeOC_6H_4$	H	Me	NO_2	66	375
C_6H_5	H	Me	NO_2	58	375
p-ClC_6H_4	H	Me	NO_2	80	375
p-$NO_2C_6H_4$	H	Me	NO_2	85	375
C_6H_5	H	CN	$PO(OEt)_2$	18	375, 376
p-$NO_2C_6H_4$	H	CN	$PO(OEt)_2$	50	375, 376
p-$NO_2C_6H_4$	H	CN	Ts	20	375, 376
p-$MeOC_6H_4$	H	CN	C_6H_5	25	376
C_6H_5	H	CN	C_6H_5	43	376
C_6H_5	H	H	CO_2Me	75	375, 376
C_6H_5	H	H	COC_6H_5	70	375, 376

(*cont.*)

TABLE VIII (*cont.*)

Substrate	Product	% Yield	Reference
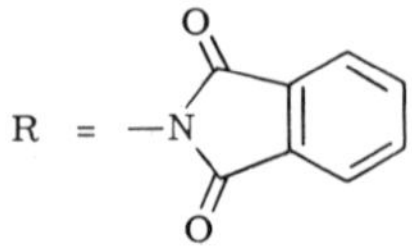			
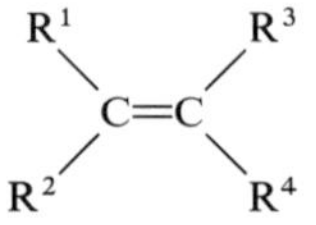	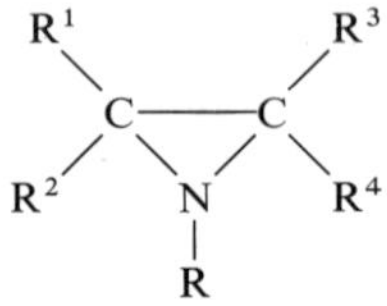		
R^1 R^2 R^3 R^4			
C_6H_5 H H CHO		50	375, 376
C_6H_5 H H CN		35	375, 376
OEt H H H		95	375, 376
H C_6H_5 C_6H_5 H		50	377
C_6H_5 C_6H_5 H H		32	378
	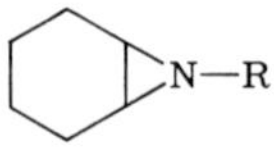	30–40	374, 375, 376
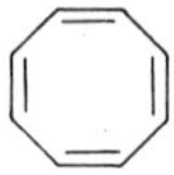	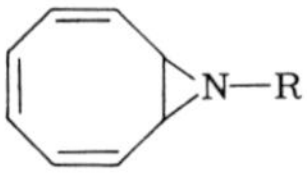	42	379
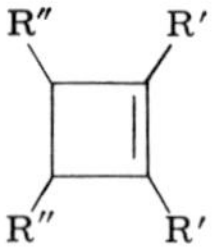	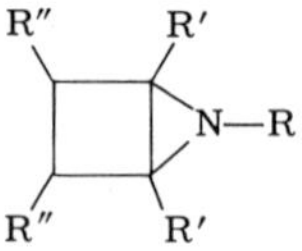	R′ R″	
		CO_2Me H 35	380
		H H 22	381
		H Cl 12	381
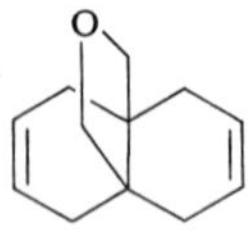	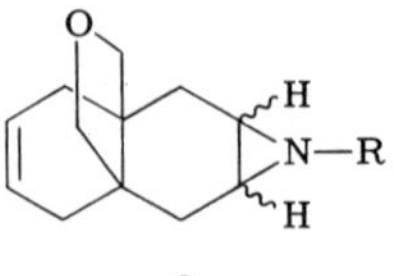	18 endo/exo mixture	382
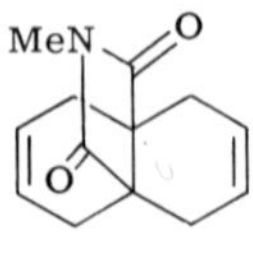	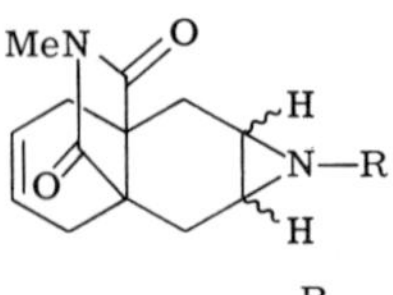	22 endo/exo mixture	382
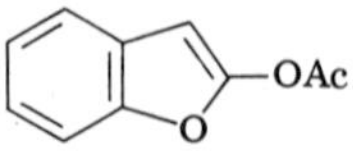	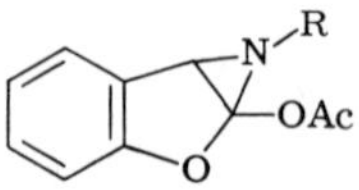	89	383

Phthalomidonitrene generated from LTA treatment of *N*-aminophthalimide reacts with 1,3-dimethoxybenzene to give 28–37% of **231** along with a trace of azepine **232**.[384] Pyrolysis of **233** affords a 2 to 1 mixture rich in

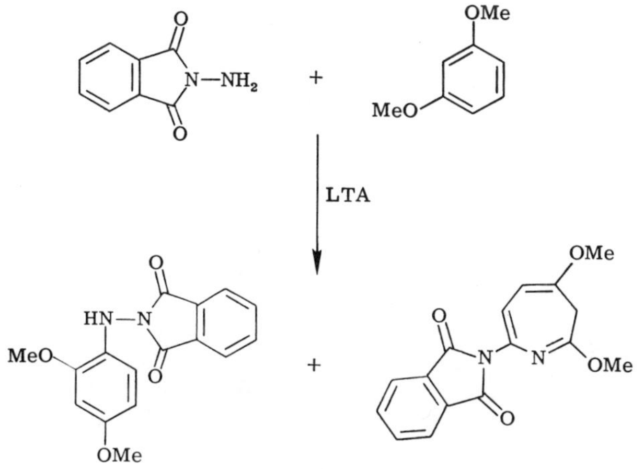

(ref. 384)

231 (28–37%) **232** (trace)

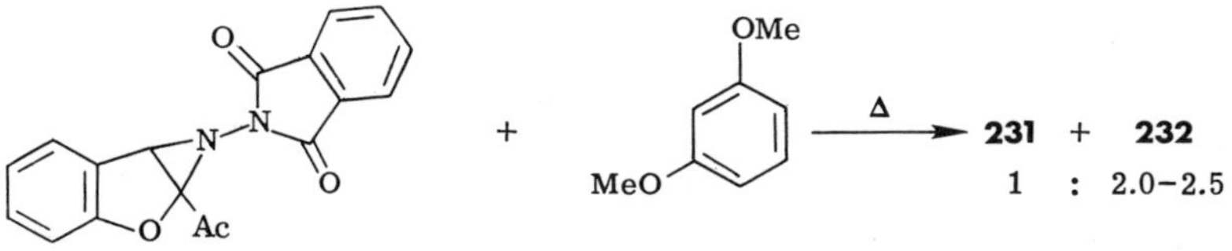

233

azepine. The difference in product ratios is attributed to acetic acid present in the LTA reaction.

In one interesting extension of the method, *N*-aminophthalimide was oxidized in the presence of alkynes in an attempt to generate the antiaromatic

---

[377] L. A. Carpino and R. K. Kirkley, *J. Am. Chem. Soc.* **92**, 1784 (1970).

[378] R. Annunziata, R. Fornasier, and F. Montanari, *J. Org. Chem.* **39**, 3195 (1974).

[379] G. C. Tustin, C. E. Monken, and W. H. Okamura, *J. Am. Chem. Soc.* **94**, 5112 (1972).

[380] G. R. Meyer and J. Stavinoha, Jr., *J. Heterocycl. Chem.* **12**, 1085 (1975).

[381] A. G. Anderson, Jr. and D. R. Fagerburg, *Tetrahedron* **29**, 2973 (1973).

[382] A. Rüttimann and D. Ginsburg, *Tetrahedron* **32**, 1009 (1976).

[383] D. W. Jones, *J. Chem. Soc. Chem. Commun.*, 884 (1972).

[384] D. W. Jones, *J. Chem. Soc. Chem. Commun.*, 67 (1973).

1*H*-azirine system. However, the only product isolated was the 2*H*-azirine, derived perhaps from rearrangement of the transiently stable 1*H*-system **234**.[385]

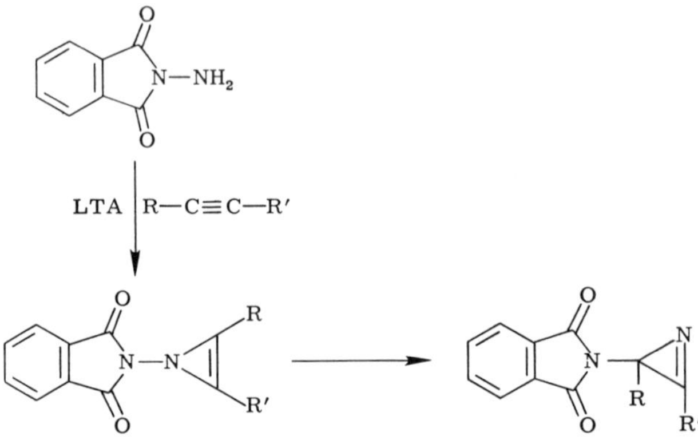

(ref. 385)

234 (1–15%)

Diazenes can also be trapped by allyl aryl sulfides to give **235** which subsequently afford **236** by a [2,3] sigmatropic shift.[386] Trapping with

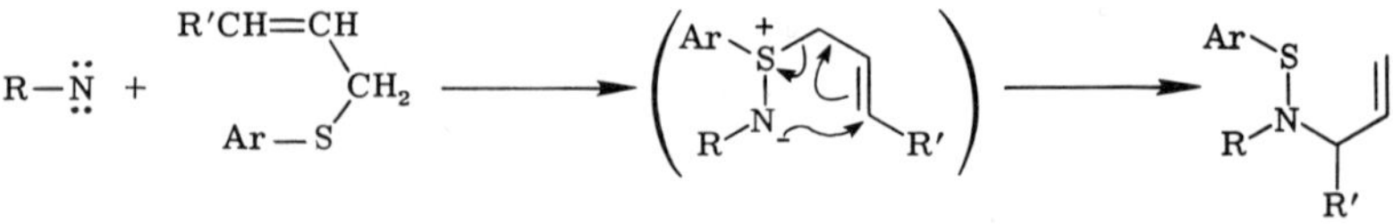

235 **236** (54–64%)

R =

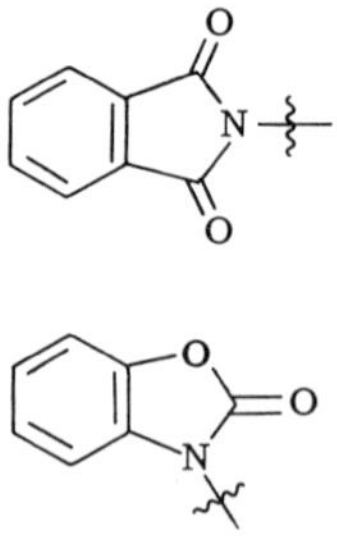

(ref. 386)

sulfoxides gives high yields of the corresponding sulfoximides[354,387,388] in a reaction that occurs with retention of configuration at sulfur.[388a,388b] The

[385] D. J. Anderson, T. L. Gilchrist, G. E. Gymer, and C. W. Rees, *J. Chem. Soc. Perkin Trans. 1*, 550 (1973).

[386] R. S. Atkinson and S. B. Awad, *J. Chem. Soc. Perkin Trans. 1*, 651 (1975).

[387] T. L. Gilchrist, C. W. Rees, and E. Stanton, *J. Chem. Soc. Chem. Commun.*, 801 (1971).

formation of **237** is one interesting application of the process.[388b] Pyrolysis

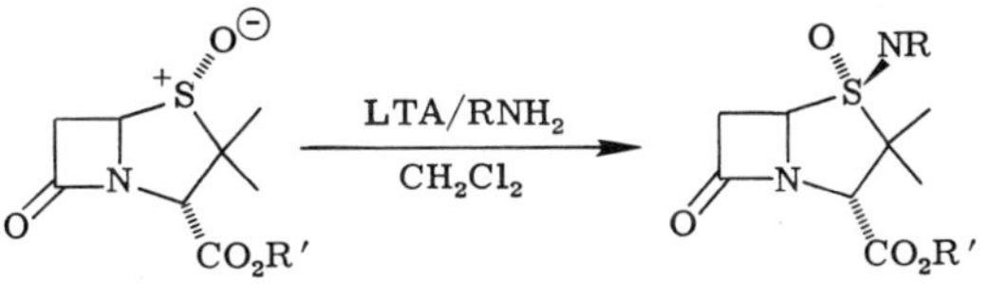

237 (76%)

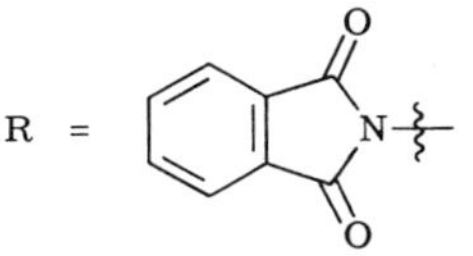

(ref. 388b)

of **238** at low pressure has been shown to give benzocyclobutene-1,2-dione[387,388] whereas photolysis of **238** in the presence of cyclohexene affords aziridine **239**.[388] Thermal decomposition of the readily available

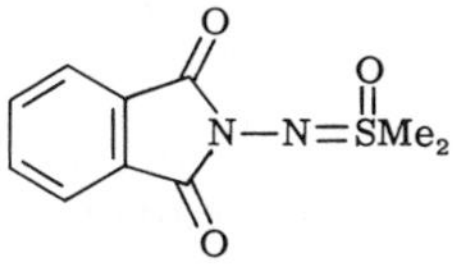

238

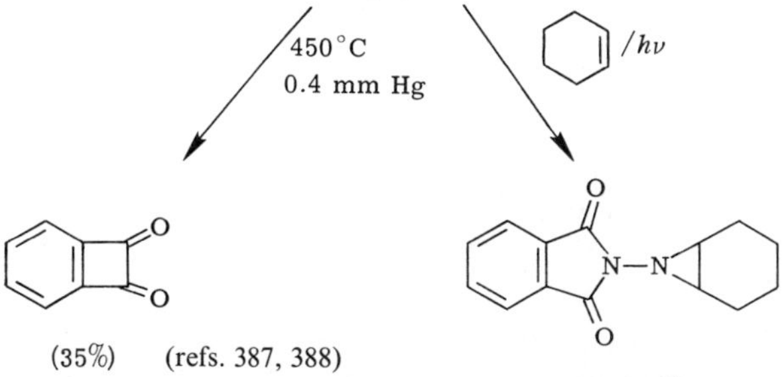

(35%) (refs. 387, 388)

239 (20%)

(ref. 388)

3-amino-2-oxazolidone system **240** results in concomitant loss of N_2 and

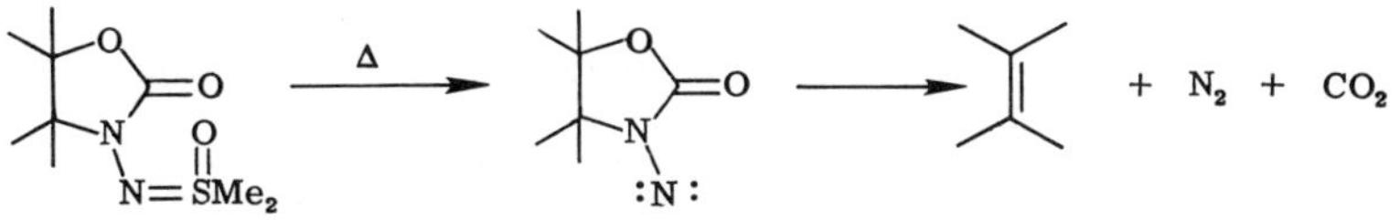

240

---

[388] D. J. Anderson, D. C. Horwell, E. Stanton, T. L. Gilchrist, and C. W. Rees, *J. Chem. Soc. Perkin Trans. 1*, 1317 (1972).

[388a] S. Colonna and C. J. M. Stirling, *J. Chem. Soc. Perkin Trans. 1*, 2120 (1974).

[388b] J. E. G. Kemp, M. D. Closier, and M. H. Stefaniak, *Tetrahedron Lett.*, 3785 (1979).

CO_2 from an intermediate diazene to afford excellent yields of alkene.[389,390] This predominantly syn-elimination allows the preparation of interesting strained alkenes such as **241**[389] and **242**.[389,390]

241 (ref. 389)

242 (refs. 389, 390)

D. AZOMETHINES

The chemistry of the oxidation of azomethines (**243**) with LTA has been studied extensively. Two excellent comprehensive reviews on the subject have

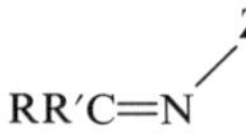

243

appeared[9,391] along with several specific reviews dealing with oxidation of oximes,[392] substituted hydrazones,[393] heteroallylic systems,[394] and oxidative cyclization of carbonyl derivatives.[395] The depth of coverage of this interesting area of LTA chemistry noted above precludes further discussion except to note several recent reports.

Hydrazone **244** is cyclized in good yield,[396] while treatment of **245** with

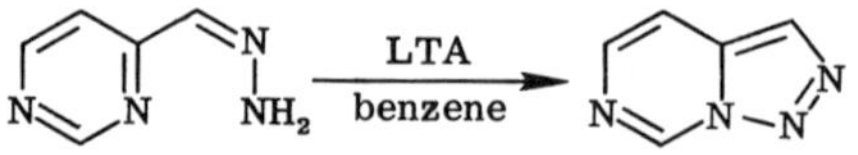

(ref. 396)

244 (60%)

LTA in the presence of acrylonitrile gives **246**.[397] Under the same conditions, **247** affords the corresponding triazolopyridine.[397] Both **248**[397a] and

[389] M. Kim and J. D. White, *J. Am. Chem. Soc.* **99**, 1172 (1977).
[390] M. Kim and J. D. White, *J. Am. Chem. Soc.* **97**, 451 (1975).
[391] R. N. Butler, F. L. Scott, and T. A. F. O'Mahony, *Chem. Rev.* **73**, 93 (1973).
[392] R. N. Butler, *Chem. Ind. (London)*, 523 (1972).
[393] R. N. Butler, *Chem. Ind. (London)*, 437 (1968).
[394] R. N. Butler, *Chem. Ind. (London)*, 499 (1976).
[395] J. Warkentin, *Synthesis*, 279 (1970).
[396] G. Maury, J.-P. Paugam, and R. Pougam, *J. Heterocycl. Chem.* **15**, 1041 (1978).
[397] B. Stanovnik and M. Tisler, *Croat. Chem. Acta* **49**, 135 (1977).
[397a] T. Tsuchiya and J. Kurita, *J. Chem. Soc. Chem. Commun.*, 803 (1979).

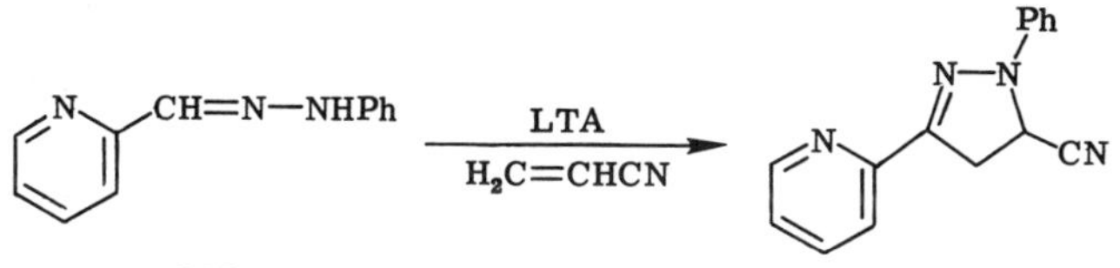

245 **246**

(ref. 397)

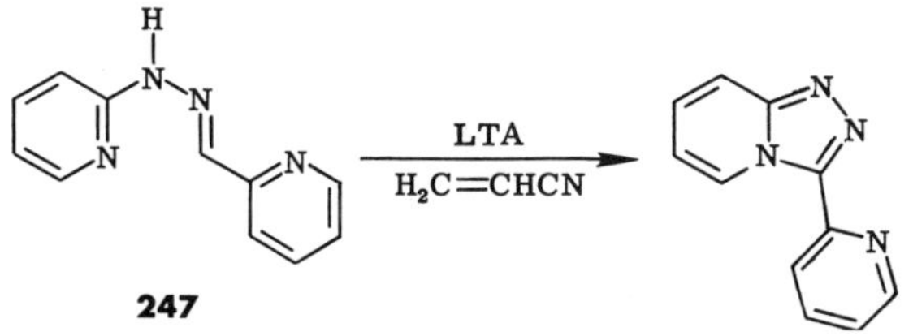

247

249[397b] give interesting ring-contracted products when allowed to react with LTA.

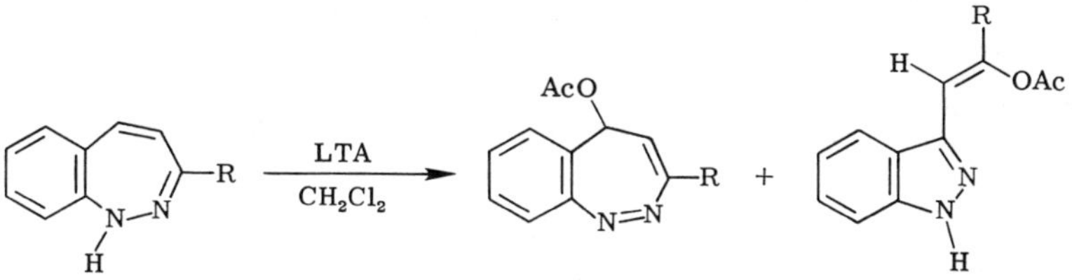

248 (60–65%) (25%) (ref. 397a)

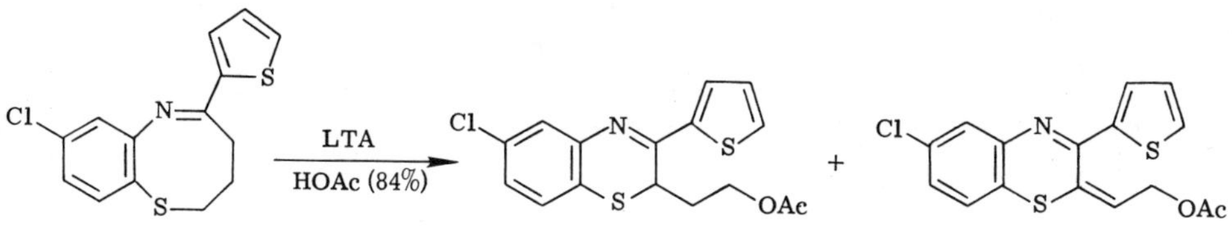

249 1.4 : 1 (ref. 397b)

Studies on the reaction of LTA with a series of aliphatic aldehyde phenylhydrazones **250** containing ortho substituents in the *N*-phenyl ring indicate that formation of **251** is favored by both the ortho substituent on the *N*-phenyl ring and an alkyl group (R) on the methine carbon of **250**. Production of **252** is enhanced by replacement of alkyl with aryl or by increased

[397b] J. B. Press, N. H. Eudy, F. M. Lovell, and N. A. Perkinson, *Tetrahedron Lett.*, 1705 (1980).

complexity in the alkyl group. A partitioning of **253** accounts for both products.[398,398a] The acyl hydrazone **254** is oxidized by LTA to afford

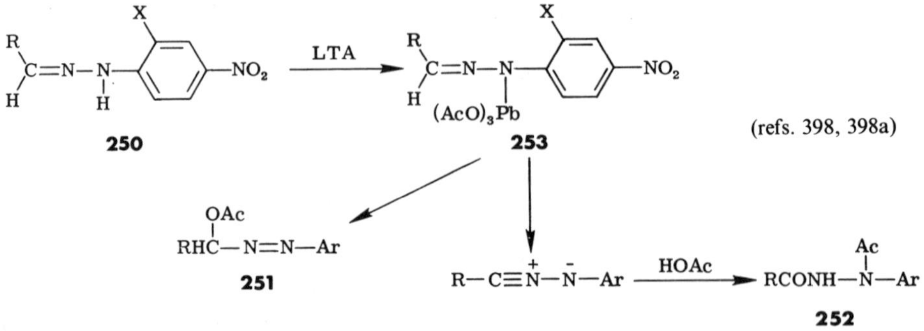

stereoselective ring formation.[398b]

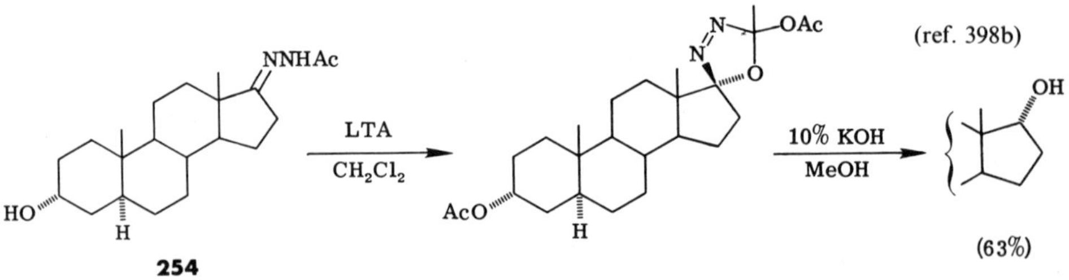

Oxime **255** gives a mixture of nitrosoacetates upon treatment with LTA,[399] while oximes **256** produce mixtures of nitrosoacetates and nitroacetates in moderate yields.[399a] A series of dioximes were found to give mainly

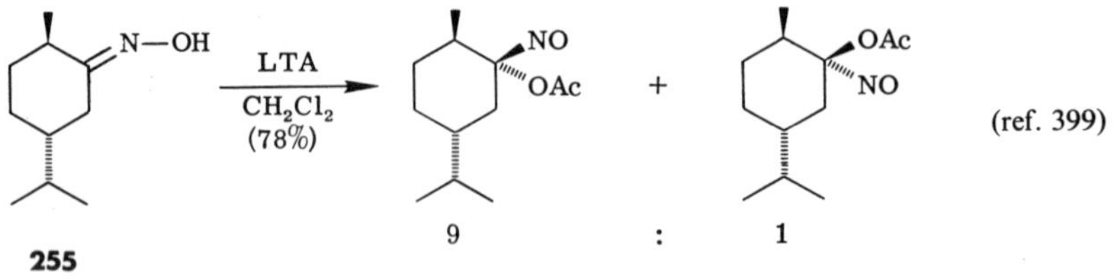

---

[398] R. N. Butler and W. B. King, *J. Chem. Soc. Perkin Trans. 1*, 282 (1977).
[398a] R. N. Butler and A. M. O'Donohue, *Tetrahedron Lett.* 4583 (1979).
[398b] I. R. McDermott and C. H. Robinson, *J. Chem. Soc. Chem. Commun.*, 28 (1979).
[399] T. Bosch, G. Kresze, and J. Winkler, *Liebigs Ann. Chem.*, 1009 (1975).
[399a] D. Shafiullah and H. Ali, *Synthesis*, 124 (1979).
[400] A. Ohsawa, H. Arai, and H. Igeta, *Heterocycles* **9**, 1367 (1978).
[400a] H. Arai, A. Ohsawa, and H. Igeta, *Heterocycles* **12**, 204 (1979).

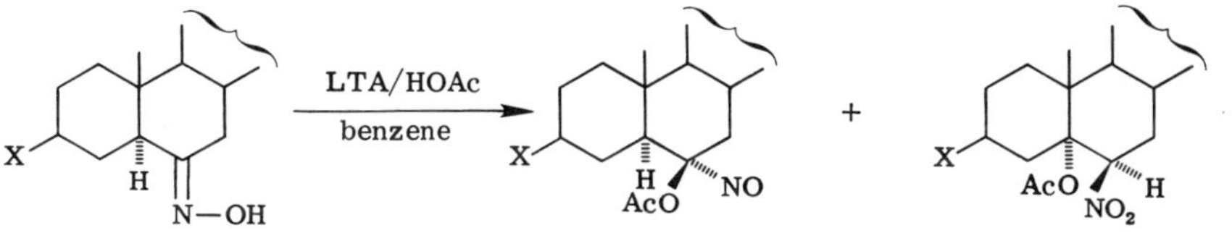

256 (15–18%) (18–22%)

(ref. 399a)

257,[400–400b] but the analogous bis-hydrazones give mainly butadienes[400b] when treated with LTA in methylene chloride.

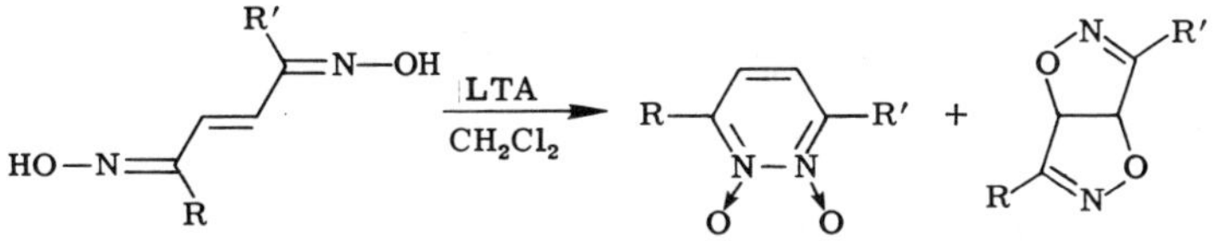

257

(refs. 400–400b)

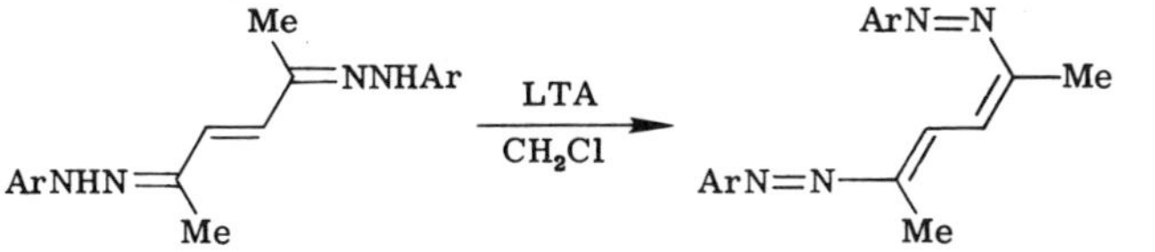

(ref. 400b)

The oxidation of bis-tosylhydrazones[401,402] and bis-semicarbazones[403,404] with LTA gives rise to the cyclization noted. The process is favored when X in **258** is a good leaving group. The proposed diazo intermediate **259** (Y=O) can be isolated when **258** (Y=O) is oxidized using methylene chloride–triethylamine as solvent. Subsequent treatment of **259** (Y=O) with acetic

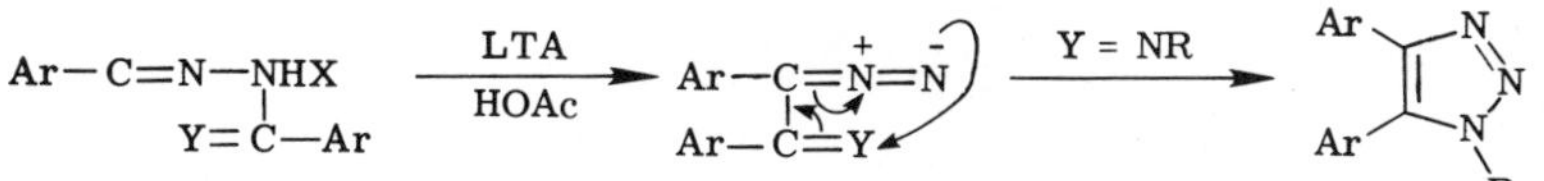

258 **259**

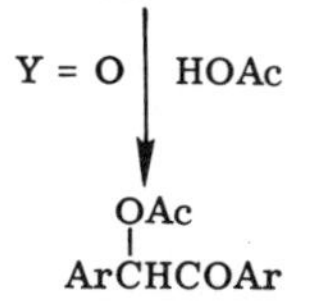

X	Y
$CONH_2$	$NNHCONH_2$
Ts	NNHTs
Ts	O
$CONH_2$	O

(refs. 401, 404, 405)

R	Refs.
$NHCONH_2$	403, 404
NHTs	401, 402

[400b] A. Ohsawa, H. Arai, H. Igeta, T. Akimoto, A. Tsuji, and Y. Iitaka, *J. Org. Chem.* **44**, 3524 (1979).

[401] R. N. Butler, A. B. Hanahoe, and W. B. King, *J. Chem. Soc. Perkin Trans. 1*, 881 (1978).

[402] R. N. Butler and A. B. Hanahoe, *J. Chem. Soc. Chem. Commun.*, 622 (1977).

acid gives *O*-acetylbenzoin.[401,404,405] Systems of type **260** also cyclize upon treatment with LTA.[406,407]

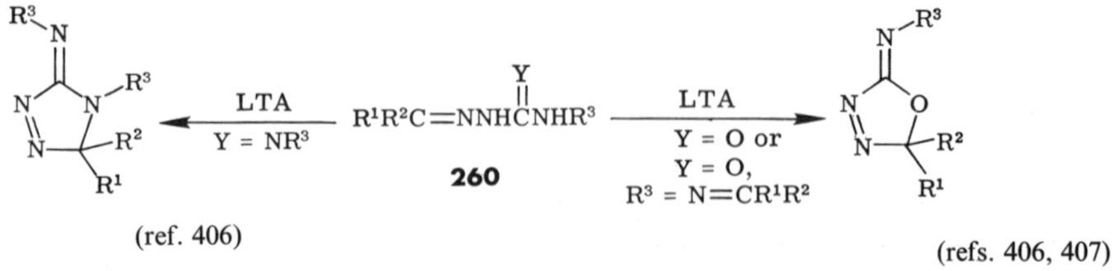

IV. LTA Reactions with Hydrocarbons

A. ALKANES

Acetoxylation of saturated hydrocarbons by LTA received little attention prior to 1970.[122] It was reported that oxidation of polycyclic alkanes such as adamantane was complicated by further reaction of primary products.[408] Further, it has been shown that regioselectivity is not particularly high with adamantyl derivatives.[17,18,409] However, recent studies have discovered that

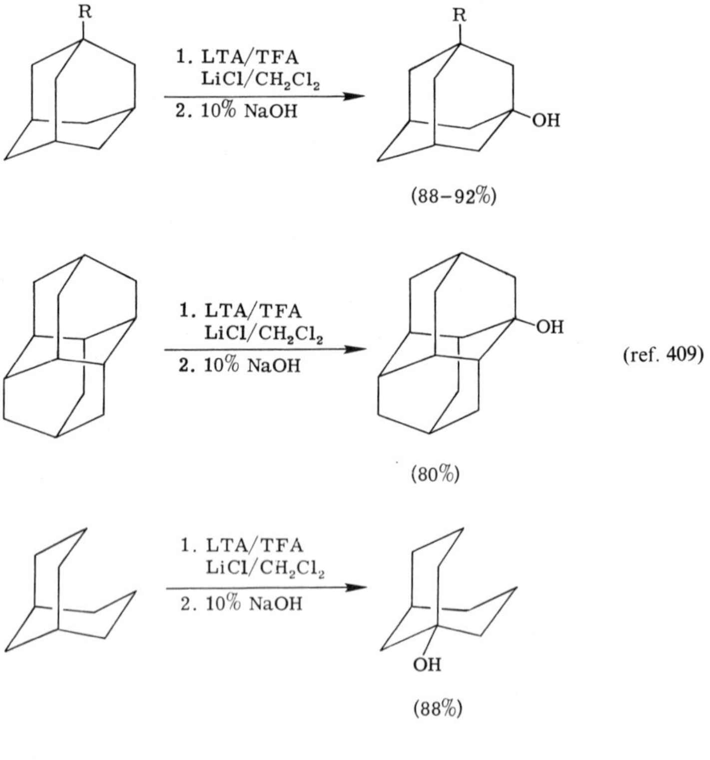

the use of LTA/TFA in the presence of chloride ion gives excellent yields of trifluoroacetoxylated polycyclic hydrocarbons isolated as the corresponding alcohols.[409] The reaction was less successful with methylcyclohexane and 3-methylhexane, but with **261**, a regioisomer and a product with rearranged carbon skeleton were isolated.[409] A very useful extension of the reaction

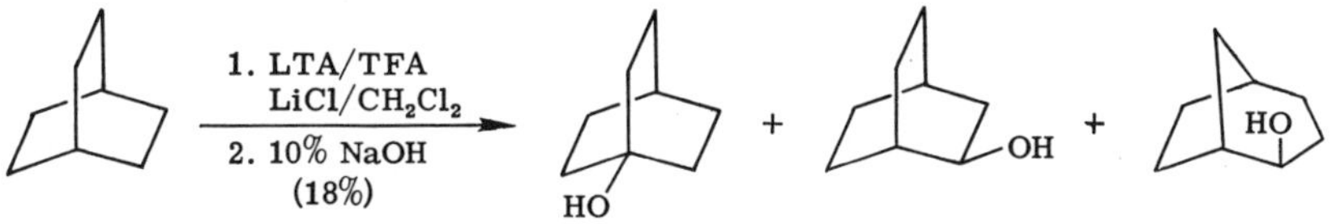

261 45 : 26 : 24

(ref. 409)

involves sequential treatment of hydrocarbon with LTA/TFA and then with a nucleophile. In this manner, a number of bridgehead functionalized hydrocarbons are obtained in excellent yield.[409,410] The identity of the oxidizing

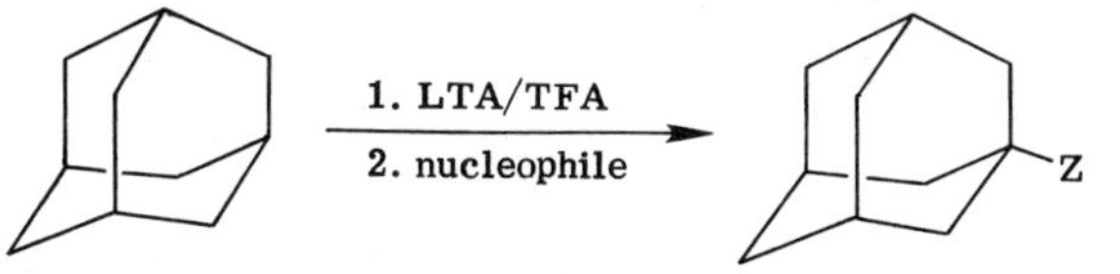

Z	% yield
—NHCOMe	85
—NHCHO	82
—C_6H_4OMe-*p*	64
—C_6H_4OH-*p*	81
—$CH(CO_2Et)COMe$	87
—S-*n*-Bu	97

(ref. 410)

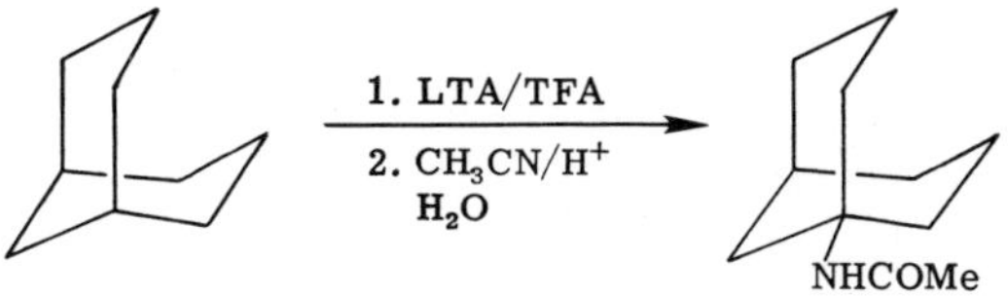

(78%)

agent in the LTA/TFA/LiCl reaction is not clear, but mechanistic studies have shown that oxidation does not occur via radical-cation intermediates.

[403] N. E. Alexandrou and S. Adamopoulos, *Synthesis*, 482 (1976).

[404] R. N. Butler, A. B. Hanahoe, and A. M. O'Donohue, *J. Chem. Soc. Perkin Trans. 1*, 1246 (1979).

[405] R. N. Butler and A. B. Hanahoe, *Chem. Ind. (London)*, 39 (1978).

[406] L. M. Cabelkova-Taguchi and J. Warkentin, *Can. J. Chem.* **56**, 2194 (1978).

[407] K. Ramakrishnan, J. B. Fulton, and J. Warkentin, *Tetrahedron* **32**, 2685 (1976).

[408] W. H. W. Lunn, *J. Chem. Soc. C*, 2124 (1970).

[409] S. R. Jones and J. M. Mellor, *J. Chem. Soc. Perkin Trans. 1*, 2576 (1976).

[410] S. R. Jones and J. M. Mellor, *Synthesis*, 32 (1976).

At this point, a mechanism proceeding by electrophilic attack at the carbon–hydrogen bond is favored.[411,412]

The LTA acetoxylation of hydrocarbons, as noted above, is not particularly useful as a synthetic technique. Acetoxylation of the benzylic position of aromatic hydrocarbons, however, is frequently employed.[122] The reaction of **262** with LTA is typical.[413] The reaction of toluene with

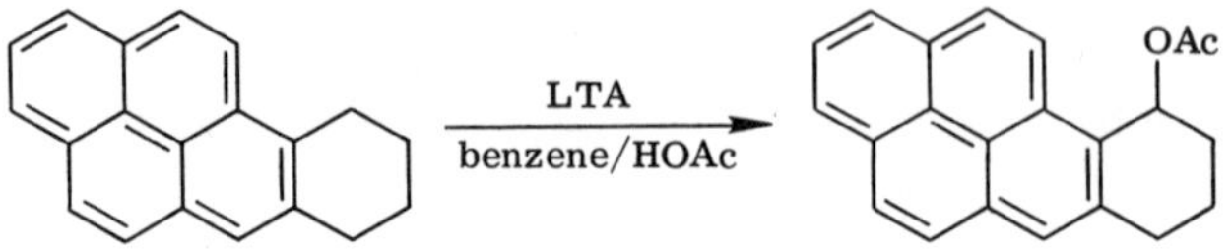

(ref. 413)

262 (82%)

LTA has been rationalized by a free-radical mechanism in which the radicals have a slight positive character.[414,415] The fact that cumene reacts twice as fast as toluene with LTA is taken as supportive evidence,[416] as is the isolation of products resulting from ring substitution by methyl and carboxymethyl radicals.[414] Both short-lived radicals[415] and concerted carbon–lead bond formation[416a] have been suggested in the acetoxylation of fluorene derivatives with LTA.

A mechanistic change occurs, however, in aromatic systems with low ionization potentials. In these systems an electron-transfer process pertains.[417] Thus, **263** is initially oxidized to **264** followed by production of **265**; with excess LTA **266** is obtained.[418]

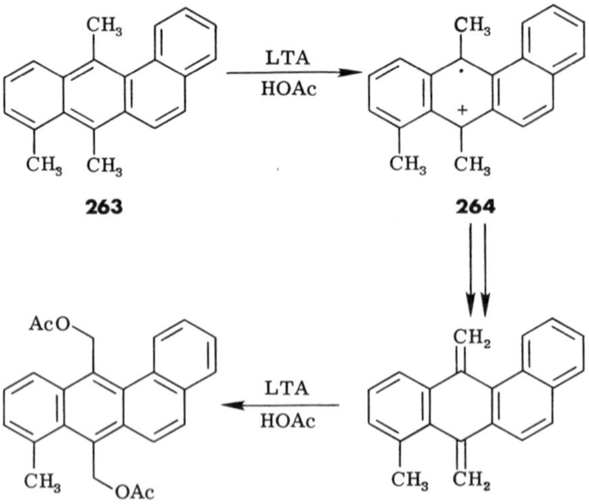

(ref. 418)

263 **264**

266 (50%) **265** (37%)

[411] S. R. Jones and J. M. Mellor, *J. Chem. Soc. Perkin Trans. 2*, 511 (1977).

A similar process may be occurring in the LTA oxidations of both 7-methyl- and 12-methylbenz(a)anthracene.[419]

A very useful adaptation of the electron-transfer pathway occurs with the oxidation of methoxy-substituted alkyl benzenes. Due to the shift in mechanism away from the radical route, **267** reacts in preference to **268** as shown

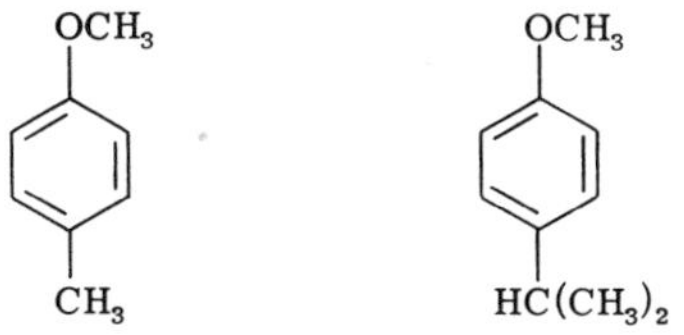

267 **268**

by a competition experiment. Therefore, methyl functionalization by LTA is possible in the presence of tertiary benzylic hydrogen with oxygenated aromatics.[420]

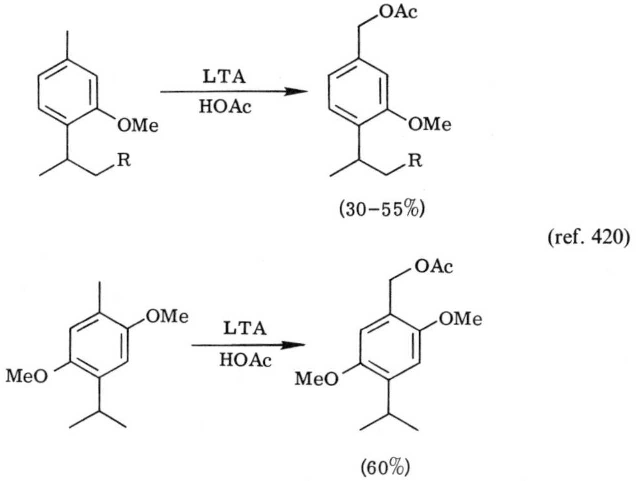

---

[412] S. R. Jones and J. M. Mellor, *J. Chem. Soc. Chem. Commun.*, 385 (1976).
[413] H. Yagi and D. M. Jerina, *J. Am. Chem. Soc.* **97**, 3185 (1975).
[414] E. I. Heiba, R. M. Dessau, and W. J. Koehl, Jr., *J. Am. Chem. Soc.* **90**, 1082 (1968).
[415] P. S. Radhakrishnamurti and S. N. Mahapatro, *Indian J. Chem. Sect. A* **14**, 478 (1976).
[416] E. I. Heiba, R. M. Dessau, and W. J. Koehl, Jr., *J. Am. Chem. Soc.* **91**, 6830 (1969).
[416a] S. Narasimhan and N. Venkatasubramanian, *Indian J. Chem. Sect. B* **17**, 143 (1979).
[417] E. I. Heiba, R. M. Dessau, and W. J. Koehl, Jr., *J. Am. Chem. Soc.* **91**, 138 (1969).
[418] J. Pataki and R. Balick, *Tetrahedron Lett.*, 3447 (1974).
[419] P. Jacquignon and M. Croisy-Delcey, *C. R. Acad. Sci. Ser. C* **276**, 955 (1973).
[420] V. V. Dhekne and A. S. Rao, *Synth. Commun.* **8**, 135 (1978).

The oxidation reaction of benzylic methyl groups with LTA has also been applied to pyrroles.[421,421a,422] Use of 2 molar equivalents of LTA with **269**

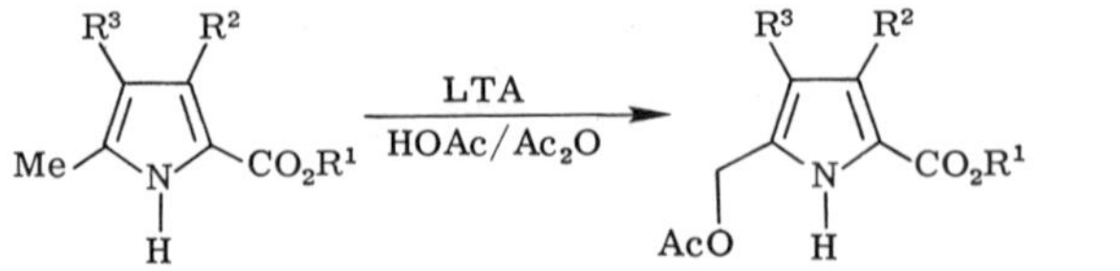

(53–98%)

(refs. 421, 421a, 422)

yields large amounts of the corresponding aldehydes upon hydrolysis.[422] In a manner analogous to pyrrole oxidation, indoles can also be transformed

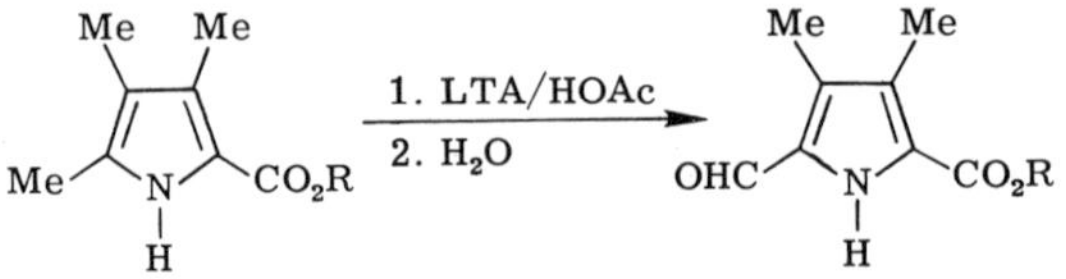

269

(57–81%)

(ref. 422)

into the corresponding acetoxy derivatives by treatment with LTA.[422a] Attempts at bis-acetoxylation were unsuccessful.

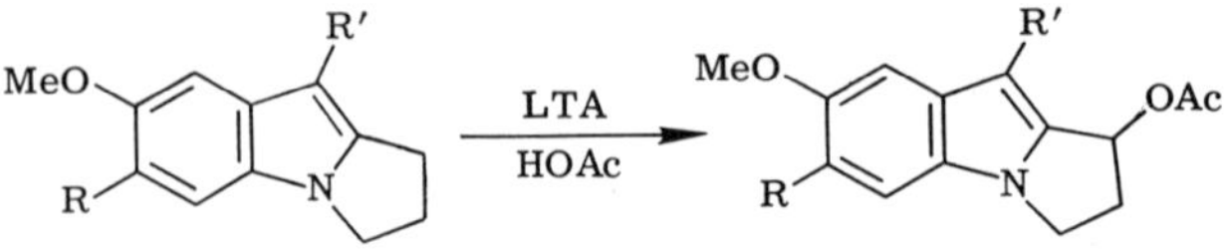

(ref. 422a)

R	R′	% yield
OMe	CN	76
OMe	CHO	65
Me	CHO	—
Me	CN	—
Me	CO_2Me	—

Dehydrogenation was observed when **270** was treated with LTA,[423] and

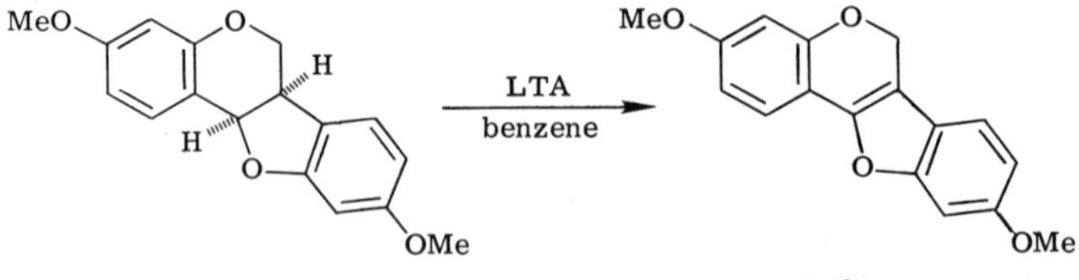

270

(30%)

(ref. 423)

[421] A. H. Jackson, G. W. Kenner, K. M. Smith, and C. J. Suckling, *Tetrahedron* **32**, 2757 (1976).
[421a] L. Diaz, G. Buldain, and B. Frydman, *J. Org. Chem.* **44**, 973 (1979).
[422] J. B. Paine III, R. B. Woodward, and D. Dolphin, *J. Org. Chem.* **41**, 2826 (1976).
[422a] K. Takahashi and T. Kametani, *Heterocycles* **13**, 411 (1979).
[423] P. J. M. Gunning, P. J. Kavanagh, M. J. Meegan, and D. M. X. Donnelly, *J. Chem. Soc. Perkin Trans. 1*, 691 (1977).

the interesting bromo derivative **271** was obtained when **272** was allowed to react with LTA/NBS.[424] In the latter case, no product of benzylic oxidation was isolated.

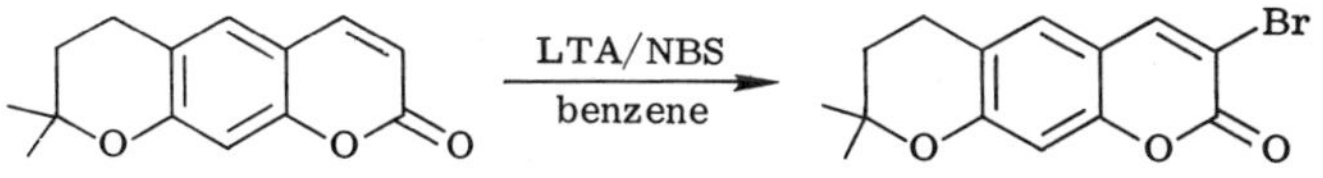

272 **271** (10%) (ref. 424)

In rare instances, LTA cleavage of carbon–carbon σ-bonds has been observed for the most part in systems containing cyclopropyl rings.[425,426]

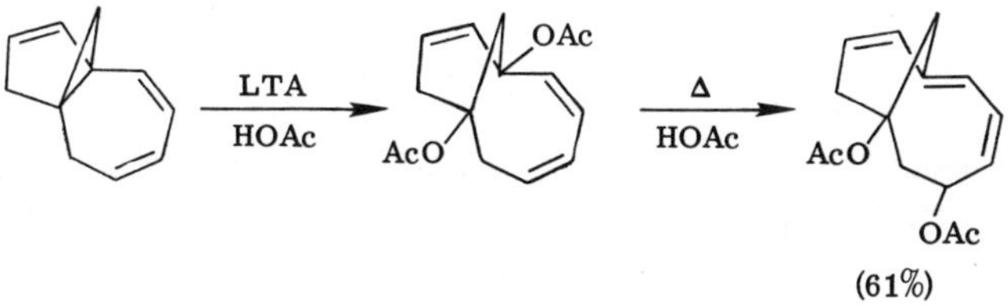

(ref. 425)

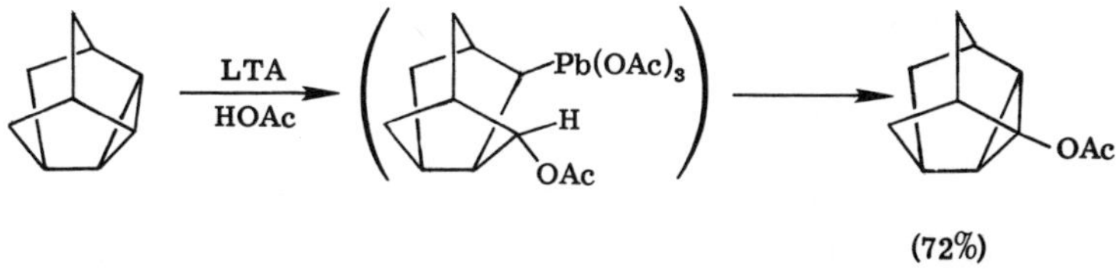

(ref. 426)

Mechanistic studies on a series of substituted arylcyclopropanes reveal that the strained ring is attacked by LTA and not by $(AcO)_3Pb^+AcO^-$. A concerted mechanism involving coordination of the cyclopropane with lead as acetate ion departs is favored.[427] Lead(IV) diacetate difluoride also cleaves phenylcyclopropane in a manner analogous to LTA.[428]

B. AROMATIC HYDROCARBONS

Reactive polynuclear aromatics and aromatic compounds containing electron-donating groups can undergo nuclear acetoxylation with LTA.

[424] B. Talapatra, S. K. Mukhopadhyay, M. K. Chaudhuri, and S. K. Talapatra, *Indian J. Chem. Sect. B* **14**, 129 (1976).
[425] L. T. Scott and W. R. Brunsvold, *J. Am. Chem. Soc.* **100**, 4320 (1978).
[426] D. F. Covey and Alex Nickon, *J. Org. Chem.* **42**, 794 (1977).
[427] R. J. Ouellette, D. Miller, A. South, Jr., and R. D. Robins, *J. Am. Chem. Soc.* **91**, 971 (1969).
[428] J. Bornstein and L. Skarlos, *J. Chem. Soc. Chem. Commun.*, 796 (1971).

When benzylic oxidation is not a competitive process, yields of this type product are moderate.[122] The synthesis of **273** is an example.[429]

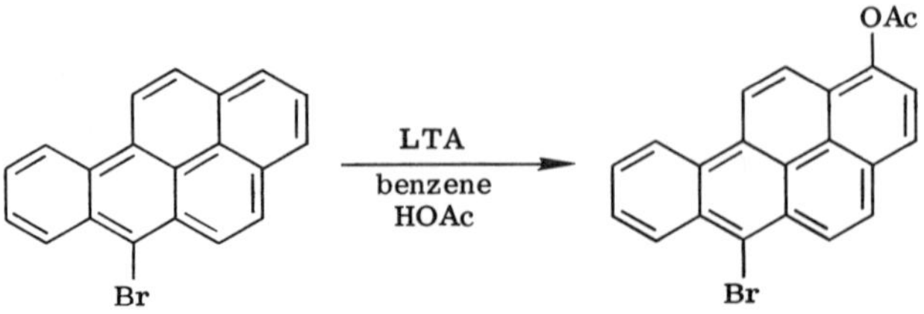

273 (40%)

(ref. 429)

LTA oxidation of anthracene in benzene, benzene–pyridine, benzene–cyclohexane, or chloroform gives a mixture containing approximately equal amounts of *cis*- and *trans*-9,10-diacetoxy-9,10-dihydroanthracene.[430] When benzene–methanol is used as solvent, the corresponding dimethoxy derivatives are formed showing a preference for formation of the *trans*-isomer **274**. In the former case, products are believed to arise from a carbonium ion intermediate, while in the latter, concerted methoxy transfer from a methoxylated lead species to afford **275** is followed by S_N2 displacement of lead to give **274**.[430] Results consistent with this second pathway were also

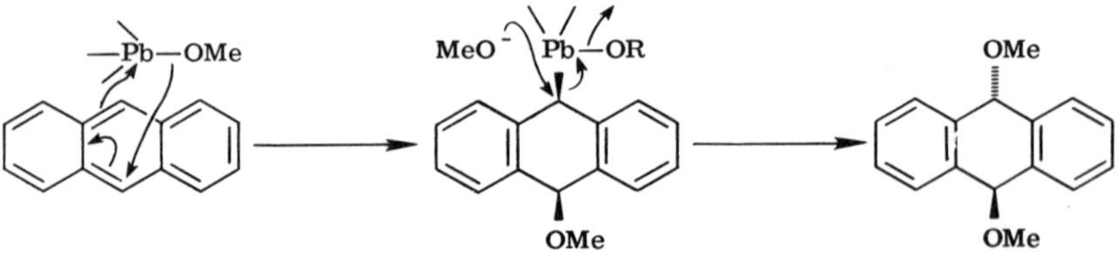

275 **274**

(ref. 430)

obtained in the LTA/benzene–methanol oxidation of acenaphthylene.[431] The bis-acetoxylation referred to above is also observed with furan derivatives as illustrated by the conversions of **276**[432] and **277**.[432a]

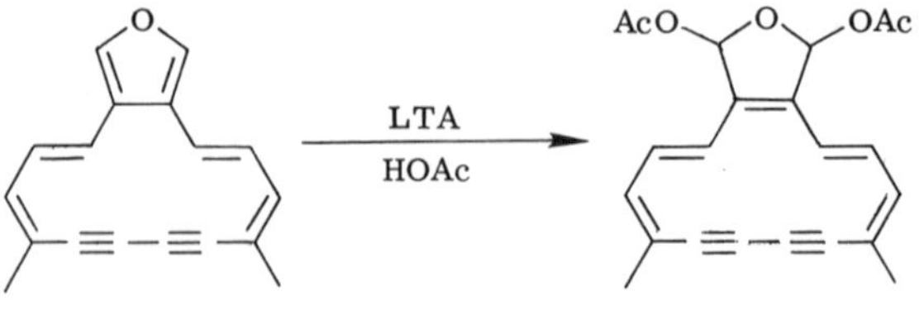

276

(ref. 432)

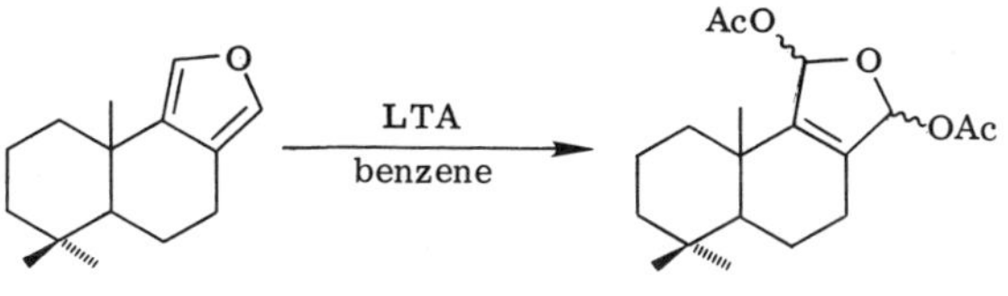

(ref. 432a)

277 (90%)

The LTA oxidation of methoxy-substituted benzenes has proven to be a complex reaction. In general, methoxy-substituted benzenes react with LTA via S_E2 plumbylation to afford aryl lead compounds which either homolyze or heterolyze.[433,434] Alternatively, electron transfer occurs to afford a radical cation that is intercepted by a nucleophile. The radical thus formed is oxidized further to give products.[433,435] Radical initiators such as Perkadox[435] enhance the latter process, whereas monomethyl oxalate facilitates both types of reactions.[436] The use of acetic acid as solvent in the oxidation of anisole increases the amounts of nuclear acetoxylated products observed at the expense of ring methylation, acetoxymethylation, and attack of the alkyl methyl group.[434,435] Oxidation studies using lead tetrabenzoate[437] and a series of naphthalene ethers[437a] have also been carried out.

An interesting synthetic application of the reaction of methoxylated aromatics with LTA involves the oxidative cyclization of diarylidene succinic anhydrides into 1-phenylnaphthalenes. The reaction occurs in yields of over 60% and can be illustrated by the conversion of **278** into **279**.[438]

---

[429] R. G. Harvey and H. Cho, *J. Chem. Soc. Chem. Commun.*, 373 (1975).
[430] B. Rindone and C. Scolastico, *J. Chem. Soc.* C, 3983 (1971).
[431] B. Rindone and C. Scolastico, *Tetrahedron Lett.*, 1479 (1973).
[432] T. M. Cresp and F. Sondheimer, *J. Am. Chem. Soc.* **97**, 4412 (1975).
[432a] H. Akita, T. Naito, and T. Oishi, *Chem. Lett.*, 1365 (1979).
[433] R. O. C. Norman and C. B. Thomas, *J. Chem. Soc. B*, 421 (1970).
[434] L. C. Willemsens, D. de Vos, J. Spierenburg, and J. Wolters, *J. Organomet. Chem.* **39**, C61 (1972).
[435] R. A. McClelland, R. O. C. Norman, and C. B. Thomas, *J. Chem. Soc. Perkin Trans. 1*, 562 (1972).
[436] R. A. McClelland, R. O. C. Norman, and C. B. Thomas, *J. Chem. Soc. Perkin Trans. 1*, 578 (1972).
[437] R. A. McClelland, R. O. C. Norman, and C. B. Thomas, *J. Chem. Soc. Perkin Trans. 1*, 570 (1972).
[437a] E. R. Cole, G. Crank, and B. J. Stapleton, *Aust. J. Chem.* **32**, 1749 (1979).
[438] A. S. R. Anjaneyulu, V. K. Rao, P. Satyanarayana, and L. R. Row, *Indian J. Chem.* **11**, 203 (1973).

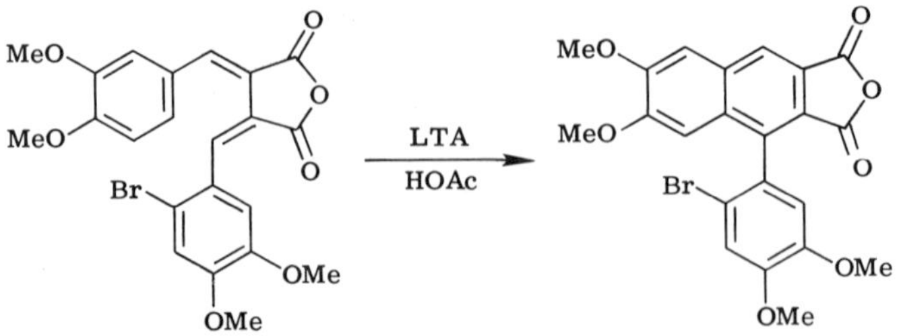

278 **279** (70%) (ref. 438)

As with alkanes, the oxidation of aromatic hydrocarbons is facilitated by the use of either LTA/TFA or lead tetrakistrifluoroacetate (LTTFA) in TFA. Both electrophilic plumbylation[439,440,441] and electron-transfer processes[439,442] are thought to occur. Plumbylation affords **280** which then converts to **281** in a reaction thought to involve aryl cations.[439,443] Since **281** is readily hydrolyzed, the sequence represents a high yield phenol synthesis.[443] Intermediates such as **280** can also be conveniently prepared via metathesis involving TFA and the corresponding aryl lead triacetates.[443]

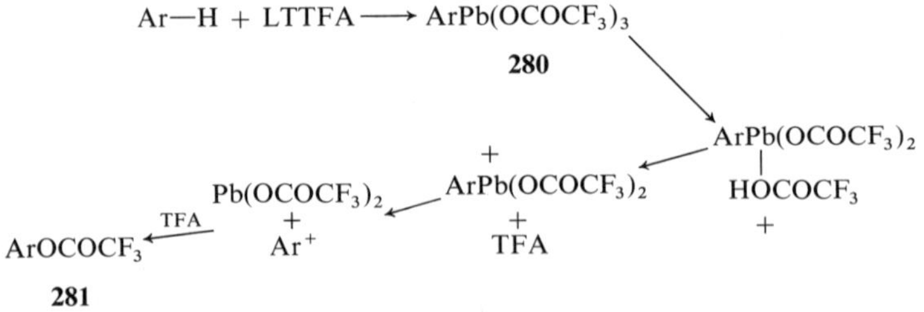

(ref. 443)

The proposed intermediate, $\mathrm{Ar\overset{+}{Pb}(OCOCF_3)_2}$ is also trapped by reactive aromatic compounds to afford high yields of biphenyl derivatives.[444,444a] Combinations of TFA/polymethylbenzenes and $AlCl_3$ or $[Al(OCOC_nF_{2n+1})_3]$/benzene or toluene give the best results.[444,444a]

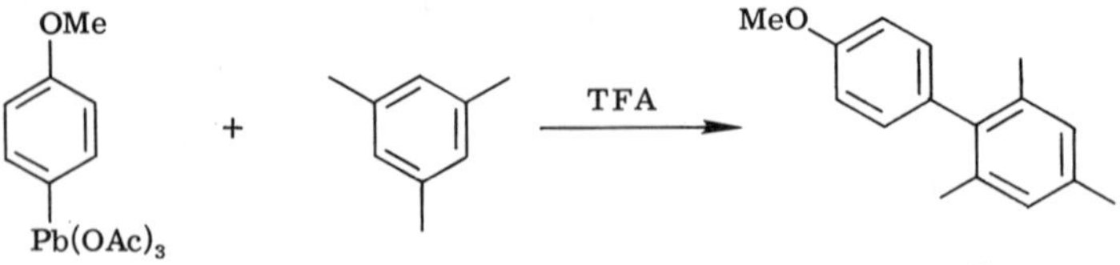

(84%) (ref. 444)

[439] R. O. C. Norman, C. B. Thomas, and J. S. Willson, *J. Chem. Soc. B*, 518 (1971).
[440] J. R. Campbell, J. R. Kalman, J. T. Pinhey, and S. Sternhell, *Tetrahedron Lett.*, 1763 (1972).
[441] D. de Vos, J. Wolters, and A. van der Gen, *Recl. Trav. Chim. Pays-Bas* **92**, 701 (1973).
[442] R. O. C. Norman, C. B. Thomas, and J. S. Willson, *J. Chem. Soc. Perkin Trans. 1*, 325 (1973).

Plumbylation also occurs readily with LTA in the presence of monohalogenoacetic acids,[445] dichloroacetic acid,[443,446] and trichloroacetic acid.[439,443,446] A wide range of aryl lead triacetates can therefore be prepared by using the appropriate combination of aromatic substrate, LTA, and halogenated acetic acid followed by metathesis with acetic acid.[443]

C. ALKENES

1. ACYCLIC ALKENES

Oxidation of acyclic alkenes by LTA[52] is believed to occur via the sequence shown in Scheme 4.[12] Recent studies using methanol serve to confirm this

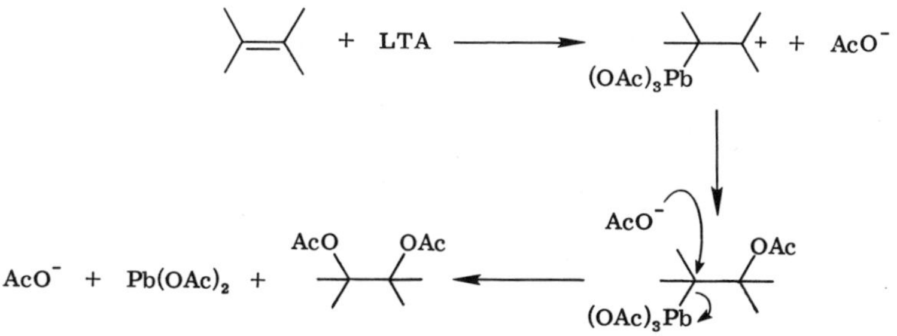

SCHEME 4

premise,[447,448] and evidence has been presented favoring the decomposition of the organometallic adduct **282** with concomitant formation of an acetoxonium ion **283**.[447] The presence of water then gives high yields of **284** and **285**. With internal alkenes evidence is presented which implicates **286**

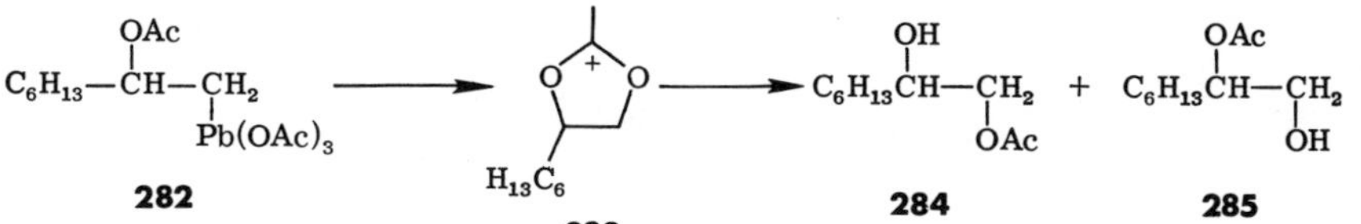

282 **283** **284** **285**

(ref. 447)

---

[443] H. C. Bell, J. R. Kalman, J. T. Pinhey, and S. Sternhell, *Tetrahedron Lett.*, 853 (1974).

[444] H. C. Bell, J. R. Kalman, J. T. Pinhey, and S. Sternhell, *Tetrahedron Lett.*, 857 (1974).

[444a] H. C. Bell, J. R. Kalman, G. L. May, J. T. Pinhey, and S. Sternhell, *Aust. J. Chem.* **32**, 1531 (1979).

[445] D. de Vos, J. Spierenburg, and J. Wolters, *Recl. Trav. Chim. Pays-Bas* **91**, 1465 (1972).

[446] D. de Vos, F. E. H. Boschman, J. Wolters, and A. van der Gen, *Recl. Trav. Chim. Pays-Bas* **92**, 467 (1973).

[447] A. Lethbridge, R. O. C. Norman, C. B. Thomas, and W. J. E. Parr, *J. Chem. Soc. Perkin Trans. 1*, 231 (1975).

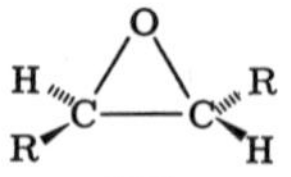

286

as the source of at least part of the diacetates formed.[447] Allylic products arise from homolytic pathways.[447]

When LTA/TFA is used in the oxidation of a series of styrene-related compounds, preparatively worthwhile yields of carbonyl compounds are formed,[449,450] and a similar transformation was observed with 1-octene.[447]

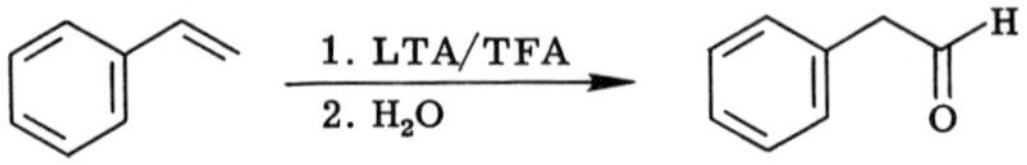

(refs. 449, 450)

(98%)

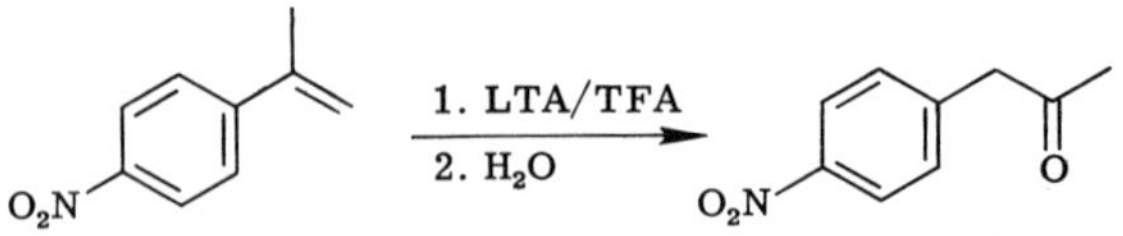

(ref. 449)

(91%)

The generalized mechanism for the reaction is illustrated for the conversion of **287** to **288**.[449,450] Similar mechanistic considerations would seem to apply in the reaction of LTA/HF with 1,2-diphenylethylene.[451]

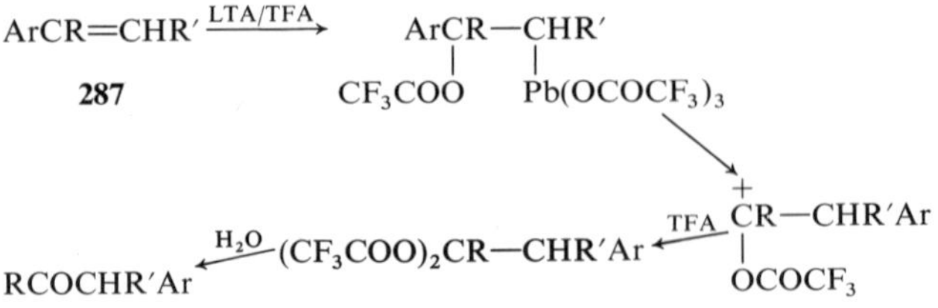

(refs. 449, 450)

288

The reaction of biallyl with LTA leads to a partitioning of the initially formed **289** to give both **290** and **291**.[452] Allenes react with LTA in acetic

[448] A. Lethbridge, R. O. C. Norman, and C. B. Thomas, *J. Chem. Soc. Perkin Trans. 1*, 1929 (1974).
[449] A. Lethbridge, R. O. C. Norman, and C. B. Thomas, *J. Chem. Soc. Perkin Trans. 1*, 35 (1973).
[450] D. Westphal and E. Zbiral, *Monatsh. Chem.* **106**, 679 (1975).
[451] D. D. Tanner and P. Van Bostelen, *J. Am. Chem. Soc.* **94**, 3187 (1972).
[452] I. Tabushi and R. Oda, *Tetrahedron Lett.*, 2487 (1966).

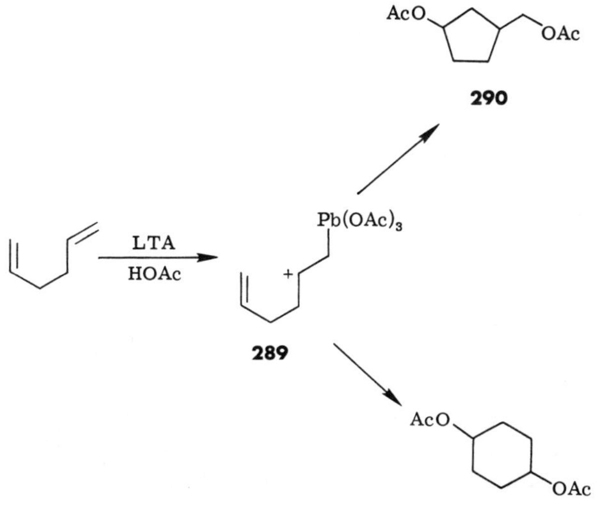

290

289

291

(ref. 452)

acid to give the corresponding 3-acetoxyalkyne.[453,454] Use of optically active allene reveals that the addition of LTA occurs predominantly by a suprafacial pathway. The reaction of (*S*)-(+)-1,3-dimethylallene to give (*S*)-(+)-4-acetoxy-2-pentyne is illustrative[453] and similar results pertain with 1,2-cyclononadiene.[454]

(ref. 453)

2. Cyclic Alkenes

The oxidation of cyclic alkenes with LTA has been widely studied.[52] In general, products can be rationalized by invoking **292** which can lead to **293** and **294**. Intermediate **293** can also arise from syn addition of LTA to

[453] R. D. Bach, R. N. Brummel, and J. W. Holubka, *J. Org. Chem.* **40**, 2559 (1975).
[454] R. D. Bach, U. Mazur, R. N. Brummel, and L.-H. Lin, *J. Am. Chem. Soc.* **93**, 7120 (1971).

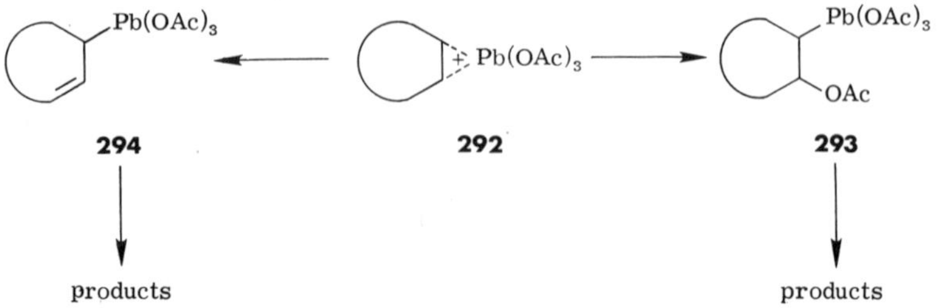

294 **292** **293**

the alkene,[455] and **293** and **294** can undergo homolysis of the lead–carbon bond followed either by product formation from the ensuing radical or by oxidation to the corresponding carbonium ion which then gives products.[52]

When the reaction is applied to strained alkenes or alkenes capable of carbonium ion rearrangement, novel results often occur. For example, treatment of 1,3,3-trimethylcyclopropene with LTA affords **295** and **296** in yields of 40% and 5% respectively.[456,457] The interesting intermediate **297** is

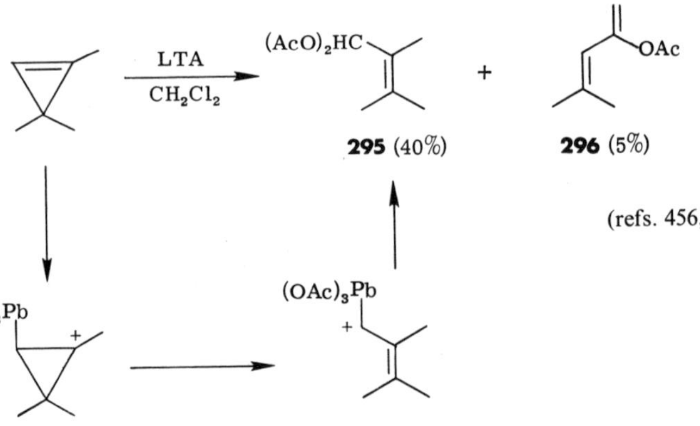

295 (40%) **296** (5%) **297**

(refs. 456, 457)

postulated to account for the formation of **295**. In a related case, the series **298** reacts with LTA to selectively produce **299**. The selectivity is explained by invoking **300**.[458] More remote cyclopropyl functionality can remain

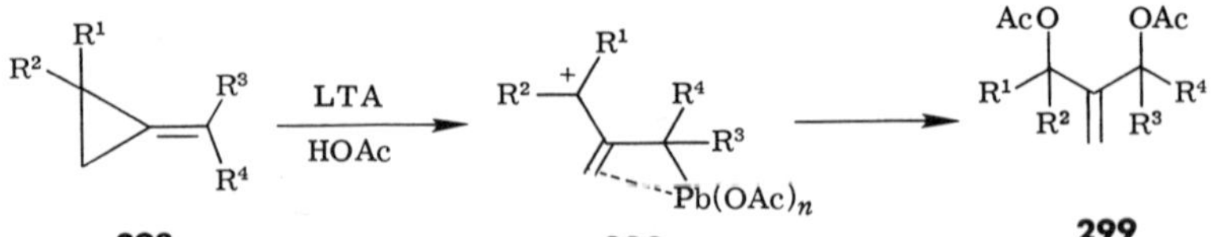

298 **300** **299**

(ref. 458)

[455] D. D. Tanner, P. B. Van Bostelen, and M. Lai, *Can. J. Chem.* **54**, 2004 (1976).
[456] T. Shirafuji and H. Nozaki, *Tetrahedron* **29**, 77 (1973).
[457] T. Shirafuji, Y. Yamamoto, and H. Nozaki, *Tetrahedron Lett.*, 4713 (1971).
[458] R. Noyori, Y. Tsuda, and H. Takaya, *J. Chem. Soc. Chem. Commun.*, 1181 (1970).

intact during oxidation, as in the preparation of **301** and **302** from **303**.[459] This behavior is in contrast to the oxidation of 4-carene by LTA/benzene in

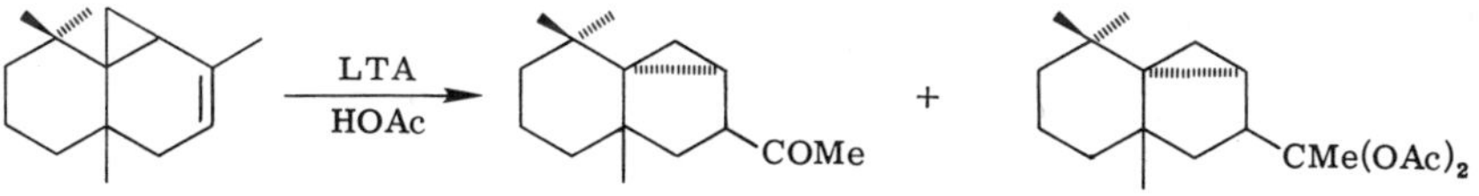

303 **301** (60–65%) **302** (16–20%) (ref. 459)

which a number of products were isolated which are consistent with homoallylic rearrangement of the three-membered ring.[460] With 3-carene, the major product of LTA oxidation was **304**.[461]

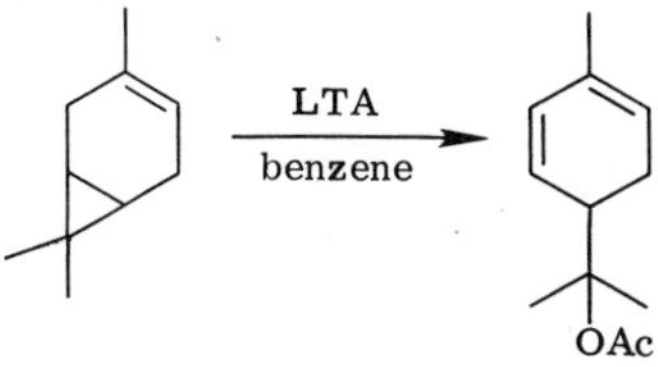

(ref. 461)

304

Ring contraction was observed when **305** or the corresponding anhydride was treated with LTA, and the intermediate **306** is implicated in the rearrangement.[462] Analogous to the reaction of 4-carene with LTA mentioned

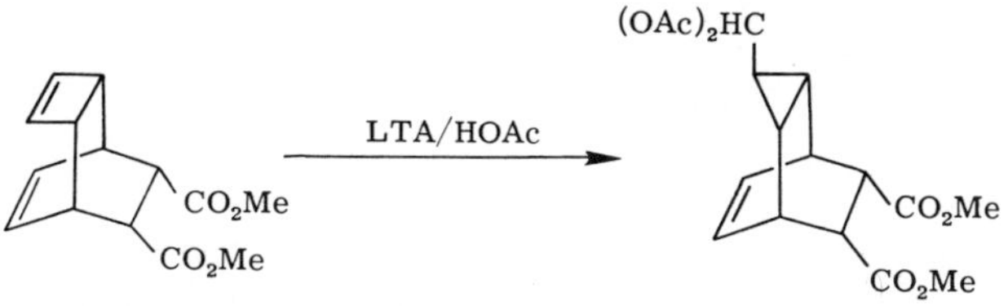

305 (67%) (ref. 462)

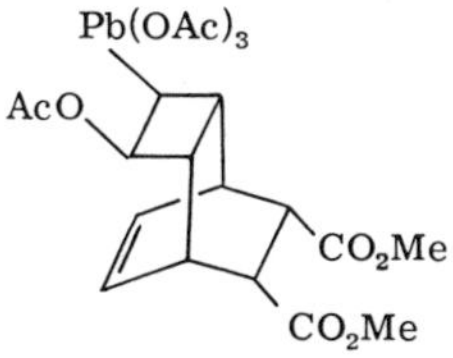

306

[459] H. Sekizaki, M. Ito, and S. Inoue, *Bull. Chem. Soc. Jpn.* **51**, 3663 (1978).
[460] B. A. Arbuzov, V. V. Ratner, Z. G. Isaeva, and M. G. Belyaeva, *Dokl. Akad. Nauk SSSR* **204**, 1115 (1972).
[461] B. A. Arbuzov, V. V. Ratner, and Z. G. Isaeva, *Izv. Akad. Nauk SSSR, Ser Khim*, 45 (1973).
[462] T. Sasaki, K. Kanematsu, A.Kondo, and K. Okada, *J. Org. Chem.* **41**, 2231 (1976).

above, β-pinene affords products involving participation of the neighboring four-membered ring.[52,463]

With **307**, a mixture containing the diacetate **308** and allylic acetate **309** was formed.[464] The results are less straightforward with alkenes containing

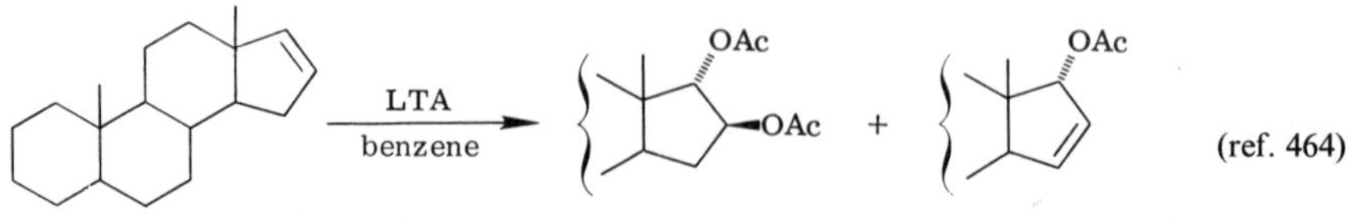

307 **308** **309**

a double bond exocyclic to a five-membered ring. The oxidation of **310** is a case in point.[465] A pronounced solvent effect has been reported for the

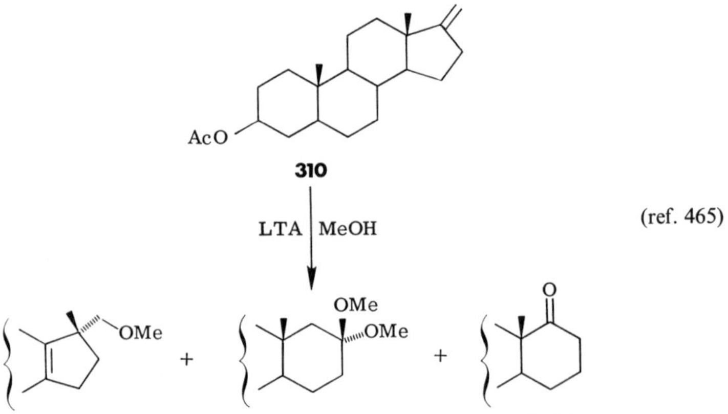

oxidation of both longifolene, **311**, and camphene, **312**.[466] In each case, the use of acetic acid as solvent results in ring expansion whereas the use of

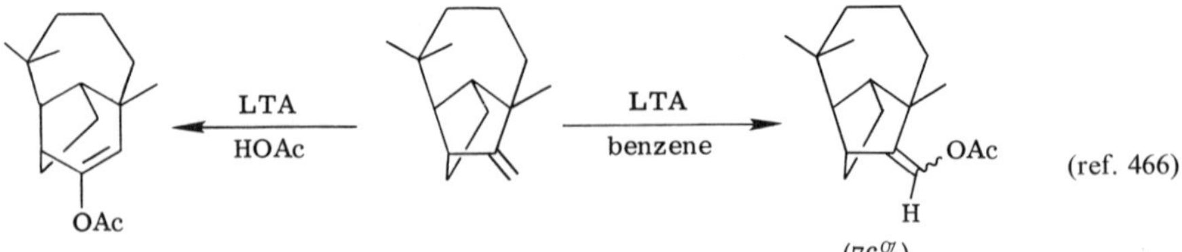

(61%) **311** (76%) *E*/*Z*-mixture

[463] H. Ohue, M. Matsushita, T. Ikeda, and H. Miki, *Osaka Kogyo Daigaku Kiyo Rikohen* **22**, 175 (1978).

[464] V. Balogh, J.-C. Beloeil, and M. Fetizon, *Tetrahedron* **33**, 1321 (1977).

[465] G. Ortar and I. Torrini, *Tetrahedron* **33**, 859 (1977).

[466] S. N. Suryawanshi, P. K. Jadhav, and U. R. Nayak, *Indian J. Chem. Sect. B* 16, 446 (1978).

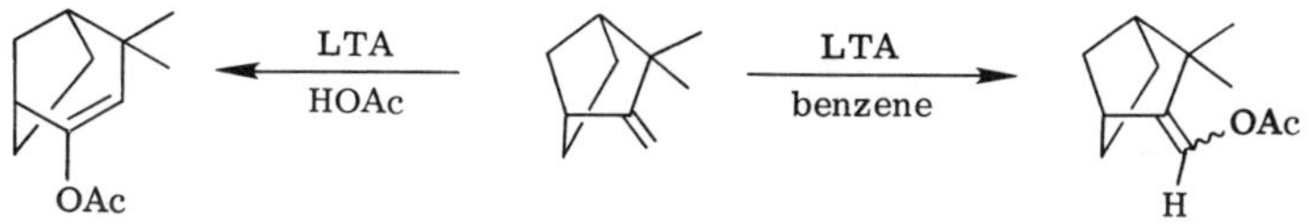

(66%) 312 (59%) *E/Z*-mixture (ref. 466)

benzene gives the corresponding aldehyde enol acetate. This dramatic change in product composition is attributed to the dielectric constant of the solvent used.

Solvent effects also play a role in the oxidation of both norbornadiene and norbornene, a fact alluded to in some detail in reference 52. Norbornene was originally reported to give an 85% yield of **313** upon treatment with LTA in acetic acid.[467] However, improved techniques of product analysis have re-

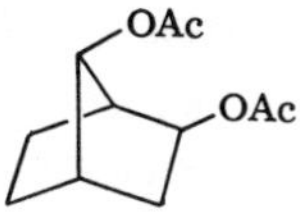

313

vealed a much more complex set of products.[52,468] The mode of formation of **313** is an area of some controversy.[52] A recent comparison of the behavior of norbornene[451] and benzobicyclo[2.2.2]octatriene[455] toward both LTA and LTA/HF has resulted in the formulation of a mechanism involving initial syn-addition of LTA to the alkene in question. Displacement of lead then gives a cation **314**, which then results in the production of **313**.[451,455]

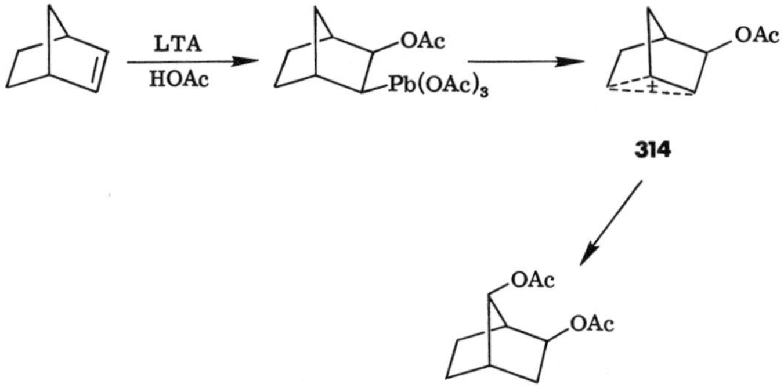

314

313

The LTA/HF oxidant and $PbF_2(OAc)_2$ both have been used to fluorinate cyclohexene rings contained in steroids, and a spectrum of products is

[467] K. Alder, F. H. Flock, and H. Wirtz, *Chem. Ber.* **91**, 609 (1958).
[468] J. Kagan, *Helv. Chim. Acta* **55**, 2356 (1972).

obtained.[469,470] LTTFA in methylene chloride–nitromethane reacts with steroidal cyclohexenes to yield products arising from 1,2-ditrifluoroacetoxylation and allylic trifluoroacetoxylation.[450] Analogous to the reaction of conjugated dienes with LTA,[52] LTTFA reacts with both 1,3-cyclohexadiene and cyclopentadiene to give products of 1,4-addition.[450]

(ref. 450)

(33%)

Extensive studies have been carried out concerning the reactions of lead(IV) acetate azides [$Pb(OAc)_{4-n}(N_3)_n$] with alkenes and much of the work in this area has been reviewed.[471] For the reaction run at low temperatures (−20°C), a mechanism involving addition of "positive azide" to the alkene has been postulated. Products then arise from the carbonium ion thus formed.[471,472] Alternatively, products might evolve from a sequence involving the cyclo addition of azide to the double bond to give **315** which then reacts further.[472,473] An interesting and typical example of the synthetic

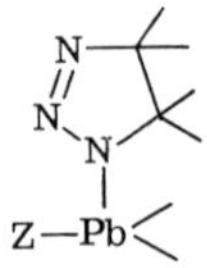

Z = OAc or N_3

315

use of the reaction occurs when lead(IV) acetate azide and norbornene interact to give **316** in a yield of 75%.[474] A series of Δ^5-steroidal alkenes has

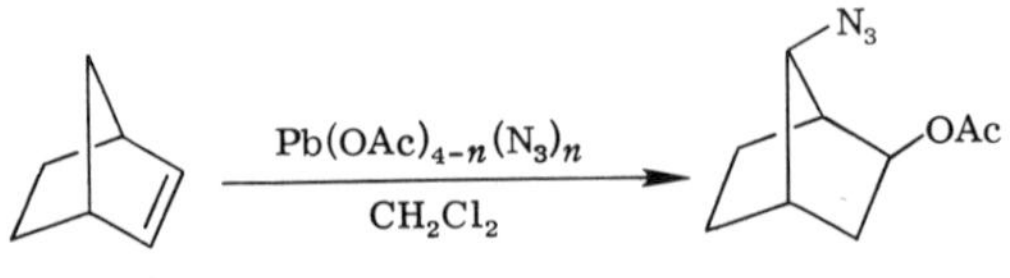

(ref. 474)

316 (75%)

[469] M. Ephritikhine and J. Levisalles, *Bull. Soc. Chim. Fr.*, 339 (1975).
[470] M. Ephritikhine and J. Levisalles, *J. Chem. Soc. Chem. Commun.*, 429 (1974).
[471] E. Zbiral, *Synthesis*, 285 (1972).
[472] E. Zbiral and A. Stütz, *Monatsh. Chem.* **104**, 249 (1973).
[473] A. Stütz and E. Zbiral, *Liebigs Ann. Chem.* **765**, 34 (1972).
[474] E. Zbiral and A. Stütz, *Tetrahedron* **27**, 4953 (1971).

been found to fragment when treated with lead(IV) acetate azide,[471,475] whereas steroidal alkenes unsubstituted at the double bond in question, afford α-azido ketones.[471,476] In the former case, temperature is crucial

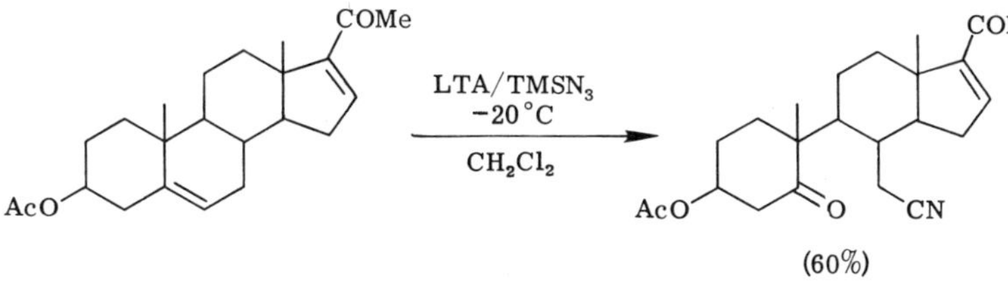

(ref. 475)

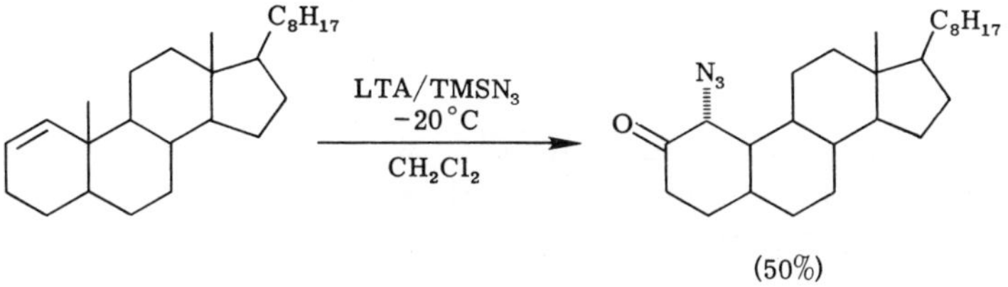

(ref. 476)

since allyl azides[475] and other oxidation products[477] arise when the reaction is carried out at +20°C. Dienes react with lead(IV) acetate azide to give

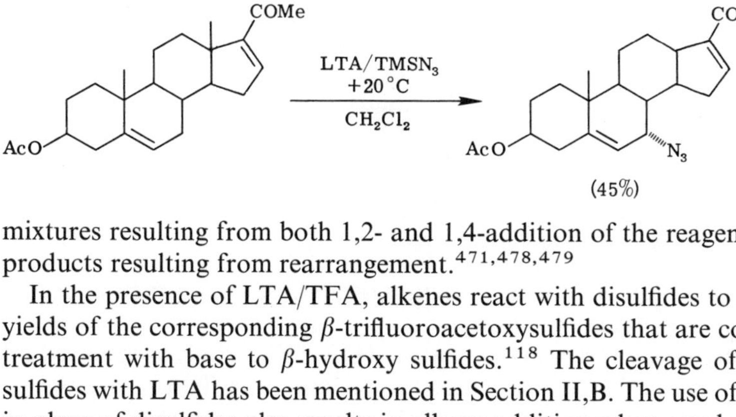

(ref. 475)

mixtures resulting from both 1,2- and 1,4-addition of the reagent as well as products resulting from rearrangement.[471,478,479]

In the presence of LTA/TFA, alkenes react with disulfides to afford high yields of the corresponding β-trifluoroacetoxysulfides that are converted by treatment with base to β-hydroxy sulfides.[118] The cleavage of β-hydroxy sulfides with LTA has been mentioned in Section II,B. The use of diselenides in place of disulfides also results in alkene addition when employed in conjunction with LTA.[479a]

[475] H. Hugl and E. Zbiral, *Tetrahedron* **29**, 759 (1973).
[476] E. Zbiral and G. Nestler, *Tetrahedron* **27**, 2293 (1971).
[477] E. Zbiral and H. Hugl, *Tetrahedron* **29**, 769 (1973).
[478] A. Wolloch, E. Zbiral, and E. Haslinger, *Liebigs Ann. Chem.*, 2339 (1975).
[479] H. Hugl and E. Zbiral, *Tetrahedron* **29**, 753 (1973).
[479a] N. Miyoshi, Y. Ohno, K. Kondo, S. Murai, and N. Sonoda, *Chem. Lett.*, 1309 (1979).

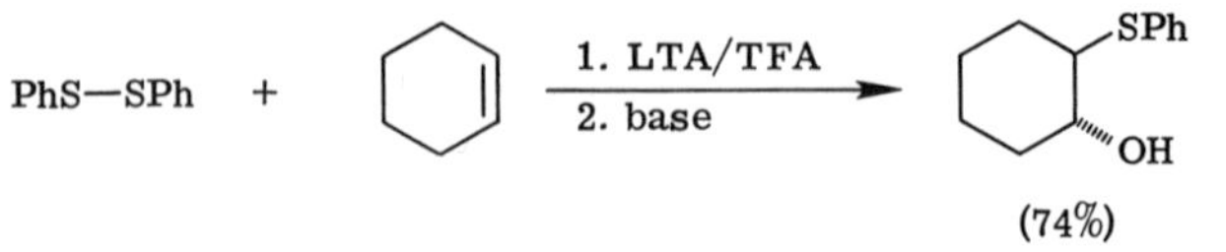

(ref. 118)

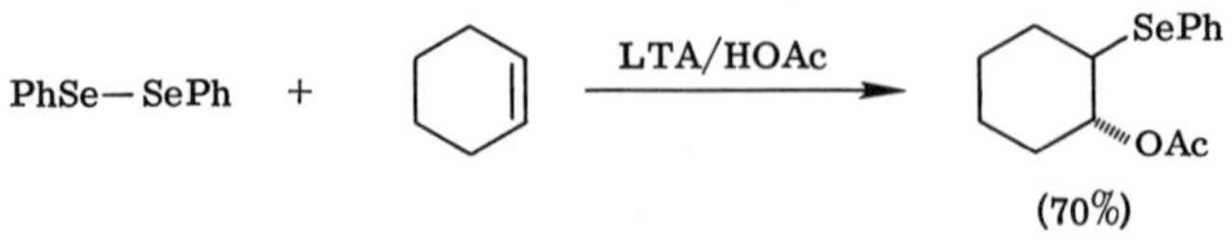

(ref. 479a)

V. LTA Reactions with Organometallics

Interest in the reactions of organometallics with LTA has led to the development of a number of useful synthetic procedures. Oxidation of trialkylboranes with LTA gives moderate yields of the corresponding alkyl acetates.[480] With mixed trialkylboranes containing both primary and

$$R_3B + LTA \xrightarrow{\text{benzene}} 3ROAc \qquad \text{(ref. 480)}$$

secondary alkyl groups on boron, preferential formation of secondary alkyl acetate occurs. The reaction has also been extended to the oxidation of 1-bromo-1-alkenyl dialkyl boranes as a means of obtaining the corresponding 1-bromo-1,2-dialkylethenes. In this case, alkyl migration occurs instead of acetate formation.[481] Use of low temperature ($-50°C$) and CH_2Cl_2 as

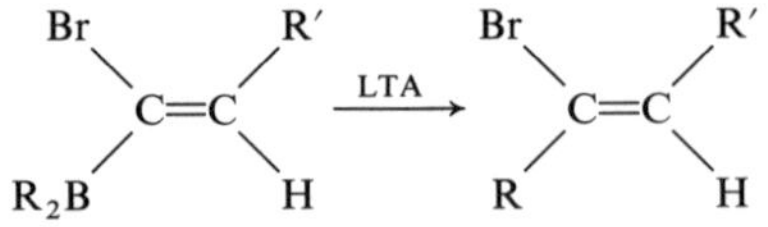

(ref. 481)

solvent favors production of the *Z*-isomer while *E*-isomer formation occurs at 0°C in benzene–hexane mixtures.

Hydroalumination of 1-alkenes with LAH by $TiCl_4$ catalysis followed by LTA treatment provides another excellent method for the preparation of primary alkyl acetates.[482] The use of **317** is crucial since organoaluminates

$$2RCH{=}CH_2 + LAH \xrightarrow[\text{THF}]{TiCl_4} \underset{\mathbf{317}}{LiAl(CH_2CH_2R)_2H_2} \xrightarrow{\text{LTA}} 2AcOCH_2CH_2R \qquad \text{(ref. 482)}$$

[480] Y. Masuda and A. Arase, *Bull. Chem. Soc. Jpn.* **51**, 901 (1978).
[481] Y. Masuda, A. Arase, and A. Suzuki, *Chem. Lett.*, 665 (1978).
[482] F. Sato, Y. Mori, and M. Sato, *Tetrahedron Lett.*, 1405 (1979).

of the type $LiAlR_4$ lose only two of the R-groups in the LTA reaction. The selective monohydroalumination of unconjugated dienes also allows for the preparation of unsaturated acetates.[482]

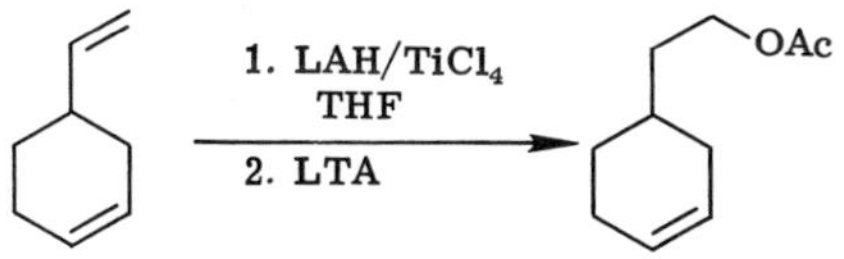

(ref. 482)

(64%)

Both arylthallium ditrifluoroacetates[483] and diarylthallium trifluoroacetates[484] react with LTA–triphenylphosphine in TFA to give excellent yields of aryl trifluoroacetates. Subsequent alkaline hydrolysis leads to the production of the corresponding phenols. The aryllead tri(trifluoroacetate) **318** is proposed as a likely intermediate.[484]

$$ArTl(OCOCF_3)_2 \xrightarrow[\text{2. } Ph_3P]{\text{1. LTA/TFA}} ArOCOCF_3 \xrightarrow[CH_2Cl_2]{NaOH/H_2O} \underset{39\text{–}78\%}{ArOH} \qquad \text{(ref. 483)}$$

$$Ar_2TlOCOCF_3 \xrightarrow[\text{2. } Ph_3P]{\text{1. LTA/TFA}} 2ArOCOCF_3 \xrightarrow[CH_2Cl_2]{NaOH/H_2O} \underset{21\text{–}69\%}{ArOH} \qquad \text{(ref. 484)}$$

$(ArPb(OCOCF_3)_3)$

318

When arylthallium ditrifluoroacetates are treated with LTTFA in TFA, formation of **318** is actually observed, with subsequent decomposition to the corresponding aryl trifluoroacetates.[443,485] An extension of the reaction allows preparation of **319** from both arylmercury(II) trifluoroacetates and

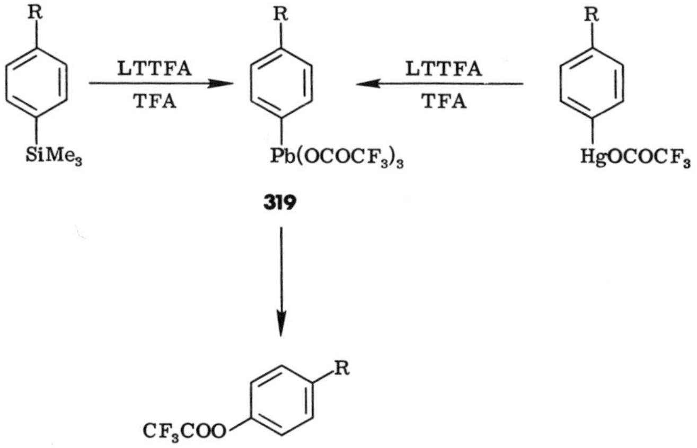

arylsilicon compounds.[443,485,485a] In each case, metal–metal exchange and not protonation followed by plumbylation prevails.[485,485a]

A number of organotin compounds containing Sn—H, Sn—C, Sn—Sn, and Sn—O groups are effectively acetoxylated with LTA,[486] and dialkyldiarylstannanes give 70–90% yields of diacetoxydiarylplumbanes along with the corresponding diacetoxyldialkylstannanes.[487] *E*-1-Alkenyl-*n*-butylstannanes are converted smoothly into terminal alkynes upon treatment with

$$R_2SnAr_2 \xrightarrow[\text{HOAc}]{\text{LTA/Hg(OAc)}_2} R_2Sn(OAc)_2 + \underset{(70\text{–}90\%)}{Ar_2Pb(OAc)_2} \qquad \text{(ref. 487)}$$

LTA in acetonitrile.[488] Intermediate **320** leading to **321** and thus the alkyne is proposed.

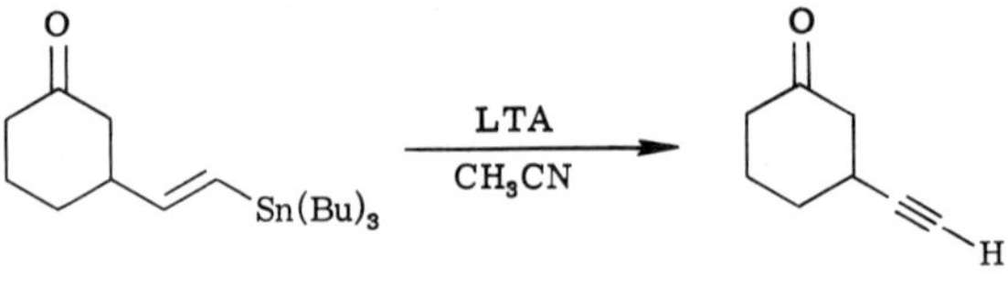

(96%)

(ref. 488)

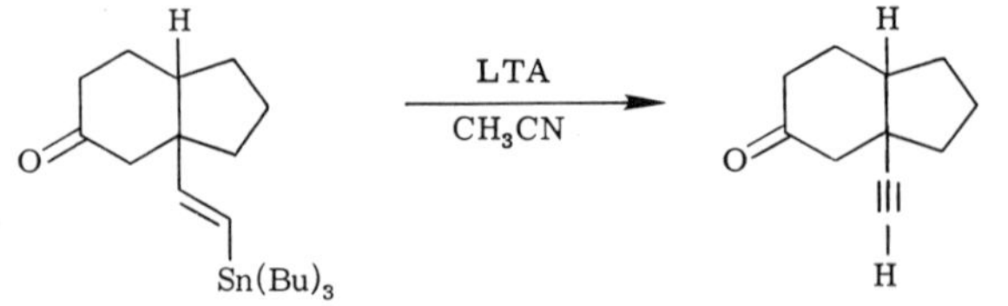

(64%)

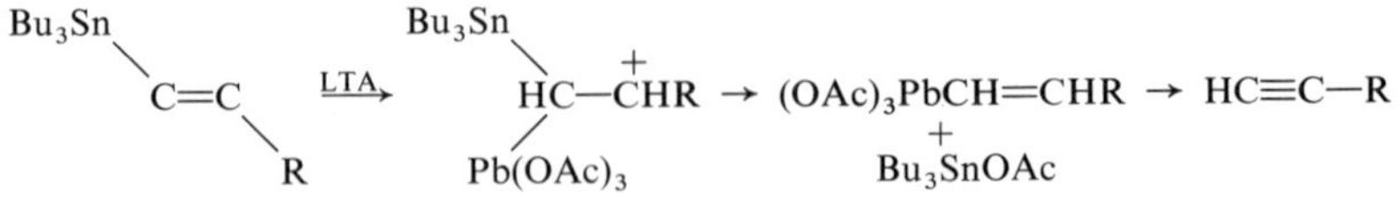

320 **321**

---

[483] E. C. Taylor, H. W. Altland, R. H. Danforth, G. McGillivray, and A. McKillop, *J. Am. Chem. Soc.* **92**, 3520 (1970).

[484] E. C. Taylor, H. W. Altland, and A. McKillop, *J. Org. Chem.* **40**, 2351 (1975).

[485] J. R. Kalman, J. T. Pinhey, and S. Sternhell, *Tetrahedron Lett.*, 5369 (1972).

[485a] H. C. Bell, J. R. Kalman, J. T. Pinhey, and S. Sternhell, *Aust. J. Chem.* **32**, 1521 (1979).

[486] U. Christen and W. P. Neumann, *J. Organomet. Chem.* **39**, C58 (1972).

[487] O. P. Syutkina, E. M. Panov, and K. A. Kocheshkov, *Zh. Obshch. Khim.* **43**, 1322 (1973).

[488] E. J. Corey and R. H. Wollenberg, *J. Am. Chem. Soc.* **96**, 5581 (1974).

Allylic mercuric acetates are transformed into mixtures of allylic acetates by LTA. The major reaction pathway is thought to involve S_E' (probably S_Ei') formation of a σ-allylic lead derivative which then demetalates.[489] With the dimeric π-complex **322** direct attack by LTA is postulated.[489] Oxidation of

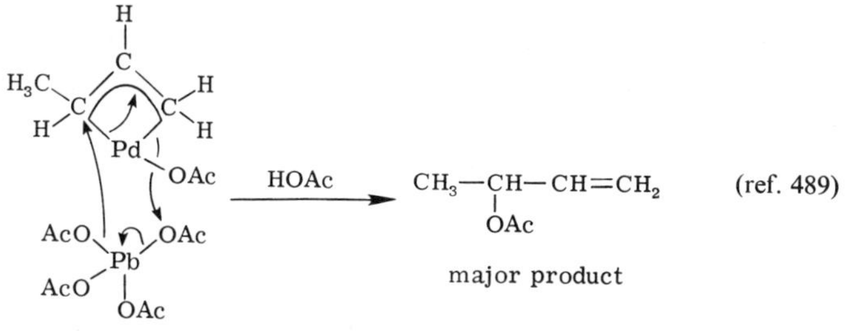

322

cycloocta-1,5-diene with LTA/$PdCl_2$ gives a 70% yield of **323**,[490] and arylbutenylacetates are formed in high yield from the reaction of aryl-

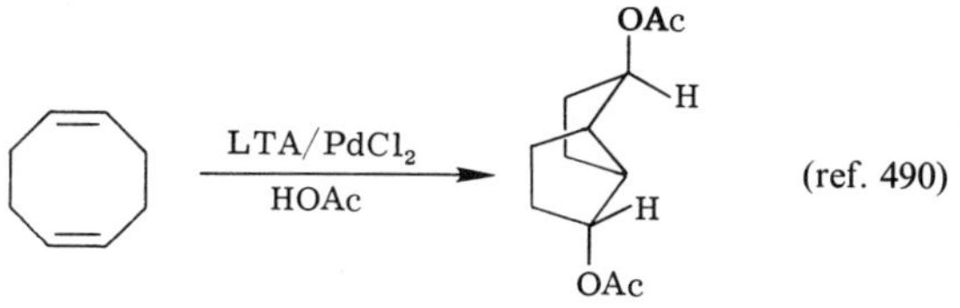

323 (70%)

mercuric salts, butadiene, and LTA in the presence of a catalytic amount of palladium acetate.[491]

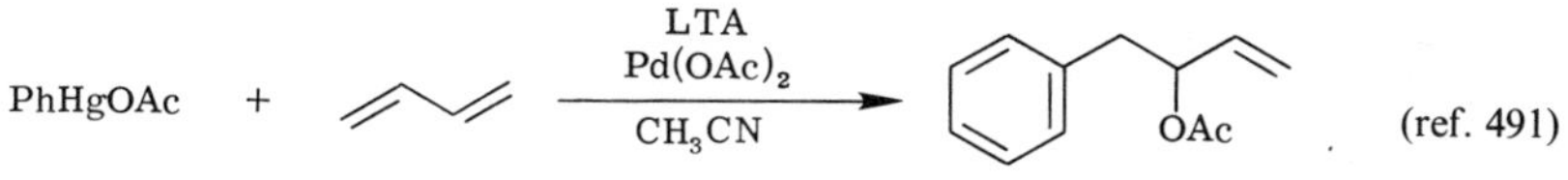

(78%)

A series of β-aminopalladium compounds **324** were treated with LTA to afford the corresponding β-aminoacetates. Hydrolysis or LAH reduction gives β-aminoalcohols.[492] Stereochemical studies show that β-aminoacetate

[489] W. Kitching, T. Sakakiyama, Z. Rappoport, P. D. Sleezer, S. Winstein, and W. G. Young, *J. Am. Chem. Soc.* **94**, 2329 (1972).

[490] P. M. Henry, M. Davies, G. Ferguson, S. Phillips, and R. Restivo, *J. Chem. Soc. Chem. Commun.*, 112 (1974).

[491] R. F. Heck, *J. Am. Chem. Soc.* **90**, 5542 (1968).

[492] J.-E. Bäckvall, *Tetrahedron Lett.*, 2225 (1975).

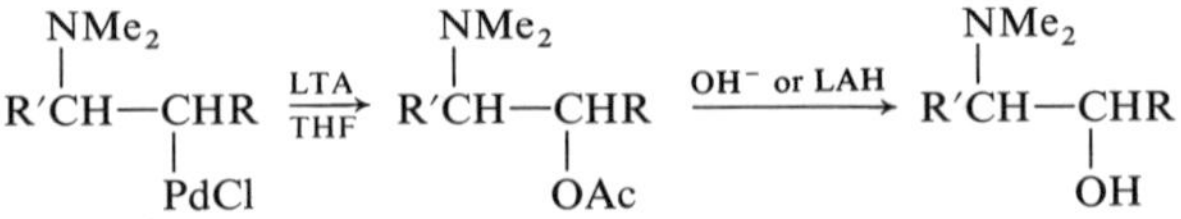

324

(ref. 492)

formation occurs with inversion at the site containing palladium. Alkyl transfer to lead with retention, followed by acetate ion attack with inversion, accounts for the observed results.[492]

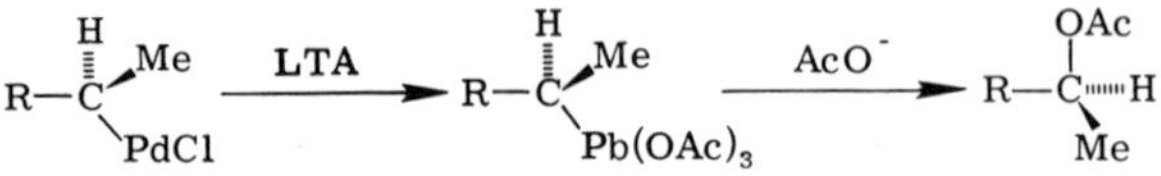

(ref. 492)

The reaction of LTA in pyridine to free cyclobutadiene from cyclobutadieneiron tricarbonyl is a method which complements the use of ceric ammonium nitrate (CAN) for the same purpose.[493–495] LTA is especially useful in systems where acidic conditions can be deleterious. The examples given are illustrative.

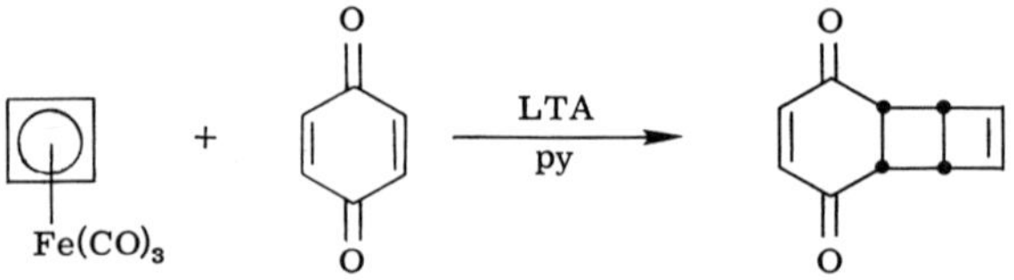

(40–44%)

(ref. 493)

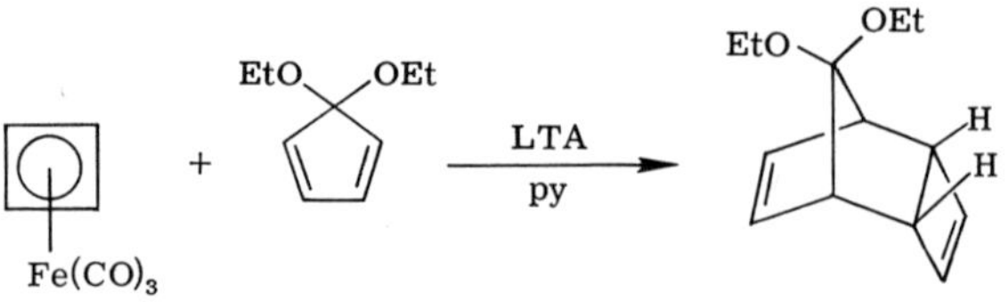

(ref. 494)

(ref. 495)

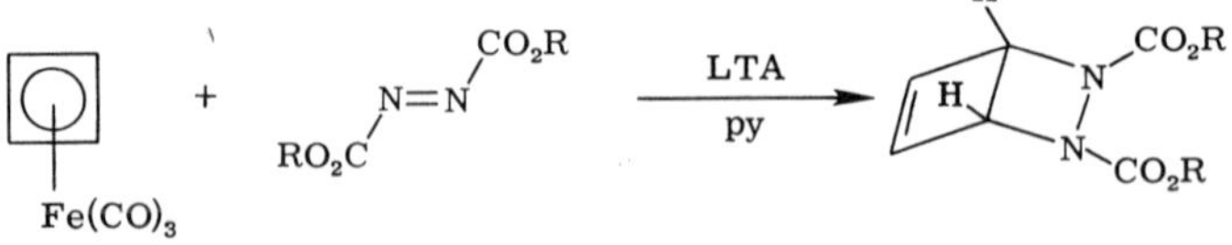

R	% yield
$-CH_2CH_2Ts$	45
$-CH_2CH_3$	50–60

[493] L. Brener, J. S. McKennis, and R. Pettit, *Org. Synth.* **55**, 43 (1976).
[494] J. C. Barborak and R. Pettit, *J. Am. Chem. Soc.* **89**, 3080 (1967).
[495] S. Masamune, N. Nakamura, and J. Spadaro, *J. Am. Chem. Soc.* **97**, 918 (1975).

Benzocyclobutadiene is generated by the LTA oxidation of **325** and when **325** is used in conjunction with cyclobutadieneiron tricarbonyl and LTA the mixed adduct **326** is obtained in 75% yield.[496] The latter intriguing result stems from the fact that oxidation of **325** and cyclobutadieneiron tricarbonyl

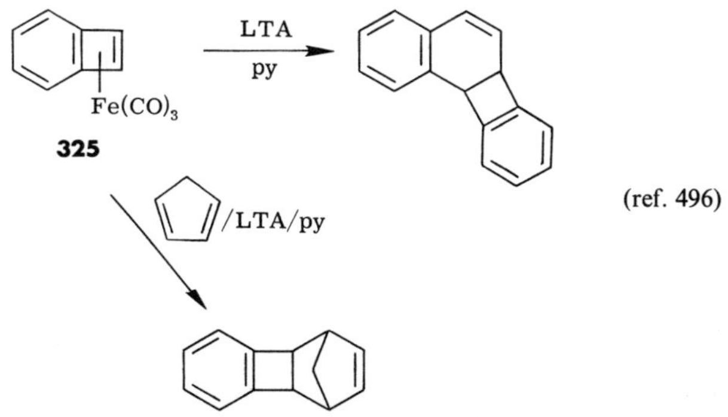

occurs at comparable rates. This being the case, the benzocyclobutadiene thus produced can only function in the role of dienophile in the subsequent [4 + 2]cycloaddition with cyclobutadiene.[496]

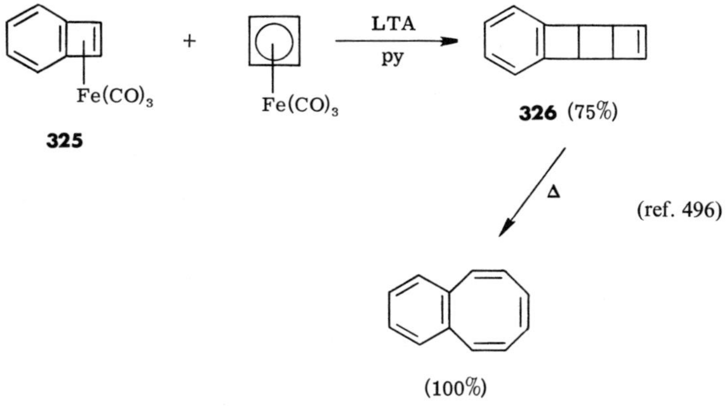

ACKNOWLEDGMENT

The author gratefully acknowledges H. D. Juve, Jr., D. K. Heckendorn, W. D. Boyce, and L. A. Rubottom for their help in preparing the manuscript, and Professor J. H. Cooley for his most useful comments.

[496] W. Merk and R. Pettit, *J. Am. Chem. Soc.* **89**, 4787 (1967).

CHAPTER **II**

Phase Transfer Assisted Permanganate Oxidations

DONALD G. LEE

I. Introduction . . . 147
II. The Phase Transfer Process . . . 152
A. Phase Transfers from Aqueous Permanganate . . . 152
B. Phase Transfer Utilizing Solid Permanganate . . . 162
C. Phase Transfer Mechanisms . . . 166
D. Use of Solid Supports for Permanganate Oxidations . . . 167
E. Reactions in Micelles . . . 167
III. Oxidation of Alkenes . . . 168
A. Terminal Alkenes . . . 168
B. Nonterminal Alkenes . . . 173
IV. Oxidation of Alkynes . . . 182
A. Nonterminal Alkynes . . . 182
B. Terminal Alkynes . . . 185
V. Oxidation of Arenes . . . 186
VI. Oxidation of Alkanes . . . 190
VII. Oxidation of Polycyclic Aromatic Hydrocarbons . . . 192
VIII. Oxidation of Alcohols . . . 193
IX. Oxidations of Phenols . . . 196
X. Oxidation of Ethers . . . 199
XI. Oxidation of Aldehydes . . . 201
XII. Oxidation of Sulfur Compounds . . . 201
XIII. Oxidation of Halides . . . 203
XIV. Oxidation of Amines . . . 204

I. Introduction

The use of permanganate as an oxidant for organic transformations has a long and extensive history.[1] It has been used both as a selective oxidant and as a scavenger to remove small amounts of organic material present as contaminants in either water or air.

[1] J. W. Ladbury and C. F. Cullis, *Chem. Rev.* **58**, 403 (1958), and references therein.

Oxidation in Organic Chemistry, Part D

ISBN 0-12-697253-2

The use of permanganate as a selective oxidant for a variety of reactions has been reviewed by Stewart in a previous volume in this series.[2] Other reviews include those by Arndt,[3] Freeman,[4] Waters,[5] and Lee.[6]

The purification of water by treatment with permanganate has been discussed by a number of authors[7] and Posselt and Reidies[8] have described the removal of malodorus compounds from factory off-gases by use of a permanganate scrubber.

The use of aqueous permanganate in syntheses is limited to the oxidation of organic compounds that are at least partially soluble in water. If the solubility is too low there is not sufficient contact between the oxidant and the reductant at the interface and the rate of reaction is drastically reduced. This problem has been illustrated by the work of Eastman and Quinn.[9] They showed that the alkane, 2,2,4-trimethylpentane, did not undergo an appreciable amount of oxidation at the tertiary hydrogen over a period of several months in contact with aqueous permanganate, while the corresponding alcohol, 2,4-dimethylpentan-2-ol, which contains a similar tertiary hydrogen, was oxidized with a half-life of about 5 days under similar conditions. The alcohol, because of its ability to hydrogen-bond with water, has greater solubility in the aqueous phase and therefore came into more intimate contact with the oxidant.

Consequently aqueous permanganate has found its greatest application in the oxidation of organic compounds containing polar groups that provide them with at least partial solubility in water. Stewart[2] has tabulated these reactions and described the mechanisms involved.

The classical way of overcoming the solubility problem has been by use of polar organic solvent systems that will dissolve both reactants. Examples of solvents that have been used include ethanol, *tert*-butyl alcohol, acetone, pyridine, acetic acid, acetic anhydride, and trifluoroacetic acid. Some representative reactions have been summarized in Table I.

[2] R. Stewart, *in* "Oxidation in Organic Chemistry" (K. B. Wiberg, ed.), Part A, pp. 1–68. Academic Press, New York, 1965.

[3] D. Arndt, *in* "Methoden der Organischen Chemie, Houben-Weyl" (E. Müller, ed.). Thieme, Stuttgart, 1975; D. Arndt, "Manganese Compounds as Oxidizing Agents in Organic Chemistry." Open Court, La Salle, 1981.

[4] F. Freeman, *Rev. React. Species Chem. React.* **2**, 179 (1973).

[5] W. A. Waters, "Mechanisms of Oxidation of Organic Compounds." Methuen, London, 1964.

[6] D. G. Lee, "Oxidation of Organic Compounds by Permanganate Ion and Hexavalent Chromium." Open Court, La Salle, 1980.

[7] R. G. Spicher and R. T. Skrinde, *J. Am. Water Works Assoc.* **57**, 472 (1965), and references therein.

[8] H. S. Posselt and A. H. Reidies, *I & EC Prod. Res. Dev.* **4**, 48 (1965).

[9] R. H. Eastman and R. A. Quinn, *J. Am. Chem. Soc.* **82**, 4249 (1960).

TABLE I

USE OF ORGANIC SOLVENT SYSTEMS FOR PERMANGANATE OXIDATIONS

Reactant	Solvent system	Product (% yield)	Ref.
1,4-Cyclohexadiene	Aq. ethanol	4,5-Dihydroxycyclohexene (7)	10
Cyclohexene	Aq. ethanol	*cis*-1,2-Dihydroxycyclohexane (33)	11
Methyl cinnamate	Aq. ethanol	Phenylglyceric acid (67)	12
2,5-Dimethoxy-2,5-dihydrofuran	Aq. ethanol	2,5-Dimethoxy-3,4-dihydroxytetrahydrofuran (37)	13
endo-Dicyclopentadiene	Aq. ethanol	5-*exo*,6-*exo*-Dihydroxy-*endo*-3a,4,5,6,7,7a-hexahydro-4,7-methanoindene (28)	13a
Norbornene	Aq. *t*-butyl alcohol	2,3-*exo*,*cis*-Dihydroxynorbornane (40)	14
Butylamine	Aq. *t*-butyl alcohol	Butanol (46)	15
Norbornene	Aq. acetone	1,3-Diformylcyclopentane (54–66)	14
4-Oxo-4-(2-xanthenyl)butyric acid	Aq. acetone	4-Oxo-4-(2-xanthonyl)butyric acid (93)	16
2-(4-Methoxyphenyl)cyclohexanone	Aq. acetone	6-Oxo-6-(4-methoxyphenyl)hexanoic acid (70)	17
3-Methyl-5-carbomethoxypentanal	Aq. acetone	3-Methyl-5-carbomethoxypentanoic acid (65)	18
2-Amino-2,4,4-trimethylpentane	Aq. acetone	2-Nitro-2,4,4-trimethylpentane (77)	19
1,2-Dimethylcyclopentene	Aq. acetone	*cis*-1,2-Dimethyl-1,2-dihydroxycyclopentane (45)	20
(*Z*,*Z*)-2,6-Octadiene	Aq. acetone	*cis*-2,5-Bis(hydroxyethyl) tetrahydrofuran (32)	20a
3β-Acetoxy-20-ketopregna-5,16-diene	Aq. acetone	16α,17α-Dihydroxy-3β-acetoxy-20-ketopregn-5-ene (40)	21
Methyl levopimarate	Aq. acetone	Methyl dehydroabietate (76)	22

(*Continued*)

[10] K. A. Powell, A. L. Hughes, H. Katchian, J. F. Jerauld, and H. Z. Sable, *Tetrahedron* **28**, 2019 (1972).

[11] M. F. Clarke and L. N. Owen, *J. Chem. Soc.*, 315 (1949).

[12] C. N. Riiber, *Chem. Ber.* **48**, 823 (1915).

[13] J. C. Sheehan and B. M. Bloom, *J. Am. Chem. Soc.* **74**, 3825 (1952).

[13a] D. Brewster, M. Myers, J. Ormerod, P. Otter, A. C. B. Smith, M. E. Spinner, and S. Turner, *J. Chem. Soc. Perkin 1*, 2796 (1973).

[14] K. Wiberg and K. A. Saegebarth, *J. Am. Chem. Soc.* **79**, 2822 (1957).

[15] S. S. Rawalay and H. Shechter, *J. Org. Chem.* **32**, 3129 (1967).

[16] A. M. El-Abbady, S. Ayoub, and F. G. Baddar, *J. Chem. Soc.*, 2556 (1960).

[17] E. R. Clark and J. G. B. Howes, *J. Chem. Soc.*, 1152 (1956).

[18] V. R. Mandapur, P. P. Pai, K. K. Chakravarti, U. G. Nayak, and S. C. Bhattacharyya, *Tetrahedron* **20**, 2601 (1964).

[19] N. Kornblum, R. J. Clutter, and W. J. Jones, *J. Am. Chem. Soc.* **78**, 4003 (1956).

[20] P. D. Bartlett and A. Bavley, *J. Am. Chem. Soc.* **60**, 2416 (1938).

[20a] D. M. Walba, M. D. Wand, and M. C. Wilkes, *J. Am. Chem. Soc.* **101**, 4396 (1979).

[21] G. Cooley, B. Ellis, F. Hartley and V. Petrow, *J. Chem. Soc.*, 4373 (1955).

[22] B. Marchand, *Chem. Ber.* **91**, 405 (1958).

TABLE I (*cont.*)

Reactant	Solvent system	Product (% yield)	Ref.
Heptafluoroisoquinoline	Aq. acetone	2,5,6-Trifluoropyridine-3,4-dicarboxylic acid (52)	23
Phenylmethanesulfinamide	Aq. acetone	Phenylmethanesulfonamide	24
2,5-Dihydro-5-methyl-3,6-diphenyl-1,2,4-triazine	Aq. acetone	5-Methyl-3,6-diphenyl-1,2,4-triazine (73)	25
4-Methyl-1-(4-nitrophenyl)-5-morpholino-4,5-dihydro-*v*-triazole	Aq. acetone	4-Methyl-1-(4-nitrophenyl)-5-(2-oxo-morpholino)-4,5-dihydro-*v*-triazole (40)	26
8-Hexadecyne	Aq. acetone	8,9-Hexadecanedione (81)	27
5-Decene	Aq. acetone	5-Hydroxy-6-decanone (73)	28
3-Phenoxytoluene	Aq. pyridine	3-Phenoxybenzoic acid (71)	29
2,5-Dimethylbiphenyl	Aq. pyridine	2,5-Biphenyldicarboxylic acid (75)	30
2-Acetylbiphenylene	Aq. pyridine	2-Biphenylenylglyoxylic acid (49)	31
4-(2-Phenylvinyl)-6-quinolinecarboxylic acid	Aq. pyridine	4,6-Quinolinedicarboxylic acid (76)	32
2,8-Dimethyl-10-acetylphenoxazine	Aq. pyridine	10-Acetylphenoxazine-2,8-dicarboxylic acid (75)	33
Pregnenolone acetate	Aq. pyridine	Pregnane-3β,5,6-triol-7,20-dione 3-acetate (26)	34
3,5-Di-*t*-butyltoluene	Aq. pyridine	3,5-Di-*t*-butylbenzoic acid (73)	35
1-Phenyl-2-methyl-1-propanol	Acetic acid	Phenyl isopropyl ketone (71)	36
1,2-Diphenyl-1,2-bis(3,4-diethoxyphenyl)ethylene	Acetic acid	2,3,6,7-Tetraethoxy-9,10-diphenylphenanthrene (65)	37
2-Chloromethyl 4-methylphenyl sulfide	Acetic acid	2-Chloromethyl 4-methylphenyl sulfone (83)	38
Bis(2-keto-2-phenylethyl) sulfide	Acetic acid	Bis(2-keto-2-phenylethyl) sulfone	39
4-Aminophenyl 4-nitrophenyl sulfide	Acetic acid	4-Aminophenyl 4-nitrophenyl sulfone	40
Oleic acid	Acetic anhydride	9,10-Diketostearic acid (46)	41
Cyclohexane	Trifluoroacetic acid	Adipic acid (75)	42

[23] R. D. Chambers, M. Hole, W. K. R. Musgrave, A. A. Storey, and B. Iddon, *J. Chem. Soc. C*, 2331 (1966).

[24] H. Seiler and H. Erlenmeyer, *Helv. Chim. Acta* **40**, 88 (1957).

[25] C. M. Atkinson and H. D. Cossey, *J. Chem. Soc.*, 1805 (1962).

[26] L. M. Rossi and P. Trimarco, *Synthesis*, 743 (1978).

[27] N. S. Srinivasan and D. G. Lee, *J. Org. Chem.* **44**, 1574 (1979).

[28] N. S. Srinivasan and D. G. Lee, *Synthesis*, 520 (1979).

[29] French Pat. 1398558 (1964), The British Petrolium Co. Ltd., C. F. Forster.

[30] E. K. Weisburger and J. H. Weisburger, *J. Org. Chem.* **23**, 1193 (1958).

[31] J. F. W. McOmie and S. D. Thatte, *J. Chem. Soc.*, 5298 (1962).

[32] A. P. Shroff, H. Jaleel, and F. M. Miller, *J. Pharm. Sci.* **55**, 844 (1966).

[33] R. Hazard, J. Cheymol, P. Chabrier, A. Sekera, and J. De Antoni, *Compt. Rend.* **252**, 4166 (1961).

[34] H. R. Nace and A. L. Rieger, *J. Org. Chem.* **35**, 3846 (1970).

Obviously, the use of organic solvent systems is limited to the oxidation of those compounds that react much more readily with permanganate than does the solvent itself.

Another more recent approach involves the use of salts such as tetrabutylammonium permanganate[43] or benzyltriethylammonium permanganate[44] which, because of the organophilicity of the quaternary ammonium cations, are soluble in nonpolar solvents. These salts may be prepared by adding a solution of the corresponding quaternary ammonium halide to a solution of potassium permanganate. However, caution should be exercised because there have been reports of explosions associated with their use.[45] Above a certain temperature most quaternary ammonium permanganates spontaneously ignite.[45a]

It is unnecessary, however, to prepare and isolate the quaternary ammonium permanganates. They can be prepared *in situ* (without danger of ignition) by reacting a quaternary ammonium halide (Q^+X^-) with potassium permanganate. Anion exchange [Eq. (1)] then produces a quaternary ammonium ion pair that is soluble in nonpolar solvents.

$$Q^+X^- + K^+MnO_4^- \rightleftharpoons Q^+MnO_4^- + K^+X^- \qquad (1)$$

This exchange usually takes place with a change of phase by the permanganate ion. Typically, the quaternary ammonium halide dissolved in an organic solvent such as methylene chloride, is added to either solid $KMnO_4$ or an aqueous solution of potassium permanganate. During the exchange, the permanganate ion migrates to the organic phase and exists there as an ion pair.

Quaternary phosphonium and arsonium salts can also be used as phase transfer agents, as can both cyclic and acyclic polyethers. In the latter cases, the polyether complexes the potassium ion causing it to become organophilic

---

[35] M. G. J. Beets, W. Meerburg, and H. van Essen, *Rec. Trav. Chim. Pays-Bas* **78**, 570 (1959).

[36] H. A. Neidig, D. L. Funck, R. Uhrich, R. Baker, and W. Kreiser, *J. Am. Chem. Soc.* **72**, 4617 (1950).

[37] J. Gardent, *Bull. Soc. Chim. Fr.*, 1049 (1962).

[38] D. Klamann and H. Bertsch, *Chem. Ber.* **88**, 201 (1955).

[39] J. D. Loudon and L. B. Young, *J. Chem. Soc.*, 5496 (1963).

[40] T. Naito, S. Nakagawa, and K. Takahashi, *Chem. Pharm. Bull.* **16**, 148 (1968).

[41] H. P. Jensen and K. B. Sharpless, *J. Org. Chem.* **39**, 2314 (1974).

[42] R. Stewart and U. A. Spitzer, *Can. J. Chem.* **56**, 1273 (1978).

[43] T. Sala and M. V. Sargent, *J. Chem. Soc. Chem. Comm.*, 253 (1978).

[44] H. J. Schmidt and H. J. Schafer, *Angew. Chem. Int. Ed. Engl.* **18**, 68 (1979).

[45] J. A. Morris and D. C. Mills, *Chem. Ind.* 446 (1978); H. Jager, L. Lutolf, and M. W. Meyer, *Angew. Chem. Int. Ed. Engl.* **18**, 786 (1979); H. J. Schmidt and H. J. Schafer, *Angew. Chem. Int. Ed. Engl.* **18**, 787 (1979).

[45a] H. Karaman and D. G. Lee, unpublished results.

and thereby produces an ion pair that is soluble in organic solvents [Eqs. (2) and (3)].

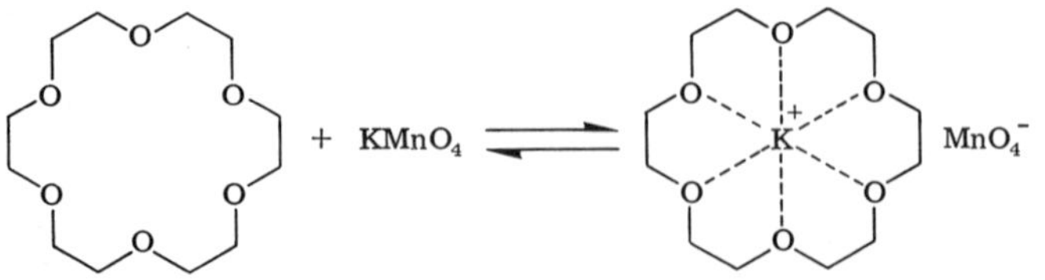
(2)

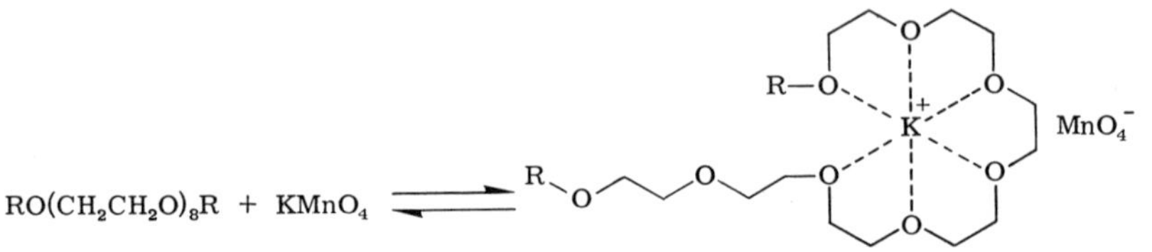
(3)

This phase transfer technique has been applied extensively to a variety of synthetic procedures in addition to oxidation reactions. Numerous review articles[46] and three books[47] on the subject are available. In the next section we will examine the actual phase transfer process, with particular attention to permanganate. In succeeding sections the use of permanganate ion (dissolved in organic solvents) as an oxidant for a variety of organic compounds will be considered.

II. The Phase Transfer Process

A. PHASE TRANSFERS FROM AQUEOUS PERMANGANATE

The effectiveness of any particular phase transfer system will be dependent on the ability of that system to bring the permanganate ion into solution in the organic phase. (It should be noted that, whereas most reactions will probably be more efficient when the concentration of permanganate in the nonaqueous phase is high, some selective oxidations may require a low, but constant concentration of oxidant.)

[46] E. V. Dehmlow, *Chem. Technol.* **5**, 210 (1975); *Angew. Chem. Int. Ed. Engl.* **13**, 170 (1974), *ibid.* **16**, 493 (1977); R. A. Jones, *Aldrichimica Acta* **9**, 35 (1976); G. W. Gokel and H. D. Durst, *Synthesis*, 168 (1976); G. W. Gokel and W. P. Weber, *J. Chem. Ed.* **55**, 350, 429 (1978).

[47] W. P. Weber and G. W. Gokel, "Phase Transfer Catalysis in Organic Synthesis." Springer-Verlag, New York, 1977; C. M. Starks and C. Liotta, "Phase Transfer Catalysis, Principles and Techniques." Academic Press, New York, 1978. E. V. Dehmlow and S. S. Dehmlow, "Phase Transfer Catalysis." Verlag Chemie, Weinheim, West Germany, 1980.

It is well known that potassium permanganate exists in an ionic form when dissolved in water. However, when a quaternary ammonium permanganate salt dissolves in an organic solvent it exists predominantly as an ion pair. This can be seen from the calculations described in Brändström.[48] He has shown that for two spherical ions, Q^+ and X^-, situated in a large sphere of solvent with dielectric constant D, the probability (P) of finding Q^+ and X^- separated by a distance between r and $r + dr$ is given by Eq. (4), where R is the radius of the solvent sphere, k is the Boltzmann constant, and T is the temperature.[49]

$$dP = \frac{3}{4\pi R^3} 4\pi r^2 \exp\left(-\frac{e^2}{DkTr}\right) dr \tag{4}$$

Assuming that the sphere is large enough to give ideal behavior even when the concentration of ions is increased, the total probability (dH) of finding the center of an ion, X^-, at a distance between r and $r + dr$ from the center of an ion Q^+ when more ions are present is given by Eq. (5).

$$dH = [Q^+][X^-]\left(\frac{4\pi N}{1000}\right)^2 \frac{R^3}{3} r^2 \exp\left(-\frac{e^2}{DkTr}\right) dr \tag{5}$$

A plot of dH/dr against r passes through a minimum at $r = e^2/2DkT$ (Fig. 1). The significance of this figure can be understood in the following way: if the ions can approach each other sufficiently close so that the distance between their centers is less than $e^2/2DkT$, the attractive forces will be strong enough to cause ion pair formation. If the ions cannot approach each other sufficiently close, the attractive forces will be smaller and solvation (resulting in the formation of individual ions) will occur. The fact that the point at which ion pair formation occurs is inversely dependent on the dielectric constant of the solvent means that ion pairing is much more likely to occur in nonpolar solvents with low dielectric constants. The values of $e^2/2DkT$ for several solvents have been collected in Table II.

From this table one can determine if a particular quaternary ammonium permanganate will exist in a particular solvent as an ion pair or as free ions. For example, since the ionic radii of tetramethyl and tetraoctyl ammonium ions have been found to be about 3 Å and 6 Å, respectively,[50] it can be seen that the value of r for the ammonium permanganates will be substantially

[48] A. Brändström, *Adv. Phys. Org. Chem.* **15**, 267 (1977).

[49] N. Bjerrum, *K. Dan. Vibensk. Selsk. Mat.-Fys.* Medd. **7**, 1 (1926); *C. A.* **22**, 1263 (1928); see also, M. Szwarc, *in* "Ions and Ion Pairs in Organic Reactions" (M. Szwarc, ed.), Vol. 1. pp. 1–26. Wiley (Interscience), New York, 1972.

[50] S. R. C. Hughes and D. H. Price, *J. Chem. Soc. A*, 1093 (1967); B. S. Krumgal'z, *Russ. J. Phys. Chem.* **45**, 1448 (1971).

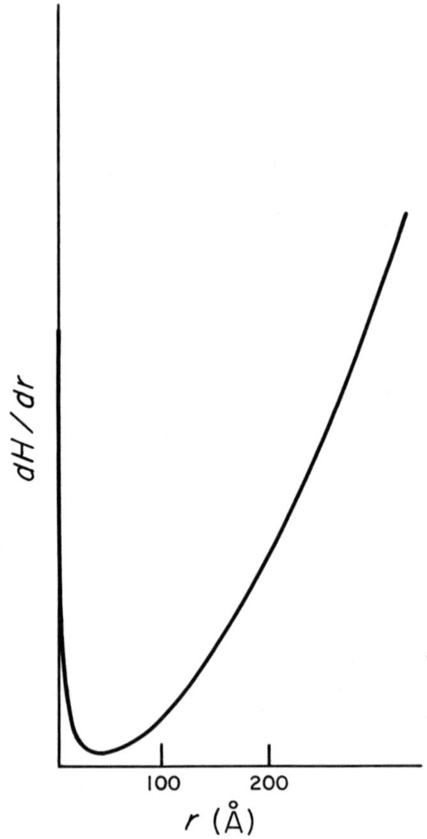

FIG. 1. Plot of dH/dr against r where dH is the probability of finding the center of an ion, X^-, at a distance between r and $r + dr$ from the center of an ion, Q^+. (Reprinted with permission from Brändström[48]. Copyright by Academic Press, London.)

less than e^2/DkT in solvents with low dielectric constants (accepting the value of about 2.5 Å for the ionic radius of the permanganate ion[51]). Consequently, it can be assumed that when permanganate ion is transferred into an organic solvent with the aid of a phase transfer agent, it exists for all practical purposes as an ion pair.

The concentration of permanganate ion pair that can be obtained in the nonaqueous phase will be dependent on at least three factors: (1) the structure of the organic cation; (2) the polarity of the organic phase; and (3) the nature of the anions involved.

[51] G. J. Palenik, *Inorg. Chem.* **6**, 503 (1967).

TABLE II

SOLVENT PARAMETERS[a]

Solvent	Dielectric constant	$e^2/2DkT$ (Å)
Water	78.5	3.57
Methanol	32.5	6.8
Ethanol	24.3	11.5
Acetone	20.5	13.7
1,2-Dichloroethane	10.17	27.6
Methylene chloride	8.9	31.5
Chloroform	4.7	60
Diethyl ether	4.2	67
Benzene	2.27	123
Carbon tetrachloride	2.22	126

[a] Reprinted with permission from Brändström.[48] Copyright by Academic Press, London.

1. STRUCTURE OF THE CATIONS

A high concentration of anion in the organic phase is promoted by use of a phase transfer agent which provides a highly organophilic cation. This was illustrated clearly by the work by Gibson and Weatherburn[52] when they studied the distribution of a number of quaternary salts between water and four different organic solvents (chloroform, methylene chloride, 1,2-dichloroethane, and 2,2′-dichlorodiethyl ether). The concentrations of quaternary salt in each phase was measured spectrophotometrically and expressed as a ratio (α) of the concentration in the organic phase over the concentration in the aqueous phase, i.e., $\alpha = [QX]_{org}/[QX]_{aq}$. Some of the results which they obtained using methylene chloride as the solvent have been summarized in Table III.

An examination of these distribution ratios indicates that they increase with the length of the alkyl chain. Eq. (6), where n is the number of methylene groups in the alkyl chain and α' is the ratio for the methyl compound, was found to correlate the data.

$$\log \alpha = 0.5n + \log \alpha' \tag{6}$$

Somewhat similar studies on the effectiveness of various phase transfer agents with potassium permanganate have also been reported. Okimoto and Swern[53] have published data from which the ratios summarized in Table

[52] N. A. Gibson and D. C. Weatherburn, *Anal. Chim. Acta* **58**, 149, 159 (1972).
[53] T. Okimoto and D. Swern, *J. Am. Oil Chem. Soc.* **54**, 862A (1977).

TABLE III

DISTRIBUTION RATIOS FOR VARIOUS QUATERNARY SALTS BETWEEN METHYLENE CHLORIDE AND WATER[a,b]

Cation	α-Values[c]					
	Cl^-	ClO_3^-	ClO_4^-	Br^-	BrO_3^-	I^-
$Ph_3PCH_3^+$	0.02	0.40	9.5	0.08	0.04	1.7
$Ph_3AsCH_3^+$	0.02	0.53	11.0	0.24	0.04	1.9
$Ph_3PCH_2CH_3^+$	0.01	0.90	40	0.25	0.08	4.9
$Ph_3PCH_2CH_2CH_3^+$	0.03	2.1	60	0.68	0.22	4.3
$Ph_3PCH_2CH_2CH_2CH_2CH_3^+$	0.22	5.6	70	2.3	1.0	8.5

[a] Data from Gibson and Weatherburn.[52]

[b] Temp. = 25°C, ionic strength = 0.1, initial concentration of quaternary salt in aqueous phase = 5×10^{-3} *M*.

[c] $\alpha = [QX]_{org}/[QX]_{aq}$.

TABLE IV

EXTRACTION OF PERMANGANATE ION FROM WATER INTO BENZENE OR METHYLENE CHLORIDE USING PHASE TRANSFER AGENTS[a]

Phase transfer agent	$[QMnO_4]_{C_6H_6}/[MnO_4^-]_{aq}$	$[QMnO_4]_{CH_2Cl_2}/[MnO_4^-]_{aq}$
Tetrabutylphosphonium chloride	99	252
Hexadecyltrimethylammonium bromide	0.03[b]	105
Tetrabutylammonium bromide	14	100
Propyltrioctylammonium bromide	96	43
Benzyltriethylammonium chloride	0.02[b]	9.4
Tetraethylammonium bromide	0	0.26[c]

[a] From Okimoto and Swern.[53]

[b] A precipitate formed in the aqueous phase; 40–80% of the permanganate unaccounted for.

[c] About 20% of the permanganate unaccounted for.

IV may be calculated. Their approach was to extract 25 ml of 0.02*M* aqueous $KMnO_4$ with 25 ml of a 0.03*M* solution of phase transfer agent in benzene or methylene chloride, and to estimate the concentration in each phase iodometrically.

Herriott and Picker[54] have also reported results obtained when solutions of water containing 1.00 mmol of $KMnO_4$ were extracted with an equal volume of benzene containing 1.5 mmol of various phase transfer agents. Their results are summarized in Table V.

[54] A. W. Herriott and D. Picker, *Tetrahedron Lett.*, 1511 (1974).

TABLE V

EXTRACTION OF WATER CONTAINING 1.00 MMOL OF PERMANGANATE ION WITH AN EQUAL VOLUME OF BENZENE CONTAINING VARIOUS PHASE TRANSFER AGENTS[a]

Phase transfer agent (mmol)	Amount of permanganate transferred (mmol)
Tetramethylammonium chloride (1.51)	0.00
Triethylbenzylammonium chloride (1.63)	0.84
Tetrabutylammonium bromide (1.54)	0.95
Tetrabutylphosphonium chloride (1.54)	0.96
Cetyltrimethylammonium bromide (1.53)	0.86
Aliquat 336[b] (1.52)	0.93
Sodium dodecyl sulfate (1.50)	0.00

[a] From Herriott and Picker.[54]
[b] Trioctylmethylammonium chloride (General Mills).

In a somewhat more thorough study of the distribution constants associated with permanganate phase transfer reactions, Karaman[55] prepared solutions of permanganate salts in various organic solvents by stirring a solution of phase transfer agent over finely powdered potassium permanganate, removed an aliquot and extracted it with an equal volume of water. The equilibrium involved in this process would be defined by Eqs. (7) and (8).

$$Q^+_{aq} + MnO_4{}^-_{aq} \overset{K_D}{\rightleftharpoons} Q^+MnO_4{}^-_{org} \tag{7}$$

$$K_D = \frac{[Q^+MnO_4{}^-]_{org}}{[Q^+]_{aq}[MnO_4{}^-]_{aq}} = \frac{[Q^+MnO_4{}^-]_{org}}{[MnO_4{}^-]^2_{aq}} \tag{8}$$

The values of K_D (Table VI) were then obtained by plotting $[Q^+MnO_4{}^-]_{org}$ against $[MnO_4{}^-]^2_{aq}$ for several concentrations of phase transfer agents; both concentrations were determined spectrophotometrically.

The results obtained from all of these studies indicate that to a first approximation the most important factor is the size of the quaternary ion. In fact, there is a rough correlation between the total number of carbons in the quaternary ion and the corresponding equilibrium constant.

It may also be noted that the two most readily available, and inexpensive, commercial phase transfer agents, Adogen 464 (Ashland Chemicals) and Aliquat 336 (General Mills) exhibited excellent transferability with respect to permanganate. Both of these agents are quaternary ammonium chlorides containing three large alkyl groups (8 to 10 carbons) and a methyl group.

[55] H. Karaman, M. S. Thesis, Univ. of Regina, Regina, Canada, 1979.

TABLE VI

EQUILIBRIUM CONSTANTS FOR THE DISTRIBUTION OF QUATERNARY AMMONIUM PERMANGANATES BETWEEN WATER AND METHYLENE CHLORIDE.[a]

Phase transfer agent	Number of carbons	K_D[b]
Tetramethylammonium chloride	4	0
Tetraethylammonium bromide	8	54
Phenyltrimethylammonium bromide	9	24
Benzyltrimethylammonium bromide	10	66
Benzyltriethylammonium bromide	13	957
Tetrabutylammonium bromide	16	∞
Benzyltributylammonium chloride	19	1489
Cetyltrimethylammonium bromide	19	∞
Adogen 464	25–31	∞
Aliquat 336	25–31	∞

[a] From Karaman.[55]

[b] $K_D = [Q^+MnO_4^-]_{org}/[MnO_4^-]_{aq}^2$, 0 indicates that all of the permanganate was in the aqueous phase while ∞ indicates that it was all in the organic phase.

It is possible that their effectiveness is due to the high organophilicity imparted by the three long alkyl chains combined with the small methyl group which would allow the permanganate ion to be closely associated with the positive nitrogen.[48]

Karaman[55] also examined the distribution of permanganate ion between aqueous and organic phases in the presence of polyethers. His results (Table VII) indicate that crown ethers are substantially more effective than the corresponding linear polyethers in promoting phase transfers. This is likely true because the energy requirement associated with the process would be decreased by use of a preformed cyclic ether. Such an assumption is consistent with the results reported by Cram[56] and his co-workers; they found that preorganization of the binding sites in 18-crown-6 caused it to form a stronger complex with cations than the corresponding acyclic polyether, dimethylhexaethylene glycol. With the *tert*-butylammonium cation, the difference in free energy was found to be about 6 kcal/mol.

Although the linear polyethers in Table VII contain three different types of terminal groups (CH_3O—, HO— and $(CH_3)_3Si$—), the most important factor appears to be the length of the chain. From Fig. 2, where a plot of

[56] E. P. Kyba, R. C. Helgeson, K. Madan, G. W. Gokel, T. L. Tarnowski, S. S. Moore, and D. J. Cram, *J. Am. Chem. Soc.* **99**, 2564 (1977). J. M. Timko, S. S. Moore, D. M. Walba, P. C. Hiberty, and D. J. Cram, *J. Am. Chem. Soc.* **99**, 4207 (1977).

TABLE VII

EQUILIBRIUM CONSTANTS FOR THE DISTRIBUTION OF POTASSIUM PERMANGANATE BETWEEN WATER AND METHYLENE CHLORIDE IN THE PRESENCE OF POLYETHERS[a]

Polyether	Average molecular weight	K_D[b]
15-Crown-5	220	47
18-Crown-6	264	128
$CH_3(OCH_2CH_2)_nOCH_3$	178	0.77
$CH_3(OCH_2CH_2)_nOCH_3$	397	1.46
$CH_3(OCH_2CH_2)_nOH$	350	1.63
$CH_3(OCH_2CH_2)_nOH$	550	2.71
$CH_3(OCH_2CH_2)_nOH$	750	4.10
$(CH_3)_3Si(OCH_2CH_2)_nOSi(CH_3)_3$	462	1.46
$(CH_3)_3Si(OCH_2CH_2)_nOSi(CH_3)_3$	487	1.45
$(CH_3)_3Si(OCH_2CH_2)_nOSi(CH_3)_3$	662	2.48

[a] From Karaman.[55]

[b] $K_D = [Q^+MnO_4^-]_{org}/[MnO_4^-]_{aq}^2$, where Q^+ is a polyether–potassium cation complex.

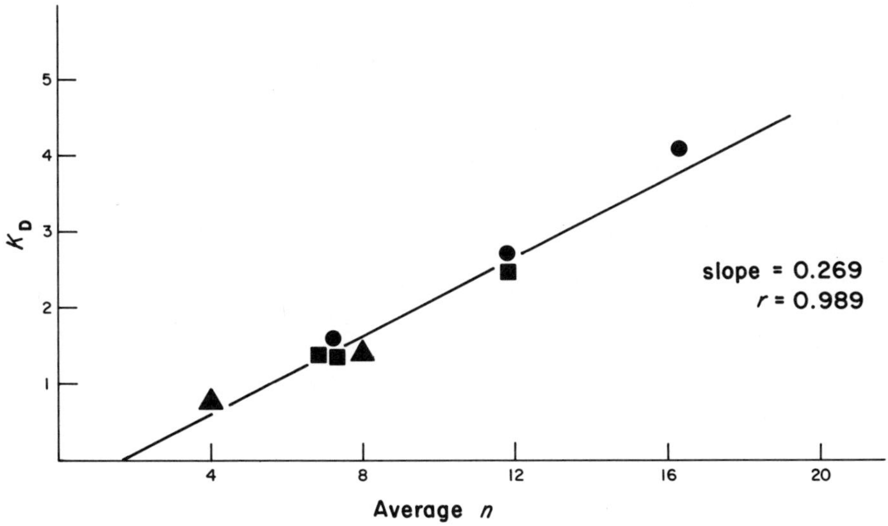

FIG. 2. Plot of K_D against chain length for various polyethers.[55] Key: ▲, $CH_3(OCH_2CH_2)_n$-OCH_3; ●, $CH_3(OCH_2CH_2)_nOH$; ■, $(CH_3)_3Si(OCH_2CH_2OSi(CH_3)_3$.

K_D against *n* (the number of ethylene glycol units) has been reproduced, it can be seen that there is a direct relationship between the two parameters. Hence it would seem that greater organophilicity or stronger binding is associated with longer chains.

2. Solvent Polarity

From a study of the distribution ratios for a number of phase transfer agents in four solvents (chloroform, methylene chloride, 1,2-dichloroethane and 2,2′-dichlorodiethyl ether) Gibson and Weatherburn[52] concluded, in part, that

> . . . the effect of the solvent on the extraction is quite small, very much smaller than the effects of the cation discussed above, and the anion discussed below. However, definite solvent effects on the extraction may be seen, although there is little correlation between extracting ability and bulk properties such as dielectric constant.

On the other hand, Brändström[48] has reported constants for the extraction of tetrabutylammonium bromide into 41 different solvents and noted that they vary by factors of greater than 10^6. He also noted that the magnitudes of these constants do not correlate with the dielectric constants of the solvents, but that the best solvents were those containing two or more donor groups on the same or adjacent carbons (e.g., $ClCH_2CN$, $ClCH_2CH_2CN$, CH_3NO_2, and $CHCl_2CHCl_2$). It was assumed that the donor groups would solvate the cations, while the acidic hydrogens adjacent to these groups would accommodate the anion and that tetrabutylammonium bromide would consequently be extracted as a solvent-separated ion pair.

Okimoto and Swern[53] have examined the extraction of permanganate ion from water into either benzene or methylene chloride with six different phase transfer agents. Their results (Table IV) indicated that permanganate ion was more readily extracted into methylene chloride than benzene in almost every case.

This conclusion was also confirmed by Karaman[55] when he examined the distribution coefficients for a variety of phase transfer agents in five different solvents: methylene chloride, chloroform, benzene, carbon tetrachloride, and pentane. His results (some of which have been summarized in Table VIII) indicate that methylene chloride is the most satisfactory solvent and that the distribution coefficients are, in general, lower in solvents with low dielectric constants.

In many cases he found that the permanganate ion did not distribute between two phases; it either remained completely in the aqueous phase or extracted completely into the organic phase (within detectable limits, at least). Again, the practical observation was made that Adogen 464 and

TABLE VIII

EQUILIBRIUM CONSTANTS FOR THE DISTRIBUTION OF PERMANGANATE SALTS BETWEEN WATER AND VARIOUS ORGANIC SOLVENTS[a,b]

Phase transfer agents	Methylene chloride	Chloroform	Benzene	Carbon tetrachloride	Pentane
Tetramethylammonium chloride	[d]	[d]	[d]	[d]	[d]
Tetraethylammonium bromide	54	0	[d]	[d]	[d]
Tetrabutylammonium bromide	∞	∞	0	0	[d]
Phenyltrimethylammonium bromide	24	0	[d]	[d]	[d]
Benzyltrimethylammonium bromide	63	28	[d]	[d]	[d]
Benzyltriethylammonium bromide	960	132	[d]	[d]	[d]
Methyltriphenylphosphonium bromide	∞	∞	[d]	[d]	[d]
Dodecyltriphenylphosphonium bromide	∞	∞	∞	∞	[d]
18-Crown-6	128	116	0	0	0
Dimethylpolyethylene glycol 178[c]	0.77	0	0	0	0
Dimethylpolyethylene glycol 397[c]	1.46	0.54	0	0	0
Methoxypolyethylene glycol 350[c]	1.63	0.41	0	0	0
Methoxypolyethylene glycol 550[c]	2.71	0.76	0	0	0
Methoxypolyethylene glycol 750[c]	4.10	0.88	0	0	0
Adogen 464	∞	∞	∞	∞	∞
Aliquat 336	∞	∞	∞	∞	∞

[a] From Karaman.[55]
[b] Equilibrium constant $= [Q^+MnO_4^-]_{org}/[MnO_4^-]_{aq}^2$. 0 indicates all of the oxidant was in the aqueous phase, ∞ that it was all in the organic phase.
[c] This number indicates the average molecular weight.
[d] Phase transfer agent insoluble.

Aliquat 336 were both effective phase transfer agents for all solvents investigated.

In summary, all evidence indicates that methylene chloride is generally the most useful solvent. Besides the fact that it solvates permanganate salts well, it is resistant to oxidation and, because of its low boiling point, it can be easily removed by vacuum distillation from the reaction products.

3. NATURE OF THE ANION

Gibson and Weatherburn[52] have reported that anions are extracted from aqueous solutions into organic solvents with the aid of phase transfer catalysts in the following general order:

$$MnO_4^- > ClO_4^- > SCN^- > I^- > ClO_3^- > NO_3^- > Br^- > BrO_3^- > Cl^-$$

This order has also been confirmed by other workers[47] and indicates that permanganate will be extracted in preference to almost all other anions. Its ease of extraction is probably associated with the fact that it has a low charge to volume ratio and that the charge is extensively delocalized. This would decrease the extent of hydration and thus reduce the energy required to transfer it from an aqueous solution into an organic solvent.

B. PHASE TRANSFERS UTILIZING SOLID PERMANGANATE

Permanganate can also be brought into solution in organic solvents by adding solvent containing a phase transfer agent to finely ground potassium permanganate.[57] An exchange takes place [Eq. (9)] and permanganate dissolves in the form of an ion pair.

$$KMnO_{4(s)} + Q^+X^-_{(org)} \rightleftharpoons Q^+MnO_4{}^-_{(org)} + KX_{(s)} \tag{9}$$

This procedure has advantages in certain cases where it is important to maintain anhydrous conditions. For example, it is known that nonterminal alkynes undergo a cleavage reaction if the organic solvent is wet from being in contact with aqueous permanganate. However, under anhydrous conditions the cleavage is suppressed and good yields of diketones may be obtained.[58]

It has also been observed that some phase transfer agents function better in the absence of water. For example, dimethyloctaethylene glycol does not extract potassium permanganate from an aqueous solution into benzene, but it will solubilize solid anhydrous potassium permanganate in the same solvent.[59]

Many of the same factors which control the effectiveness of transfers from aqueous solutions also influence the solid to solution transfers.

1. Structure of the Phase Transfer Agents

Karaman[55] has observed that both quaternary ammonium and phosphonium salts as well as polyethers can be used to solubilize anhydrous permanganate in organic solvents. He measured the concentration of permanganate that could be obtained from the treatment of finely ground potassium permanganate with limited amounts of phase transfer agent in various solvents, and calculated ratios [Eq. (10)] which indicate the effectiveness of the phase transfer agents in solubilizing permanganate.

$$K_r = \frac{[Q^+MnO_4{}^-]_{org}}{[Q^+X^-]_{initial}} \tag{10}$$

[57] A finely ground (3μ) form of potassium permanganate is available from the Carus Chemical Company under the trade name "Cairox M."

[58] D. G. Lee and V. S. Chang, *J. Org. Chem.* **44**, 2726 (1979).

[59] D. G. Lee and V. S. Chang, *J. Org. Chem.* **43**, 1532 (1978).

TABLE IX

SOLUBILITY RATIOS FOR PERMANGANATE IN METHYLENE CHLORIDE SOLUTIONS[a]

Phase transfer agents	K_r[b]
Methyltriphenylphosphonium bromide	0.86
Ethyltriphenylphosphonium bromide	0.97
n-Propyltriphenylphosphonium bromide	0.91
n-Butyltriphenylphosphonium bromide	0.89
Dodecyltriphenylphosphonium bromide	0.85

[a] From Karaman.[55]
[b] $K_r = [Q^+MnO_4^-]_{org}/[Q^+X^-]_{initial}$.

His results indicated, among other things, that there is a certain specificity of phase transfer agents for each solvent. For example, Table IX shows that the magnitude of K_r passes through a maximum as the alkyl group in alkyltriphenylphosphonium bromides is varied from methyl to dodecyl.

The K_r values for polyethers in methylene chloride are quite interesting. Crowns which contain 6 oxygens and 12 carbons in the ring were found to have K_r values close to unity, thus indicating their capacity to solubilize approximately one mol of $KMnO_4$ per mol of crown in solution. In the case of acyclic polyethers, the K_r values were found to vary from 0.007 to 5.84 depending on the chain lengths (Fig. 3). Values greater than unity suggest that these compounds are capable of complexing more than one potassium ion per mol of phase transfer agent as depicted in Fig. 4. Other polyethers are also known to be capable of complexing more than one mol of potassium ion per mol of phase transfer agent.[60]

2. SOLVENT EFFECTS

The effectiveness of phase transfer agents under these conditions was found to be markedly dependent on the nature of the solvent.[55] For example, it was noted that whereas methyltriphenylphosphonium bromide was a satisfactory phase transfer agent with methylene chloride and chloroform, dodecyltriphenylphosphonium bromide was far more effective for use in the less polar solvents such as benzene and carbon tetrachloride (Table X).

[60] F. Vogtle and V. Heiman, *Chem. Ber.* **111**, 2757 (1978); G. Weber and W. Saenger, *Angew. Chem. Int. Ed. Engl.* **18**, 227 (1979); F. Vogtle and E. Weber, *Angew. Chem. Int. Ed. Engl.* **18**, 753 (1979).

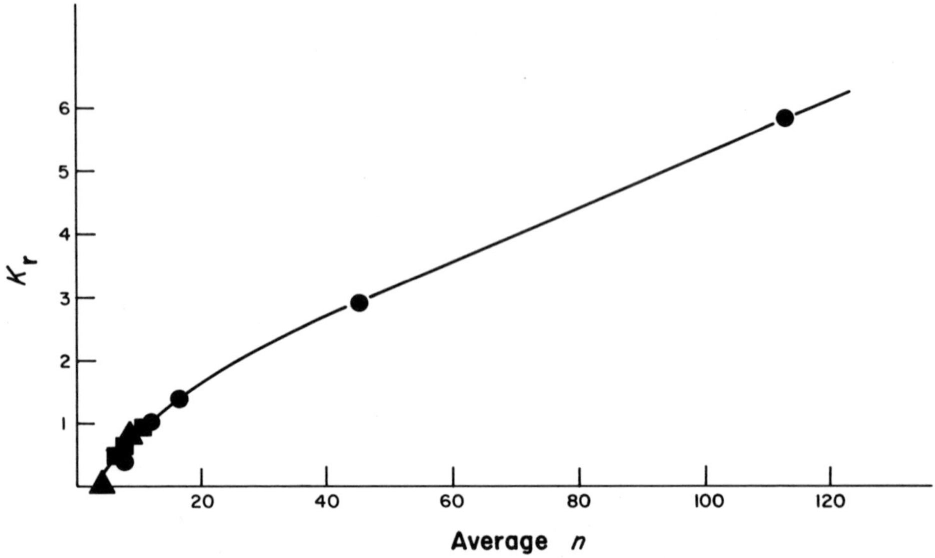

FIG. 3. Plot of K_r against chain length for various polyethers.[55] Key: ▲, $CH_3(OCH_2CH_2)_n$-OCH_3; ●, $CH_3(OCH_2CH_2)_nOH$; ■, $(CH_3)_3Si(OCH_2CH_2)_nOSi(CH_3)_3$.

Tetrabutylammonium bromide is also quite effective for solubilizing permanganate in dichloromethane, chloroform and benzene, but less satisfactory in carbon tetrachloride and pentane. Adogen 464 and Aliquat 336 were found to be quite good phase transfer agents, although the concentrations produced in pentane were rather low.

3. IDENTITY OF THE ANION

Equation (9) involves an exchange of anions between a solid phase and a solution. Hence it is not surprising that the identity of the anion in the phase

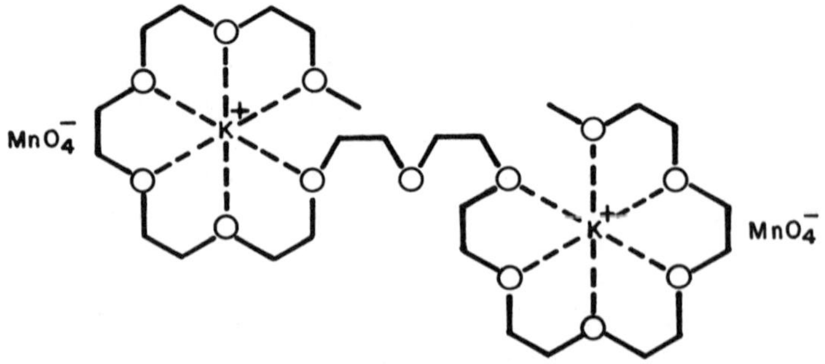

FIG. 4. Complexation of two moles of $KMnO_4$ by a single acyclic polyether.

TABLE X

SOLVENT EFFECTS ON PERMANGANATE SOLUBILITY RATIOS[a]

Phase transfer agents	K_r[b] Methylene chloride	Chloro-form	Benzene	Carbon tetra-chloride	Pentane
Methyltriphenylphosphonium bromide	0.86	0.74	[c]	[c]	[c]
Dodecyltriphenylphosphonium bromide	0.85	0.58	1.03	0.89	[c]
Tetrabutylammonium bromide	1.04	0.58	1.04	0.11	[c]
Aliquat 336[d]	0.82	0.84	0.84	0.88	0.07
Adogen 464[d]	0.82	0.88	0.90	0.91	0.08

[a] From Karaman.[55]
[b] $K_r = [Q^+MnO_4^-]_{org}/[Q^+X^-]_{initial}$.
[c] The phase transfer agent is insoluble.
[d] Methyltrialkyl (C_8–C_{10}) ammonium chloride.

transfer agent can effect the magnitude of the K_r values. Table XI contains some examples that illustrate this phenomena. Although the number of examples in this table is too small to reach any general conclusions, it appears that the greatest effects are present when chloroform is used as the solvent. On the basis of the available evidence, it seems that quaternary ammonium chlorides are substantially more effective phase transfer agents in chloroform than are the corresponding bromides.

TABLE XI

THE EFFECT OF ANIONS ON PERMANGANATE SOLUBILITY RATIOS[a]

Phase transfer agents	K_r[b] Methylene chloride	Chloro-form	Benzene	Carbon tetra-chloride	Pen-tane
Tetrabutylammonium chloride	0.92	0.97	0.78	0.11	[c]
Tetrabutylammonium bromide	1.04	0.58	1.04	0.11	[c]
Benzyltriethylammonium chloride	0.97	0.96	[c]	[c]	[c]
Benzyltriethylammonium bromide	0.93	0.45	[c]	[c]	[c]

[a] From Karaman.[55]
[b] $K_r = [Q^+MnO_4^-]_{org}/[Q^+X^-]_{initial}$.
[c] The phase transfer agent is insoluble.

C. PHASE TRANSFER MECHANISMS

In the earliest descriptions of phase transfer reactions it was assumed that the exchange of anions took place in the aqueous phase as depicted:[47]

$$\text{R—Y} + \text{Q}^+\text{X}^- \rightarrow \text{R—X} + \text{Q}^+\text{Y}^- \quad \text{organic phase}$$

$$\text{-------}\rightleftharpoons\text{------------}\rightleftharpoons\text{----------}$$

$$\text{Y}^- + \text{Q}^+\text{X}^- \longleftarrow \text{X}^- + \text{Q}^+\text{Y}^- \quad \text{aqueous phase}$$

Alternatively, the exchange of anions could take place across the interface without migration of the phase transfer agent into the aqueous phase.

$$\text{R—Y}_{(org)} + \text{Q}^+\text{X}^-{}_{(org)} \rightarrow \text{R—X}_{(org)} + \text{Q}^+\text{Y}^-{}_{(org)}$$

$$\text{Q}^+\text{Y}^-{}_{(org)} + \text{X}^-{}_{(aq)} \rightarrow \text{Q}^+\text{X}^-{}_{(org)} + \text{Y}^-{}_{(aq)}$$

The possibility that the latter mechanism may be important in many reactions has been demonstrated by Landini, Maia, and Montanari.[61] Using liquid membranes, they showed that the transport of anions from one phase into another did not require the simultaneous transfer of a quaternary ammonium or phosphonium ion.

The experiments were carried out in a "U" system in which two organic phases (A) and (B) were separated by an aqueous solution of sodium bromide. When quaternary cations such as tetrabutylphosphonium bromide and tetrapropylammonium bromide, which are partly soluble in water, were dissolved in solution A and the system stirred, partitioning between A and B was observed. However, if more organophilic cations such as hexadecyltributylphosphonium bromide or tetraoctylammonium bromide were used, no transfer from A to B could be detected. Hence, it would appear that these two cations lack sufficient solubility in water to be transferred. They were found, however, to be effective phase transfer agents for the following reaction:

$$C_8H_{17}OSO_2CH_3 + KY \xrightarrow[\text{PhCl/H}_2\text{O}]{\text{Q}^+\text{Y}^-} C_8H_{17}Y + CH_3SO_3K$$

The importance of this observation was further emphasized by the fact that the largest pseudo-first-order rate constants were obtained for those cations that had the least solubility in water. In fact, it was found that the greater rates obtained with these agents could be attributed almost entirely to their insolubility in water. An analysis of the kinetic data showed that the rates were directly dependent on the concentration of the phase transfer agent in the organic phase.[62]

[61] D. Landini, A. Maia and F. Montanari, *J. Chem. Soc. Chem. Comm.*, 112 (1977).
[62] D. Landini, A. Maia and F. Montanari, *J. Am. Chem. Soc.* **100**, 2796 (1978).

Detailed mechanisms for phase transfers associated with alkylation and dihalocarbene reactions have been considered by Dehmlow.[46]

D. USE OF SOLID SUPPORTS FOR PERMANGANATE OXIDATIONS

It has been observed that the attachment of reagents to solid supports can be used effectively in several synthetic procedures.[63] For example, Regen and his co-workers have reported that both molecular sieves and alumina coated with potassium permanganate may be used to oxidize alcohols in benzene or toluene solutions. Yields varying from 26 to 100% were obtained.[64]

Menger and Lee have subsequently shown that moisture is essential for these reactions.[64a] For example, they found that drying of the molecular sieve/permanganate reagent over phosphorous pentoxide greatly reduced its ability to oxidize secondary alcohols. Furthermore they showed that the solid hydrate, $CuSO_4 \cdot 5H_2O$, also acted as an excellent support for the oxidation of secondary alcohols by potassium permanganate in benzene.[64a]

E. REACTIONS IN MICELLES

Certain quaternary cations act as phase transfer agents by forming micelles that bring organic substrates into the aqueous phase where they can react with water-soluble reagents. Micelles are usually formed with quaternary cations that contain one large and three small alkyl groups. Aggregates (10–50 molecules) of these compounds form with the large alkyl groups providing an hydrophobic center in which organophilic substrates also dissolve. The polar exterior of this droplet promotes its dispersion into the aqueous phase. In this way the organic substrates are brought into contact with reagents dissolved in the aqueous phase.[65]

Certain oxidation reactions are believed to proceed in this way. For example, piperonal has been oxidized to piperonylic acid in 65% yield by treatment with aqueous potassium permanganate containing a small amount of cetyltrimethylammonium bromide.[65] (In the absence of the phase transfer agent, the yield was reduced to about 35%.)

It has also been demonstrated that the rate of oxidation of 1-octene by aqueous potassium permanganate is accelerated by addition of emulsifying agents including sodium stearate and sodium laurate.[66]

[63] C. C. Leznoff, *Chem. Soc. Rev.* **3**, 65 (1974).

[64] S. L. Regen and C. Koteel, *J. Am. Chem. Soc.* **99**, 3837 (1977); S. Quici and S. L. Regen, *J. Org. Chem.* **44**, 3437 (1979).

[64a] F. M. Menger and C. Lee, *J. Org. Chem.* **44**, 3446 (1979).

[65] F. M. Menger, J. U. Rhee, and H. K. Rhee, *J. Org. Chem.* **40**, 3803 (1975); F. M. Menger, *Acc. Chem. Res.* **12**, 111 (1979).

[66] F. Yamashita, A. Atsumi, and H. Inoue, *C. A.* **83**, 113378 (1975).

III. Oxidation of Alkenes

A. TERMINAL ALKENES

One of the first examples of the use of phase transfer assisted permanganate oxidations involved the conversion of 1-decene into nonanoic acid.[67] The reaction of terminal alkenes was subsequently investigated thoroughly by Krapcho, Larson, and Eldridge[68] and has now become a standard procedure for the oxidation of long-chain terminal alkenes that lack solubility in water.[69] As the results summarized in Table XII indicate, a variety of phase transfer agents and solvents have been successfully used with Adogen 464 (or Aliquat 336) in methylene chloride the most economical and generally convenient combination.

When terminal alkenes are oxidized a certain amount of over oxidation occurs; e.g., the oxidation of 1-octene gives about 10% hexanoic acid in addition to the main product, heptanoic acid. This problem has been investigated extensively by Krapcho *et al.*[68] with the conclusion that over oxidation is promoted by hydroxide ions which are produced during the reduction of permanganate. They found, for example, that the over oxidation of allylbenzene could be greatly increased by adding base, but substantially reduced by adding acetic acid. In fact, the addition of acetic acid to the reaction mixture was found to almost completely eliminate over oxidation in most cases. As a consequence, most synthetic preparations are now carried out in solvents containing 5–10% acetic acid. Table XIII contains some examples that illustrate the control of over oxidation reactions.

Whereas it is clear that hydroxide ion promotes over oxidation, the mechanism of the reaction is still somewhat obscure. Krapcho *et al.*[68] observed that the product carboxylic acids were subject to further oxidation under reaction conditions particularly in the presence of base. Phenylacetic acid, for example, was oxidized to benzoic acid with yields higher (65%) in 0.1 *M* KOH than in the presence of acetic acid. Similarly, the treatment of heptanoic acid (0.03 mol) with $KMnO_4$ (0.12 mol in 0.1 *M* KOH) for 6 hr gave a 4% yield of hexanoic acid.

It is possible that the presence of base promotes the formation of enolates that would be readily cleaved by permanganate [Eq. (11)]. Such an explanation is consistent with the greater ease of oxidation of phenylacetic acid where

[67] C. M. Starks, *J. Am. Chem. Soc.* **93**, 195 (1971).

[68] A. P. Krapcho, J. R. Larson, and J. M. Eldridge, *J. Org. Chem.* **42**, 3749 (1977).

[69] D. G. Lee, S. E. Lamb, and V. S. Chang, *Org. Synth.*, **60**, 11 (1981).

TABLE XII

PHASE TRANSFER ASSISTED OXIDATION OF TERMINAL ALKENES BY AQUEOUS PERMANGANATE

Substrate	Phase transfer agent	Solvent[a]	Products (% yield)	Ref.
1-Octene	Tetrabutylammonium bromide	Benzene	Heptanoic acid (81)[d]	54
1-Octene	Benzylhexadecyldimethyl-ammonium chloride	Benzene	Heptanoic acid (97) Hexanoic acid (3)	68
1-Octene	Adogen 464[b]	Methylene chloride	Heptanoic acid (73)[d]	70
1-Decene	Aliquat 336[c]	Benzene	Nonanoic acid (90)[d]	67
1-Decene	Benzylhexyldimethyl ammonium chloride	Benzene	Nonanoic acid (87) Octanoic acid (3)	68
1-Decene	Dimethyloctaethylene glycol	Methylene chloride	Nonanoic acid (95)[d]	59
1-Undecene	Benzylhexyldimethyl ammonium chloride	Benzene	Decanoic acid (86)[d]	68
1-Dodecene	Benzylhexyldimethyl ammonium chloride	Benzene	Undecanoic acid (90)[d]	68
1-Tetradecene	Benzylhexyldimethyl ammonium chloride	Benzene	Tridecanoic acid (83)[d]	68
1-Tetradecene	Adogen 464[b]	Methylene chloride	Tridecanoic acid (83)[d]	69
1-Hexadecene	Benzylhexyldimethyl ammonium chloride	Benzene	Pentadecanoic acid (84)[d]	68
1-Hexadecene	Adogen 464[b]	Methylene chloride	Pentadecanoic acid (84)[d]	69
1-Octadecene	Benzylhexyldimethyl-ammonium chloride	Benzene	Heptadecanoic acid (80)[d]	68
1-Octadecene	Adogen 464[b]	Methylene chloride	Heptadecanoic acid (81)[d]	69
1-Eicosene	Benzylhexyldimethyl-ammonium chloride	Benzene	Nonadecanoic acid (90)[d]	68
1-Eicosene	Adogen 464[b]	Methylene chloride	Nonadecanoic acid (80)[d]	69
1-Docosene	Adogen 464[b]	Methylene chloride	Heneicosanoic acid (84)[d]	69
Styrene	Adogen 464[b]	Methylene chloride	Benzoic acid (96)[d]	69
Allylbenzene	Benzylhexyldimethyl-ammonium chloride	Benzene	Phenylacetic acid (80) Benzoic acid (20)	68

[a] With the exception of the first and fourth entries 5–10% acetic acid was added.
[b] Methyltrialkyl (C_8–C_{10}) ammonium chloride (Ashland Chemicals).
[c] Methyltrialkyl (C_8–C_{10}) ammonium chloride (General Mills).
[d] Isolated yield.

TABLE XIII

EXAMPLES OF OVER OXIDATION[a]

Substrate	Phase transfer agent	Additives	Products (% yield)
1-Octene	Aliquat 336[b]	none	Heptanoic acid (90) Hexanoic acid (10)
1-Octene	Aliquat 336[b]	Acetic acid	Heptanoic acid (97) Hexanoic acid (3)
1-Decene	Benzylhexadecyldimethyl-ammonium chloride	none	Nonanoic acid (77) Octanoic acid (8)
1-Decene	Benzylhexadecyldimethyl-ammonium chloride	KOH	Nonanoic acid (82) Octanoic acid (14)
1-Decene	Benzylhexadecyldimethyl-ammonium chloride	Acetic acid	Nonanoic acid (87) Octanoic acid (3)
Allylbenzene	Benzylhexadecyldimethyl-ammonium chloride	none	Phenylacetic acid (50) Benzoic acid (50)
Allylbenzene	Benzylhexadecyldimethyl-ammonium chloride	KOH	Phenylacetic acid (5) Benzoic acid (95)
Allylbenzene	Benzylhexadecyldimethyl-ammonium chloride	Acetic acid	Phenylacetic acid (80) Benzoic acid (20)

[a] The solvent used in these experiments was benzene. The oxidant was aqueous potassium permanganate. See ref. 68 for details.
[b] Methyltrialkyl (C_8–C_{10}) ammonium chloride (General Mills).

formation of an intermediate enolate would be aided by conjugation of the double bond with the ring.

$$RCH_2\overset{\overset{\displaystyle O}{\|}}{C}O^- + OH^- \rightleftharpoons RCH{=}C\begin{matrix}\diagup O^-\\ \diagdown O^-\end{matrix} + H_2O \qquad (11)$$

Good yields are obtained when the oxidant is introduced either as a solid anhydrous powder or as an aqueous solution. Table XIV compares the results obtained for the oxidation of 1-decene using a variety of different phase transfer agents and conditions. Since the yields are quite good under all conditions, it appears that in most cases practical considerations such as the cost and availability of the phase transfer agent will determine the best method to be used.

Some comments concerning the use of acyclic polyethers are appropriate at this point.[59] As the results in Table XIV indicate, acyclic polyethers $[RO(CH_2CH_2O)_nR']$ function fairly well as phase transfer agents for this reaction regardless of the nature of the terminal (R and R′) groups. It was noted, however, that terminal hydroxyl groups were gradually oxidized to

TABLE XIV

OXIDATION OF 1-DECENE USING DIFFERENT PHASE TRANSFER AGENTS AND CONDITIONS[a]

Phase transfer agent	Solvent[b]	Oxidant phase	Products (% yield)
Adogen 464[c]	Benzene	Solid	Nonanoic acid (91) Octanoic acid (3)
Adogen 464[c]	Methylene chloride	Aq.	Nonanoic acid (94) Octanoic acid (1)
$CH_3(OCH_2CH_2)_8OCH_3$[d]	Benzene	Solid	Nonanoic acid (70) Octanoic acid (4)
$CH_3(OCH_2CH_2)_8OCH_3$[d]	Methylene chloride	Aq.	Nonanoic acid (92) Octanoic acid (2)
$CH_3(OCH_2CH_2)_7OH$[e]	Methylene chloride	Aq.	Nonanoic acid (88)
$CH_3(OCH_2CH_2)_7OH$[e]	Methylene chloride	Solid	Nonanoic acid (84)
$CH_3(OCH_2CH_2)_3OCH_3$[c]	Benzene	Solid	Nonanoic acid (93)
$CH_3(OCH_2CH_2)_3OCH_3$[f]	Methylene chloride	Aq.	Nonanoic acid (93)
$(CH_3)_3Si(OCH_2CH_2)_7OSi(CH_3)_3$[g]	Methylene chloride	Aq.	Nonanoic acid (85)
Dicyclohexano-18-crown-6	Benzene	Solid	Nonanoic acid (86) Octanoic acid (5)

[a] From Chang and Lee.[70]
[b] Acetic acid (5–10%) was added to the solvent.
[c] Methyltrialkyl (C_8–C_{10}) ammonium chloride (Ashland Chemicals).
[d] H. Lehmkuhl, F. Rabet and K. Hauschild, *Synthesis*, 184 (1977).
[e] Carbowax methoxy polyethylene glycol (Union Carbide).
[f] Ansul E-181 (Ansul Company).
[g] Obtained as a gift from Dr. Gerd Rossmy (Goldschmidt).

carboxyl groups. Although this process consumed some oxidant it did not appear to decrease the effectiveness of these compounds as phase transfer agents.

Because it is known (Fig. 2) that the ability of polyethers to act as phase transfer agents for potassium permanganate decreases with the number of ethylene glycol units in the chain, it is surprising (but convenient) that dimethyl triethylene glycol proved to be as effective as most of the other phase transfer agents for these purposes. (The fact that the concentration of permanganate in the organic solvent was not high did not prevent the reaction from going to completion.)

It was also noted in Table VIII that dimethyl octaethylene glycol could not be used to transfer potassium permanganate from water to benzene, however it does bring solid permanganate into a benzene solution. This observation can sometimes be used advantageously in the following way: If the reaction

[70] V. S. Chang and D. G. Lee, unpublished results.

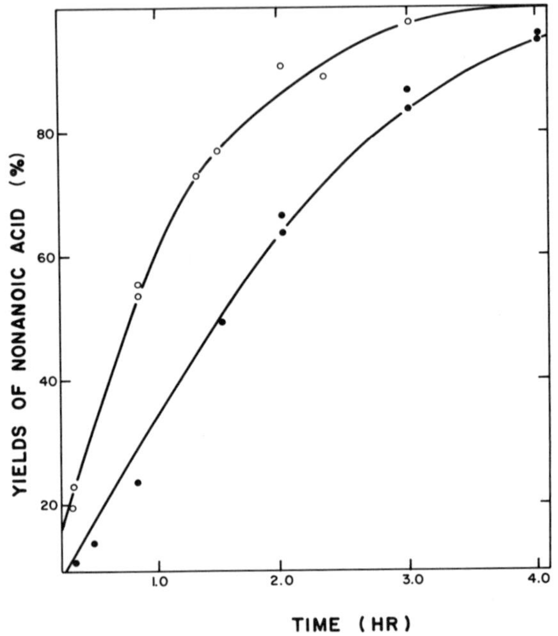

FIG. 5. Comparison of the use finely powdered (○) and reagent grade (●) $KMnO_4$ for the phase transfer assisted oxidation of 1-decene.

is carried out using solid $KMnO_4$, solubilized by dimethyl octaethylene glycol in benzene, the phase transfer agent can be removed easily from the final product by a simple aqueous extraction. Evaporization of the solvent then leaves a product that can be used for many purposes without further purification.

It should also be noted that when solid oxidant is employed, it is advantageous to use a finely divided form.[57] This is illustrated by Fig. 5 where it is apparent that the reaction in which powdered permanganate was used proceeded at a faster rate.

A Typical Procedure

Preparation of Tridecanoic Acid by Oxidation of 1-*Tetradecene*.* A 1-l round-bottomed flask was charged with $KMnO_4$ (32 g, 0.20 mol) and 300 ml

* Reprinted with permission from Krapcho *et al.*, *J. Org. Chem.* **42,** 3749. Copyright (1977) American Chemical Society.

of water. The flask was immersed in an ice bath and stirred vigorously with an egg-shaped magnet (1 in.). A solution of 1-tetradecene (11.8 g, 0.06 mol), 300 ml of benzene (or toluene), 60 ml of glacial acetic acid, and benzylhexadecyldimethylammonium chloride (0.2 g, 0.5 mmol) was added in one portion. Stirring was continued without any further addition of ice to the bath for about 4 hr. A total of 35 g of Na_2SO_3 was added to the cooled reaction mixture followed by the slow addition of a solution of 35 ml of concentrated HCl in 35 ml of water. Two clear layers resulted. The layers were separated and the benzene layer washed once with a 100-ml portion of cold water. The benzene layer was dried over anhydrous sodium sulfate, the drying agent removed by filtration, and the bulk of the benzene removed by distillation. The residual benzene was removed on a rotary evaporator to yield 12.7 g (99%) of crude solid. The crude acid was dissolved in pentane (60 ml), filtered to remove traces of insoluble material, and placed in the freezer overnight. Filtration yielded 10.6 g (83%) of tridecanoic acid of mp 43°–44°C. Treatment of a sample of the crude or crystallized acid with CH_2N_2 followed by GLC analysis showed about 2% contamination of dodecanoic acid and a trace amount of a short retention time impurity.

B. NONTERMINAL ALKENES

In aqueous solutions permanganate readily oxidizes water-soluble alkenes.[2] The nature of the products is, however, dependent on the reaction conditions: in acidic solutions, cleavage reactions predominate;[71] under basic conditions, dihydroxylation is the main reaction;[72] and in a neutral medium, ketols are formed as the major products.[27] Unfortunately the conditions under which these reactions take place are not sharply defined and mixtures of products are often obtained. However, it is generally possible to predict what will be the major product of a reaction under a particular set of conditions.

There appears to be general agreement[2,71,73,74] with a suggestion made nearly a century ago[75] that the initial reaction between alkene and per-

[71] D. G. Lee and J. R. Brownridge, *J. Am. Chem. Soc.* **96**, 5517 (1974).

[72] A. Lapworth and E. N. Mottram, *J. Chem. Soc.* 1628 (1925).

[73] F. Freeman, *Rev. React. Species Chem. React.* **2**, 179 (1973); F. Freeman, C. O. Fuselier, and E. M. Karachefski, *Tetrahedron Lett.*, 2133 (1975); F. Freeman and E. M. Karchefski, *Biochim. Biophys. Acta* **447**, 238 (1976).

[74] M. Jaky and L. I. Simandi, *React. Kinet. Catal. Lett.* **3**, 397 (1975); K. Polgar, M. Jaky, and L. I. Simandi, *React. Kinet. Catal. Lett.* **5**, 489 (1976); L. I. Simandi and M. Jaky, *J. Am. Chem. Soc.* **98**, 1995 (1976).

[75] G. Wagner, *Zh. Russ. Fiz Khim. Ova* **27**, 219 (1895).

manganate results in the formation of a cyclic manganate(V) diester [Eq. (12)].

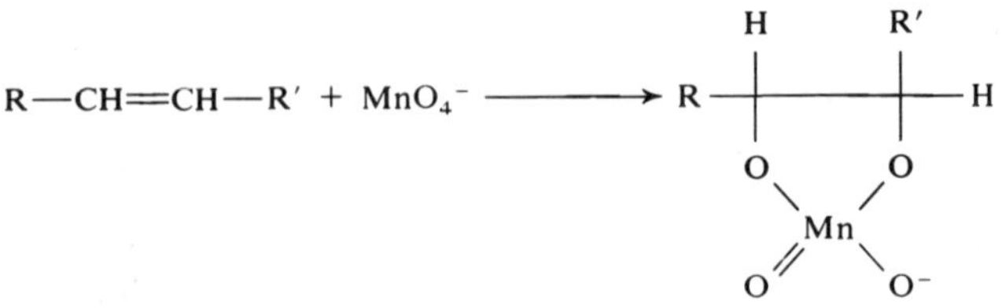

(12)

Under acidic conditions the manganate(V) diester apparently undergoes an oxidative decomposition that results in cleavage products plus manganate (III) [Eq. (13)].[71]

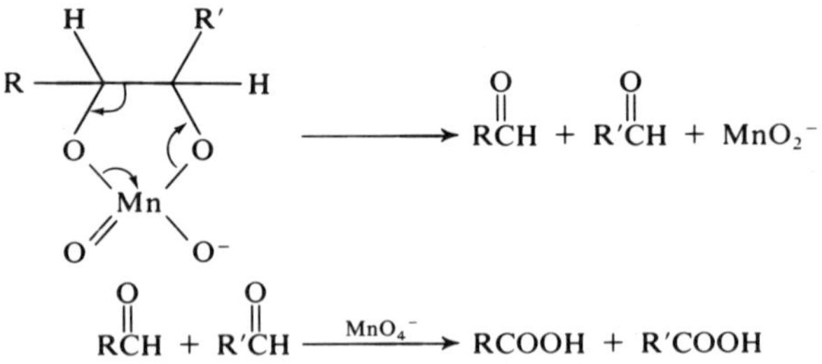

(13)

On the other hand, under basic conditions the intermediate would be hydrolyzed to a *cis*-diol [Eq. (14)][2]

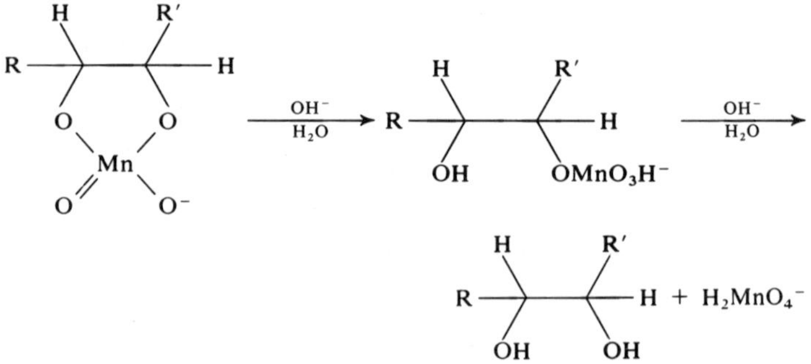

(14)

Convincing evidence for these mechanisms has been provided by the work of Ogino and Mochizuki.[76] They have reported the preparation of stable solutions of several manganate(V) diesters by reacting alkenes with permanganate in cold (0°–3°C) methylene chloride solutions while using

[76] T. Ogino and K. Mochizuki, *Chem. Lett.*, 443 (1979).

benzyltriethylammonium chloride as a phase transfer agent. The homogeneous dark brown solution of manganate(V) diester that formed was then treated with an aqueous solution to produce products. If the aqueous solution was basic, diols were formed; if it was acidic, dialdehydes were produced:

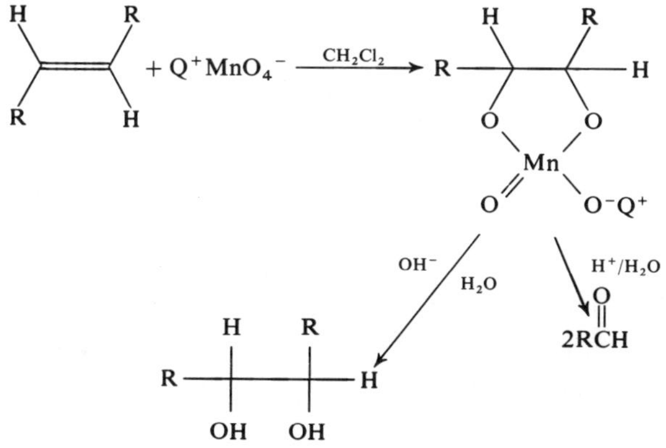

It is probable that the intermediate manganate(V) diester could be more stable under these conditions because it would be complexed with the quaternary ammonium ion and because the hydrolysis reaction would be suppressed in the absence of water.

This procedure has provided a simple route to a number of diols and dialdehydes that are difficult to prepare by other methods (see Tables XV and XVI). In particular it may be noted that it is possible to prepare 1,2-diols from terminal alkenes by use of this procedure.

α-Diols are also obtained from phase transfer assisted permanganate oxidations if the reactions are carried out in contact with an aqueous solution of sodium hydroxide.[59,77,78,79] In fact, the yields of diols are often better than for the corresponding reaction in water,[79] and because one is not limited to the use of water-soluble alkenes, a greater range of products is available.

In order for the reaction to be effective, it is probably necessary to use a phase transfer agent that is capable of transferring both the hydroxide ion and the permanganate ion into the organic phase. Several examples of such dihydroxylation reactions have been summarized in Table XV.

[77] T. Okimoto and D. Swern, *J. Am. Oil Chem. Soc.* **54**, 867A (1977).

[78] T. A. Foglia, P. A. Barr, and A. J. Malloy, *J. Am. Oil Chem. Soc.* **54**, 858A (1977).

[79] W. P. Weber and J. P. Shepherd, *Tetrahedron Lett.*, 4907 (1972).

TABLE XV

PHASE TRANSFER ASSISTED DIHYDROXYLATION OF ALKENES IN METHYLENE CHLORIDE

Substrate	Phase transfer agent	Aqueous phase	Products (% yield)	Ref.
Oleyl alcohol (*cis*-9-Octadecen-1-ol)	Tetrabutylammonium bromide	15–20% NaOH	*erythro*-1,9,10-Octadecanetriol (60–80)	77
Elaidyl alcohol (*trans*-9-Octadecen-1-ol)	Tetrabutylammonium bromide	15–20% NaOH	*threo*-1,9,10-Octadecanetriol (40–50)	77
9-Octadecene	Dicyclohexano-18-crown-6	40% NaOH	9,10-Octadecanediol (80)	78
Methyl oleate	Dicyclohexano-18-crown-6	40% NaOH	Methyl 9,10-dihydroxyoctadeconate (25)	78
cis-Cyclooctene	Benzyltriethylammonium chloride	40% NaOH	*cis*-1,2-Cyclooctanediol (50, 70)[a]	79, 76
trans-Cyclododecene	Benzyltriethylammonium chloride	40% NaOH	*trans*-1,2-Cyclooctanediol (50)	79
Cyclohexene	Benzyltriethylammonium chloride	40% NaOH	*cis*-1,2-Cyclohexanediol (15)	79
trans-5-Decene	Adogen 464	15% NaOH	5,6-Decanediol (20)	59
endo-Dicyclopentadiene	Triethylbenzylammonium chloride	3% NaOH[b]	5-*exo*,6-*exo*-Dihydroxy-*endo*-3a,4,5,6,7,7a-hexahydro-4,7-methanoindene (83)	76
Norbornene	Triethylbenzylammonium chloride	3% NaOH[b]	*exo*-*cis*-Bicyclo[2.2.1]heptane-2,3-diol (69)	76
trans-Stilbene	Triethylbenzylammonium chloride	3% NaOH[b]	Hydrobenzoin (26)	76
1-Octene	Triethylbenzylammonium chloride	3% NaOH[b]	1,2-Octanediol (80)	76

[a] The higher yield (80%) was obtained when the reaction intermediate was allowed to form under anhydrous conditions and then hydrolyzed with water.[76]

[b] The reaction was carried out under anhydrous conditions and then treated with 3% NaOH.[83]

Under neutral aqueous conditions, the intermediate manganate(V) diesters are hydrolyzed to the corresponding acyclic manganate(V) esters which undergo an oxidative decomposition giving the observed ketols [Eq. (15)].[71]

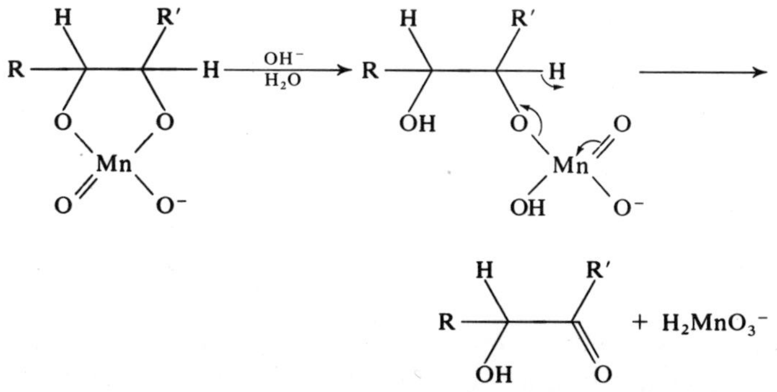

(15)

Alternatively the intermediate cyclic manganate(V) diester could be oxidized prior to hydrolysis and oxidative decomposition [Eq. (16)].

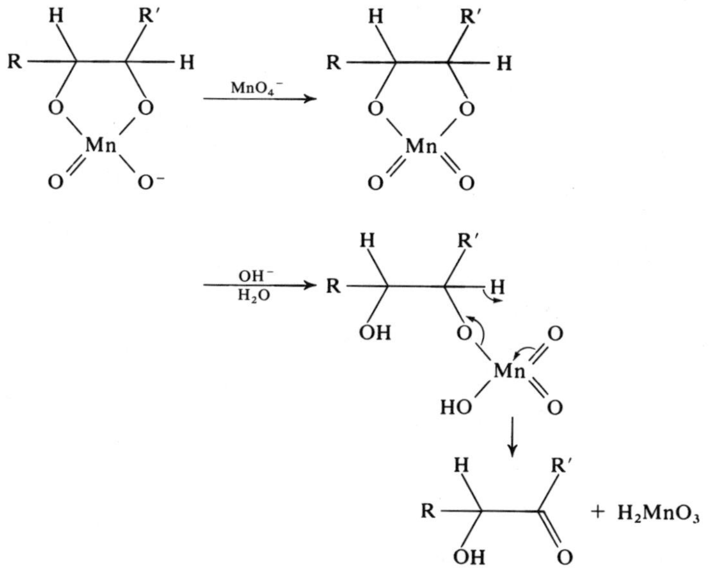

(16)

[80] N. S. Srinivasan and D. G. Lee, unpublished results.

[81] K. B. Sharpless, R. F. Lauer, O. Repic, A. Y. Teranishi, and D. R. Williams, *J. Am. Chem. Soc.* **93**, 3303 (1971); H. P. Jensen and K. B. Sharpless, *J. Org. Chem.* **39**, 2314 (1974).

[82] D. J. Sam and H. E. Simmons, *J. Am. Chem. Soc.* **94**, 4024 (1972).

[83] W. Rennie and D. G. Lee, unpublished results.

TABLE XVI

PREPARATION OF DIALDEHYDES FROM THE OXIDATION OF NONTERMINAL ALKENES[a]

Substrate	Quenching solution	Products (% yield)
endo-Dicyclopentadiene	Acetate buffer[b]	1β,5β-Bicyclo[3.3.0]oct-6-ene-2α,4α-dicarboxaldehyde (81)
Norbornene	Acetate buffer[b]	1,3-Diformylcyclopentane (63)
cis-Cyclooctene	1 M $HClO_4$	Octanedial (74)
trans-Stilbene	Acetate buffer[c]	Benzaldehyde (79)

[a] From Ogino and Mochizuki.[76]
[b] pH = 3.
[c] pH = 5.

Experiments in which the products obtained from the oxidation of *E*-5-decene in aqueous acetone were monitored by GLC, indicated that the diones were more rapidly formed from the oxidation of the ketols when the water content of the solvent was decreased.[77,80] Thus it is found that in the absence of water (e.g., in acetic anhydride solutions) diones are observed as the major products.[81]

It has also been found that diones can be obtained in good yield from the phase transfer assisted oxidation of nonterminal alkenes. For example, 5,6-decanedione and 1,2-cyclodecanedione have been obtained in 53% and 69% yields, respectively, from the oxidation of the corresponding alkenes under specific conditions.[59] If sufficient oxidant is used, oxidative cleavages accompany these reactions (Table XVII).

Phase transfer assisted permanganate reactions have also been used to oxidize polymers with unsaturated side chains. These compounds, formed from the copolymerization of terminal alkenes and nonconjugated dienes, are readily oxidized to oil-soluble polymers with polar pendent groups.[84] For example:

[84] U.S. Patent, 4,152,276 (1979). P. F. Jackisch, Ethyl Crop., Richmond, Virginia.

TABLE XVII

OXIDATIVE CLEAVAGES OF NONTERMINAL ALKENES

Substrate	Phase transfer agent	Solvent	Oxidant phase	Additives	Products (% yield)	Ref.
9-Octadecene	Tetrabutylammonium bromide	Methylene chloride	Aqueous		Nonanoic acid (80)	78
9-Octadecene	Cetyltrimethylammonium bromide	Benzene	Aqueous		Nonanoic acid (50) 9,10-Diketooctadecane (27) 9-Hydroxy-10-octadecanone (5)	78
trans-5-Decene	Dimethylpolyethylene glycol	Benzene	Solid	Acetic acid	Valeric acid (69) 5,6-Decanedione (28)	59
trans-2-Heptene	Dimethylpolyethylene glycol	Benzene	Solid	Acetic acid	Valeric acid (67)	70
Methyl oleate	Tetrabutylammonium bromide	Methylene chloride	Aqueous		Nonanoic acid (67) Monomethyl azelate (72)	78
Cyclohexene	Dicyclohexano-18-crown-6	Benzene	Solid		Adipic acid (100)	82
Cyclododecene	Adogen 464	Methylene chloride	Solid	Acetic acid	Dodecanedioic acid (27–83) 1,2-Cyclododeanedione (7–19) 2-Hydroxycyclododecanone (6–7)	59
α-Pinene	Dicyclohexano-18-crown-6	Benzene	Solid		Pinonic acid (90)	82
trans-Stilbene	Dicyclohexano-18-crown-6	Benzene	Solid		Benzoic acid (100)	82
trans-Stilbene	Tetrabutylammonium bromide	Benzene	Aqueous		Benzoic acid (95)	54
cis-Stilbene[a]	Tetrabutylammonium bromide	Pyridine	Solid		Benzoic acid (98)	43
trans-Stilbene	Dicyclohexano-18-crown-6	Benzene	Aqueous		Benzoic acid (92) Benzaldehyde (7)	83
Triphenylethylene	Adogen 464	Benzene	Solid		Benzophenone (50) Benzoic acid (17)	83

[a] Tetrabutylamonium permanganate, prepared from the reaction of aqueous permanganate with tetrabutylammonium bromide, was used in a pyridene solution.

If the diene is cyclic, oxidation gives two alkanoic acid groups on adjacent carbons, i.e.,

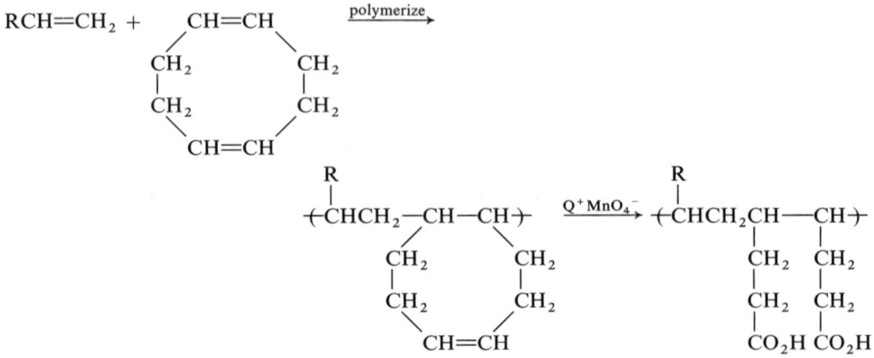

These products may be used as ashless dispersants and viscosity lift additives for lubricating oils.[84]

TYPICAL PROCEDURES

*Oxidation of Nonterminal Alkenes by Solid Potassium Permanganate.** Alkene (0.054 mol) was dissolved in 130 ml of solvent (methylene chloride or benzene) and 25 ml of acetic acid in a 500-ml, three-necked, round-bottomed flask equipped with a mechanical stirrer. About 3 g of phase transfer agent (Adogen 464, dicyclohexano-18-crown-6, or dimethyl poly-ethylene glycol) dissolved in 20 ml of solvent was added, followed by addition of powdered potassium permanganate (0.177 mol) in small portions for a period of 2 hr. An ice bath was used to maintain the temperature below 20°C. The mixture was then stirred vigorously overnight, cooled, and treated with 100 ml of water and 5 g of sodium bisulfite to reduce any excess oxidant. After 20 min the solution was acidified (concentrated HCl) and the manga-nese dioxide was reduced by addition of the required amount of sodium bisulfite in small portions. Any solid carboxylic acids that precipitated were collected by filtration, and the nonaqueous layer was separated. The aqueous layer was saturated with sodium chloride and extracted with 2 × 50 ml of ether. The combined organic layers were extracted with 2 × 100 ml of 5% sodium hydroxide solution to remove any additional carboxylic acids, dried over anhydrous magnesium sulfate, and concentrated by rotary evaporation. The resulting yellow oil could then be analyzed directly by GLC (when benzene and dimethylpolyethylene glycol had been used as solvent and phase transfer agent) or distilled under vacuum and then analyzed.

* Reprinted with permission from Lee and Chang, *J. Org. Chem.* **43,** 1532. Copyright (1978) American Chemical Society.

The precipitated carboxylic acids were dissolved in a 5% sodium hydroxide solution and combined with the basic solutions from previous extractions. The solution was filtered to remove any residual manganese dioxide, acidified with concentrated hydrochloric acid, and extracted with 2 × 250 ml of ether. This solution was dried over magnesium sulfate and evaporated to give purified carboxylic acids.

Oxidation of Nonterminal Alkenes by Aqueous Potassium Permanganate.† Alkene (0.054 mol) was dissolved in a mixture of methylene chloride (130 ml), acetic acid (25 ml), and water (100 ml) in a 500-ml, three-necked, round-bottomed flask equipped with a mechanical stirrer. About 3.5 g of phase transfer agent (Adogen 464 or dimethyl polyethylene glycol) dissolved in 20 ml of methylene chloride was added. The mixture was cooled in an ice bath and powdered potassium permanganate (0.177 mol) was added in small portions over a 1-hr period. The mixture was stirred vigorously for another 6 hr, cooled, and treated with 5 g of sodium bisulfite to reduce any excess oxidant. After 20 min the solution was acidified (concentrated HCl) and the manganese dioxide reduced by addition, in small portions, of the required amount of sodium bisulfite. The products were then isolated as described above.

Oxidation of endo-Dicyclopentadiene.** The oxidant solution was prepared by treating pulverized potassium permanganate (3.41 mmol) with triethylbenzylammonium chloride (3.41 mmol) in methylene chloride (40 ml) and added dropwise to a stirred solution of *endo*-dicyclopentadiene (2.27 mol) in methylene chloride (20 ml) at such a rate that the temperature was maintained at 0°–3°C under cooling with an ice bath (40–50 min). After addition was complete, stirring was continued until the permanganate ion was completely consumed (30–40 min). The homogeneous dark brown mixture was then treated with either acidic or basic aqueous solutions.

When the reaction mixture was treated with 3% NaOH solution (30 ml) under a nitrogen atmosphere at room temperature for 18 hr, a crystalline product, mp 47°–52°C, was obtained in 83% yield from the organic layer upon usual work-up. This product was identified as the *exo,cis*-diol from IR and NMR data. No other products were detected on TLC and GLC. On the other hand, when the reaction mixture was treated with an acetate solution (30 ml) at pH 3, the dialdehyde, mp 42°–44°C, was obtained in 81% yield as the single product (TLC, GLC).[76]

† Reprinted with permission from Lee and Chang, *J. Org. Chem.* **43**, 1532. Copyright (1978) American Chemical Society.

** Reprinted with permission from Ogino and Mochizuki,[76] courtesy of The Chemical Society of Japan.

IV. Oxidation of Alkynes

A. NONTERMINAL ALKYNES

Carbon–carbon triple bonds are also succeptable to oxidation by permanganate although the rate of reaction is somewhat slower when compared to the corresponding alkenes. Nonterminal alkynes containing substituents that impart some solubility in water are oxidized to diketones by neutral aqueous permanganate. A well-known example is the oxidation of stearolic acid which has been described by Khan and Newman[85] [Eq. (17)].

$$CH_3(CH_2)_7C{\equiv}C(CH_2)_7CO_2H \xrightarrow[pH = 7.5]{MnO_4^-} CH_3(CH_2)_7COCO(CH_2)_7CO_2H \quad (17)$$

On the other hand, nonsubstituted low molecular weight alkynes are cleaved by aqueous permanganate, whereas the higher molecular weight compounds are insoluble in water and thus unreactive. This is illustrated by the results summarized in Table XVIII.[58]

The use of phase transfer agents overcomes the solubility problems encountered with large alkynes and permits the preparation of diones in good yields.[58] Some examples are found in Table XIX.

It is of interest to note that the results presented in Table XIX indicate that alkynes undergo a nonsymmetrical cleavage when oxidized in organic solvents. For example, 8-hexadecyne, when it is cleaved, gives approximately equal amounts of octanoic and heptanoic acids rather than the expected two moles of octanoic acid.[58] This observation has been rationalized by assuming cleavage takes place by reaction of the intermediate dione

TABLE XVIII

THE OXIDATION OF ALKYNES BY AQUEOUS POTASSIUM PERMANGANATE[a]

Alkyne	Recovered substrate (%)	Products (% yield)
8-Hexadecyne	94	—
7-Tetradecyne	91	—
1-Phenyl-1-pentyne	90	Benzoic acid (8)
		1-Phenyl-1,2-pentanedione (trace)
5-Decyne	56	Valeric acid (19)
		5,6-Decanedione (trace)
4-Octyne	6	Butyric acid (73)
		Propionic acid (6)

[a] From Lee and Chang.[58]

[85] N. A. Khan and M. S. Newman, *J. Org. Chem.* **17**, 1063 (1952).

TABLE XIX

THE PHASE TRANSFER ASSISTED OXIDATION OF ALKYNES BY PERMANGANATE ION IN METHYLENE CHLORIDE[a]

Alkyne	Oxidant[b] phase	Products (% yield)
Diphenylacetylene	Aqueous	Benzil (93)
		Benzoic acid (2)
1-Phenyl-1-pentyne	Solid	1-Phenyl-1,2-pentanedione (81)
1-Phenyl-1-hexyne	Aqueous	1-Phenyl-1,2-hexanedione (41)
		Benzoic acid (32)
		Butyric acid (27)
7-Tetradecyne	Aqueous	7,8-Tetradecanedione (54)
		Heptanoic acid (25)
		Hexanoic acid (17)
8-Hexadecyne	Solid	8,9-Hexadecanedione (80)
		Octanoic acid (5)
		Heptanoic acid (trace)

[a] From Lee and Chang.[58]
[b] The phase transfer agent used was Adogen 464.

in its enol form [Eq. (18)]. Such a suggestion is not unreasonable since permanganate is known to react readily with carbon–carbon double bonds and it has been established that diones are intermediates in the cleavage reaction.[58]

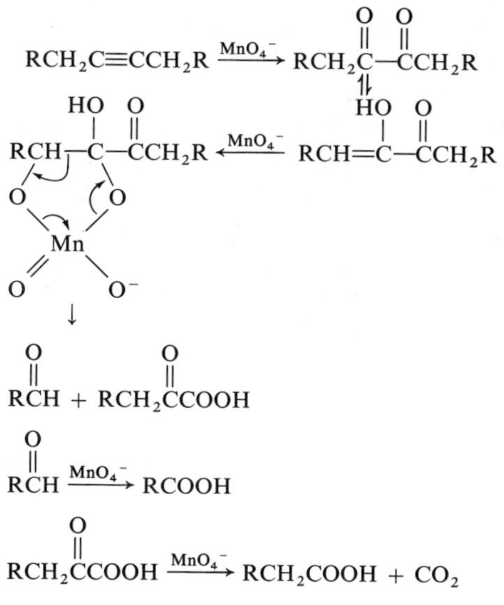

(18)

Typical Experimental Procedures

Preparation of 1-Phenyl-1,2-pentanedione by Oxidation of 1-Phenyl-1-pentyne.†† A 200-ml Erlenmeyer flask, equipped with a reflux condenser, was charged with dichloromethane (100 ml), acetic acid (5 ml), and 1-phenyl-1-pentyne (2.0 g, 0.014 mol). The solution was stirred magnetically and heated to reflux temperature before powdered potassium permanganate[57] (5.85 g, 0.037 mol) and the phase transfer agent (Adogen 464, 1.6 g) were added. After being stirred vigorously for 4 hr the precipitated manganese dioxide was collected and washed with dichloromethane (2 × 50 ml). Residual manganese dioxide was reduced by addition (to the combined filtrates) of 20% hydrochloric acid (40 ml) followed by small portions of sodium hydrogen sulfite until all of the brown color had disappeared. The organic phase was separated, washed with water, and dried with anhydrous magnesium sulfate. Most of the solvent was removed using a rotary evaporator and the remaining yellow oil distilled under vacuum to give unreacted starting material (0.14 g) and 1-phenyl-1,2-pentanedione; yield: 1.98 g (81%); bp 108°–110°C/5.5 torr.

*Preparation of 8,9-Hexadecanedione by Oxidation of 8-Hexadecyne.** Potassium permanganate (5.85 g, 0.037 mol) was dissolved in water (100 ml) in a 500-ml Erlenmeyer flask and a solution consisting of 8-hexadecyne (2.0 g, 0.009 mol), dichloromethane (100 ml), acetic acid (5 ml), and phase transfer agent (Adogen 464, 1.5 g) was added. The solution was stirred magnetically and heated under reflux for 6 hr. After cooling, sodium hydrogen sulfite (2 g) was added to reduce any unreacted permanganate. After 15 min the solution was acidified (conc. hydrochloric acid) and the precipitated manganese dioxide reduced by addition, in small portions, of the required amount of sodium hydrogen sulfite. The aqueous phase was separated, saturated with sodium chloride, and extracted with dichloromethane (3 × 75 ml). The organic layers were combined and extracted with 5% aqueous sodium hydroxide (3 × 75 ml), dried over anhydrous magnesium sulfate, and concentrated by rotary evaporation to give a yellow liquid which solidified on cooling. This yellow solid was recrystallized from methanol (15 ml) to give 8,9-hexadecanedione; yield: 1.55 g (68%); mp 51°–52°C.

†† Reprinted with permission from Lee and Chang[86], courtesy of Thieme, Stuttgart.

* Reprinted with permission from Lee and Chang[86], courtesy of Thieme, Stuttgart.

[86] D. G. Lee and V. S. Chang, *Synthesis*, 462 (1978).

B. TERMINAL ALKYNES

Terminal alkynes are smoothly oxidized to the corresponding carboxylic acids under phase transfer conditions. The yield is slightly reduced by over oxidation,[68] which probably arises from the oxidation of enols as described in the previous section. The reaction seems most applicable for high molecular weight alkynes that are not soluble in water. Krapcho *et al.*[68] have shown that 1-hexyne reacts quantitatively with aqueous permanganate; however, the yield drops off progressively with 1-octyne and 1-decyne, presumably because of their decreased solubilities in water.

Some examples of the phase transfer assisted reaction are presented in Table XX.

A Typical Experimental Procedure

*Preparation of Heptanoic Acid from Oxidation of 1-Octyne.** Potassium permanganate (28 g, 0.18 mol) and 200 ml of tap water were placed into a 1-l round-bottomed flask fitted with a 1-in. egg-shaped spinbar. The mixture was stirred and immersed in an ice bath. A solution of 1-octyne (5.0 g, 0.045 mol), 120 ml of pentane, 60 ml of acetic acid, and 0.2 g of Aliquat 336 was added in one portion. The mixture is stirred for 5 hr without replenishing the ice. The black–brown mixture was cooled in an ice bath and

TABLE XX

Phase Transfer Assisted Oxidation of Terminal Alkynes by Permanganate[a,b]

Alkyne	Phase transfer agent	Solvent	Oxidant phase	Products (% yield)
1-Hexyne	Aliquat 336	Pentane	Aqueous	Pentanoic acid (99) Butanoic acid (1)
1-Octyne	Aliquat 336	Pentane	Aqueous	Heptanoic acid (88) Hexanoic acid (2)
1-Decyne	Aliquat 336	Pentane	Aqueous	Nonanoic acid (76) Octanoic acid (3)
Phenylacetylene	Dimethylpolyethylene glycol	Methylene chloride	Solid	Benzoic acid (91)

[a] From Lee and Chang.[58]
[b] From Krapcho *et al.*[68]

* Reprinted with permission from Krapcho *et al.*, *J. Org. Chem.* **42**, 3749. Copyright (1977) American Chemical Society.

Na_2SO_3 (30 g) was added in several portions. A solution of 60 ml of concentrated HCl in 60 ml of water was then cautiously added. The top pentane layer was separated and extracted once with 50 ml of cold water, dried over Na_2SO_4, decanted from the drying agent, and concentrated on a rotary evaporator to yield 5.4 g of crude product (90% recovery). Vacuum distillation yielded 4.1 g (70%) of heptanoic acid (98% pure by GLC of the methyl esters). Trace amounts of short retention-time impurities were also present.

V. Oxidation of Arenes

There are numerous examples in the literature of the use of aqueous permanganate for the oxidation of toluene derivatives to the corresponding benzoic acids.[6] The evidence indicating that these reactions are likely initiated by hydrogen atom abstraction at the benzylic position has been reviewed in a previous volume in this series.[2]

Under phase transfer conditions it has been observed that it is possible to oxidize toluene and xylene when they are used as solvents for the reaction.[82] However, the phase transfer assisted oxidation of toluene dissolved in methylene chloride gave a very low yield of benzoic acid.[70] Similarly it was observed that 1,2-diphenylethane and acenaphthene were not oxidized when treated in a methylene chloride solution with potassium permanganate and Adogen 464.[70]

On the other hand, Sala and Sargent[43] have reported that *p*-nitrotoluene is oxidized to *p*-nitrobenzoic acid by tetrabutylammonium permanganate in pyridine at 65°C. In this procedure, tetrabutylammonium permanganate was first prepared and isolated by reacting tetrabutylammonium bromide with aqueous potassium permanganate. The resulting precipitate was collected and purified by crystallization from a dichloromethane–benzene solution before being used as an oxidant. This process, which has also been applied to the oxidation of alkenes, alcohols and aldehydes provides an important new oxidation procedure. It should be noted, however, that some caution must be exercised when handling solid quaternary ammonium permanganates; on at least one occasion tetrabutylammonium permanganate was observed to undergo a violent spontaneous ignition while being transferred from a bottle to weighing paper.[45]

In a similar series of experiments Schmidt and Schafer[44] have described the preparation of benzyltriethylammonium permanganate and its use for the oxidation of arenes in either methylene chloride or glacial acetic acid. Their results may be summarized by the following equations where the yields given are based on the amount of arene consumed. (The extent of conversion, based on the amount of arene initially present, was approximately 29%, 16%, 49%, and 70%, respectively, for these four reactions.)

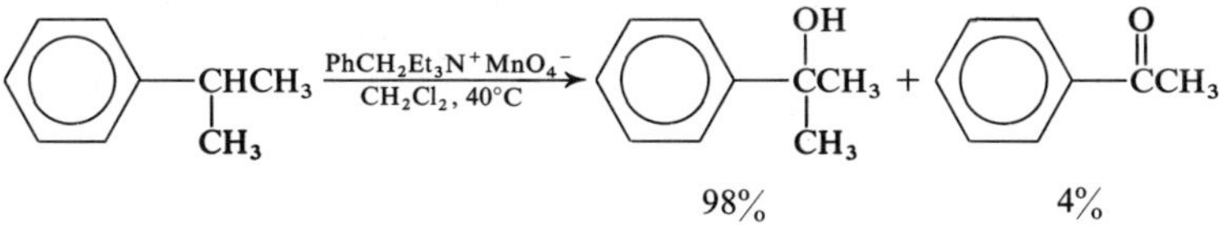

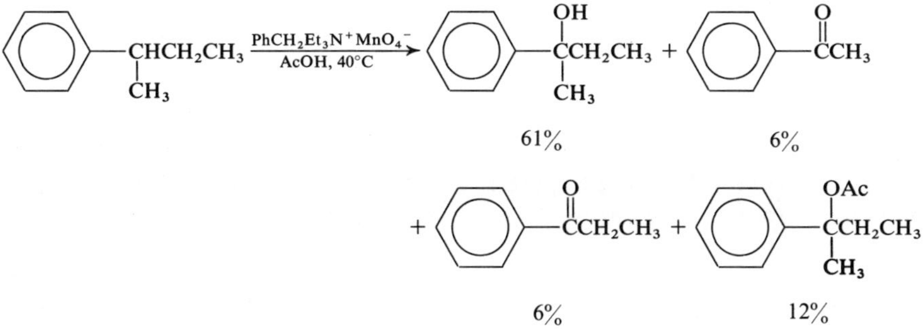

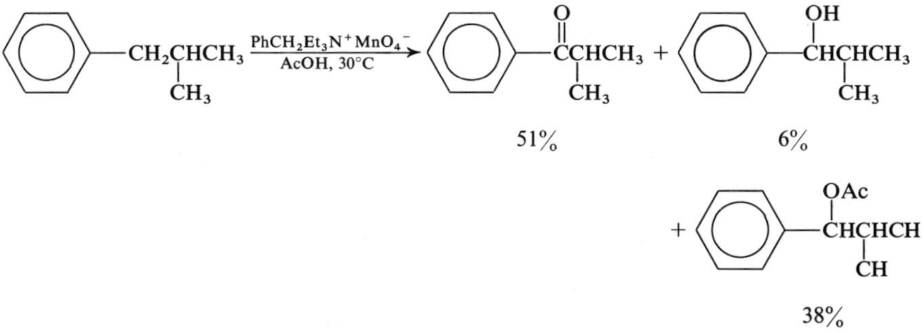

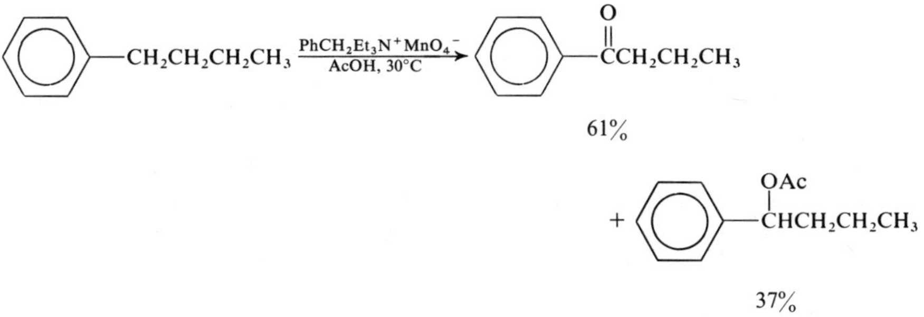

These products suggest that the reaction proceeds initially to the corresponding benzyl alcohol. If the initial product is a secondary alcohol, it may be further oxidized to the corresponding phenyl ketone or react with acetic acid to give an ester as in the following reaction.

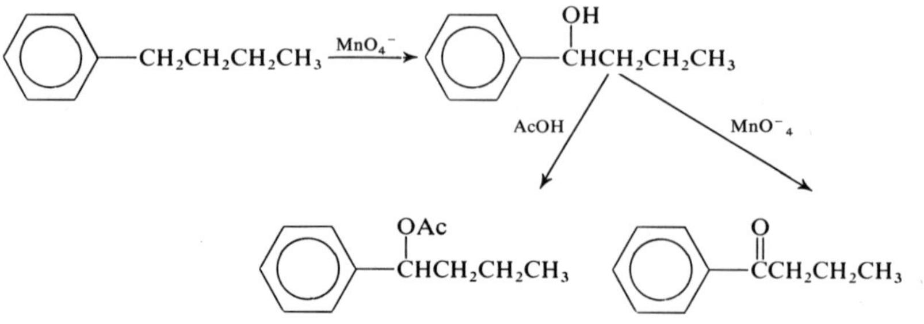

If the initial product is a tertiary alcohol, it may undergo dehydration followed by oxidative cleavage of the carbon–carbon double bond as well as ester formation. For example, the products obtained from *sec*-butylbenzene may be formed as in the following reaction sequence.

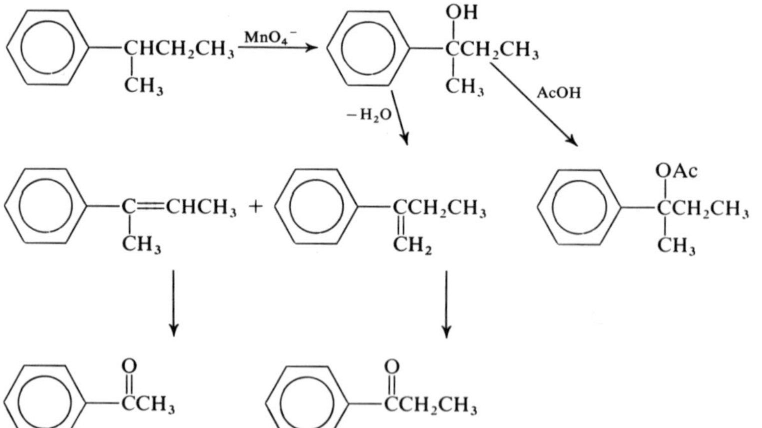

Sangaiah and Krishna Rao[86a] have used a modification of the procedure developed by Schmidt and Schafer[44] to oxidize the side chains of arenes bearing alkoxy and acetoxy substituents. When compounds such as 4-acetoxytoluene were treated with a suspension of benzyltriethylammonium permanganate (1 part) and potassium permanganate (10 parts) in glacial acetic acid the corresponding acetoxybenzoic acids were obtained in moderate yields (25–50%). Under similar conditions benzylic methylene groups

[86a] R. Sangaiah and G. S. Krishna Rao, *Synthesis*, 1018 (1980).

were converted into carbonyl groups as in the syntheses of espeleton shown below.

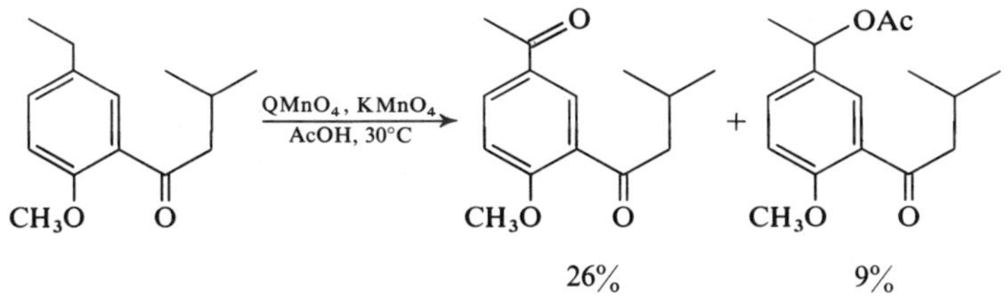

The formation of a benzyl acetate as a side product suggests that this reaction (as well as those reported by Schmidt and Schafer) may proceed via a benzyl carbonium ion intermediate. However, Sangaiah and Krishna Rao also reported that the reaction of thymol acetate (below) under similar conditions results in oxidation at the primary rather than the tertiary benzyl carbon (which would be expected to give a more stable carbonium ion). The results are unfortunately inconclusive because of the lack of a mass balance for the reaction; i.e., 72% of the starting material was unaccounted for.

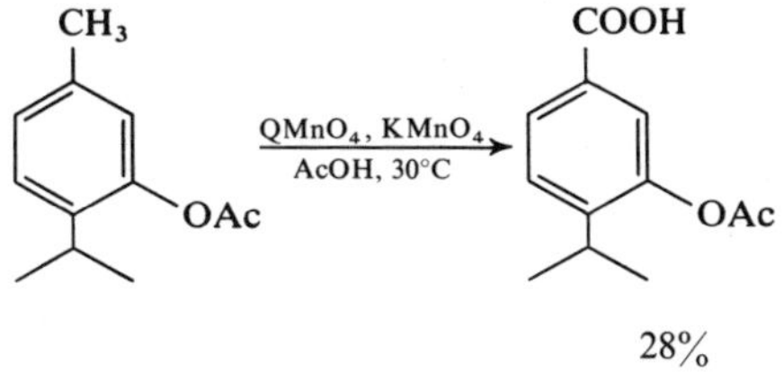

Phase transfer assisted oxidations (not involving prior preparation and isolation of quaternary ammonium permanganates) were found to be useful for the preparation of diaryl ketones from the corresponding diarylmethanes.[70] As the examples summarized in Table XXI indicate this reaction is particularly appropriate for the oxidation of a number of compounds that are not readily soluble in aqueous solutions.

TYPICAL EXPERIMENTAL PROCEDURE

Preparation of 1-Tetralone by Oxidation of 1,2,3,4-Tetrahydronaphthalene.[70] A 500-ml Erlenmeyer flask, equipped with a reflux condenser, was charged with methylene chloride (150 ml), acetic acid (10 ml), sulfuric acid (10 ml), water (100 ml), 1,2,3,4-tetrahydronaphthalene (10.0 g, 0.076 mol), and Adogen 464 (2.0 g). The solution was stirred magnetically and heated at reflux temperature for 2 hr while potassium permanganate (25.0 g, 0.16 mol) was added in small portions. After cooling, sufficient sodium

TABLE XXI

PHASE TRANSFER ASSISTED OXIDATION OF DIARYLMETHANES BY AQUEOUS POTASSIUM PERMANGANATE[a]

Substrate[b]	Products (% yield)
Diphenylmethane	Benzophenone (80–85%)
4-Benzylpyridine	4-Benzoylpyridine (70–80%)
Fluorene	9-Fluorenone (80–82%)
9,10-Dihydroanthracene	Anthraquinone (96%)
Tetralin	Tetralone (70–75%)
Indane	Indanone (70–75%)

[a] From Chang and Lee.[70]

[b] The phase transfer agent was Adogen 464, the solvent was methylene chloride containing 5% acetic acid, the oxidant was used as an aqueous solution containing 10% H_2SO_4.

hydrogen sulfite to reduce the precipitated manganese dioxide was gradually added. The aqueous phase was separated, saturated with sodium chloride, and extracted with methylene chloride (3 × 50 ml). The organic layer was extracted with a 5% sodium hydroxide solution (2 × 50 ml). The organic layers were combined, dried over anhydrous magnesium sulfate, and concentrated by rotary evaporation to give a yellow liquid. Distillation of this product under vacuum (6 torr) gave unreacted starting material (1.0 g) and 1-tetralone (7.7 g, 0.053 mol, 70%), bp 113°–116°C.

VI. Oxidation of Alkanes

Schmidt and Schafer[44] have reported the products obtained from the oxidation of several alkanes by benzyltriethylammonium permanganate in glacial acetic acid at 30°–60°C. The results are summarized in the following reactions. (The yields given here are based on the amount of alkane consumed; conversions based on the amount of alkane used were 25–40%.)

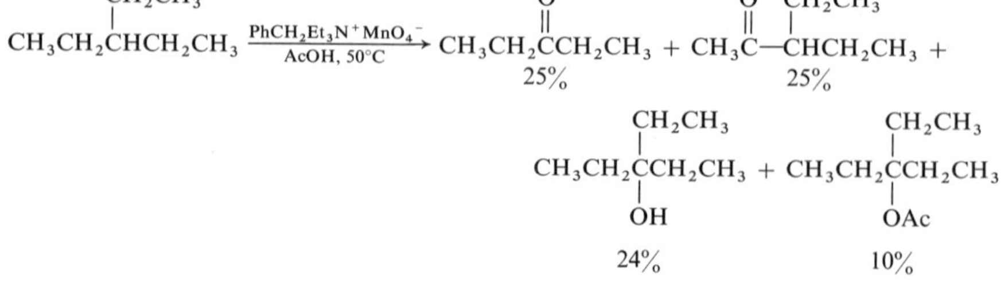

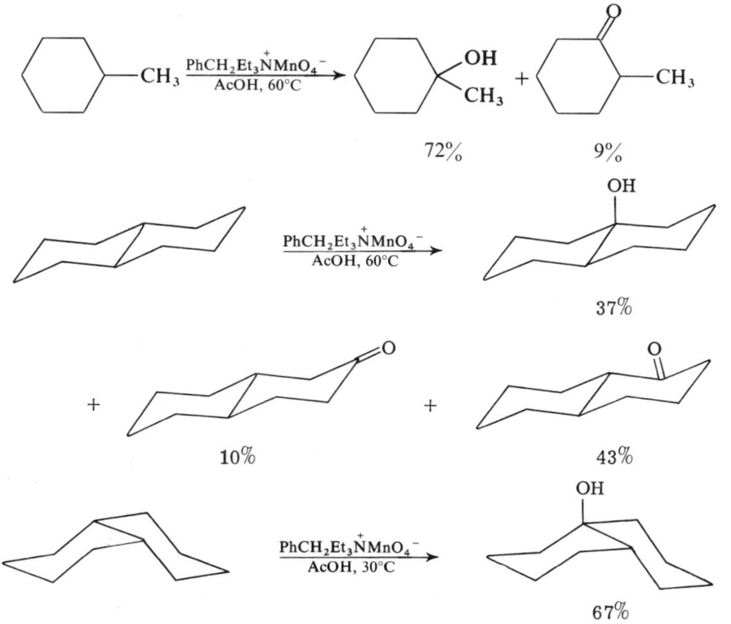

These results clearly indicate that tertiary carbon–hydrogen bonds are more easily cleaved than secondary carbon–hydrogen bonds. When statistical factors are taken into account, the former appear to be 15–25 times as reactive under these conditions.

The initial products in each of these reactions appear to be either secondary or tertiary alcohols. For example, the products obtained from 3-ethylpentane may be produced in the following sequence of reactions.

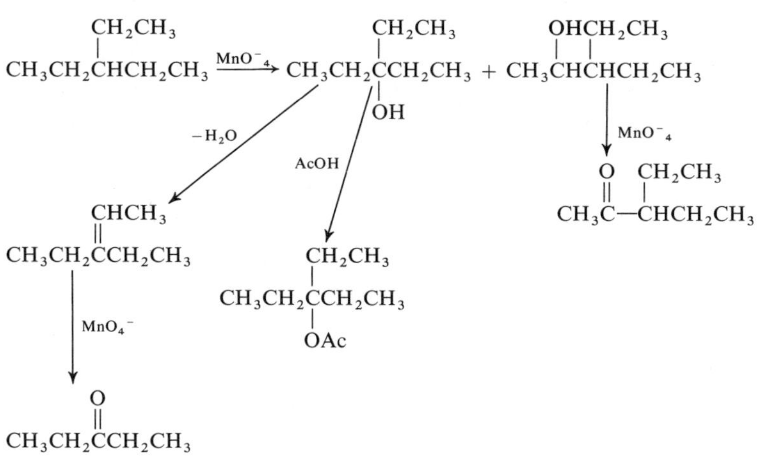

Leddy, McKervey and McSweeney[87] have noted that the stability of quaternary ammonium permanganates can be increased by adsorbing them on alumina. For example, they have reported the hydroxylation of diamantane and triamantane by benzyltriethylammonium permanganate dispersed on alumina. As illustrated by the following reaction, oxidation was observed to take place preferentially at the tertiary positions.

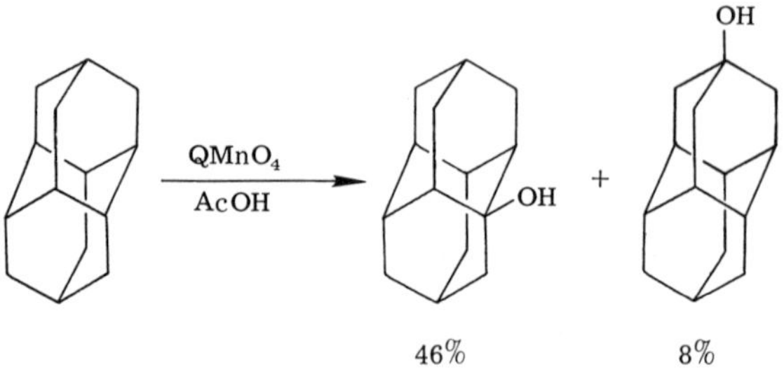

VII. Oxidation of Polycyclic Aromatic Hydrocarbons

Although the yield of phthalic acid that may be obtained from the oxidation of naphthalene under phase transfer conditions is only 20%, anthracene and phenanthracene are converted to the corresponding quinones in about 75% yields.[70]

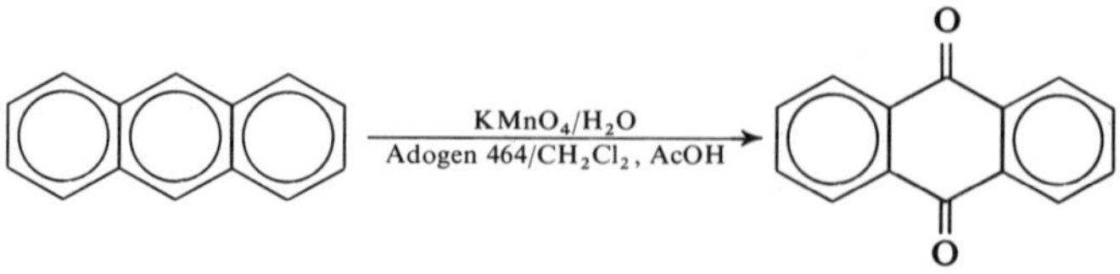

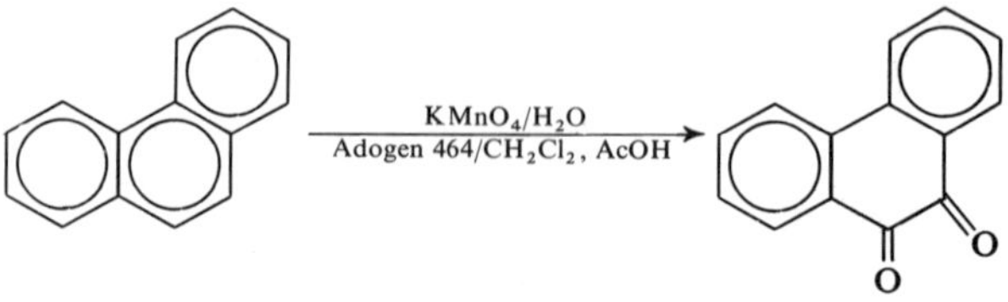

TYPICAL EXPERIMENTAL PROCEDURE

Preparation of 9,10-Phenanthraquinone by Oxidation of Phenanthrene.[70] A 500-ml Erlenmeyer flask, equipped with a reflux condenser, was charged with methylene chloride (150 ml), acetic acid (5 ml), sulfuric acid (10 ml), water (100 ml), phenanthrene (5.0 g, 0.27 mol), and Adogen 464 (1.5 g).

[87] B. P. Leddy, M. A. McKervey, and P. McSweeney, *Tetrahedron Lett.*, 2261 (1980).

Powdered potassium permanganate (28.0 g, 0.18 mol) was added slowly with stirring and the solution was gently refluxed for 4 hr. The solution was cooled in an ice bath and the precipitated manganese dioxide was reduced by addition of the required amount of sodium hydrogen sulfite in small portions. The aqueous phase was separated, saturated with sodium chloride, and extracted with ether (2 × 40 ml). The organic layers were combined, dried over anhydrous magnesium sulfate, and concentrated by rotary evaporation to give a pale yellow liquid which solidified on cooling. It was further purified by recrystallization from a water–acetone solution to give 9,10-phenanthraquinone (4.2 g, 0.02 mol, 75%) mp 206°–208°C.

VIII. Oxidation of Alcohols

Alcohols are readily oxidized by permanganate ion in both aqueous and nonaqueous solutions. The use of phase transfer agents to solubilize the oxidant in nonpolar solvents is particularly useful for high molecular weight compounds. The employment of volatile organic solvents also simplifies product isolation somewhat.

Table XXII contains several examples of the oxidation of both primary and secondary alcohols. As in the corresponding aqueous reactions, primary alcohols were found to yield carboxylic acids whereas ketones were obtained from secondary alcohols.

Solid supports have also been used to assist the oxidation of alcohols by permanganate. The supports include molecular sieve,[64] alumina,[64] and hydrated copper sulfate.[64a] There is good evidence that the presence of moisture on the solid supports is essential if a high yield of product is to be realized. When the solid supports were carefully dried over phosphorous pentoxide before use, the yields decreased markedly.[64a]

Furthermore, it was shown that the low yields obtained from the oxidation of primary alcohols is due to a retardation of the reaction by carboxylic acids (which are formed from the oxidation of primary alcohols).[64a]

Table XXIII contains several examples of the use of this procedure.

Typical Experimental Procedure

Preparation of Hexadecanoic Acid by Oxidation of 1-Hexadecanol.[70] A 500-ml Erlenmeyer flask equipped with a reflux condenser was charged with methylene chloride (150 ml), acetic acid (5 ml), 1-hexadecanol (4.0 g, 0.017 mol), water (100 ml), and Adogen 464 (1.0 g). Powdered potassium permanganate (5.0 g, 0.032 mol) was added in small portions and the solution stirred under gentle reflux for 4 hr. The solution was cooled and

TABLE XXII

THE PHASE TRANSFER ASSISTED OXIDATION OF ALCOHOLS

Substrate	Phase transfer agent	Solvent	Oxidant phase	Additives	Products (% yield)	Ref.
Primary alcohols						
Benzyl alcohol	Tetrabutylammonium bromide	Benzene	Aqueous		Benzoic acid (92)	54
Benzyl alcohol	Tetrabutylammonium bromide	Pyridine	Solid		Benzoic acid (98)	43
Benzyl alcohol	Dicyclohexano-18-crown-6	Benzene	Solid		Benzoic acid (100)	82
1-Heptanol	Dicyclohexano-18-crown-6	Benzene	Solid		Heptanoic acid (70)	82
1-Octanol	Tetrabutylammonium bromide	Benzene	Aqueous		Octanoic acid (47)	54
1-Decanol	Adogen 464	Methylene chloride	Aqueous	Acetic acid	Decanoic acid	70
1-Dodecanol	Adogen 464	Methylene chloride	Aqueous	Acetic acid	Dodecanoic acid (92)	70
1-Tridecanol	Adogen 464	Methylene chloride	Aqueous	Acetic acid	Tridecanoic acid (92)	70
1-Hexadecanol	Adogen 464	Methylene chloride	Aqueous	Acetic acid	Hexadecanoic acid (95)	70
1-Docosanol	Adogen 464	Methylene chloride	Aqueous	Acetic acid	Docosanoic acid (87)	70
Secondary alcohols						
2-Propanol	Triphenylmethylarsonium chloride	Chloroform	Aqueous		Acetone (100)	88
5-Decanol	Adogen 464	Methylene chloride	Aqueous	Acetic acid	5-Decanone	80
l-Menthol	Adogen 464	Methylene chloride	Aqueous	Acetic acid	Menthone (92)	70

Benzhydrol	Tetrabutylammonium permanganate	Pyridine	Solid		Benzophenone (97)	43
Benzhydrol	Dicyclohexano-18-crown-6	Benzene	Solid		Benzophenone (100)	82
Benzhydrol	Adogen 464	Methylene chloride	Aqueous	Acetic acid	Benzophenone (93)	70
Unsaturated alcohols						
$C_6H_5CH{=}CHCH_2OH$	Adogen 464	Methylene chloride	Aqueous	Acetic acid	C_6H_5COOH (72)	70
$C_6H_5{\equiv}CCH_2CH_2OH$	Adogen 464	Methylene chloride	Aqueous	Acetic acid	C_6H_5COOH (79)	70
$CH_3(CH_2)_7CH{=}CH(CH_2)_8OH$	Adogen 464	Methylene chloride	Solid	Acetic acid	$CH_3(CH_2)_7COOH$ (62) $HOOC(CH_2)_7COOH$ (66)	70
$CH_3(CH_2)_7{\equiv}C(CH_2)_8OH$	Adogen 464	Methylene chloride	Solid	Acetic acid	$CH(CH_2)_7C{\equiv}C(CH_2)_7COOH$ (73)	70
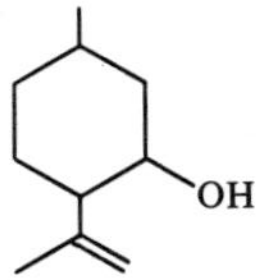	Adogen 464	Methylene chloride	Aqueous	Acetic acid	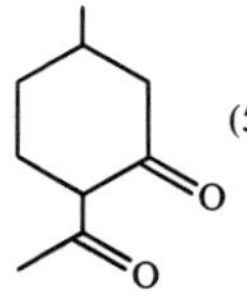 (53)	70
$CH_3(CH_2)_2CH(OH)(CH_2)_8CH{=}CH(CH_2)_7CH_3$	Adogen 464	Methylene chloride	Solid	Acetic acid	$CH_3(CH_2)_7COOH$ (19) $CH_3(CH_2)_2C({=}O)(CH_2)_8COOH$ $CH_3(CH_2)_2C({=}O)(CH_2)_8CH(OH){-}C({=}O)(CH_2)_7CH_3$ (50)	70

TABLE XXIII

OXIDATION OF ALCOHOLS USING SOLID SUPPORTS

Substrate	Solid support	Product (% yield)[a]	Ref.
2-Octanol	Molecular sieve	2-Octanone (82)[b]	64
2-Octanol	Alumina	2-Octanone (100)	64
2-Octanol	$CuSO_4 \cdot 5\ H_2O$	2-Octanone (96)	64a
2-Hexadecanol	$CuSO_4 \cdot 5\ H_2O$	2-Hexadecanone (84)[b]	64a
1-Cyclohexylethanol	$CuSO_4 \cdot 5\ H_2O$	Methyl cyclohexyl ketone (96)	64a
3-Methylcyclohexanol	$CuSO_4 \cdot 5\ H_2O$	3-Methylcyclohexanone (97)	64a
Cycloheptanol	Molecular sieve	Cycloheptanone (94)	64
Cyclooctanol	Molecular sieve	Cyclooctanone (87)[b]	64
Cyclododecanol	Molecular sieve	Cyclododecanone (90)[b]	64
Cyclododecanol	Alumina	Cyclododecanone (95)[b]	64
Cholestanol	Molecular sieve	Cholestanone (91)[b]	64
Cholestanol	$CuSO_4 \cdot 5\ H_2O$	Cholestanone (91)[b]	64a
Benzhydrol	Molecular sieve	Benzophenone (100)	64
Benzhydrol	$CuSO_4 \cdot 5\ H_2O$	Benzophenone (100)	64a
Ethyl lactate	$CuSO_4 \cdot 5\ H_2O$	Ethyl pyruvate (73)	64a
Benzyl alcohol	Molecular sieve	Benzaldehyde (80)	64
Cinnamyl alcohol	Molecular sieve	Cinnamaldehyde (94)[b]	64
1-Hexanol	Molecular sieve	Hexanal (29)	64
1-Octanol	$CuSO_4 \cdot 5\ H_2O$	Octanal (20)	64a
1-Octanol	Molecular sieve	Octanal (26)	64
1-Decanol	Molecular sieve	Decanal (26)	64
1-Dodecanol	Molecular sieve	Dodecanal (34)	64

[a] Yield determined by GLC unless indicated otherwise.
[b] Isolated yield.

the precipitated manganese dioxide was reduced by addition of the required amount of sodium hydrogen sulfite. The aqueous phase was separated, saturated with sodium chloride, and extracted with ether (2 × 40 ml). The organic layers were combined, dried over anhydrous magnesium sulfate, and concentrated by rotary evaporation to give a solid residue. The product was purified by recrystallization from aqueous ethanol (2 × 50 ml) to give hexadecanoic acid (4.0 g, 0.016 mol, 95%) mp 62°–63°C.

IX. Oxidation of Phenols

Phenols readily reduce permanganate in both aqueous and nonaqueous solutions. Although this is an important reaction in some water purification processes,[89] its application to organic syntheses is not extensive.

[88] N. A. Gibson and J. W. Hosking, *Aust. J. Chem.* **18**, 123 (1965).
[89] J. T. O'Connor and K. Y. Baliga, *J. Sanit Eng. Div. Am. Soc. Civ. Eng.* **96** (SA6), 1377 (1970).

The products obtained appear to be formed in oxidative polymerizations and/or coupling reactions. As the following reactions indicate, the amounts of coupling and polymerization obtained from 2,6-dialkylphenols is controlled by the size of the alkyl groups.[70]

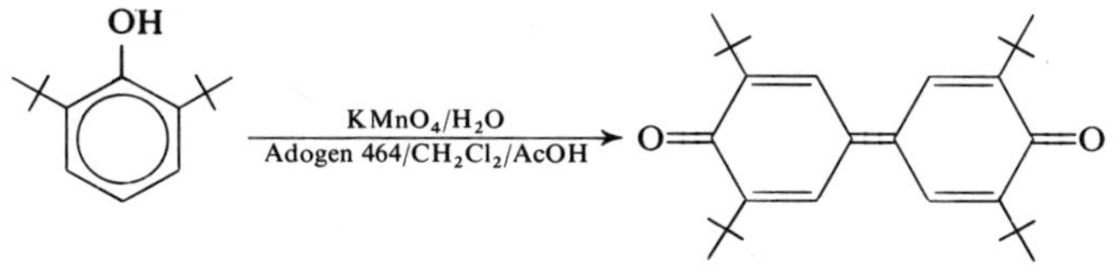

81%

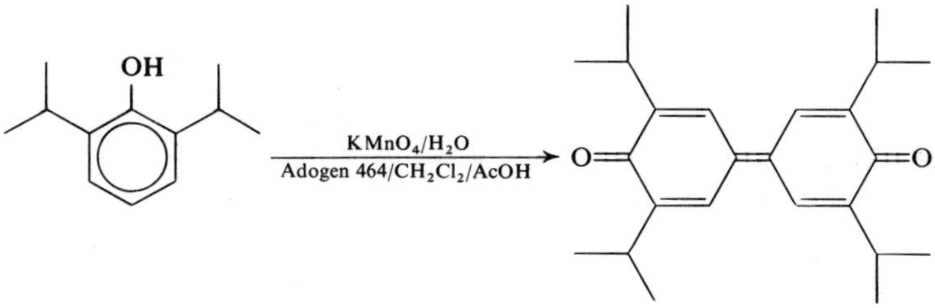

37%

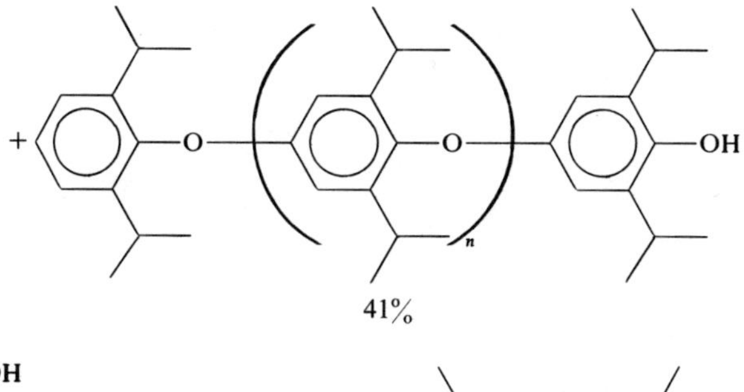

41%

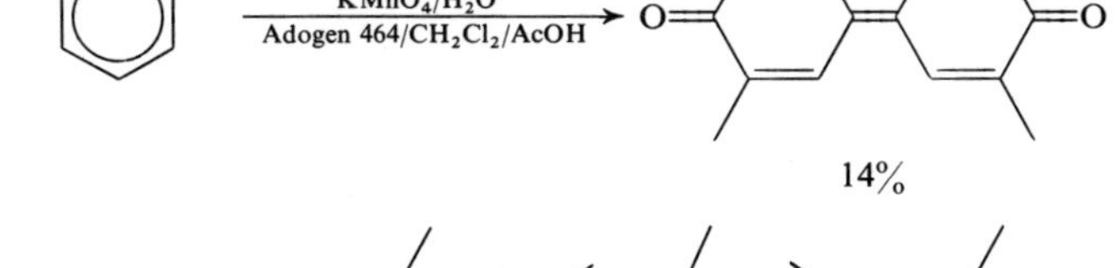

14%

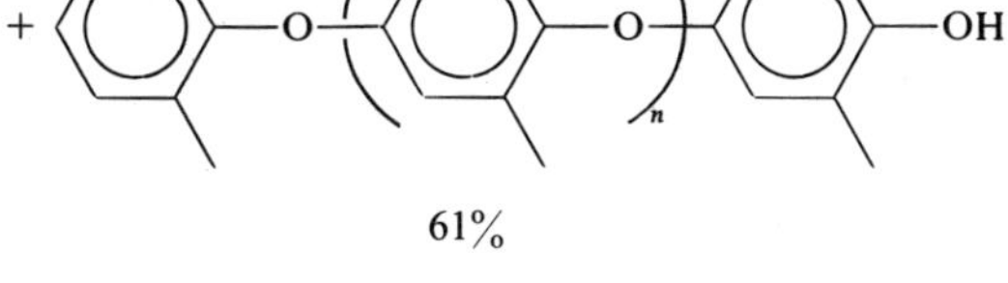

61%

3,5-Di-*tert*-butylcatechol can be converted into 3,5-di-*tert*-butyl-*o*-quinone in excellent yields in methylene chloride using either 18-crown-6 or Adogen 464 as catalysts.[70,90]

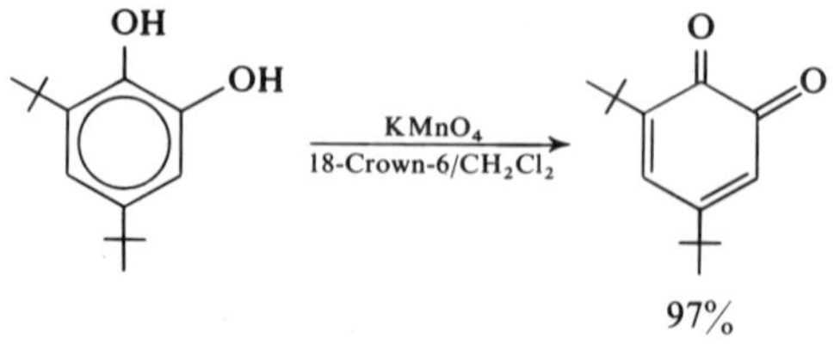

In aqueous solutions it has been found that the reaction of phenolates with permanganate ion proceeds by a mechanism in which the first step involves electron transfer with formation of a phenoxy free radical.[91,92]

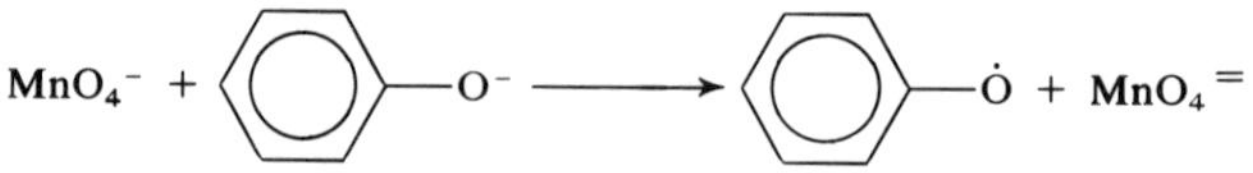

Although mechanism studies have not been reported for the phase transfer assisted oxidations, the products obtained may be accounted for if formation of a phenoxy free radical is assumed to be the initial reaction in nonpolar solvents as well. For example, the oxidation of 2,6-dimethylphenol may be visualized as proceeding as follows:

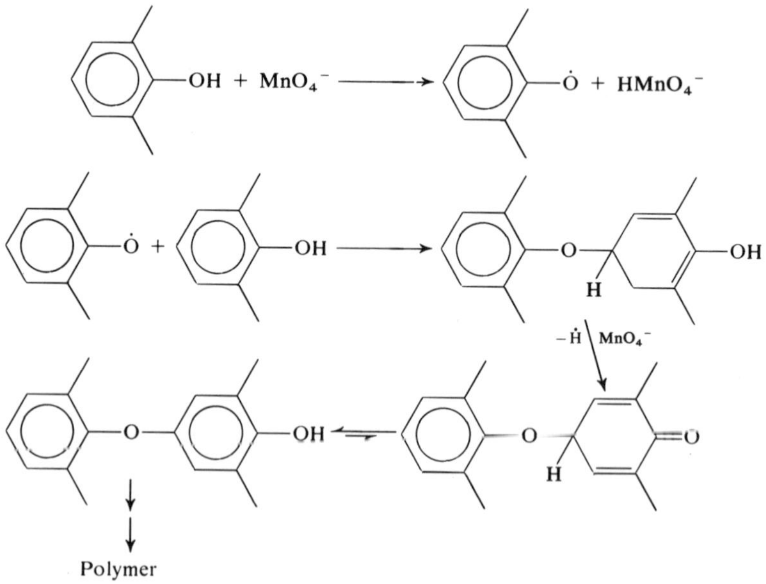

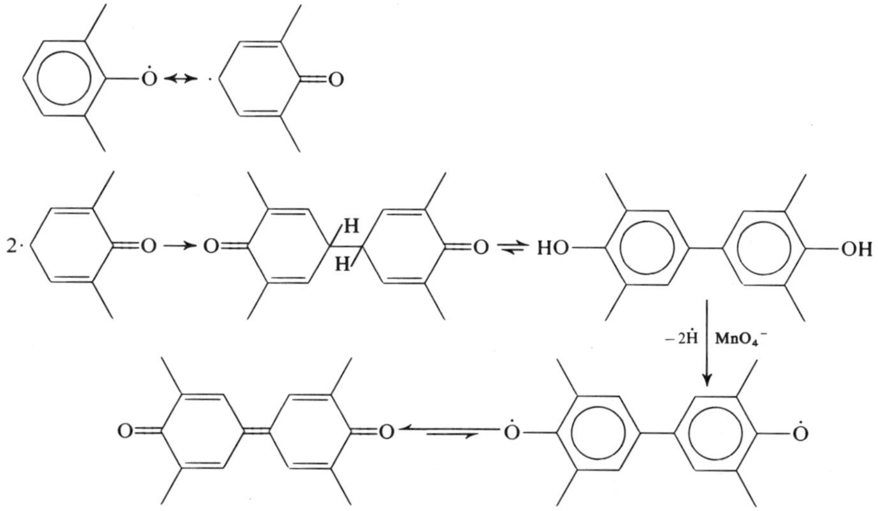

X. Oxidation of Ethers

Schmidt and Schafer[93] have reported that aliphatic ethers can be oxidized by benzyltrimethylammonium permanganate in methylene chloride solutions. The reaction, which involves conversion of an α-methylene into a carbonyl group, results in formation of an ester. In many respects the reaction is similar to the oxidation of ethers by ruthenium tetroxide.[94] However, this procedure has an advantage over the use of ruthenium tetroxide because aromatic rings are not as readily degraded by benzyltriethylammonium permanganate.[93] The results reported by Schmidt and Schafer have been summarized in Table XXIV.

Although it has been shown that ether oxidation by ruthenium tetroxide involves hydride abstraction,[95] the products summarized in Table XXIV

---

[90] G. W. Gokel and H. D. Durst, *Synthesis*, 168 (1976).
[91] M. Doré, B. Legube and N. Merlet, *J. Fr. Hydrol.* **18**, 53 (1975).
[92] C. Sebastian, M.S. Thesis, Univ. of Regina, Canada, 1979.
[93] H. J. Schmidt and H. J. Schafer, *Angew. Chem. Int. Ed. Engl.* **18**, 69 (1979).
[94] D. G. Lee and M. van den Engh, *in* "Oxidation in Organic Chemistry (W. S. Trahanovsky, ed.), Part B, pp. 222–225. Academic Press, New York, 1973.
[95] D. G. Lee and M. van den Engh, *Can. J. Chem.* **50**, 3129 (1972).

TABLE XXIV

OXIDATION OF ETHERS BY BENZYLTRIETHYLAMMONIUM PERMANGANATE IN METHYLENE CHLORIDE[a]

Ether	Time (days)	Temperature (°C)	Product	Yield (%)[b]	Conversion (%)[c]
$CH_3(CH_2)_3O(CH_2)_3CH_3$	14	30	$CH_3(CH_2)_3O\overset{O}{\overset{\|}{C}}(CH_2)_2CH_3$	80	58
$C_6H_5CH_2OCH_2C_6H_5$	6	−5 to 20	$C_6H_5CH_2O\overset{O}{\overset{\|}{C}}C_6H_5$	84	80
			$C_6H_5\overset{O}{\overset{\|}{C}}—OCH_2O—\overset{O}{\overset{\|}{C}}C_6H_5$		8
$C_6H_5CH_2O(CH_2)_3CH_3$	8	0	$C_6H_5\overset{O}{\overset{\|}{C}}O(CH_2)_3CH_3$	99	90
$CH_3O(CH_2)_3CH_3$	10	42	$CH_3O\overset{O}{\overset{\|}{C}}(CH_2)_2CH_3$	40	23
$CH_3O(CH_2)_7CH_3$	14	42	$CH_3O\overset{O}{\overset{\|}{C}}(CH_2)_6CH_3$	79	41
$C_6H_5CH_2OCH(CH_2CH_3)_2$	7	−5	$C_6H_5\overset{O}{\overset{\|}{C}}OCH(CH_2CH_3)_2$	85	84
$C_6H_5CH_2OCH_3$	7	−5	$C_6H_5\overset{O}{\overset{\|}{C}}OCH_3$	84	83
$CH_3(CH_2)_3OCH(CH_2CH_3)_2$	14	30–42	$CH_3(CH_2)_2\overset{O}{\overset{\|}{C}}OCH(CH_2CH_3)_2$	73	45
			$CH_3CH_2\overset{O}{\overset{\|}{C}}CH_2CH_3$		3
$C_6H_5CH_2O\underset{C_6H_5}{\underset{\|}{C}}HCH_3$	9	5	$C_6H_5\overset{O}{\overset{\|}{C}}O\underset{C_6H_5}{\underset{\|}{C}}HCH_3$	83	71
			$C_6H_5\overset{O}{\overset{\|}{C}}CH_3$		6
$C_6H_5CH_2OC_6H_5$	9	0	$C_6H_5\overset{O}{\overset{\|}{C}}OC_6H_5$	94	42
$CH_3(CH_2)_3O\underset{C_6H_5}{\underset{\|}{C}}HCH_3$	8	25	$C_6H_5\overset{O}{\overset{\|}{C}}CH_3$	67	38
			$CH_3(CH_2)_2\overset{O}{\overset{\|}{C}}O\underset{C_6H_5}{\underset{\|}{C}}HCH_3$		3
			$CH_3(CH_2)_3O\overset{O}{\overset{\|}{C}}C_6H_5$		2
$CH_3(CH_2)_3OC_6H_5$	14	42	$CH_3(CH_2)_2\overset{O}{\overset{\|}{C}}OC_6H_5$	92	11

[a] From Schmidt and Schafer.[93]
[b] Yield is based on the amount of ether consumed as determined by gas chromatography.
[c] Conversion is based on the amount of ether initially present.

could also result from a hydrogen atom transfer mechanism which would be more likely in a nonpolar solvent.

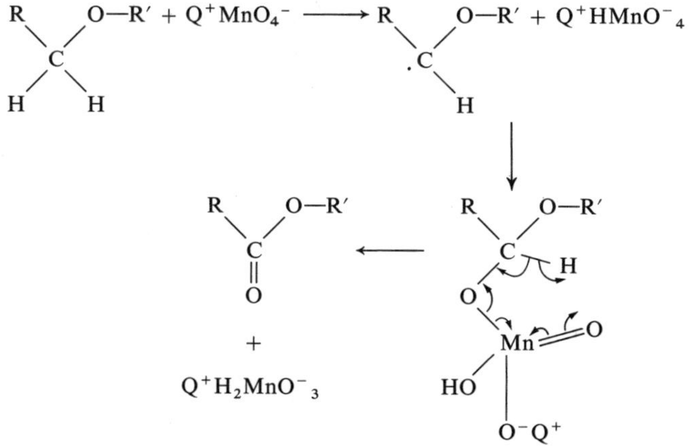

XI. Oxidation of Aldehydes

Aldehydes are easily oxidized to the corresponding carboxylic acids under phase transfer conditions. Several examples are summarized in Table XXV. The reactions proceed in good yield to give a single product. It should be noted, however, that the oxidation of cinnamaldehyde and phenylpropargyl aldehyde resulted in cleavage of the carbon–carbon double and triple bonds respectively. Hence, it is apparent that formyl groups cannot be selectively oxidized in the presence of double bonds.

XII. Oxidation of Sulfur Compounds

Sulfides and sulfoxides are oxidized by permanganate to the corresponding sulfones in both aqueous and organic solvents.[2] The phase transfer assisted reaction has also been found to produce pure products in good yields.[70,80] Table XXVI contains several examples.

Under similar conditions alkyl sulfides are converted to sulfonic acids.[70]

General Procedure for Phase Transfer Assisted Permanganate Oxidations of Sulfides.[80] Potassium permanganate (45.5 mmol), dissolved in 40 ml of water and Adogen 464 (1 g) dissolved in 10 ml of methylene chloride were placed in an Erlenmeyer flask fitted with a condenser and immersed in a water bath at room temperature.

TABLE XXV

THE OXIDATION OF ALDEHYDES

Substrate	Phase transfer agent	Solvent	Oxidant phase	Products (% yield)	Ref.
m-Nitrobenzaldehyde	Tetrabutylammonium permanganate	Pyridine	Solid	*m*-Nitrobenzoic acid (95)	43
p-Chlorobenzaldehyde	Tetrabutylammonium permanganate	Pyridine	Solid	*p*-Chlorobenzoic acid (99)	43
p-Anisaldehyde	Tetrabutylammonium permanganate	Pyridine	Solid	*p*-Anisic acid (94)	43
4-Acetoxy-2-methoxy-benzaldehyde	Tetrabutylammonium permanganate	Pyridine	Solid	4-Acetoxy-2-methoxy-benzoic acid (85)	43
Piperonal	Tetrabutylammonium permanganate	Pyridine	Solid	Piperonylic acid (99)	43
Piperonal	Cetyltrimethylammonium bromide	Water	Aqueous	Piperonylic acid (64–74)	65
Benzaldehyde	Dicyclohexano-18-crown-6	Benzene	Solid	Benzoic acid	82
Cinnamaldehyde	Adogen 464	Methylene chloride[a]	Aqueous	Benzoic acid (83)	70
Phenylpropargyl aldehyde	Adogen 464	Methylene chloride[a]	Aqueous	Benzoic acid (81)	70

[a] Acetic acid (5%) was added to the solvent.

TABLE XXVI

PHASE TRANSFER ASSISTED OXIDATION OF ORGANIC SULFUR COMPOUNDS BY AQUEOUS POTASSIUM PERMANGANATE[a]

Substrate	Product (% yield)
Phenyl sulfide	Phenyl sulfone (98)
n-Butyl sulfide	*n*-Butyl sulfone (95)
n-Octyl sulfide	*n*-Octyl sulfone (94)
t-Butyl sulfide	*t*-Butyl sulfone (82)
Dibenzothiophene	Dibenzothiophene sulfone (93)
Phenyl sulfoxide	Phenyl sulfone (95)
n-Butyl sulfoxide	*n*-Butyl sulfone (86)
1-Dodecanethiol	1-Dodecane sulfonic acid (87)

[a] The phase transfer agent was Adogen 464, the solvent was methylene chloride containing 5–10% acetic acid.[70,80]

The mixture was stirred for about 2 hr with a magnetic stirrer and the manganese dioxide that formed was reduced with sodium bisulfite and dilute sulfuric acid. The organic phase was separated and the aqueous phase was extracted with ether (3 × 40 ml). The combined organic solvents were dried (anhydrous $MgSO_4$) and the volatiles removed using a rotary evaporator. The product was purified by crystallization or distillation.

XIII. Oxidation of Halides

The oxidation of hydrocarbons can sometimes be facilitated by prior halogenation of the oxidation site. Under reaction conditions, the halide presumably undergoes hydrolysis to the corresponding alcohol which is then oxidized. The following reaction illustrates this approach.[96]

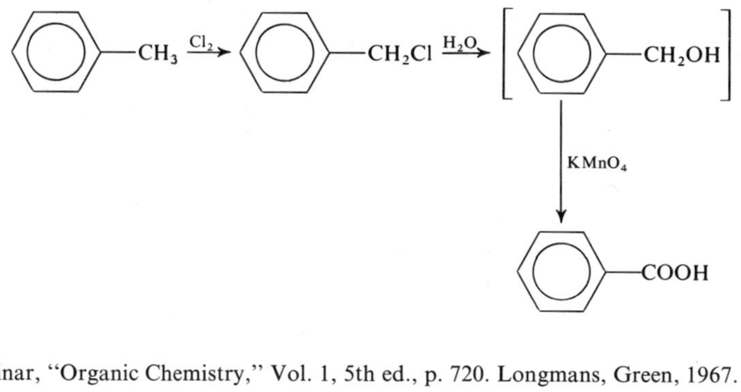

[96] I. L. Finar, "Organic Chemistry," Vol. 1, 5th ed., p. 720. Longmans, Green, 1967.

It has been observed that benzhydryl chloride can also be oxidized under phase transfer conditions. Benzophenone was obtained in 82% yield when benzhydryl chloride and tetraethylammonium bromide were dissolved in methylene chloride and treated with aqueous permanganate.[80]

TYPICAL EXPERIMENTAL PROCEDURE

Preparation of Benzophenone by Oxidation of Benzhydryl Chloride.[80] A 500-ml Erlenmeyer flask equipped with a condenser was charged with methylene chloride (120 ml), benzhydryl chloride (3.0, 0.015 mol), water (20 ml), tetraethylammonium bromide (1.0 g), sodium hydroxide (2.2 g) and potassium permanganate (2.5 g, 0.16 mol). The mixture was stirred with a magnetic stirrer and refluxed for 2 days. The precipitated MnO_2 was reduced using sodium bisulfite and dilute sulfuric acid. The organic phase was separated and the aqueous phase extracted with ether (3 × 50 ml). The combined organic solvents were dried (anhydrous $MgSO_4$) and the volatiles removed using a rotary evaporator. The crude mixture obtained was analysed by GLC.

XIV. Oxidation of Amines

The oxidation of amines under phase transfer conditions results in the formation of products similar to those found during oxidations by neutral aqueous permanganate. For example, the oxidation of dibenzylamine in a methylene chloride solution containing Adogen 464 as a phase transfer agent gave *N*-(α-dibenzylaminobenzyl)benzamide and benzaldehyde:[70]

$$PhCH_2NHCH_2Ph \xrightarrow[KMnO_4/H_2O]{CH_2Cl_2/Adogen\ 464} PhCONH\underset{\substack{|\\ Ph \\ 46\%}}{CH}(CH_2Ph)_2 + \underset{28\%}{PhCHO}$$

A similar reaction was observed when Shechter and Rawalay[97] oxidized dibenzylamine with potassium permanganate in a neutral solution of water and *tert*-butyl alcohol. They obtained *N*-(α-dibenzylaminobenzyl)benzamide as the sole product (62% yield) but noted that benzaldehyde was one of the hydrolysis products. They also described a possible reaction mechanism that leads to the formation of this rather complex product.

[97] H. Shechter and S. S. Rawalay, *J. Am. Chem. Soc.* **86**, 1706 (1964).

Rossi and Trimarco[26] also found that similar products were obtained when 1-aryl-5-morpholino-4,5-dihydro-*v*-triazoles and 1-aryl-5-dimethylamino-4,5-dihydro-*v*-triazoles were oxidized by potassium permanganate either in acetone or in benzene containing cetyltrimethylammonium bromide as the phase transfer agent.

The products were the corresponding 2-oxomorpholino or *N*-formylamino compounds:

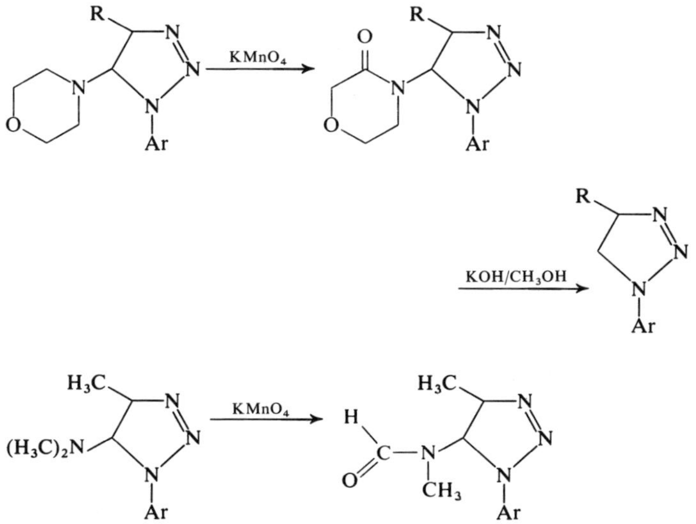

The observed products suggest that these reactions may be initiated by oxidative attack on a hydrogen adjacent to a nitrogen. However, the report did not contain any mechanistic considerations.

Typical Experimental Procedure

*Preparation of 1-(4-Fluorophenyl)-4-methyl-5-(2-oxomorpholino)-4,5-dihydro-v-triazole.** To a 250-ml round-bottomed flask fitted with a dropping funnel and an efficient mechanical stirrer was added cetyltrimethylammonium bromide (36.4 mg, 0.1 mmol), benzene (25 ml), and a solution of potassium permanganate (0.553 g, 3.5 mmol) in water (40 ml). A solution of 1-(4-fluorophenyl)-4-methyl-5-morpholino-4,5-dihydro-*v*-triazole (0.53 g, 2 mmol) in benzene (40 ml) was dropped into the flask and stirring was continued until no more starting compound could be detected by TLC. The

* Reprinted with permission from Rossi and Trimarco[26], courtesy of Thieme, Stuttgart.

reaction mixture was filtered and the organic layer separated. After concentration, the product was precipitated by adding pentane to the benzene solution and purified by chromatography over silica gel with benzene/ethyl acetate (4:1). The product was recrystallized from benzene/pentane; yield, 45%.

Acknowledgment

The author wishes to acknowledge the assistance and encouragement of several co-workers who participated in some of the work reported herein. In particular, use of previously unpublished results obtained by Dr. Victor S. Chang, Dr. N. S. Srinivasan, Mr. Hasan Karaman and Mr. William Rennie is noted with gratitude.

CHAPTER **III**

Intramolecular Oxidative Coupling of Aromatic Substrates

*OM PRAKASH DHINGRA**

I. Introduction 207
II. Oxidative Coupling of Benzyltetrahydroisoquinolines 209
A. Diphenolic Substrates 210
B. Nonphenolic Substrates 226
C. Monophenolic Substrates 238
III. Oxidative Coupling of Phenethyltetrahydroisoquinolines 241
A. Diphenolic Substrates 242
B. Nonphenolic Substrates 248
C. Monophenolic Substrates 253
IV. Oxidative Coupling of *N*-Benzylphenethylamines 257
A. Diphenolic Substrates 258
B. Monophenolic Substrates 263
C. Nonphenolic Substrates 264
V. Oxidative Coupling of Diarylalkanes 264
A. Diphenolic Substrates 264
B. Nonphenolic Substrates 267
C. Monophenolic Substrates 276

I. Introduction

Intramolecular oxidative coupling of phenols, either by a homolytic or a heterolytic mechanism,[1] is of great importance in natural products chemistry. It is widely postulated as a biosynthetic pathway to many heterocyclic

* Dedicated to the late Professor S. Morris Kupchan.

[1] For a detailed discussion of the mechanisms of phenolic oxidative coupling see: P. D. McDonald and G. A. Hamilton, *in* "Oxidation in Organic Chemistry" (W. S. Trahanovsky, ed.), Part B, p. 97. Academic Press, New York, 1973.

Oxidation in Organic Chemistry, Part D

ISBN 0-12-697253-2

products[2] including alkaloids[3] and antibiotics.[4] Since the proposal of alkaloid biosynthesis through oxidative phenol coupling presented by Barton and Cohen[5] in 1957, many biogenetic-type syntheses of alkaloids and other polycyclic compounds have been reported.[2,6] The coupling step is often replicated in the laboratory by use of inorganic oxidants such as ferric chloride, potassium ferricyanide, manganese dioxide, and lead dioxide. However, the yields of the coupling step are, in general, low and the work-up procedure is complicated by the necessity of handling polyhydric phenols in the presence of strong base.[7] Other methods of aryl–aryl coupling, such as Pschorr[7,8] or photochemical[7,9] reactions, are also commonly employed in synthesis but each presents problems either in nonselectivity or in the inaccessibility of starting materials.

Recently, a number of new reagents have been reported which appear to give consistently high yields of intramolecularly coupled products. Those which have proved particularly useful are vanadium oxytrifluoride[10–15] (VOF_3), vanadium oxytrichloride[16] ($VOCl_3$), manganese tris(acetylacetonate)[17] (MTA), thallium trifluoroacetate[16] (TTFA), ferric chloride–

[2] A. R. Battersby and W. I. Taylor (eds.), "Oxidative Coupling of Phenols." Dekker, New York, 1967.

[3] A. R. Battersby, *in* "Oxidative Coupling of Phenols" (W. I. Taylor and A. R. Battersby, eds.), p. 119. Dekker, New York, 1967.

[4] A. I. Scott, *in* "Oxidative Coupling of Phenols" (W. I. Taylor and A. R. Battersby, eds.), p. 95. Dekker, New York, 1967.

[5] D. H. R. Barton and T. Cohen, *Festchr. Prof. Dr. Arthur Stoll Siebzigsten Gerburtstag,* 129 (1957).

[6] T. Kametani, K. Fukumoto, and F. Satoh, *Bioorg. Chem.* **3**, 430 (1974).

[7] M. Shamma, "The Isoquinoline Alkaloids: Chemistry and Pharmacology." Academic Press, New York, 1972.

[8] T. Kametani and K. Fukumoto, *J. Heterocyclic Chem.* **8**, 341 (1971).

[9] T. Kametani and K. Fukumoto, *Acc. Chem. Res.* **5**, 212 (1972).

[10] S. M. Kupchan, A. J. Liepa, V. Kameswaran, and R. F. Bryan, *J. Am. Chem. Soc.* **95**, 6861 (1973).

[11] S. M. Kupchan, V. Kameswaran, J. T. Lynn, D. K. Williams, and A. J. Liepa, *J. Am. Chem. Soc.* **97**, 5622 (1975).

[12] S. M. Kupchan, Om P. Dhingra, C.-K. Kim, and V. Kameswaran, *J. Org. Chem.* **43**, 2521 (1978).

[13] S. M. Kupchan, Om P. Dhingra, and C.-K. Kim, *J. Org. Chem.* **43**, 4076 (1978).

[14a] A. S. Kende and L. S. Liebeskind, *J. Am. Chem. Soc.* **98**, 267 (1976).

[14b] K. Tomioka, T. Ishiguro, and K. Koga, *J. Chem. Soc. Chem. Commun.*, 652 (1979).

[15a] R. E. Damon, R. H. Schlessinger, and J. F. Blount, *J. Org. Chem.* **41**, 3772 (1976).

[15b] T. Biftu, B. G. Hazra, and R. Stevenson, *J. Chem. Soc. Perkin 1*, 2276 (1979).

[16] M. A. Schwartz, B. F. Rose, R. A. Holton, S. W. Scott, and B. Vishnuvajjala, *J. Am. Chem. Soc.* **99**, 2571 (1977).

[17] A. Rolan and V. D. Parker, *J. Org. Chem.* **39**, 1014 (1974).

dimethylformamide complex,[18] and lead tetraacetate (LTA).[19] These reagents have been used to couple a wide variety of diphenolic,[13,16] monophenolic,[13,16,17] and nonphenolic substrates.[10–16] In addition, controlled potential electrochemical oxidation of nonphenolic[20–22] and monophenolic[23] substrates has been developed as the method of choice for the synthesis of various complex natural products such as (±)-*O*-methylflavinantine,[20] colchicine,[24] and (±)-cryptoplurine.[25] The purpose of this chapter is to show how our understanding of biosynthetic pathways to complex alkaloids has led to the development of a number of coupling methods and to demonstrate the utility of the recently developed methods in the syntheses of complex natural products. The literature through August 1980 is covered.

II. Oxidative Coupling of Benzyltetrahydroisoquinolines

Isoquinoline alkaloids have been the subject of considerable chemical and pharmacological interest[7] since the beginning of alkaloid chemistry about 170 years ago when Serturner announced the isolation of morphine, an isoquinoline alkaloid, and demonstrated that it was chiefly responsible for the physiological effects of opium.[26] Subsequently, a large number of isoquinoline alkaloids have been reported to show a wide range of physiological activity. The structure–activity relationships for these compounds are extremely difficult to evaluate due to the diversity of their physiological properties.[7,26] These biologically active compounds can produce effects ranging from emesis, sedation, diuresis, increased endocrine secretions, and ganglionic blockade to tumor-growth inhibition.[7] Liriodenine (**1**) has shown significant cytotoxic inhibitory activity *in vitro* against cells derived from the human carcinoma of the nasopharynx.[27] Thalicarpine (**2**), a novel

[18] S. Tobinaga and E. Kotani, *J. Am. Chem. Soc.* **94**, 309 (1972).
[19] B. Umezawa and O. Hoshino, *Heterocycles* **3**, 1005 (1975).
[20] L. L. Miller, F. R. Stermitz, and J. R. Falck, *J. Am. Chem. Soc.* **95**, 2651 (1973).
[21] E. Kotani, N. Takeuchi, and S. Tobinaga, *J. Chem. Soc. Chem. Commun.*, 550 (1973).
[22] V. D. Parker and A. Rolan, *J. Am. Chem. Soc.* **97**, 4714 (1975).
[23] U. Palmquist, A. Nilsson, V. D. Parker, and A. Rolan, *J. Am. Chem. Soc.* **98**, 2571 (1976).
[24] E. Kotani, F. Mivazaki, and S. Tobinaga, *J. Chem. Soc. Chem. Commun.*, 300 (1974).
[25] E. Kotani, M. Kitazawa, and S. Tobinaga, *Tetrahedron* **30**, 3027 (1974).
[26] V. Preininger, *in* "The Alkaloids" (R. H. F. Manske and H. L. Holmes, eds.), Vol. XIV, p. 207. Academic Press, New York, 1975.
[27] D. Warthen, E. L. Gooden, and M. Jacobson, *J. Pharm. Sci.* **58**, 637 (1969).

dimeric aporphine–benzylisoquinoline, and *dl*-tetradrine (**3**), a bis-benzylisoquinoline, have been shown to possess hypotensive activity,[28–31] as well as significant inhibitory activity against the Walker 256 intramuscular carcinosarcoma in rats over a wide dosage range.[31,32] After several years of preclinical toxicological studies in higher animals, clinical trials of these compounds have recently begun.

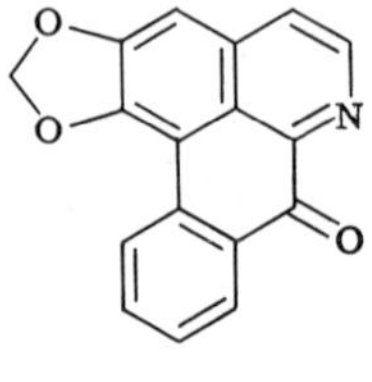

1

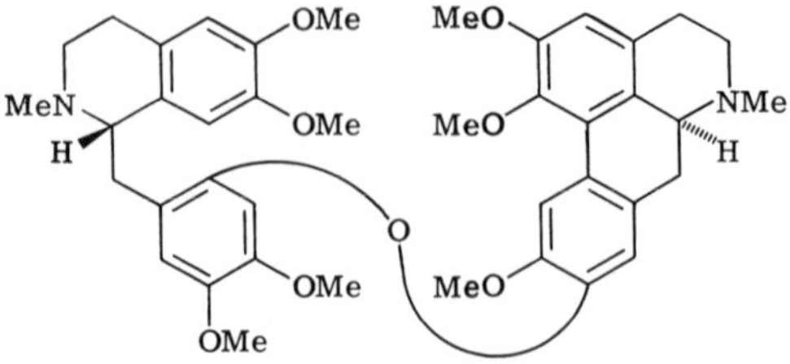

2

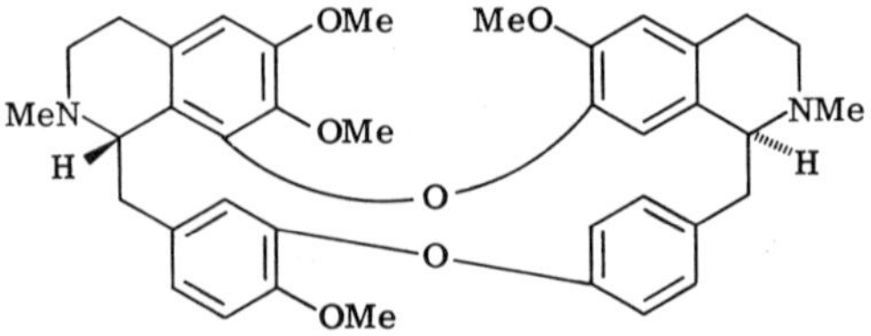

3

A. DIPHENOLIC SUBSTRATES

In a lecture in 1925, Sir Robert Robinson proposed that morphine and the aporphines in general might be formed in nature by oxidation of suitably substituted benzyltetrahydroisoquinolines.[33,34] Subsequently, in 1932, Robinson and Sugasawa attempted to oxidize laudanosoline to an aporphine with chloranil, but instead generated a dibenzopyrrocoline salt.[35] In the discussion of his results, Robinson suggested that the desired coupling to an

[28] S. M. Kupchan, K. K. Chakravarti, and N. Yokoyama, *J. Pharm. Sci.* **52**, 985 (1963).
[29] M. Tomita, H. Furukawa, S.-T. Lu, and S. M. Kupchan, *Tetrahedron Lett.*, 4309 (1965); *Chem. Pharm. Bull.* **15**, 959 (1967).
[30] S. M. Kupchan and A. J. Liepa, *Chem. Commun.*, 599 (1971).
[31] S. M. Kupchan, *Trans. N. Y. Acad. Sci.* **32**, 85 (1970).
[32] S. M. Kupchan, N. Yokoyama, and B. S. Thyagarajan, *J. Pharm. Sci.* **50**, 164 (1961).
[33] J. Gulland and R. Robinson, *Mem. Proc. Manchester Lit. Philos. Soc.* **69**, 79 (1925).
[34] R. Robinson and S. Sugasawa, *J. Chem. Soc.*, 3163 (1931).
[35] R. Robinson and S. Sugasawa, *J. Chem. Soc.*, 789 (1932).

aporphine might be brought about if the nitrogen atom was rendered nonbasic as a salt or an acyl derivative.

Robinson's ideas remained dormant until 1957, when Barton and Cohen converted Robinson's general speculations into a rational scheme based on phenolic oxidative coupling.[5] In this approach, Barton hypothesized that radical coupling ortho or para to the phenolic function of suitably substituted diphenolic benzyltetrahydroisoquinolines could yield, with further transformations where necessary, the aporphines, the morphine alkaloids, and protostephanine.

Barton proposed that the majority of aporphines could be thought of as biogenetically derived from alternative coupling modes of a hypothetical diphenolic base **4**. Ortho–ortho coupling of base **4** would yield aporphines of the corytuberine oxidation pattern (**5**), whereas ortho–para coupling of base **4** would yield aporphines of the glaucine oxidation pattern (**6**).

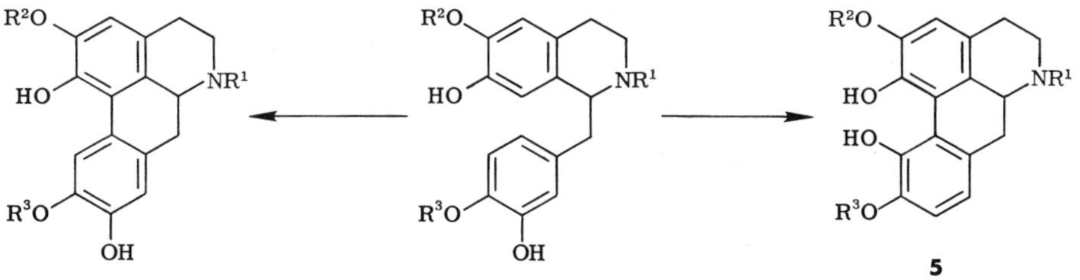

6 **4** **5**

The hypothesis that phenolic coupling of a diphenolic benzyltetrahydroisoquinoline such as **4** creates the bond between the two aromatic rings of the aporphine directly has been supported by studies with *Corydalis cava* by Blaschke.[36,37] Blaschke found that feeding (±)-reticuline (**7**) carbon-14-labeled in the *N*-methyl group to *Corydalis cava* yielded bulbocapnine (**8**) labeled in the *N*-methyl group.[36] Subsequently, feeding of (±)-reticuline triply labeled with tritium (**9**) and mixed with (±)-reticuline labeled in the *N*-methyl group with carbon-14 yielded bulbocapnine (**10**) with tritium retained at the expected site. Furthermore, the tritium:carbon-14 ratio was the same as that in the mixture of benzyltetrahydroisoquinolines with a correction made for the two tritium atoms lost.[37] Brochmann–Hassen has found that labeled reticuline is incorporated into isoboldine in opium poppies[38] and into magnofluorine in *Aqualegia*.[39]

[36] G. Blaschke, *Arch. Pharm.* **301**, 432 (1968).
[37] G. Blaschke, *Arch. Pharm.* **303**, 358 (1970).
[38] E. Brochmann-Hanssen, C.-C. Fu, and L. Misconi, *J. Pharm. Sci.* **60**, 1880 (1971).
[39] E. Brochmann-Hanssen, C.-H. Chen, H.-C. Chiang, and K. McMurtrey, *Chem. Commun.*, 1269 (1972).

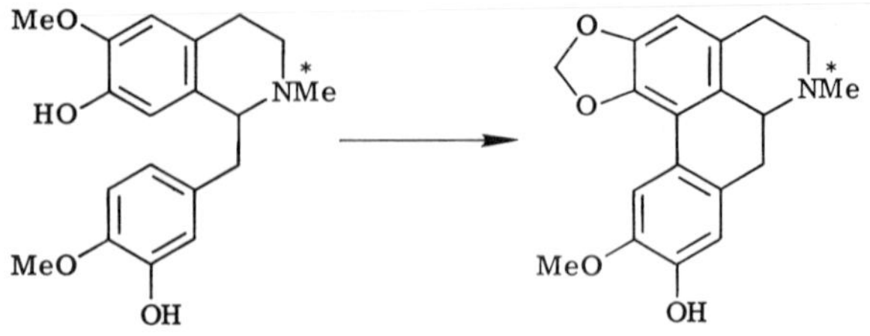

7 **8**

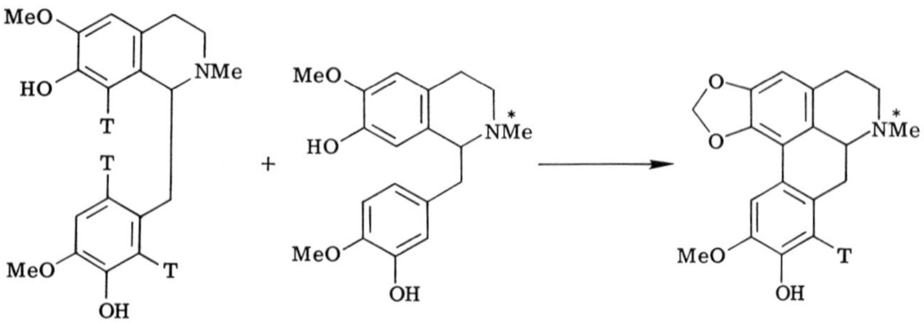

9 **7** **10**

* = [^{14}C] label

The laboratory analogy of the biogenetic direct coupling of diphenolic benzyltetrahydroisoquinolines to aporphines has, as a general rule, produced the aporphines in low yield.[40–48] Reticuline, a diphenolic benzyltetrahydroisoquinoline of great interest because of its role in aporphine and morphine biosynthesis, has been oxidized with manganese dioxide,[46] silver carbonate,[45] and potassium ferricyanide[43–45] under a variety of conditions, and in each case the yield of aporphine was less than 6%, a result which is typical for the oxidations of diphenolic benzyltetrahydroisoquinolines. There have, however, been some notable exceptions to these low yields. In 1962 Franck resurrected the hypothesis of Robinson that oxidation of an isoquinoline with a quaternary nitrogen might yield an aporphine,[35] and oxidized laudanosoline methiodide (**11**) to the aporphine **12** in 60% yield using ferric chloride.[49] This was the first successful oxidative coupling of diphenolic benzyltetrahydroisoquinoline to an aporphine in a practical sense. Jackson has since suggested that the high yield of aporphine obtained in the oxidation of laudanosoline methiodide may result perhaps from

[40] S. M. Albonico, A. M. Kuck, and V. Deulofeu, *Chem. Ind.* (*London*), 1580 (1964).
[41] T. Kametani and I. Noguchi, *J. Chem. Soc. C*, 1440 (1967).

phenolate anion addition to an *ortho*-quinone rather than from a radical pairing like the one Barton and Cohen proposed for aporphine biosynthesis.[47] While attempting to prepare a morphinandienone by oxidation of *N*-mesylnorreticuline (**13**) with manganese dioxide in chloroform, Franck isolated instead *N*-mesylnorisoboldine (**14**) in 40% yield.[50] Schwartz has oxidized reticuline perchlorate (**15**) to isoboldine (**16**) in 53% yield with vanadium oxytrichloride in dichloromethane.[51] The high yield with this reagent results possibly from oxidation via a species such as **17**.[52] A result of this mechanism would be a diminished tendency for intermolecular coupling.

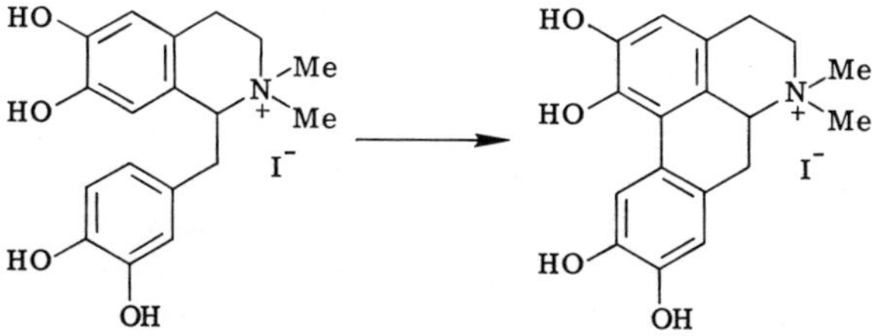

11 **12**

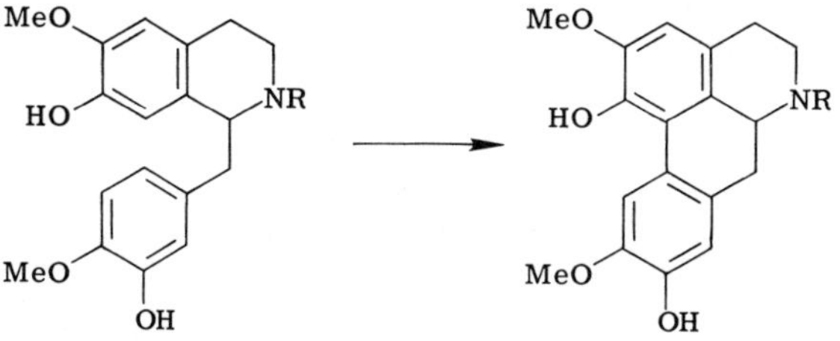

13 R = mesyl **14** R = mesyl

15 R = Me·HClO$_4$ **16** R = Me

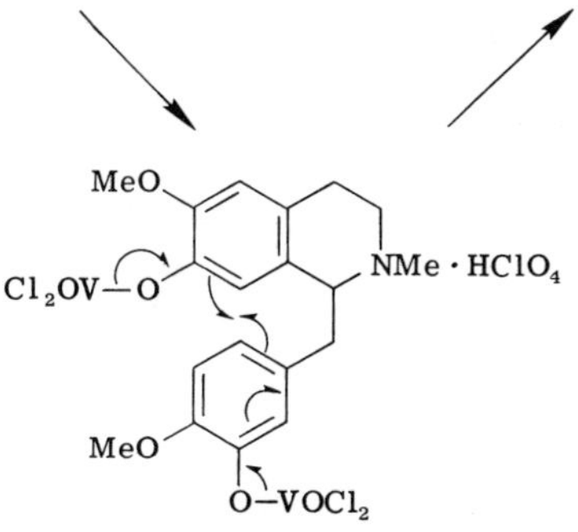

17

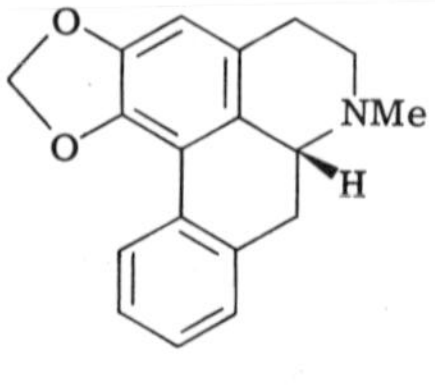

18

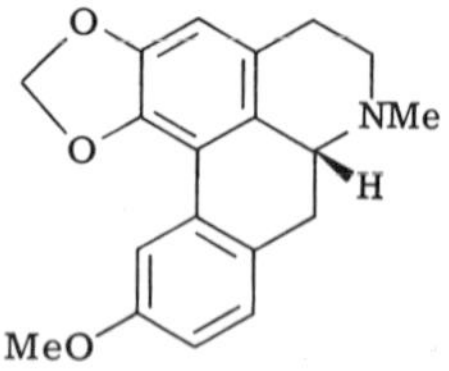

19

Although the direct oxidative coupling that Barton suggested accounts successfully for most of the aporphines found in nature, certain aporphines that are nonsubstituted, e.g., roemerine (**18**), or are monosubstituted in ring D, e.g., laureline (**19**), cannot be accounted for by this pathway. The oxygenation pattern is simply not correct for direct coupling ortho or para to a phenol in ring D to yield an aporphine. In order to account for the formation of these anomalous aporphines, Barton and Cohen suggested the intermediacy of a spirodienone (**21**) which they called a proaporphine.[5] The proaporphine could then undergo a dienone–phenol rearrangement to yield an aporphine monosubstituted in ring D meta to the final coupling site (**22**). Alternatively, the proaporphine could be reduced to the dienol (**23**) followed by a dienol–benzene rearrangement to yield an aporphine unsubstituted in ring D (**24**).

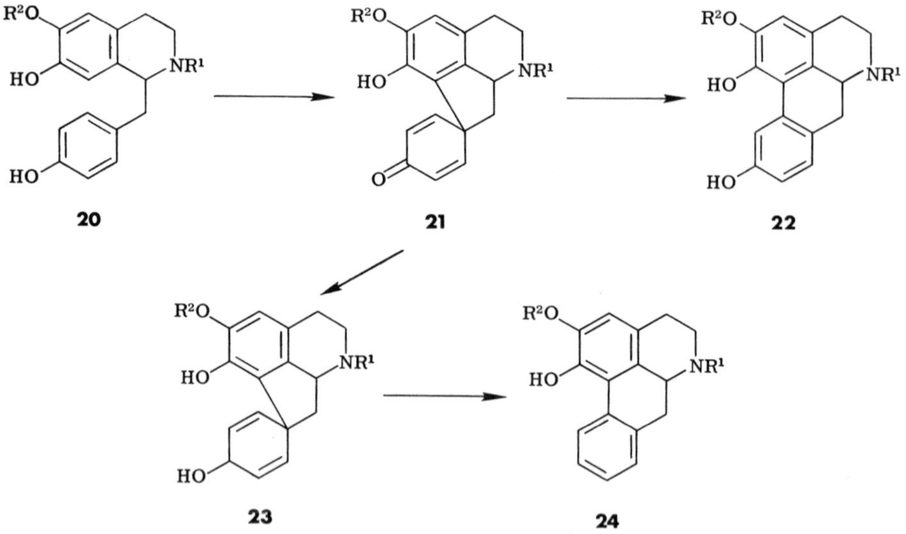

[42] T. Kametani, T. Sugahara, and K. Fukumoto, *Tetrahedron* **25**, 3667 (1969).
[43] A. H. Jackson and J. A. Martin, *Chem. Commun.*, 420 (1965).
[44] W. W. C. Chen and P. Maitland, *J. Chem. Soc. C*, 753 (1966).

Barton's speculations were rewarded in 1963 when crotonosine (**27**) was isolated and its structure elucidated.[53] Subsequently, Barton showed that *Croton linearis* converted multiply labeled (−)-coclaurine (**28**) into multiply labeled crotonosine (**29**).[54] No incorporation of (±)-[^{3}H]isococlaurine (**30**) was observed, suggesting that *Croton linearis* first oxidized coclaurine (**25**) to proaporphine **26** and then through demethylation–remethylation formed crotonosine (**27**). Using *Mecanopsis cambrica*, Barton demonstrated that tritium-labeled coclaurine (**28**) was incorporated into mecambroline (**32**) retaining the tritium labels as shown,[55] and thus, confirmed his postulated proaprophine route to aporphines.

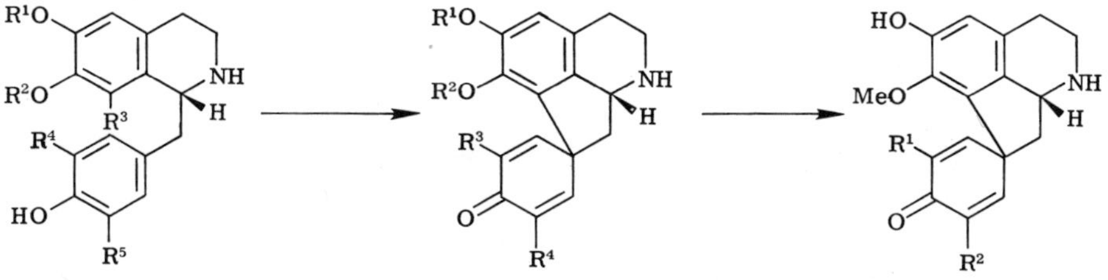

25 R^1 = Me, R^2 = R^3 = R^4 = R^5 = H
28 R^1 = Me, R^2 = H, R^3 = R^4 = R^5 = T
30 R^1 = H, R^2 = Me, R^3 = R^4 = R^5 = T

26 R^1 = Me, R^2 = R^3 = R^4 = H

27 R^1 = R^2 = H
29 R^1 = R^2 = T

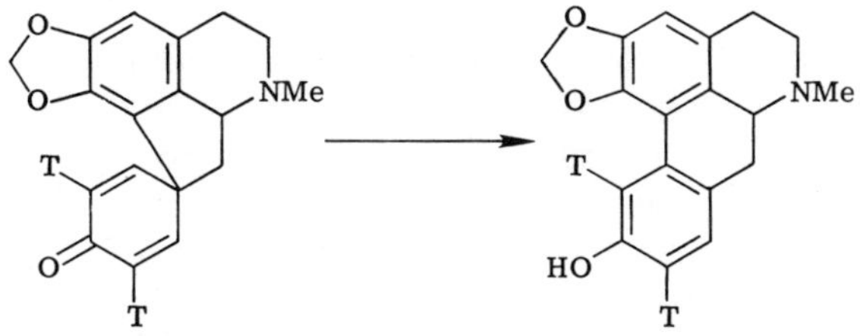

31 **32**

[45] T. Kametani, A. Kozuka, and K. Fukumoto, *J. Chem. Soc. C*, 1021 (1971).
[46] B. Franck, G. Dunkelmann, and H. J. Lubs, *Angew. Chem.* **79**, 1066 (1967).
[47] A. H. Jackson and J. A. Marten, *J. Chem. Soc. C*, 2061 (1969).
[48] T. Kametani, K. Fukumoto, K. Kigasawa, and K. Wakisaka, *Chem. Pharm. Bull.* **19**, 714 (1971).
[49] B. Franck and G. Schlingloff, *Justus Liebigs Ann. Chem.* **659**, 132 (1962).
[50] B. Franck, G. Dunkelmann, and H. J. Lubs, *Angew. Chem. Int. Ed. Eng.* **6**, 1075 (1967).
[51] M. A. Schwartz, *Synth. Commun.* **3**, 33 (1973).
[52] M. A. Schwartz, R. H. Holten, and S. W. Scott, *J. Am. Chem. Soc.* **91**, 2800 (1969).
[53] L. J. Haynes, K. L. Stuart, D. H. R. Barton, D. S. Bhakuni, and G. W. Kirby, *Chem. Commun.*, 141 (1965).

Battersby investigated the biosynthesis of isothebaine (**36**) in *Papaver orientale* concurrently with Barton's work with *M. cambrica*, and has, through feeding experiments with labeled precursors, established the sequence orientaline (**33**) → orientalinone (**34**) → orientalinol (**35**) → isothebaine (**36**).[56–58]

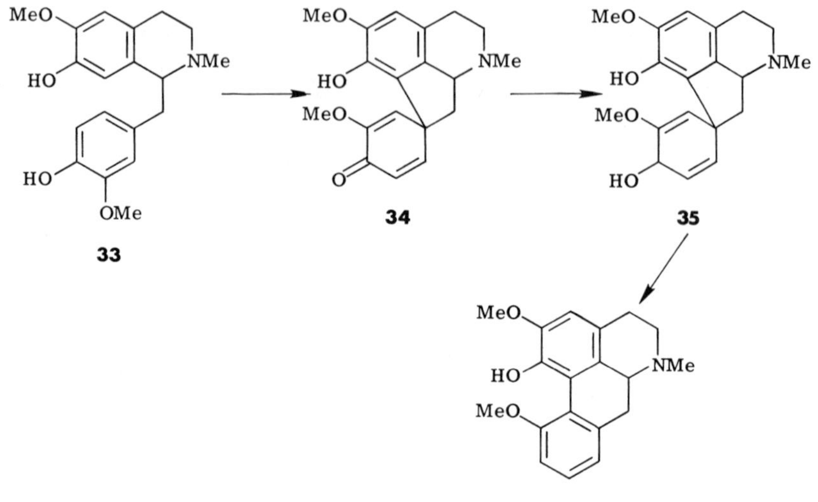

33 **34** **35** **36**

Battersby also succeeded in mimicking this biogenetic pathway in the laboratory.[59] Oxidation of orientaline (**33**) with potassium hexacyanoferrate gave orientalinone (**34**) in 4% yield. The dienone was reduced with sodium borohydride to **35** and the dienol was then rearranged with aqueous hydrochloric acid to isothebaine (**36**) in 36% yield from the dienone. Nuciferine, corydine, isocorytuberine, pseudocorydine, and other aporphines have since been synthesized by hexacyanoferrate oxidation of the appropriate diphenolic benzyltetrahydroisoquinoline to a proaporphine, followed by acid-catalyzed rearrangement to aporphines and further modifications.[60]

[54] D. H. R. Barton, D. S. Bhakuni, G. M. Chapman, and G. W. Kirby, *J. Chem. Soc. C*, 1295 (1967).
[55] D. H. R. Barton, D. S. Bhakuni, G. M. Chapman, and G. W. Kirby, *J. Chem. Soc. C*, 2134 (1967).
[56] A. R. Battersby, R. T. Brown, J. H. Clements, and G. G. Iverach, *Chem. Commun.*, 230 (1965).
[57] A. R. Battersby and T. H. Brown, *Chem. Commun.*, 171 (1966).
[58] A. R. Battersby, T. J. Brocksom, and R. R. Ramage, *Chem. Commun.*, 464 (1969).
[59] A. R. Battersby, T. H. Brown, and J. H. Clements, *J. Chem. Soc.*, 4550 (1965).
[60] T. Kametani and K. Fukumoto, *Synthesis*, 657 (1972).

In 1971 Battersby published some surprising results from labeling experiments in *Dicentra eximia*.[61] Corydine (**47**), glaucine (**43**), and dicentrine (**42**) occur in *Dicentra eximia* and their structures appear to be quite easily explained by direct ortho–ortho and ortho–para coupling of reticuline (**7**) to yield the 1,2,10,11- and 1,2,9,10-tetraoxygenated aporphines **47**, **42**, and **43**, respectively, in accordance with the routes proposed by Barton[5] and supported by Blaschke[36,37] (Scheme 1).

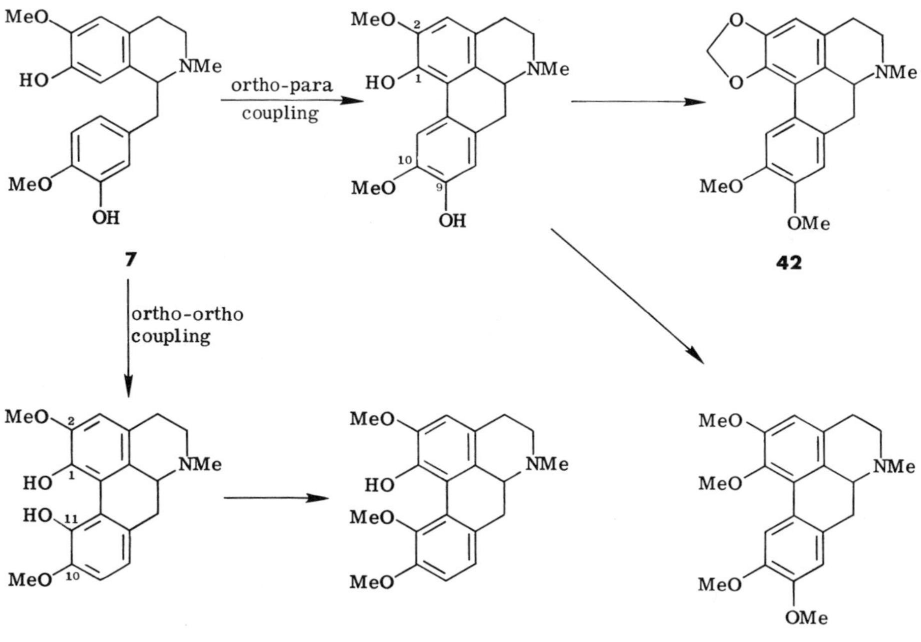

SCHEME 1

Labeled reticuline was, however, not incorporated by *Dicentra eximia* into any of the three aporphines. Battersby established that norprotosinomenine (**37**) was a precursor of the *Dicentra* aporphines. To rationalize this, Battersby proposed that oxidative coupling of protosinomenine (**38**) or norprotosinomenine (**37**) could yield the proerythrinadienones **40** and **45**, or **39** and **44**, respectively. Dienone–phenol rearrangement of **39** and **40** could yield boldine (**41**) which could then be transformed into glaucine (**43**) and dicentrine (**42**). Dienone–phenol rearrangement of **44** and **45** could yield the aporphine **46** which could be transformed into corydine (**47**). Confirmation

[61] A. R. Battersby, J. L. McHugh, J. Staunton, and M. Todd, *Chem. Commun.*, 985 (1971).

of the role of boldine in this scheme was obtained when labeled boldine was incorporated into glaucine and dicentrine in *Dicentra eximia* (Scheme 2).

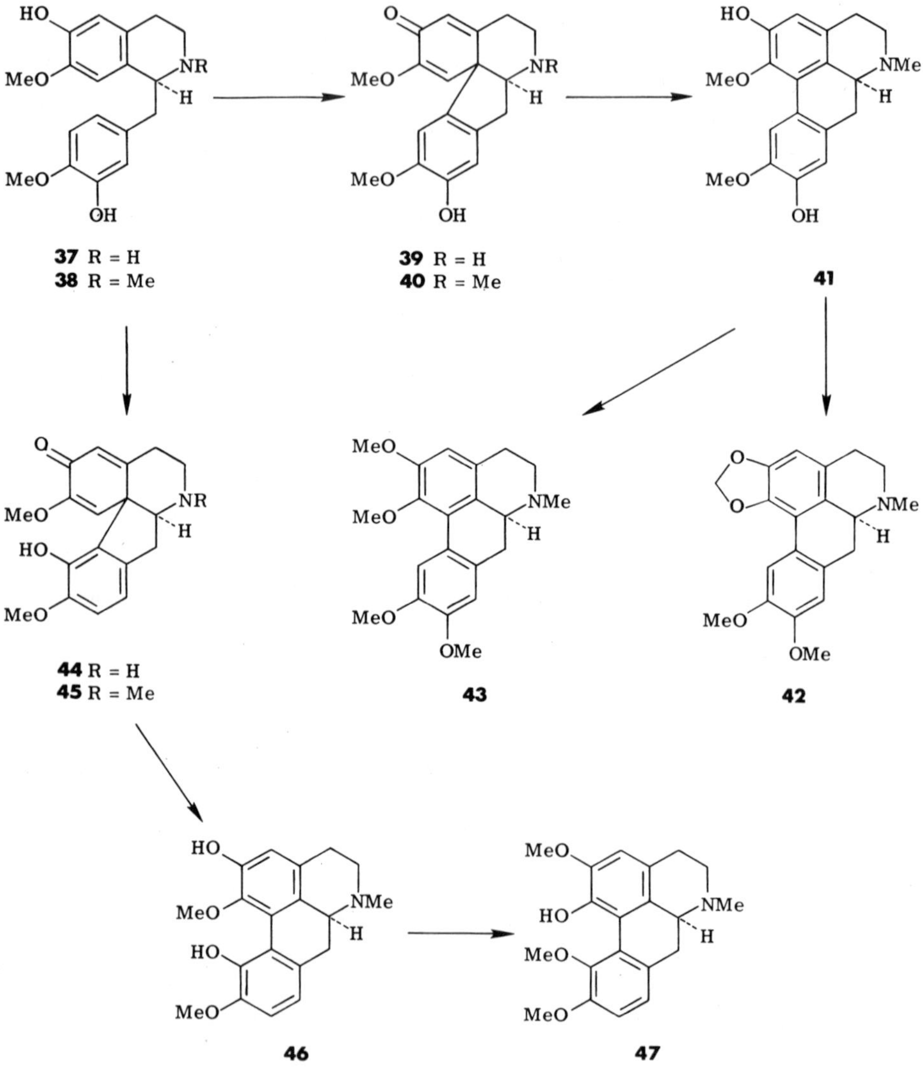

SCHEME 2

Barton and Cohen also proposed a route to morphine based on diphenolic coupling of benzyltetrahydroisoquinoline (**7a**) to a morphinandienone (**48**). Further modifications to **48** could yield morphine (**55**) as shown in Scheme 3, route 1.[5] This route has been shown to be essentially correct by extensive tracer experiments in *Papaver somniferum*, with the minor, but significant, modification shown in Scheme 3, route 2.[62]

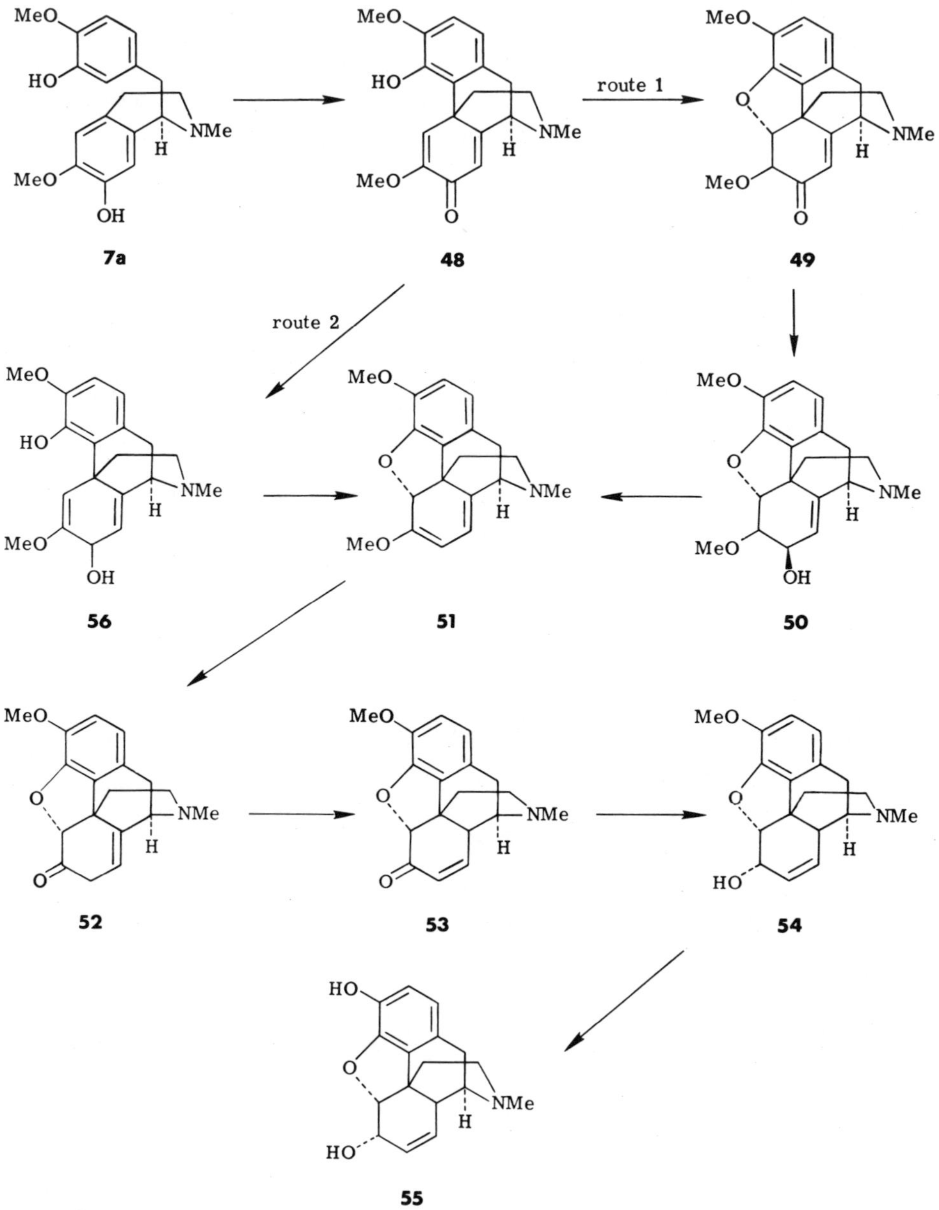

SCHEME 3

The morphinandienone structure which Barton and Cohen proposed as an intermediate in morphine biosynthesis was subsequently invoked in the biosynthesis of other alkaloids. Barton proposed that protostephanine (**62**), isolated from *Stephania japonica* Miers, is biosynthesized by oxidation of an isoquinoline such as **57** to a morphinandienone (**58**).[63] Subsequent reduction of **58** to a dienol (**59**) followed by rearrangement to the dehydrodibenzazonine **61** would yield protostephanine (**62**) on reduction. Hackett has established that the morphinandienone **58** is, indeed, incorporated into protostephanine in *S. japonica*.[64]

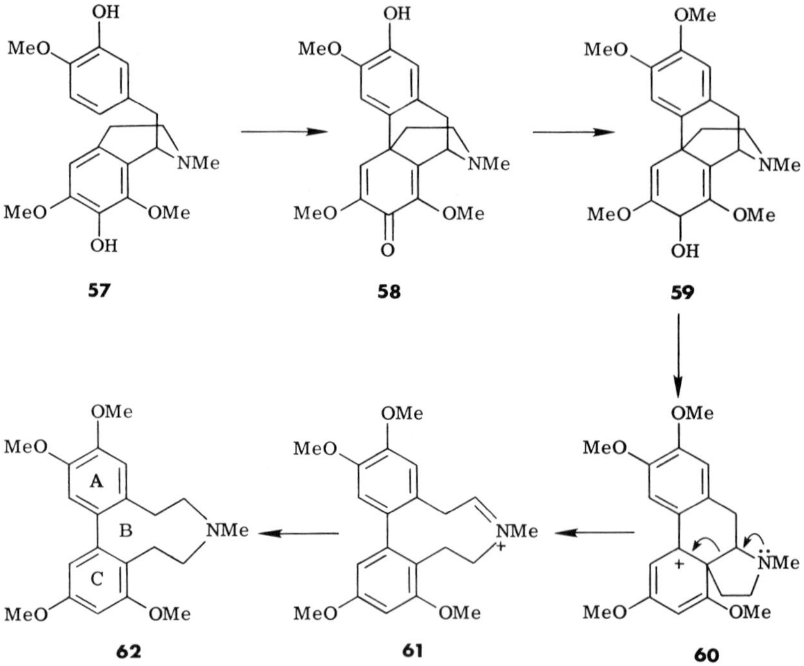

SCHEME 4

Battersby synthesized protostephanine in 1968 along Barton's proposed pathway (Scheme 4).[65] Ferricyanide oxidation of **57** yielded the morphinandienone **58** in 2% yield. O-Methylation of the phenol, followed by sodium

---

[62] T. A. Geissman and D. H. Grout, "Organic Chemistry of Secondary Plant Metabolism," p. 501. Freeman, San Francisco, California, 1969.

[63] D. H. R. Barton, *Pure Appl. Chem.* **9**, 35 (1964).

[64] P. Hackett, unpublished results cited in reference 63.

[65] A. R. Battersby, A. K. Bhatnager, P. Hackett, C. W. Thornber, and J. Staunton, *Chem. Commun.*, 1214 (1965).

borohydride reduction, yielded the dienol **59**, and acid-catalyzed rearrangement of **60** with concentrated sulphuric acid gave the neospirinedienone **63** in 80% yield. When **63** was heated with magnesium iodide and the products reduced with lithium aluminum hydride and then methylated, protostephanine (**62**) was formed.

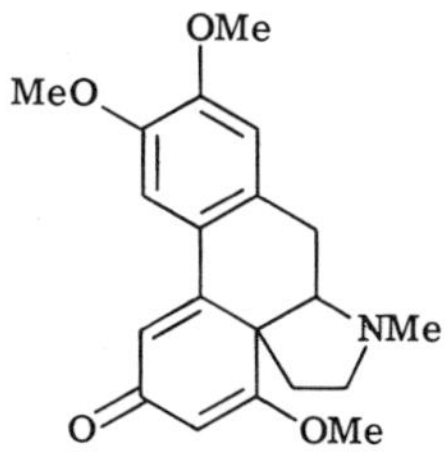

63

The intermediacy of a dibenzazonine structure (**65**) has also been proposed in the biosynthesis of the *Erythrina* alkaloids.[66] The hypothetical source of this dibenzazonine is, however, different from the source of protostephanine (**62**), with a proerythrinadienone (**39**) rather than a morphinandienol (**59**) being precursorial. Barton proposed the route shown in Scheme 5 for the biosynthesis of the *Erythrina* alkaloids in which oxidative coupling of norprotosinomenine (**37**) yields the proerythrinadienone **39**. However, rather than undergoing the dienone–phenol rearrangement to an aporphine that Battersby proposed in *Dicentra eximia*,[61] the proerythrinadienone rearranges to the ring-opened product **64**, which is reduced to the dibenazaonine **65**. Further modifications of **65** lead to erysodienone (**67**) which can be modified to other *Erythrina* alkaloids such as erythraline (**68**). Barton has shown that (+)-norprotosinomenine is, indeed, incorporated into erythraline (**68**) by feeding (+)-[5-^{3}H, 3-^{14}C]-*N*-norprotosinomenine to *Erythrina crista* galli.[67] Likewise, he showed that *O*,*O*-dimethyl[^{14}C-,1,17-$^{3}H_2$]erysodienone (**67**) was incorporated into erythraline (**68**). In both cases the tritium to carbon-14 activity ratios in the isolated product was unchanged from that in the substrate. The dibenzazonine **65** has been tested as a precursor for erythraline (**68**).[68] When **65** was labeled with tritium in C-3 and C-9 by base-catalyzed exchange in tritium oxide and fed to *E. crista*, labeled erythraline (**68**) was isolated. Furthermore, when this tritiated dibenzazonine was fed with **65** labeled in the methoxy groups with carbon-14, the tritium to

[66] D. H. R. Barton, R. James, G. W. Kirby, and D. A. Widdowson, *Chem. Commun.*, 266 (1967).
[67] D. H. R. Barton, R. B. Boar, and D. A. Widdowson, *J. Chem. Soc. C*, 1529 (1968).
[68] D. H. R. Barton, R. James, G. W. Kirby, D. W. Turner, and D. A. Widdowson, *J. Chem. Soc. C*, 1529 (1968).

carbon-14 ratio in the erythraline isolated was the same as that in the precursors fed. Thus, the intermediacy of protosinomenine (**37**), the dibenzazonine **65**, and erysodienone (**67**) in the biosynthesis of the *Erythrina* alkaloids has been established (Scheme 5).

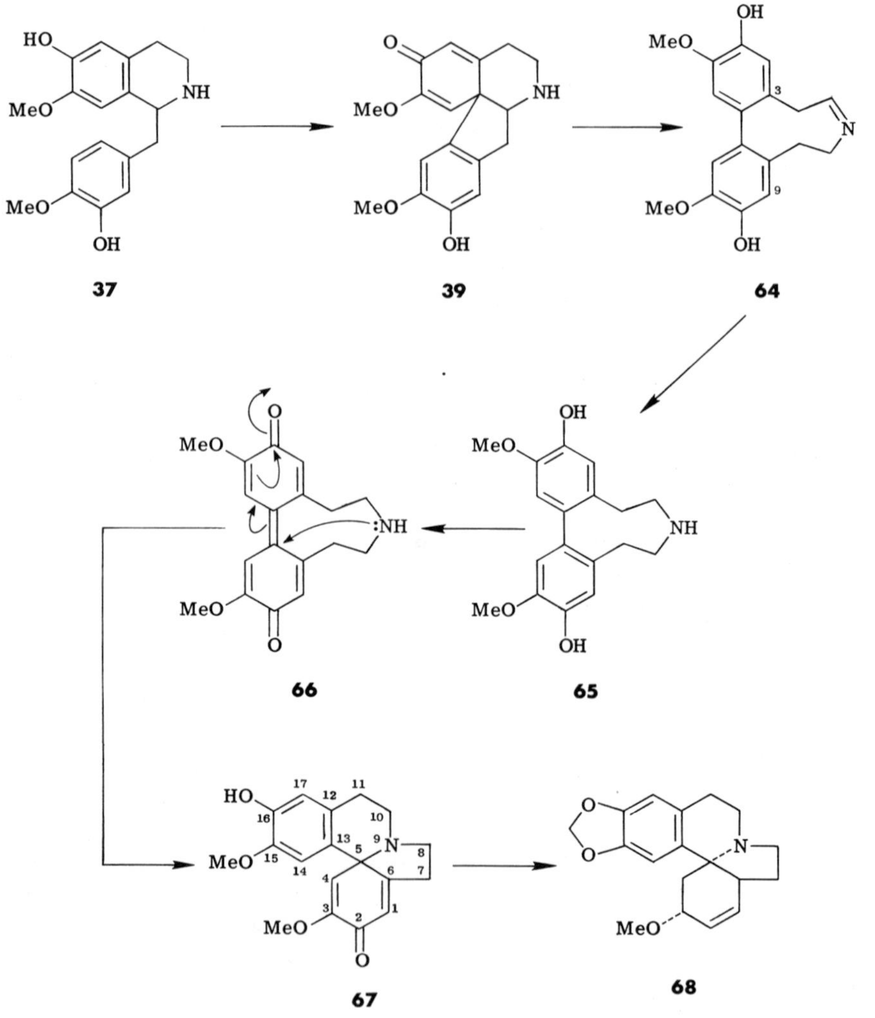

SCHEME 5

In 1971, Franck prepared the *N*-mesylmorphinandienone **71** in 34% yield by oxidation of the *N*-mesylbenzyltetrahydroisoquinoline **69** with vanadium oxytrichloride.[69] Treatment of **71** with boron trifluoride etherate gave the dehydrodibenzazonine **73** in 63% yield. Catalytic reduction of **73** followed by demesylation with lithium in ammonia gave the dibenzazonine **74** (58% yield) which was oxidized with potassium ferricyanide to give 14-methoxy-erysodienone (**75**) in 61% yield (Scheme 6). These results prompted Franck to suggest that norreticuline (**70**) and a morphinandienone intermediate (**72**) might be considered biosynthetic precursors for erysodienone (**67**). Barton has shown in *E. crista* galli that norreticuline (**70**) is not incorporated into the *Erythrina* alkaloids.[70]

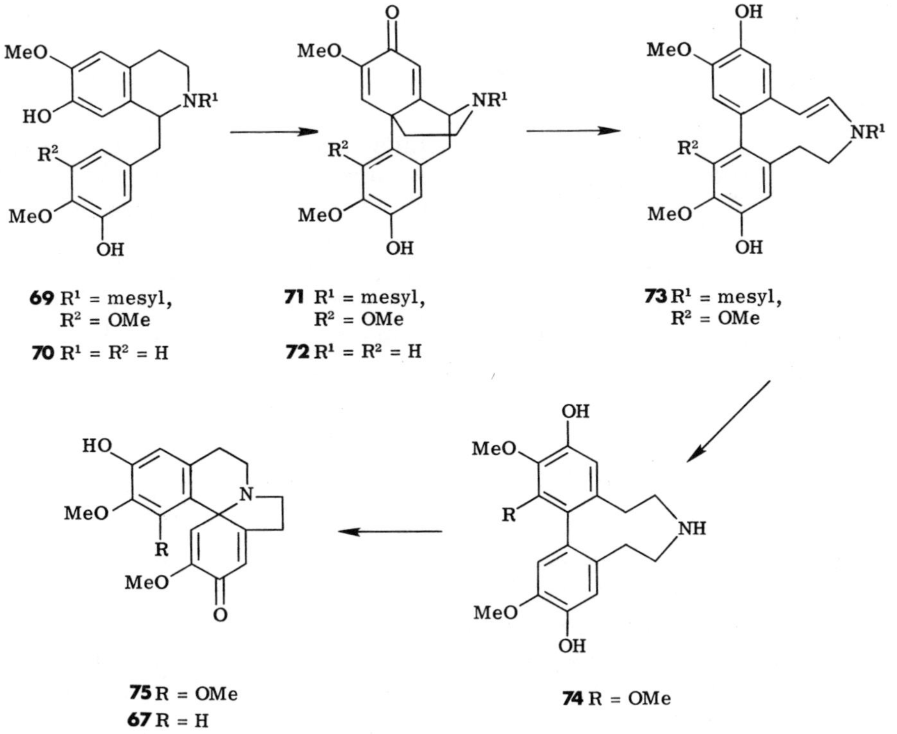

SCHEME 6

A biomimetic synthesis of the key *Erythrina* alkaloid precursor dibenzazonine **65** has been reported recently.[71] The readily accessible benzyltetrahydroisoquinoline (**76**) was oxidized with VOF_3 in CH_2Cl_2 to give the

[69] B. Franck and V. Teetz, *Angew. Chem. Int. Ed. Eng.* **10**, 411 (1971).
[70] D. H. R. Barton, C. J. Pottor, and D. A. Widdowson, *J. Chem. Soc. Perkin 1*, 346 (1974).
[71] S. M. Kupchan, C.-K. Kim, and J. T. Lynn, *J. Chem. Soc. Chem. Commun.* **86** (1976).

proerythrinadienone **77** in 40% yield. Successive base-catalyzed hydrolysis and hydride reduction afforded the dibenzazonine **65**.

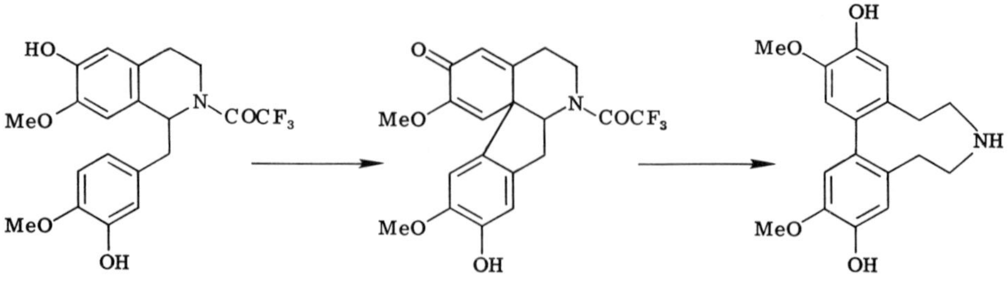

76 **77** **65**

A large number of isoquinoline alkaloids have been synthesized using diphenol oxidative coupling as the key step for forming the C—C bond between two aromatic rings.[72] The yield of the coupling step is usually very low and, in general, this methodology does not afford practical routes to alkaloids. Biogenetic methods are usually acceptable only when small amounts of alkaloids are needed to confirm structures or when one is interested in biosynthetic pathways. However, recent introduction of new reagents for nonphenolic,[10–16] monophenolic,[13,16–19] and diphenolic[16] oxidative couplings together with an emphasis on nitrogen protection

(as $\rangle$N → BH_3,[73] $\rangle NCO_2Et$,[74] and $\rangle NCOCF_3$[10–13])

offers considerable hope for the future. McDonald[75] has recently drawn attention to another factor, the removal of unfavorable conformational interactions in the transition state for cyclization by suitable design of the substrate molecule. Oxidation of the diphenols **78a**[76] and **78b**[77] with potassium ferricyanide has been shown to give a complex mixture of products from which no dienone could be isolated.

[72] T. Kametani, *in* "The Total Synthesis of Natural Products" (John ApSimon, ed.), Vol. 3, p. 1. Wiley, New York, 1977 and references cited therein.

[73] M. A. Schwartz, B. F. Rose, and B. Vishnuvajjala, *J. Am. Chem. Soc.* **95**, 612 (1973).

[74] M. A. Schwartz and I. S. Mami, *J. Am. Chem. Soc.* **97**, 1239 (1975).

[75] E. McDonald and A. Suksamrarn, *Tetrahedron Lett.*, 4421 (1975).

[76] D. H. R. Barton, R. B. Boar and D. A. Widdowson, *J. Chem. Soc. C*, 1208 (1970).

[77] E. McDonald and A. Suksamrarn, *J. Chem. Soc. C*, 440 (1978).

[78] E. McDonald and R. D. Wylie, *Tetrahedron* **35**, 1415 (1979).

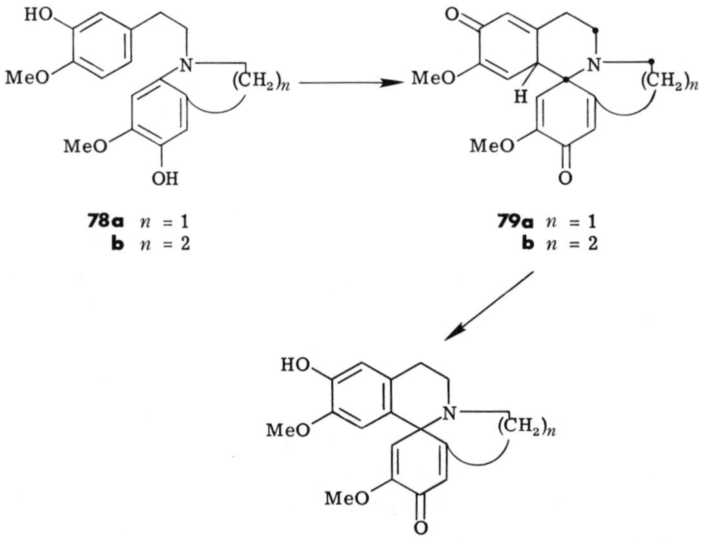

78a $n = 1$
b $n = 2$

79a $n = 1$
b $n = 2$

80a $n = 1$
b $n = 2$

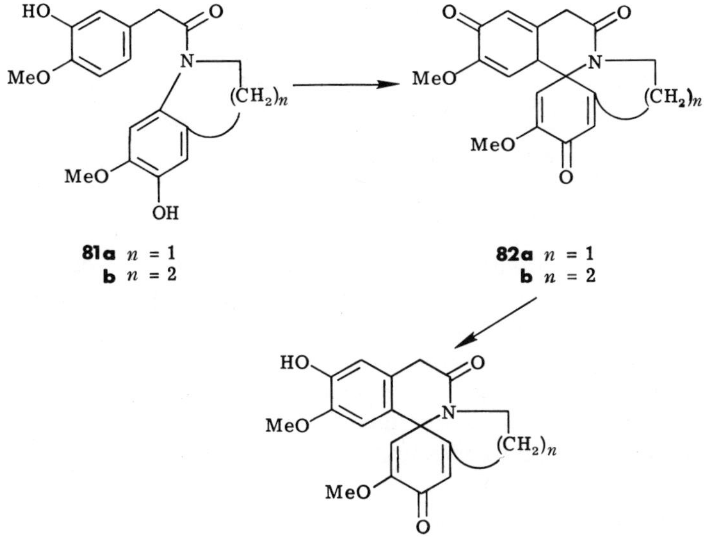

81a $n = 1$
b $n = 2$

82a $n = 1$
b $n = 2$

83a $n = 1$
b $n = 2$

However, when the diphenolic amides **81a** and **81b** were oxidized with potassium ferricyanide, dienone lactams **83a**[78] and **83b**[77] were isolated in 14% and 67% yield, respectively. McDonald has suggested that the formation of polymeric products in the case of **78a** and **78b** may be due to unfavorable 1,3-diaxial interactions between the substituents at the marked positions in the intermediate **79**. Replacement of the sp^3-hydridized carbon atom adjacent to the nitrogen atom in **78** by an sp^2-hybridized atom (>C=O) in **81** removes two of the three interactions in the intermediate **82** thus lowering the activation energy of the transition state for cyclization leading to intramolecular coupled products.

Another interesting development is the use of 1-benzyltetrahydroisoquinoline *N*-oxides for the synthesis of aporphines.[79] Thus, reaction of (±)-reticuline *N*-oxide (**84**) with cuprous chloride in methanol for 20 hr at 10°–15°C in the absence of oxygen, followed by treatment with sodium hydrosulfite, afforded (±)-corytuberine (**85**) in 61% yield. It is interesting to note that cuprous chloride favors ortho–ortho over the ortho–para coupling that is generally observed with the majority of other coupling reagents.

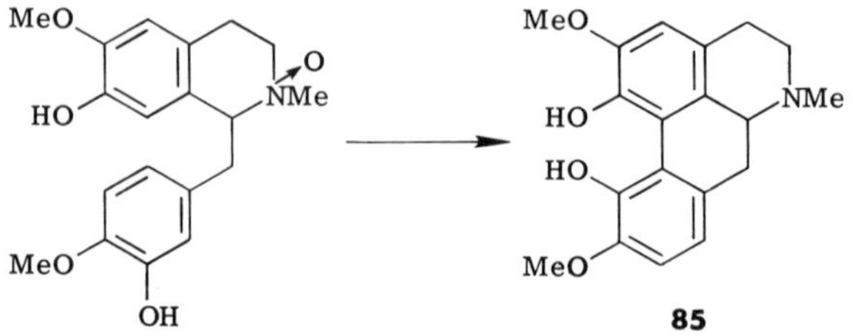

84 **85**

B. NONPHENOLIC SUBSTRATES

Recently, attention has been directed toward utilization of nonphenolic substrates in an attempt to develop effective intramolecular coupling methods for alkaloid synthesis.[80–82] Thus, Miller converted laudanosine (**86**) to *O*-methylflavinantine (**87**) in 52% yield by anodic oxidation of laudanosine in acetonitrile.[80] Tobinaga later improved the yield of (±)-*O*-methylflavinantine to 70% by anodic oxidation of laudonasine in fluoroboric acid–acetonitrile.[81] Kim and Kupchan have reported the electro-oxidative

[79] T. Kametani and M. Ihara, *Heterocycles* **12**, 893 (1979).

[80] L. L. Miller, F. R. Stermitz, and J. R. Falck, *J. Am. Chem. Soc.* **93**, 5941 (1971); L. L. Miller, F. R. Stermitz, and J. R. Falck, *J. Am. Chem. Soc.* **95**, 2651 (1973).

[81] E. Kotani and S. Tobinaga, *Tetrahedron Lett.*, 4759 (1973).

[82] C.-K. Kim and S. M. Kupchan, *J. Am. Chem. Soc.* **97**, 5623 (1975).

coupling in fluoroboric acid–acetonitrile of (±)-laudanosine (**86**) to (±)-*O*-methylflavinantine (**87**) in 94% yield.[82]

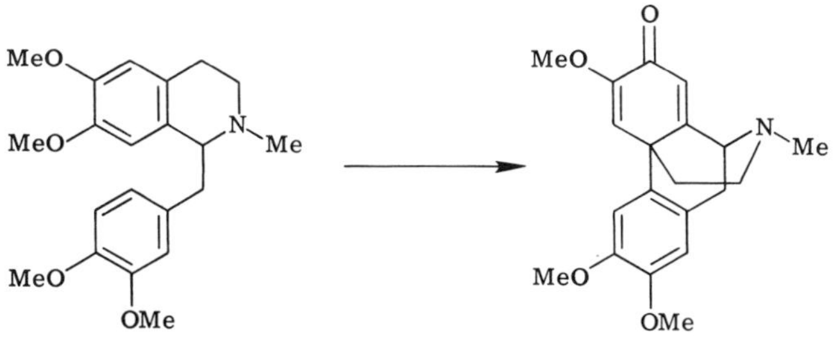

86 **87**

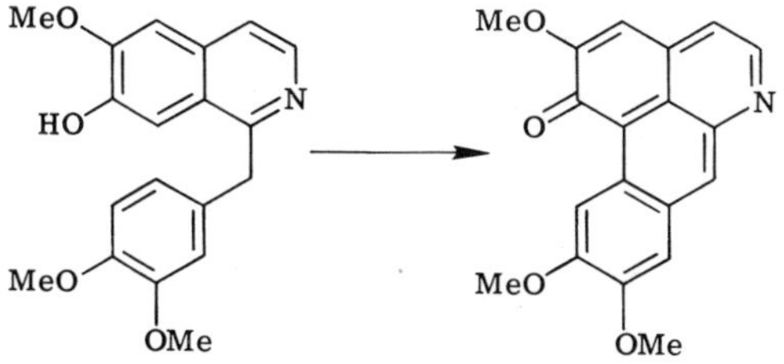

88 **89**

In 1973 Liepa and Kupchan made the fortuitous discovery that the monophenolic benzylisoquinoline **88** developed an orange-red coloration reminiscent of the quinonoid oxoaporphine **89** upon spraying an analytical silica gel plate with 3% ceric sulfate in 3 *N* sulfuric acid (a visualization aid). In a subsequent preparative experiment, conducted at 0°C by addition of **88** to ceric sulfate (4 mol equiv) in 10% sulfuric acid, **89** was obtained in 25% yield. Optimization of the yield for this oxidative coupling revealed that vanadium oxytrifluoride in trifluoroacetic acid (VOF_3–TFA) was an extremely effective system, producing yields of 59% of the conversion.[83] This system was subsequently extended to the oxidation of nonphenolic benzyltetrahydroisoquinolines.[10,11] Thus VOF_3–TFA oxidation of the nonphenolic *N*-formylnorlaudanosine (**90a**) provided in 55% yield the neospirodienone (**92a**). The formation of the neospirenedienone **92** raised questions about the mechanism of its formation since the spirodienone could not be formed directly by oxidative coupling, but rather must result from rearrangement of one or more intermediates. By using appropriately deuterated *N*-formylnorlaudanosine, it was shown[11] that neospirenedienone (**92**) formation involves morphinandienone-type intermediate **91** (Scheme 7).

[83] S. M. Kupchan and A. J. Liepa, *J. Am. Chem. Soc.* **95**, 4062 (1973).

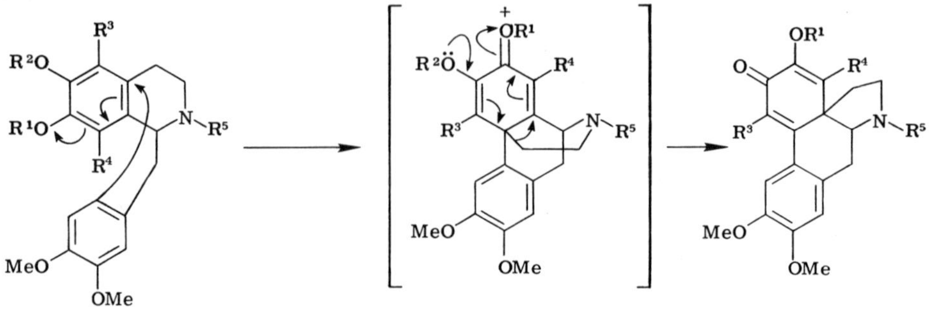

90a-g **91a-g** **92a-g**

a $R^1 = R^2 = Me$; $R^3 = R^4 = H$; $R^5 = CHO$
b $R^1 = R^2 = Me$; $R^3 = R^4 = H$; $R^5 = COCF_3$
c $R^1 = R^2 = Me$; $R^3 = R^4 = H$; $R^5 = COCH_3$
d $R^1 = CH_2Ph$; $R^2 = Me$; $R^3 = R^4 = H$; $R^5 = CHO$
e $R^1 = Me$; $R^2 = CH_2Ph$; $R^3 = R^4 = H$; $R^5 = CHO$
f $R^1 = R^2 = Me$; $R^3 = D$; $R^4 = H$; $R^5 = CHO$
g $R^1 = R^2 = Me$; $R^3 = H$; $R^4 = D$; $R^5 = CHO$

SCHEME 7

In contrast to the VOF_3–TFA oxidation of (±)-*N*-acylnorlaudanosines to neospirenedienones, oxidation of (±)-laudanosine (**86**) gives glaucine (**43**) in 43% yield.[11] Kupchan and Ramachandran have shown recently[84] that anodic oxidation of (±)-laudanosine (**86**) in trifluoroacetic acid–trifluoroacetic anhydride (TFA–TFAA) at a constant potential of 1.4 V for 130 min, followed by heating the reaction mixture on a steam bath for 2 hr, gives glaucine (**43**) in 80% yield. If heating of the anolyte after the electrolysis is avoided, the yield of glaucine drops to 28%. In view of the isolation of (±)-*O*-methylflavinantine by anodic oxidation of (±)-laudanosine in acetonitrile–fluoroboric acid, the formation of glaucine by anodic as well as VOF_3 oxidation in TFA–TFAA can best be interpreted in terms of an intermediate morphinandienone (**93**), which rearranges to proerythrinadienone (**94**) and then leads to glaucine (**91**) (Scheme 8).

Elliott has recently[85,86] reported on anodic as well as vanadium oxytrifluoride induced oxidative coupling of isochromanone **95a** and isoquinolone **95b**. Thus, anodic oxidation of **95a** in CH_2Cl_2–TFA with tetrabutyl-

[84] S. M. Kupchan and V. Ramachandran, unpublished results.
[85] I. W. Elliott, Jr., *J. Org. Chem.* **42**, 1090 (1977).
[86] I. W. Elliott, Jr., *J. Org. Chem.* **44**, 1162 (1979).

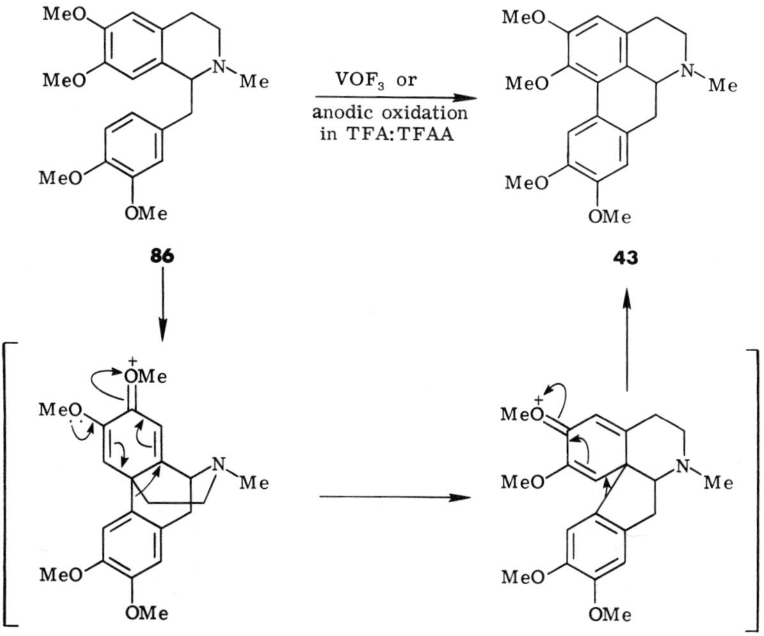

SCHEME 8

ammonium tetrafluoroborate as supporting electrolyte gave spirodienone **96a** in 33% yield. The same spirodienone **96a** was isolated in 59% yield from the oxidation of **95a** by VOF_3 in CH_2Cl_2–TFA. Oxidation of **95b**, either by anodic oxidation or VOF_3 under the above conditions, gave 40% yield of the dienone–lactam **96b**. If, on the other hand, anodic oxidation of **95a** was carried out in acetonitrile, the bridged lactone **97a** was isolated in 20% yield. The same bridged lactone **97a** was isolated in 71% yield from the oxidation of **95a** by VOF_3 in acetonitrile. Anodic oxidation of **95b** in acetonitrile did not yield the corresponding bridged lactam **97b**. However, **97b** was isolated in 40% yield from oxidation of **95b** by VOF_3 in acetonitrile. The structure of the bridged lactam **97b** was confirmed by reducing it to (±)-*O*-methylthalisopavine (**98**) with borane–THF complex.

The alklaoid thalicarpine,[28] which has the aporphine–benzylisoquinoline structure **2**,[29] exhibits a significant inhibitory activity against the Walker 256 intramuscular carcinosarcoma in rats over a wide dosage range.[87] In order to gain some insight into the structure–tumor inhibitory activity relationship

[87a] S. M. Kupchan, *Proc. Pan American Congr. Pharm. Biochem., 7th Buenos Aires*, December, 1966; *Chem. Eng. News* **44**, 64 (1966).

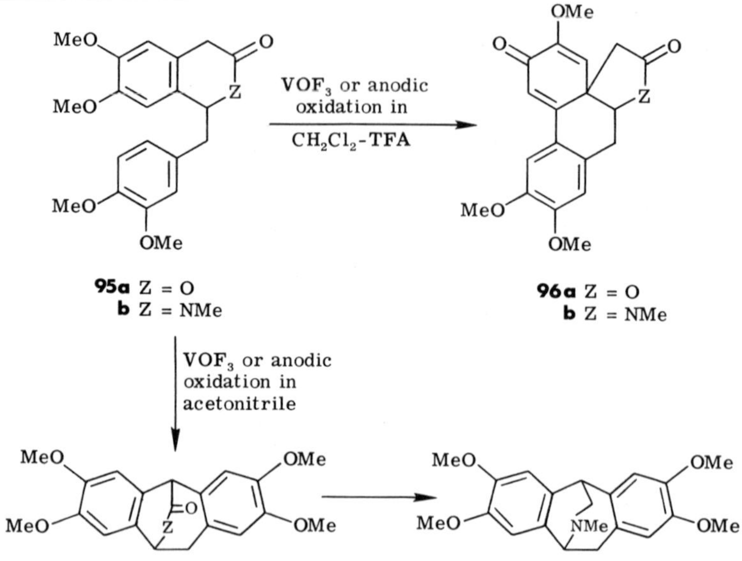

95a Z = O
b Z = NMe

96a Z = O
b Z = NMe

97a Z = O
b Z = NMe

98

in this alkaloid series, Kupchan *et al.* undertook studies directed toward structural modifications of thalicarpine.[88] On the basis of the results of chemical as well as anodic oxidative coupling reactions of nonphenolic benzyltetrahydroisoquinolines (*vide supra*), VOF_3 oxidation of thalicarpine would be expected to give the bisaporphine **99**, and anodic oxidation, the morphinandienoneaporphine dimer **100**.

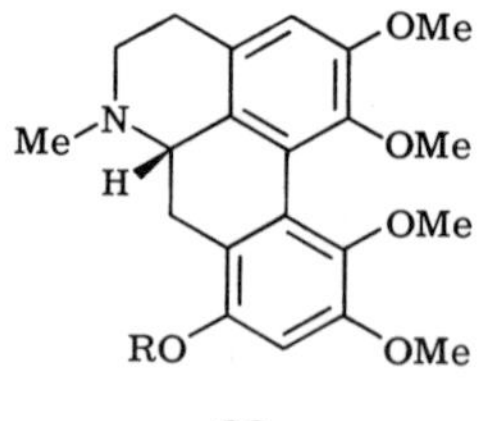

99

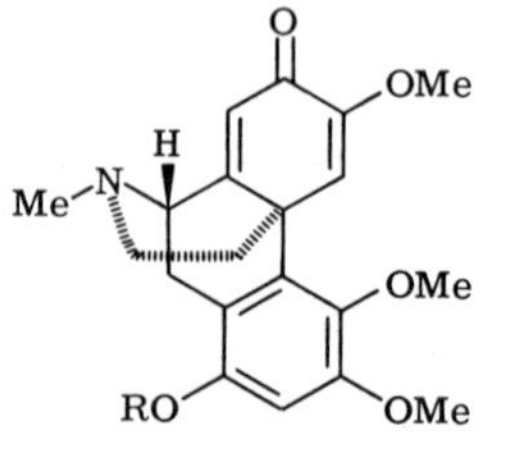

100

Quite unexpectedly, VOF_3–TFA:TFAA oxidation of thalicarpine (**2**) gave a mixture of dienones **101** and **102**, diastereomers at the spiro ring

[87b] J. L. Hartwell and B. J. Abbott, *Adv. Pharmacol. Chemother.* **7**, 117 (1969).
[88] S. M. Kupchan, Om P. Dhingra, V. Ramachandran, and C.-K. Kim, *J. Org. Chem.* **43**, 105 (1978).

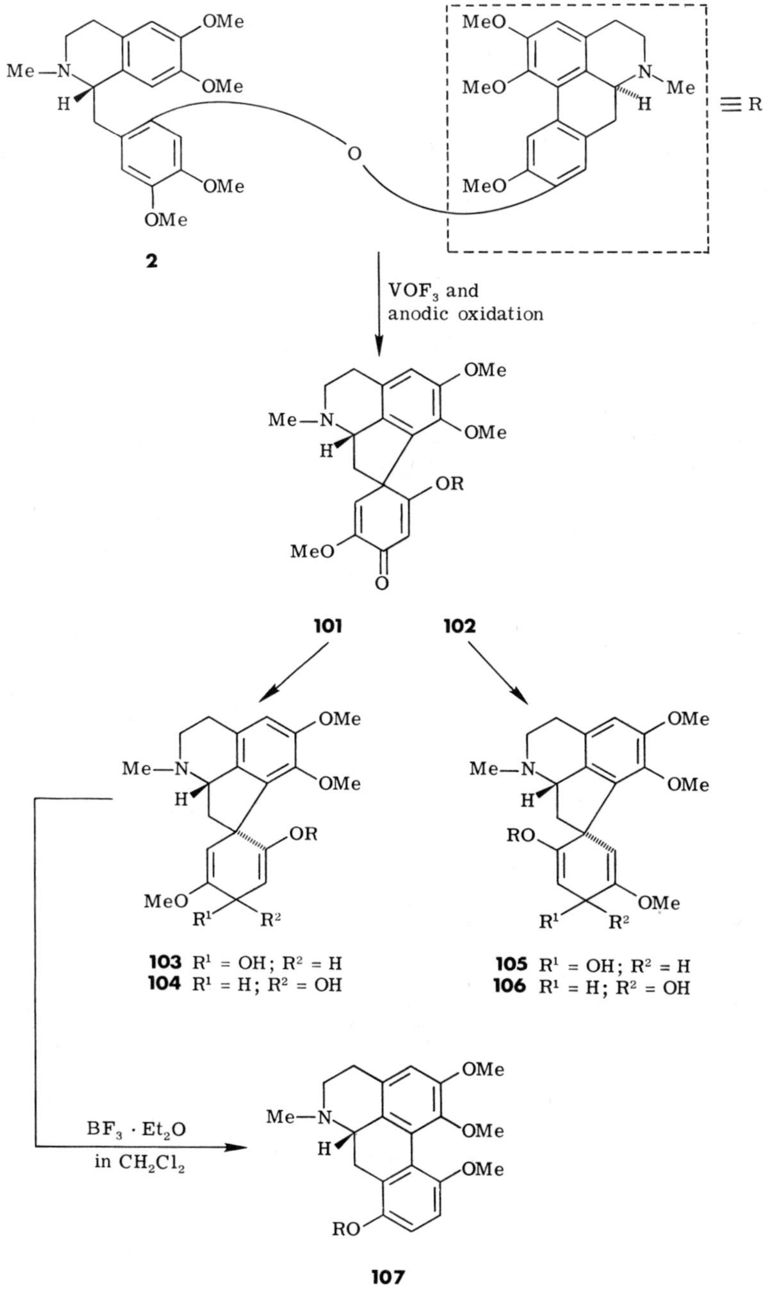
2
≡ R
VOF₃ and
anodic oxidation
101
102
103 R¹ = OH; R² = H
104 R¹ = H; R² = OH
105 R¹ = OH; R² = H
106 R¹ = H; R² = OH
BF₃ · Et₂O
in CH₂Cl₂
107

SCHEME 9

junction, in 72% and 5% yield, respectively. On the other hand, anodic oxidation of thalicarpine (**2**) in TFA–TFAA using tetraethylammonium tetrafluoroborate as supporting electrolyte give the dienones **101** and **102** in yields of 40% and 5%, respectively. The major isomer **101** was converted to the epimeric dienols **103** and **104** by sodium borohydride reduction. Treatment of **103** or **104** with boron trifluoride etherate in dichloromethane gave a new type of alkaloid bisaporphine **107** (Scheme 9). The dienols **105** and **106** obtained by borohydride reduction of the minor dienone **102**, give only cleavage products upon treatment with $BF_3 \cdot Et_2O$ in CH_2Cl_2.

Since morphinandienones were postulated as intermediates in the aforementioned transformations (Schemes 7 and 8), it was appropriate that their acid-catalyzed rearrangement also be investigated. Indeed, it was established by Kim and Kupchan[89] that they too lead to aporphines via proerythrinadienones. Thus treatment of (±)-*O*-methylflavinantine (**87**) with concentrated hydrochloric acid gave aporphine **109** in 98% yield.

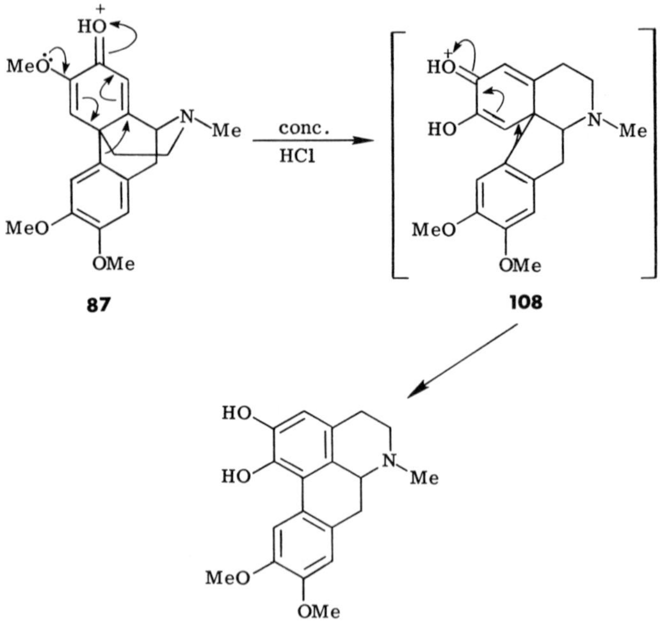

87 **108** **109**

On the other hand, if, boron trifluoride etherate is used instead of concentrated hydrochloric acid, alkyl rather than aryl migration occurs, the

[89] S. M. Kupchan and C.-K. Kim, *J. Am. Chem. Soc.* **97**, 5623 (1975).

product after catalytic reduction being the dibenzazonine alkaloid erybidine **112**.

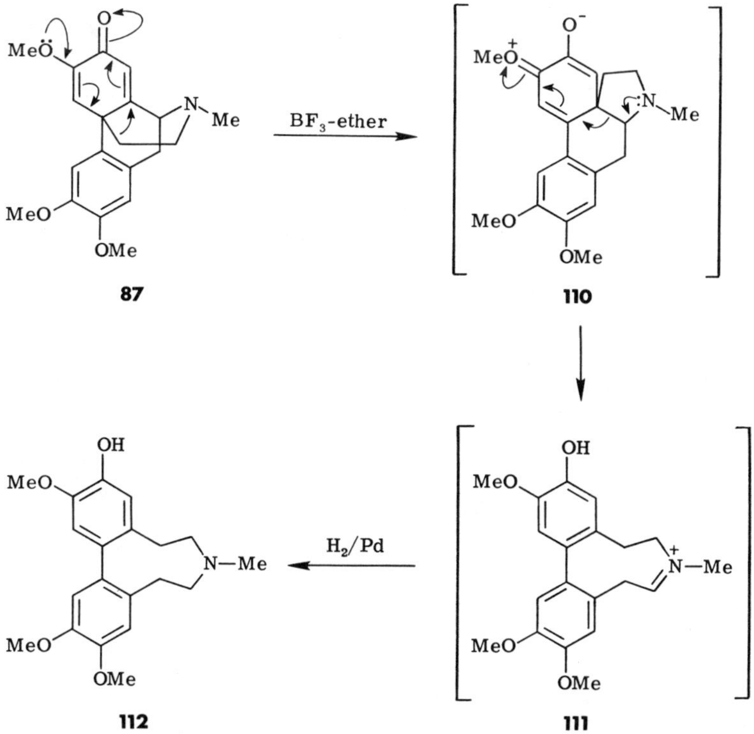

The acid-catalyzed rearrangement of morphinandienones can thus lead either to aporphines or dibenzazonines. The rearrangement to aporphines appears favored in reactions involving substrates and conditions that may enhance the participation of the nitrogen free electron pair, possibly through the intermediacy of a species such as **113**. The Lewis acid, boron trifluoride

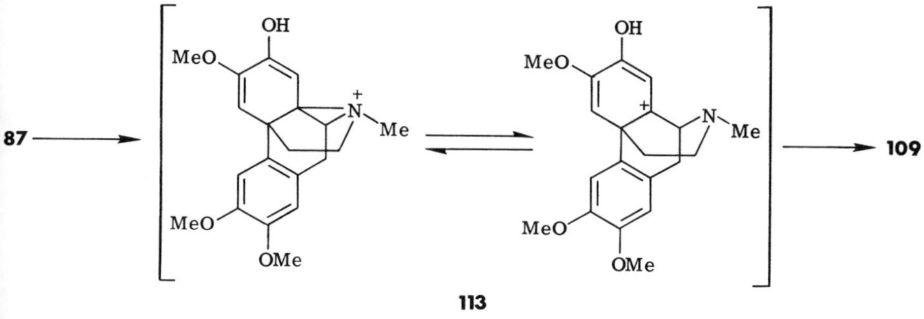

etherate, on the other hand, complexes with the nitrogen free electron pair and encourages alkyl migration, with the formation of neospirenedienones which then lead to dibenzazonines.[89]

Proerythrinadienones, proposed as intermediates in the aforementioned transformation of morphinandienones to aporphines, have also been proposed as biosynthetic precursors of aporphines and *Erythrina* alkaloids

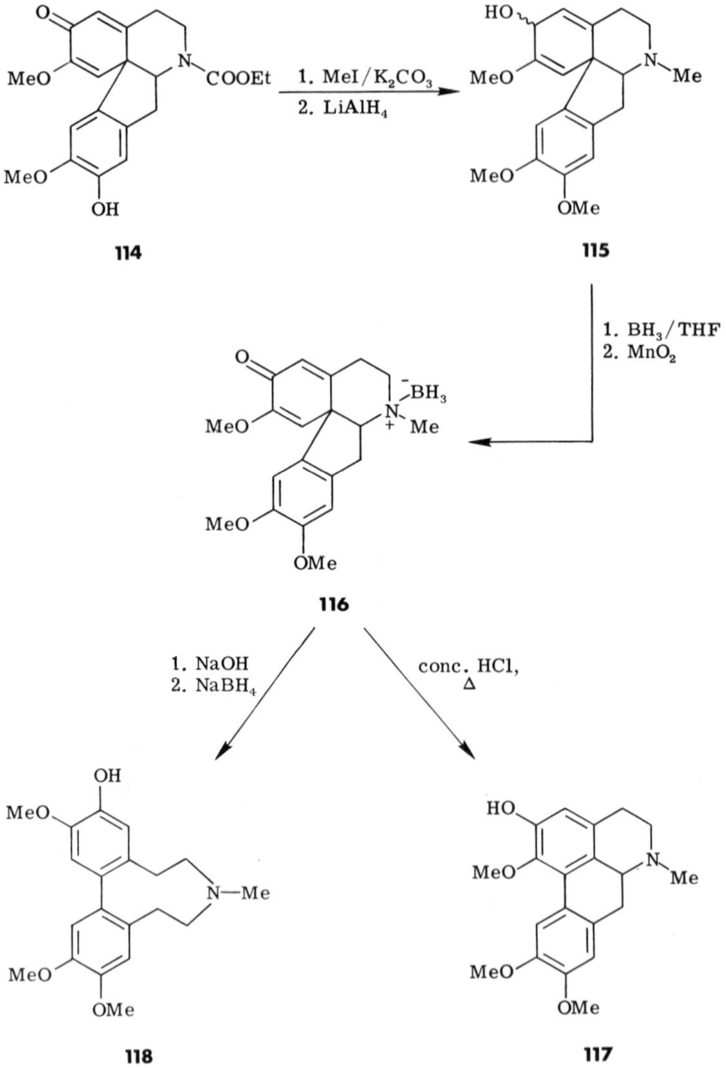

SCHEME 10

(Schemes 2 and 5). Attempts to synthesize such dienones have failed when the nitrogen atom is unprotected, and attempted transformations of *N*-acylproerythrinadienones to aporphines and dibenzazonines have also been unsuccessful.[90,91] Support for the intermediacy of proerythrinadienones and neosprienedienones in the *in vitro* conversion of morphinandienones to aporphines and dibenzazonines came when Kim and Kupchan demonstrated that *N*-methylproerythrinadienone (**116**) and *N*-methylneospirenedienone (**119**), synthesized as borane complexes, are genuine precursors of aporphine and dibenzazonines[92] (Schemes 10 and 11).

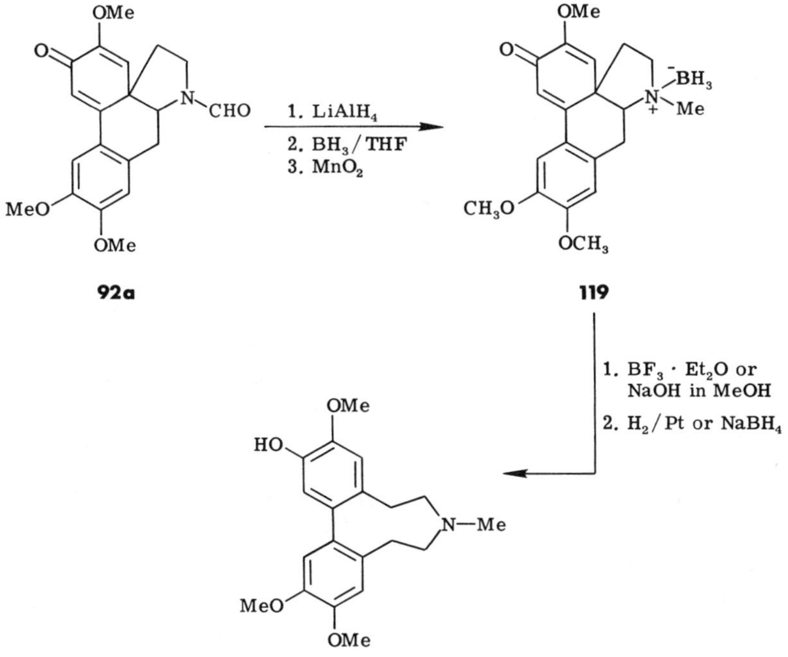

120

SCHEME 11

The results of rearrangements of spirodienones under various conditions are summarized in Table I. The results are much more interesting at elevated temperature. Thus when (±)-*O*-methylflavinantine (**87**) or its borane complex **121** was treated with boron trifluoride etherate in benzene under

[90] D. H. R. Barton, R. B. Boar, and D. A. Widdowson, *J. Chem. Soc. C*, 1213 (1970).

[91] T. Kametani, K. Takahashi, T. Sugahara, M. Koizumi, and K. Fukumoto, *J. Chem. Soc. C*, 1032 (1971).

[92] S. M. Kupchan and C.-K. Kim, *J. Org. Chem.* **41**, 3210 (1976).

TABLE I

REARRANGEMENTS OF SPIRODIENONES

	Spirodienones				Product yield (%)	
Reaction conditions	Exp. no.	Mor.	Neo.	Pro.	Aporphines	Dibenzazonines
Conc. HCl, steam bath	1	**87**			**109** (89)	
	2	**121**			**109** (78)	
	3			**116**	**117** (75)	
(1) $BF_3 \cdot Et_2O$,rt(2)H_2/Pt,MeOH	4	**87**				**118** (85)
	5	**121**				**118** (83)
	6		**119**			**120** (78)
	7			**116**	**117** (44)	**118** (35)
(1) 1 *N* NaOH,MeOH,rt(2)$NaBH_4$	8			**116**		**118** (76)
	9		**119**			**120** (73)
(1) $BF_3 \cdot Et_2O$,PhH,reflux(2)H_2/Pt,MeOH	10	87			**122** (28), **117** (8), **123** (36)	**118** (8)
	11	**121**			**122** (26), **117** (7), **123** (33)	**118** (9)
	12			**116**	**117** (47), **123** (8)	**118** (24)
	13		**119**		**122** (8) **124** (25)	**120** (35)

reflux, followed by hydrogenation over platinum in methanol (Table I, expts. 10 and 11), four products (±)-thalicmidine (**122**), (±)-predicentrene (**117**), erybidine (**118**), and (±)-3-hydroxy-2,8,10-trimethoxyaporphine (**114**) were obtained.

Under the reaction conditions used, the formation of these products can best be rationalized in terms of an equilibrium between morphinandienone, neospirenedienone, and proerythrinandienone (Scheme 12).

The (±)-*N*-methylproerythrinadienone–borane complex (**116**) under the same reaction conditions (Table I, expt. 12) gave predicentrene (**117**), erybidine (**118**), and (±)-3-hydroxy-2,8,10-trimethoxyaporphine (**123**), indicating that the reaction proceeds as shown in Scheme 13.

(±)-*N*-Methylneospirenedienone–borane complex (**119**) under the same reaction conditions (Table I, expt. 13) yielded (±)-thalicmidine (**122**), erybidine isomer (**120**), and (±)-2-hydroxy-3,9,10-trimethoxyaporphine (**124**). The genesis of these products is shown in Scheme 14.

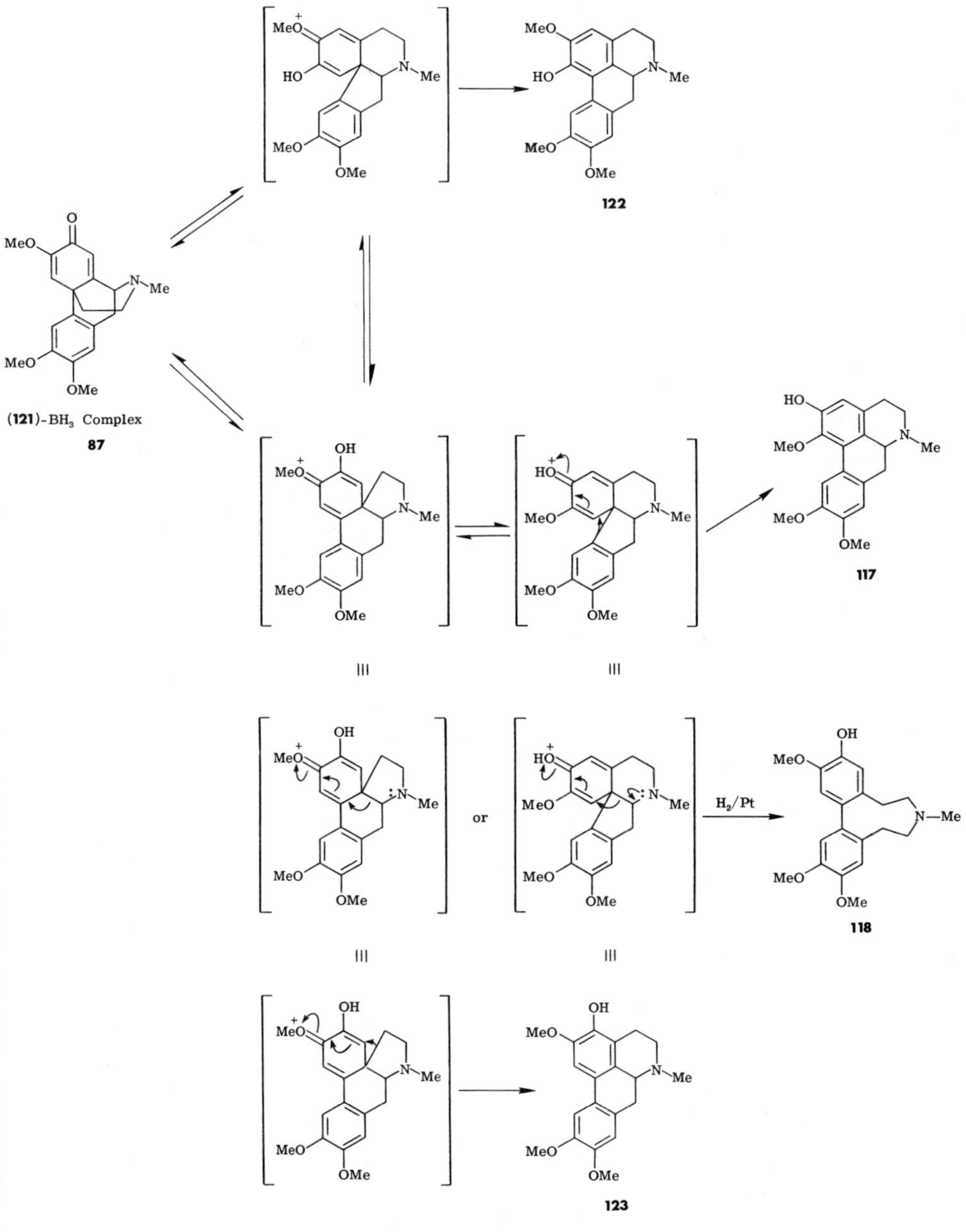

SCHEME 12

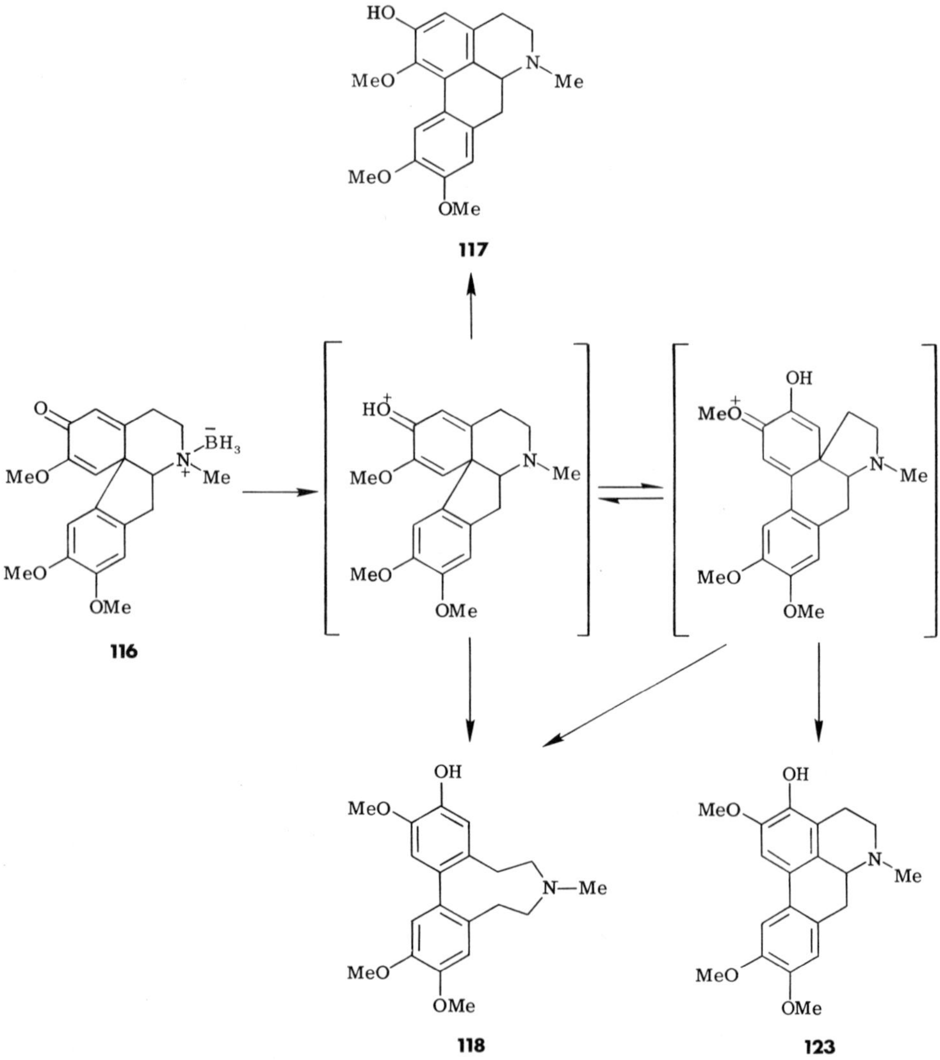

SCHEME 13

C. MONOPHENOLIC SUBSTRATES

Oxidative coupling of appropriately substituted monophenolic benzyltetrahydroisoquinolines using VOF_3–TFA:TFAA has been shown recently to be one of the simplest and most efficient routes to aporphines.[13] Thus, oxidation of (±)-*N*-trifluoroacetylnorcodamine (**125**) with the above

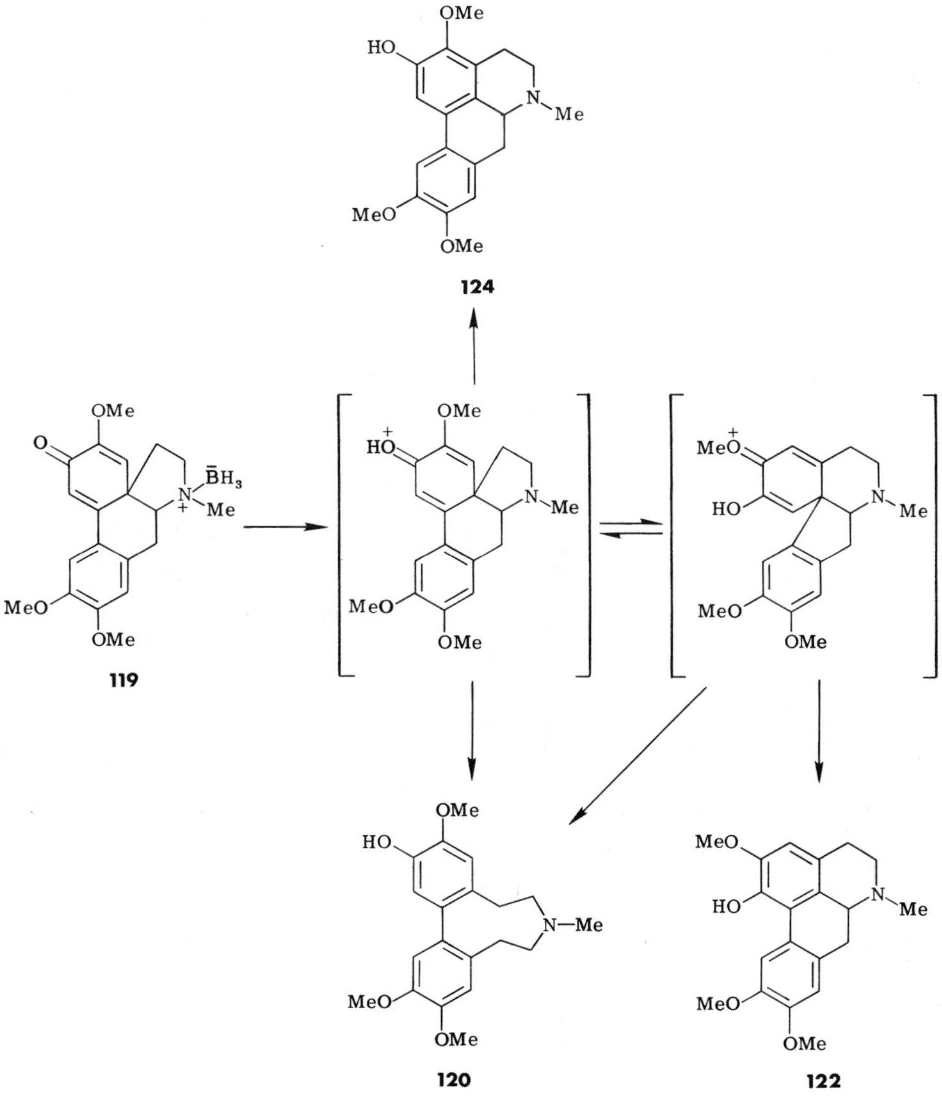

SCHEME 14

reagent gave ($\pm$)-*N*-trifluoroacetylwilsonirine **129** (70%) along with morphinandienone **130** (8%). However, oxidation of codamine (**126**) gave a complex mixture from which only ($\pm$)-thalicmidine **122** (38%) was isolable. Interestingly, when the borane complex of codamine (**127**) was oxidized with VOF_3, thalicmidine (**122**) was obtained in 80% yield.

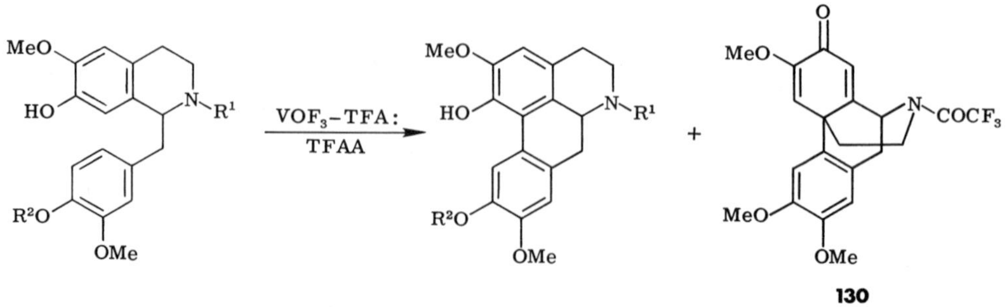

130

125 $R^1 = COCF_3$, $R^2 = Me$
126 $R^1 = R^2 = Me$
127 $R^1 = \langle {}^{\nearrow BH_3}_{\searrow Me}$, $R^2 = Me$
128 $R^1 = \langle {}^{\nearrow BH_3}_{\searrow Me}$, $R^2 = CH_2Ph$

129 $R^1 = COCF_3$, $R^2 = Me$
122 $R^1 = R^2 = Me$
131 $R^1 = Me$, $R^2 = CH_2Ph$
132 $R^1 = Me$, $R^2 = H$

The general applicability of the VOF_3–TFA:TFAA coupling method for the synthesis of 1,2,9,10-tetrasubstituted aporphines was demonstrated by the synthesis of the 1,10-dihydroxy-2,9-di-methoxyaporphine, (±)-bracteoline (**132**). Thus, oxidation of the protected amine **128** gave (±)-10-benzyloxy-1-hydroxy-2,9-dimethoxyaporphine (**131** in 75% yield. Catalytic debenzylation of **131** afforded (±)-bracteoline (**132**) in 78% yield.

Aporphines can also be obtained in fairly good yield by oxidizing appropriately substituted monophenolic benzyltetrahydroisoquinolines with lead tetraacetate in acetic acid at room temperature followed by treatment with trifluoroacetic acid of the resulting *p*-quinol acetate (**133** → **134** → **135**). An excellent review on the subject has recently appeared.[19]

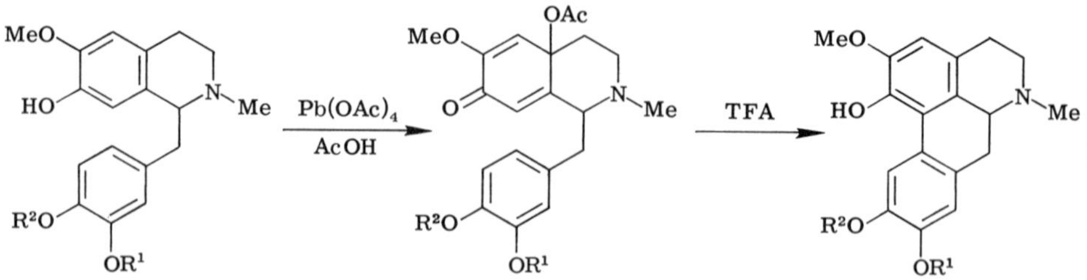

133 a $R^1 = R^2 = Me$
b $R^1 + R^2 = CH_2$
c $R^1 = Me$, $R^2 = CH_2Ph$
d $R^1 = CH_2Ph$, $R^2 = Me$

134 a–d

	Yield (%)
135 a	96
b	84
c	48
d	44

III. Oxidative Coupling of Phenethyltetrahydroisoquinolines

In recent years a number of alkaloids have been identified that can be regarded as derivatives of 1-phenethylisoquinoline. The types of structures so far recognized are simple 1-phenethyltetrahydroisoquinoline, bisphenethyltetrahydroisoquinoline, homoaporphine, homoproaporphine, homomorphinandienones, and homoerythrina alkaloids. The basic structure of each type of alkaloid is shown in Scheme 15. In contrast to the large number

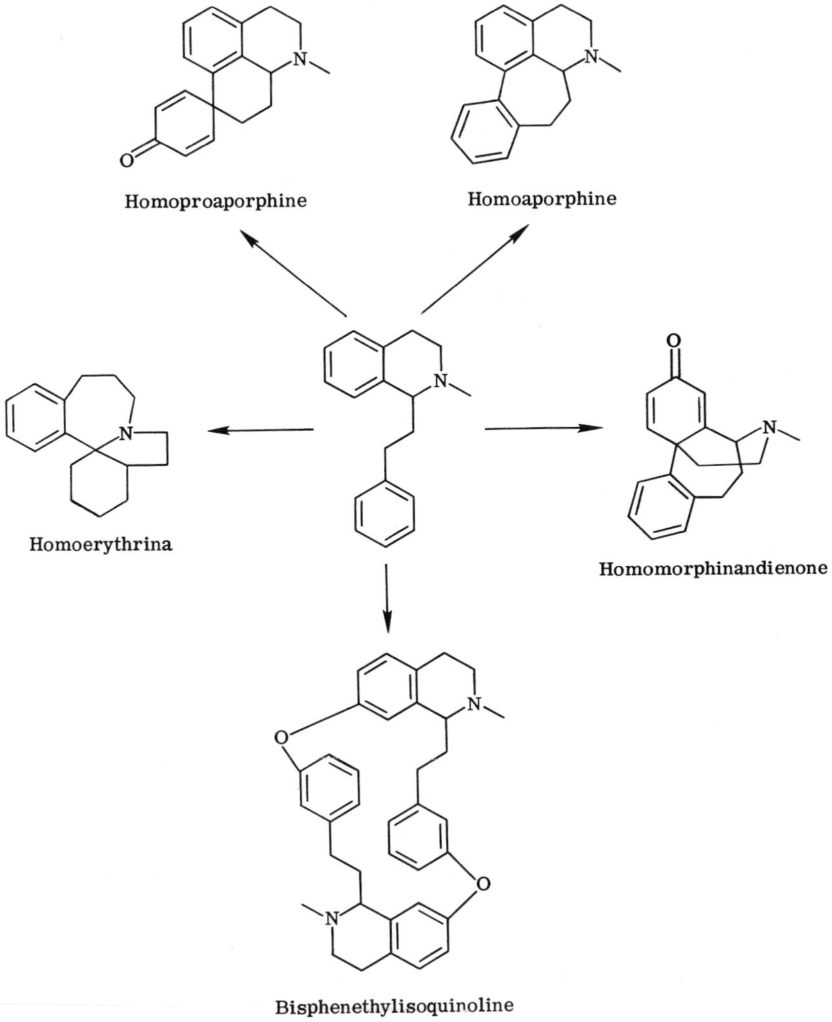

SCHEME 15

of benzylisoquinoline alkaloids, relatively few phenethylisoquinoline alkaloids are known.

A. DIPHENOLIC SUBSTRATES

Kreysiginone (**137**), a homoproaporphine alkaloid, is probably biosynthesized from homoorientaline (**136**) because kreysiginone corresponds to a homolog of orientalinone (**34**) biosynthesized from orientaline (**33**) by phenolic oxidation. Phenolic oxidation of homoorientaline (**136**) has been reported by many groups[93–98] to give a mixture of two dienones that differ in configuration at the spiro center.[99] One of these is identical with kreysiginone **137**. Battersby *et al.*[95] oxidized **136** with potassium ferricyanide and obtained homoproaporphines **137** and **138** in 31% overall yield. Kametani *et al.*[94] obtained an overall yield of 36.6% using ferric chloride as oxidizing agent. Bobbit *et al.*[98] obtained an overall yield of 36% by electro-oxidation on a graphite felt anode in water using tetraethylammonium perchlorate as an electrolyte. Kupchan and co-workers recently reported an overall yield of 68% using VOF_3–TFA:TFAA as the coupling reagent.[12]

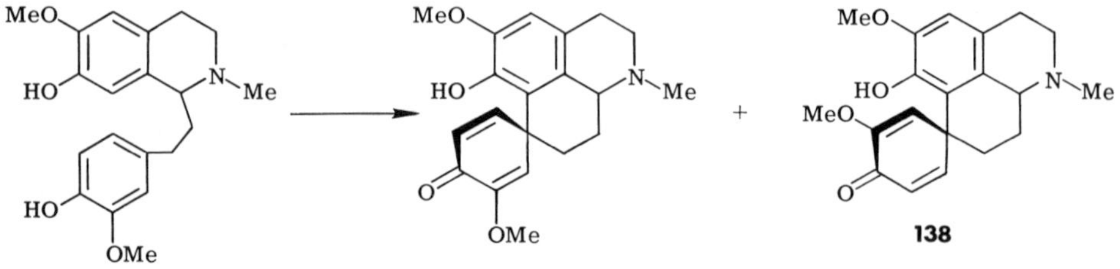

136 **137** **138**

Androcymbine (**140**), a homomorphinandienone alkaloid, is biosynthesized by phenolic oxidation of autumnaline (**139**) and transformed biogenetically into colchicine (**141**), as shown by tracer experiments.[100] Attempts to transform autumnaline (**139**) into androcymbine by *in vitro*

[93] T. Kametani, K. Fukumoto, H. Yagi, and F. Satoh, *Chem. Commun.*, 878 (1967).

[94] T. Kametani, F. Satoh, H. Yagi, and K. Fukumoto, *J. Org. Chem.* **33**, 690 (1968).

[95] A. R. Battersby, M. McDonald, M. H. G. Monro, and R. Ramage, *Chem. Commun.*, 934 (1967).

[96] A. Brossi, J. O'Brien, and S. Teitel, *Helv. Chim. Acta* **52**, 678 (1969).

[97] R. E. Hermon and B. L. Hensen, *J. Heterocycl. Chem.* **7**, 1077 (1970).

[98] J. M. Bobbitt, I. Noguchi, R. S. Ware, Kaolin Ng Chiong, and S. T. Huang, *J. Org. Chem.* **40**, 2924 (1975).

[99] T. Kametani, F. Satoh, H. Yagi, and K. Fukumoto, *Chem. Comm.*, 1103 (1967); *J. Chem. Soc. C*, 1003 (1968); *J. Chem. Soc. C* 382 (1970).

[100] A. C. Baker, A. R. Battersby, M. McDonald, R. Ramage, and J. H. Clements, *Chem. Commun.*, 390 (1967) and references cited therein.

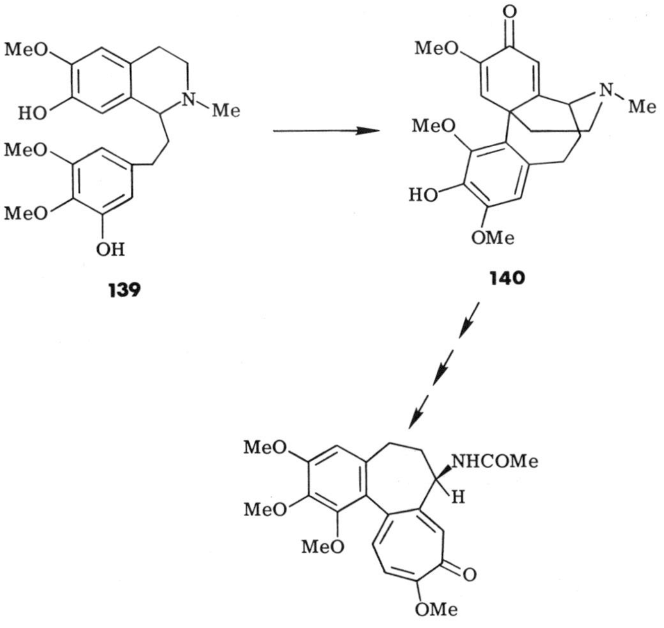

139 **140** **141**

phenolic oxidation has so far been unsuccessful.[101] However, homomorphinandienone **143** has been synthesized in low yield from homoreticuline **142** by ferricyanide oxidation.[101]

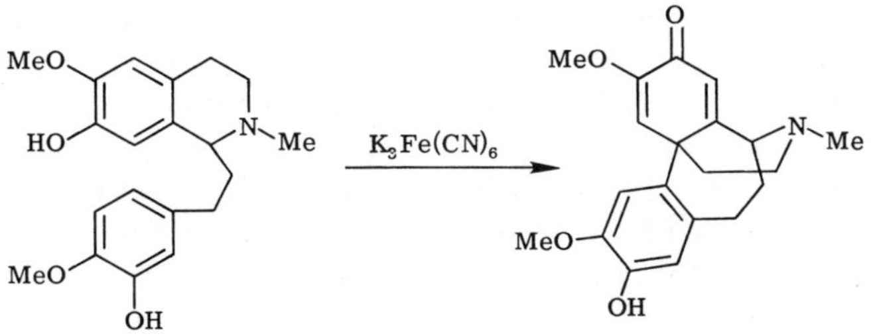

142 **143**

Homoerythrina alkaloids such as schelhammeridine (**144**) are congeners[102] of cephlotaxine (**145**) and its esters: harringtonine (**146**), isoharringtonine

[101] T. Kametani, K. Fukumoto, M. Koizumi, and A. Kozuka, *Chem. Commun.*, 1605 (1968); *J. Chem. Soc. C*, 1295 (1969).

[102] R. G. Powell, and K. L. Mikolajczak, *Phytochemistry* **12**, 2987 (1973); W. W. Paudler and J. McKay, *J. Org. Chem.* **38**, 2210 (1973); S. Asada, *Yakugaku Zasshi* **93**, 916 (1973).

(**147**), homoharringtonine (**148**), and deoxyharringtonine (**149**). The tumor-inhibitory activity of several esters of cepalotaxine (**145**)[103] has generated considerable interest in both the synthesis of these alkaloids[104] and the elucidation of their biogenesis.[105,106]

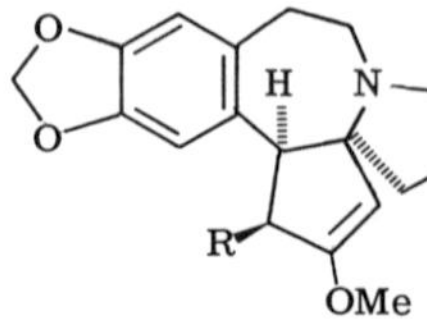

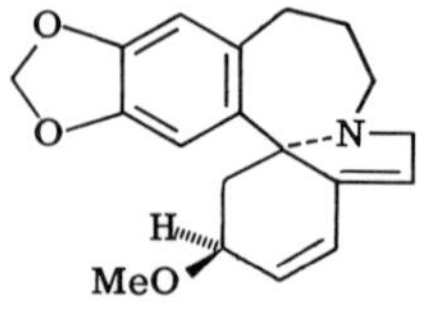

145 R = OH

144

146 R = $CH_3—C(OH)(CH_3)—(CH_2)_2—C(OH)(CO_2^-)—CH_2CO_2CH_3$

147 R = $CH_3—CH(CH_3)—(CH_2)_2—C(HO)(CO_2^-)—CH(OH)—CO_2CH_3$

148 R = $CH_3—C(OH)(CH_3)—(CH_2)_3—C(OH)(CO_2^-)—CH_2CO_2CH_3$

149 R = $CH_3CH(CH_3)—(CH_2)_2—C(OH)(CO_2^-)—CH_2CO_2CH_3$

The presence of *Schelhammera*-type alkaloids[102], e.g., **144**, in *Cephalotaxus* species has led some researchers[107–110] to propose that both the *Schelhammera*-type and *Cephalotaxus* alkaloids are biogenetically related and may be classified as homoerythrina alkaloids. Earlier investigations by Barton *et al.*,[68] have established 1-benzyltetrahydroisoquinoline **37** as the precursor of *Erythrina* alkaloids in a sequence involving phenol oxidative coupling (Scheme 5). Fitzgerald and co-workers[107] extended this concept to the *Homoerythrina* series by postulating a 1-phenethyltetrahydroisoquinoline precursor (Scheme 16, route 1). By an appropriate change in the mode of ring closure, Powell *et al.*, extended Fitzgerald's scheme to the *Cephalotaxus* series[108] (Scheme 16, route 2). The pattern of labeling expected from

[103] K. L. Mikolajczad, R. G. Powell, and C. R. Smith, Jr., *Tetrahedron* **28**, 1995 (1972).

[104] S. M. Weinreb and J. Auerbach, *J. Am. Chem. Soc.* **97**, 2503 (1975); M. F. Semmelhack, B. P. Chong, R. D. Stauffer, T. D. Rogerson, A. Chong, and L. D. Jones, *ibid.* **97**, 2507 (1975); S. M. Weinreb and M. F. Semmelhack, *Acc. Chem. Res.* **8**, 158 (1975).

[105] R. J. Parry and J. M. Schwab, *J. Am. Chem. Soc.* **97**, 2555 (1975).

[106] J. M. Schwab, M. N. T. Chang, and R. J. Parry, *J. Am. Chem. Soc.* **99**, 2368 (1977); R. J. Parry, M. N. T. Chang, J. M. Schwab, and B. M. Foxman, *ibid.* **102**, 1099 (1980).

[107] J. S. Fitzgerald, S. R. Johns, J. A. Lamberton, and A. A. Sioumis, *Aust. J. Chem.* **22**, 2187 (1969).

[108] R. G. Powell, *Phytochemistry* **11**, 1467 (1972).

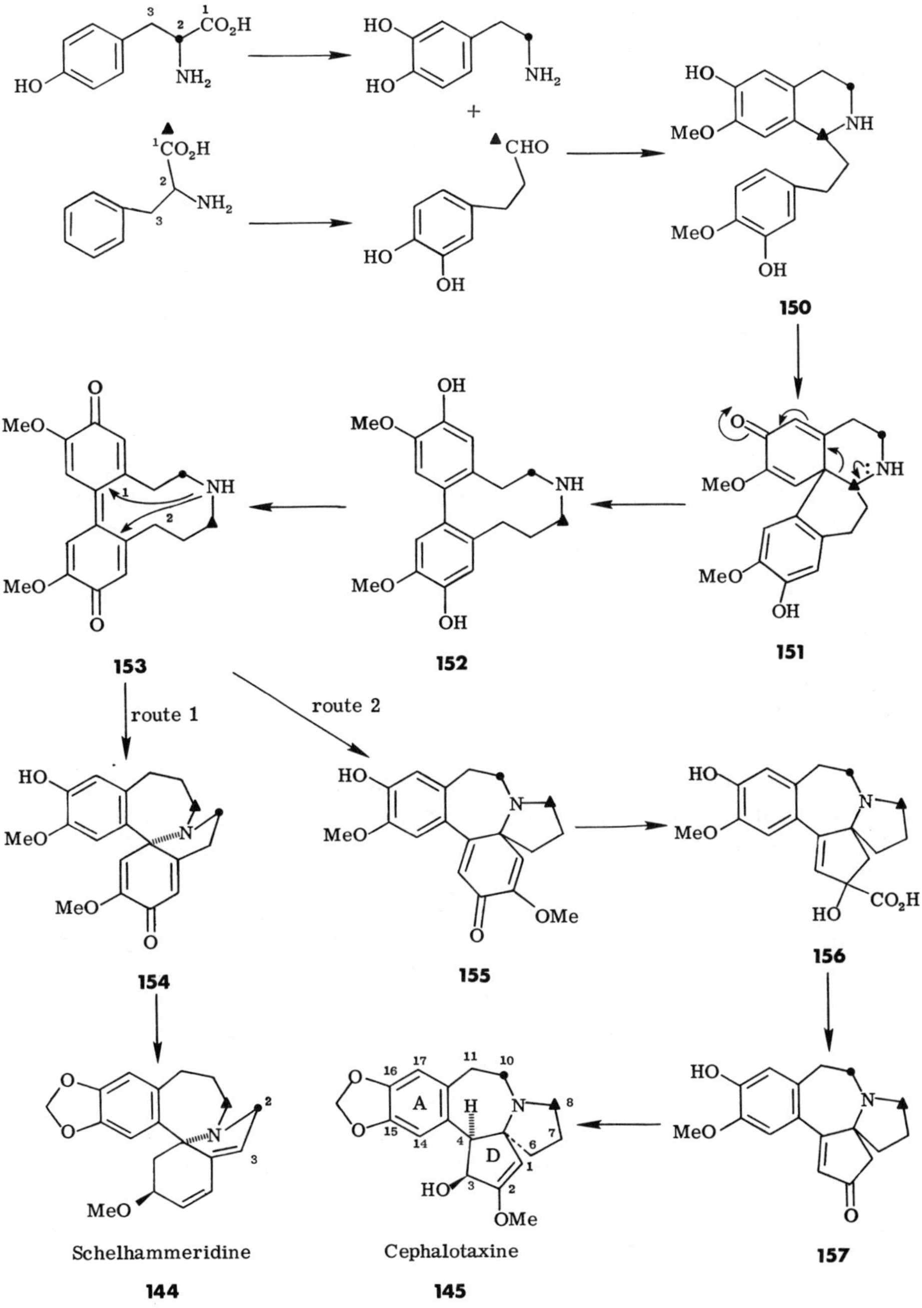

SCHEME 16: ● and ▲ represents ^{14}C labels. (Reprinted with permission from Powell,[108] courtesy of Pergamon Press.)

tyrosine labeled on the 2-position and phenylalanine labeled on the 1-position is indicated in Scheme 16. In preliminary work,[111] Battersby has indeed established that [2-^{14}C]tyrosine, when fed to *Schelhammera pedunculata* plants, yielded schelhammeridine labeled solely at C-2. However, preliminary work by Parry and Schwab[105] pointed to the biosynthesis of cephalotaxine from tyrosine in an altogether different manner which involved two molecules of this amino acid. Recently,[106] however, tyrosine labeled in the aromatic ring was examined as a cephalotaxine precursor and was found to label ring A of cephalotaxine (**145**) almost exclusively indicating that only one unit of tyrosine is used for biosynthesis. The involvement of two molecules of tyrosine, suggested on the basis of previous labeling studies, is now attributed, in part, to tyrosine catabolism in such a way that side-chain label but not ring label enters the pathway to cephalotaxine. By shortening the time of the feeding experiments, it could be shown that *dl*-[2-^{14}C]tyrosine primarily labels C-10, consistent with the ring-labeled tyrosine results; in prolonged experiments label appeared elsewhere. Thus ring A and B of cephalotaxine arise from tyrosine. On the basis of the proposed biosynthesis of cephalotaxine (Scheme 16, route 2), ring C and D would be expected to arise from a molecule of phenylalanine. Indeed, when phenylalanine labeled at C-1 was fed to *C. harringtonia* plants, label appeared largely at C-8 of cephalotaxine establishing that phenylalanine provides a C_6–C_3 unit for cephalotaxine biosynthesis and is consistent with passage through a phenethyltetrahydroisoquinoline intermediate **150**.

Marino and Samanen[109] have examined *in vitro* a unified approach to *Homoerythrina* skeletons via the pivotal diphenolic dibenz(*d,f*)azecine **159**. The compound **159** was proposed as a common precursor of the cephalotaxine-type skeleton **161**, homoerysodienone **163**, and *Schelhammera*-type skeleton **165** and, indeed, **159** was cleanly transformed into two cyclized *Homoerythrina* skeletons using potassium ferricyanide in methylene chloride–sodium bicarbonate solution. The *Schelhammera*-type skeleton **165** was isolated in 45% yield along with the homoerysodienone **163** (15%) and unreacted starting material (35%). That no cephalotaxine precursor **161** was observed in the above oxidation is suggestive of the absence of either diphenoquinone intermediate **160** or a suitably disposed *p*-hydroxy group (Scheme 17).

Subsequently, Kupchan and co-workers**112** synthesized monophenolic dibenz(*d,f*)azecines **167**, **168**, and **169** by taking advantage of the VOF_3–

[109] J. P. Marino and J. M. Samanen, *J. Org. Chem.* **41**, 179 (1976).

[110] E. McDonald and A. Suksamrarn, *Tetrahedron Lett.*, 4425 (1975).

[111] A. R. Battersby, E. McDonald, J. A. Milner, S. R. Johns, J. A. Lamberton, and A. A. Sioumis, *Tetrahedron Lett.*, 3419 (1975).

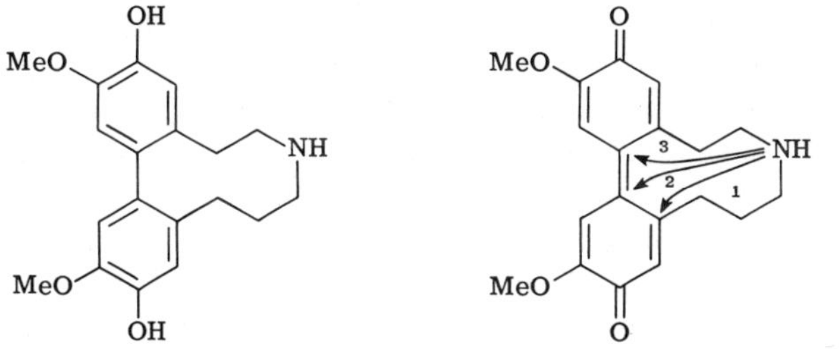

159 **160**

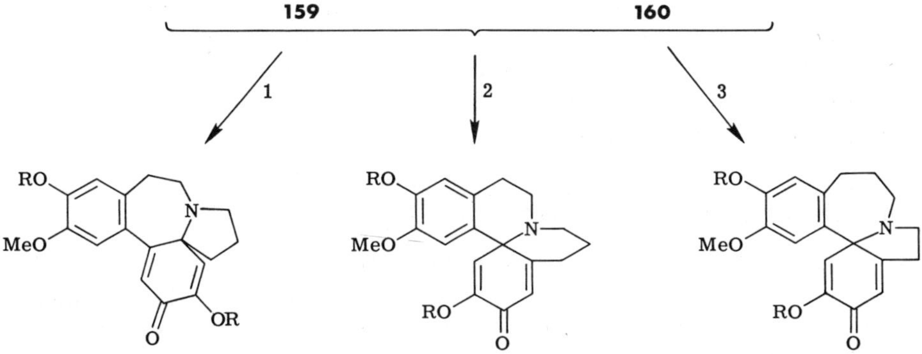

161 R = H
162 R = Me

163 R = H
164 R = Me

165 R = H
166 R = Me

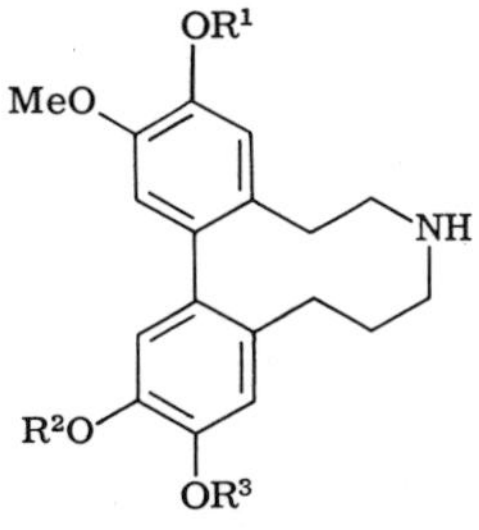

167 $R^1 = R^3$ = Me; R^2 = H
168 $R^1 = R^2$ = Me; R^3 = H
169 R^1 = H; $R^2 = R^3$ = Me

SCHEME 17

TFA:TFAA-induced oxidative coupling of nonphenolic and monophenolic phenethyltetrahydroisoquinolines to homoneospirenedienone and homoproerythrinadienones.[12]

The oxidation of 2-hydroxydibenz(*d,f*)azecine **167** with potassium ferricyanide yielded the cephalotaxine-type compound **162** (10%). Similar oxidation of 3-hydroxydibenz(*d*,*f*)azecine **168** gave the naturally unknown homoerysodienone **164** in 15% yield. However, when 12-hydroxydibenz-(*d*,*f*)azecine **169** was oxidized, no *Schelhammera*-type homoerythrinadienone **166** was detectable.[112] It follows that both phenolic groups are required for the formation of *Schelhammera*-type homoerythrinadienone **166**, implying that the diphenoquinone **160** may be an intermediate in the reaction as was proposed in the *Erythrina* alkaloid series by Barton, *et al.*[68]

The successful transformation of the dibenz(*d*,*f*)azecine **167** into the *Cephalotaxus*-type skeleton **162** represents the first reported synthesis of a *Cephalotaxus* alkaloid precursor along the biogenetic-type route, *i.e.*, substituted phenethyltetrahydroisoquinolines → spirodienones → dibenz(*d*,*f*)azecines → cephalotaxine precursor. Further elaboration of the dienone ring of **162** would result in the synthesis of a *Cephalotaxus* alkaloid similar to **145**.

B. NONPHENOLIC SUBSTRATES

The remarkable success achieved in the synthesis of morphinandienones by anodic oxidation of nonphenolic benzyltetrahydroisoquinolines prompted workers in the alkaloid field to study the electro-oxidation of nonphenolic phenethyltetrahydroisoquinolines. Thus anodic oxidation of (±)-homolaudanosine (**170**) would be expected to yield demethoxy-*O*-demethyl-

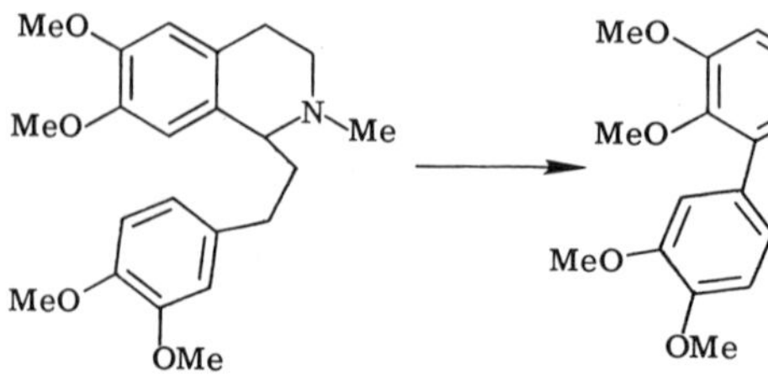

170 **171**

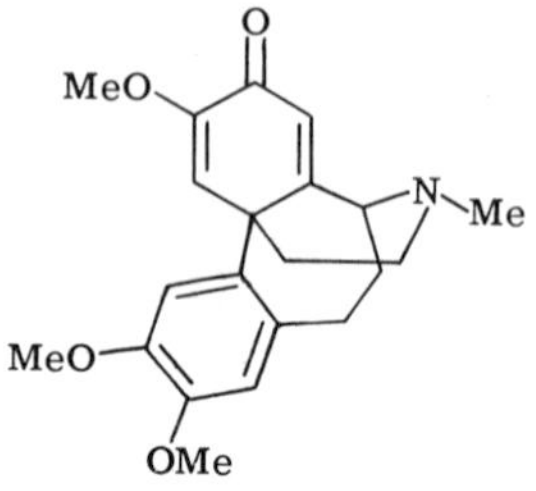

172

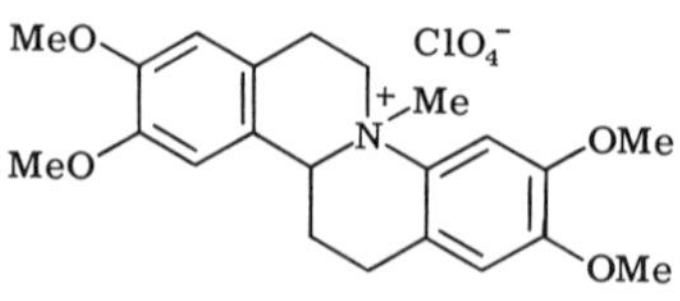

173

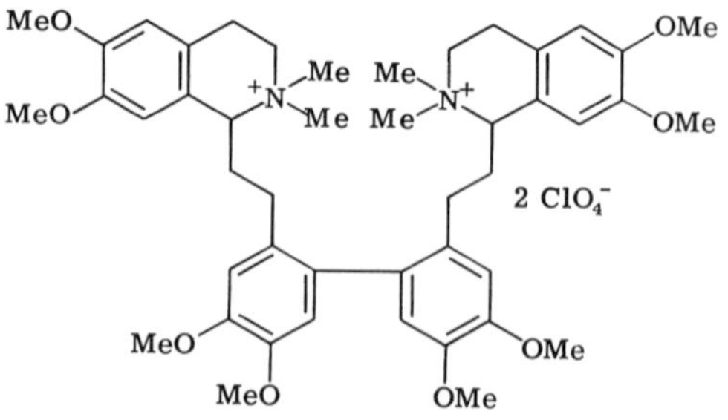

174

androcymbine (**172**). Quite unexpectedly, anodic oxidation of (±)-homolaudanosine (**170**) and its methoperchlorate salt in acetonitrile containing 10% sodium perchlorate as supporting electrolyte resulted in the formation of the N—C coupled product (**173**) and the dimer (**174**), respectively.[113] However, anodic oxidation of homolaudanosine (**170**) in a mixture of trifluoroacetic acid (TFA) and trifluoroacetic anhydride (TFAA) containing tetraethylammonium tetrafluoroborate as the supporting electrolyte at a constant potential of 1.3 V gave homoaporphine **171** in 34% yield.[12] The homoaporphine **171** was also obtained in 40% yield by VOF_3–TFA:TFAA oxidation of **170**.

In contrast to the oxidation of **170** to homoaporphine **171**, VOF_3–TFA: TFAA oxidation of (±)-*N*-trifluoroacetylhomonorlaudanosine (**175a**) yielded homoneospirenedienone **176a** (64%), homoaporphine **177** (2%), homoproerythrinadienone **178a** (5%), and the aldehydeamide **179** (22%), respectively.

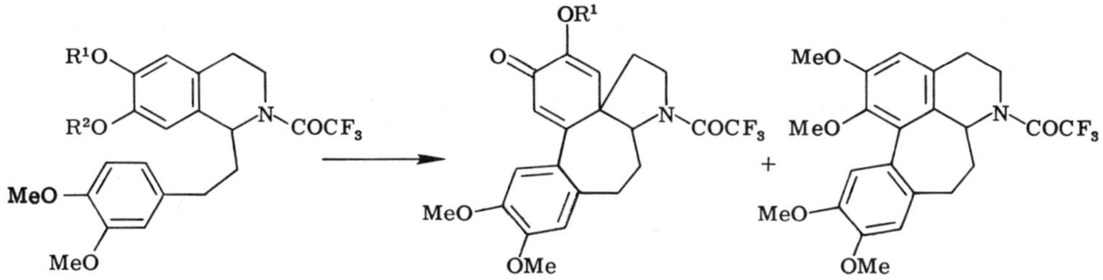

175a $R^1 = R^2 = Me$
b $R^1 = CH_2Ph$; $R^2 = Me$
c $R^1 = Me$; $R^2 = CH_2Ph$

176a-c

177

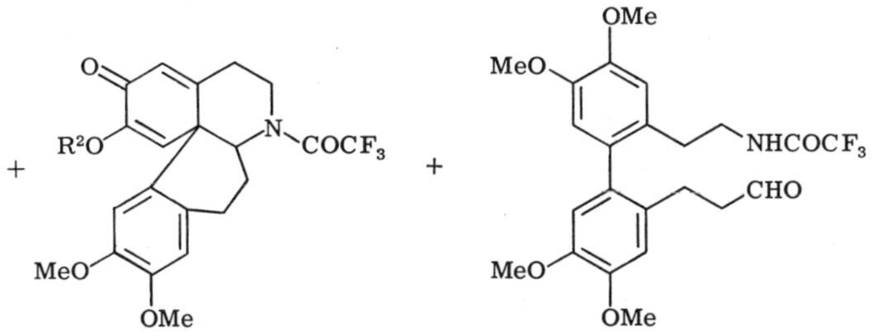

178a-c

179

[112] S. M. Kupchan, Om P. Dhingra, and C.-K. Kim, *J. Chem. Soc. Chem. Commun.*, 847 (1977); *J. Org. Chem.* **43**, 4464 (1978).

[113] A. Najafi and M. Sainsbury, *Heterocycles* **6**, 459 (1977).

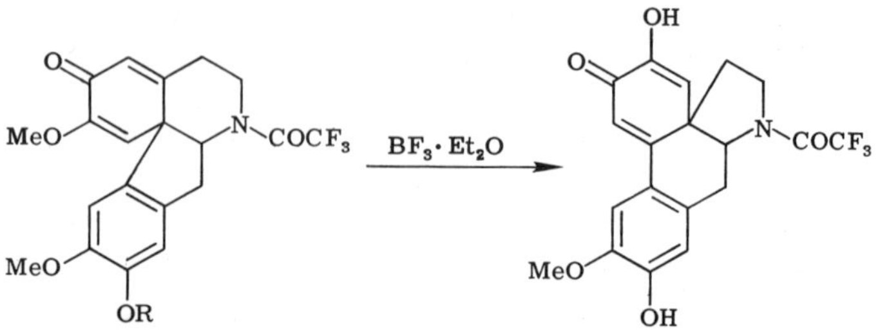

180 R = Me
177 R = H

181

Homoproerythrinadienone **178a** is a homolog of the proerythrinadienone **180**. Proerythrinadienones are intermediates in the biosynthesis of aporphine and *Erythrina* alkaloids (Schemes 3 and 5) and preliminary results[111] on the biosynthesis of *Homoerythrina* alkaloids indicate that homoproerythrinadienones may also play an important role in the biosynthesis of these metabolites. Kametani has found that treatment of proerythrinadienone **77** with boron trifluoride etherate or concentrated sulfuric acid gives neospirenedienone **181**.[114] In contrast, homoproerythrinadienones rearrange to homoaporphines on treatment with boron trifluoride etherate.[115] Thus, treatment of **178a** with boron trifluoride etherate in dichloromethane at room temperature for 24 hr yielded homoaporphine **182** which, on methylation with diazomethane, gave **177** (Scheme 18).

Interestingly, the homoneospirenedienone **176a** also rearranged to a monophenolic homoaporphine upon treatment with boron trifluoride etherate. The monophenolic homoaporphine product was different from homoaporphine **182** yet, upon methylation with diazomethane, gave the same tetramethoxy homoaporphine **177** as was obtained by methylation of **182** indicating that the structure of this monophenolic homoaporphine must be **184**. The formation of **184** and **176a** may be rationalized if homoneospirenedienone (**176a**) first rearranges to a homoproerythrinadienone-type intermediate **183**, which then rearranges to the homoaporphine **184** (Scheme 18).

Homoneospirenedienone **176a** could not be formed directly by oxidative coupling from *N*-trifluoroacetylhomonorlaudanosine (**175a**). Rather, the formation of **176a** must result from the rearrangement of either the homoproerythrinadienone-type intermediate (Scheme 19, route 1), as in the acid-catalyzed rearrangement of proerythrinadienone to neospirenedienones,[114]

[114] T. Kametani, K. Takahashi, T. Honda, M. Ihava, and K. Fukumoto, *Chem. Pharm. Bull.* **20**, 1793 (1972).

[115] J. P. Marino and J. M. Samenen, *Tetrahedron Lett.*, 4553 (1973).

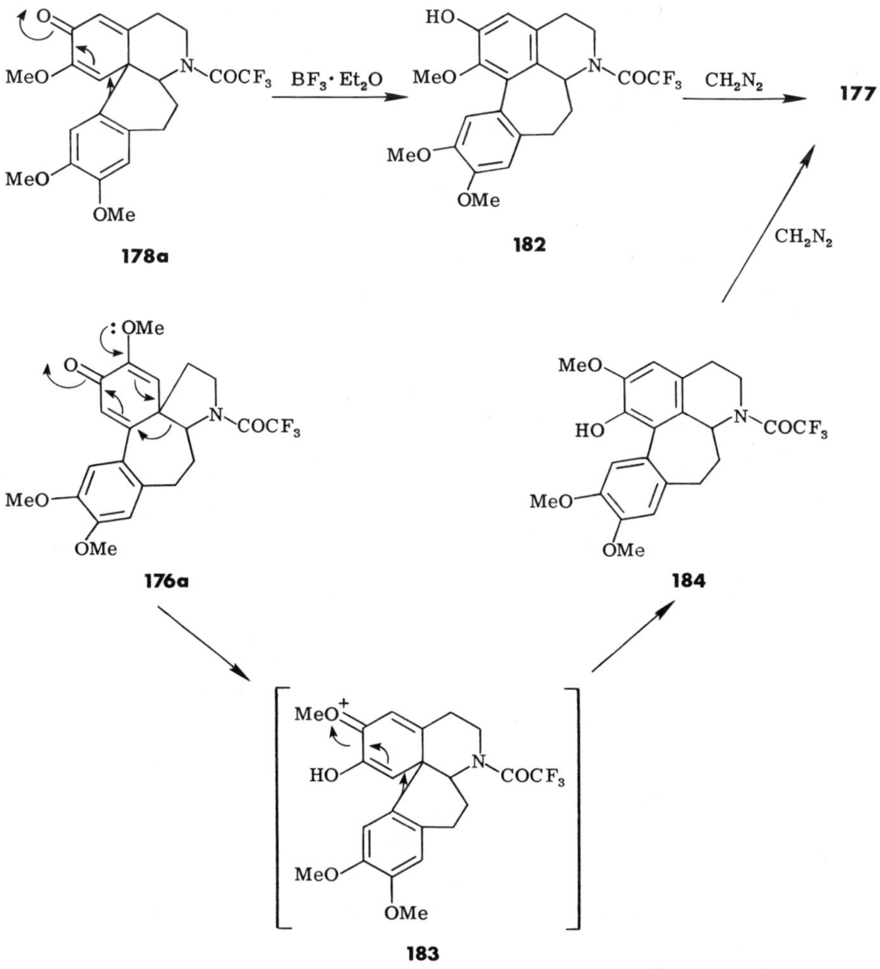

SCHEME 18

or of a homomorphinanedienone-type intermediate (route 2), as in the conversions of (±)-*N*-acylnorlaudanosines to (±)-*N*-acylmorphinedienones and then to (±)-*N*-acylneospirenedienones in nonphenol oxidative coupling of benzyltetrahydroisoquinolines.[10] By using 6- and 7-benzyloxy analogs of (±)-*N*-trifluoroacetylhomonorlaudanosine (**175b,c**), it was shown conclusively that homoneospirenedienone (**176**) formation involves a homoproerythrinadienone-type intermediate (**186**). The results of oxidation of various phenethyltetrahydroisoquinolines are summarized in Table II.

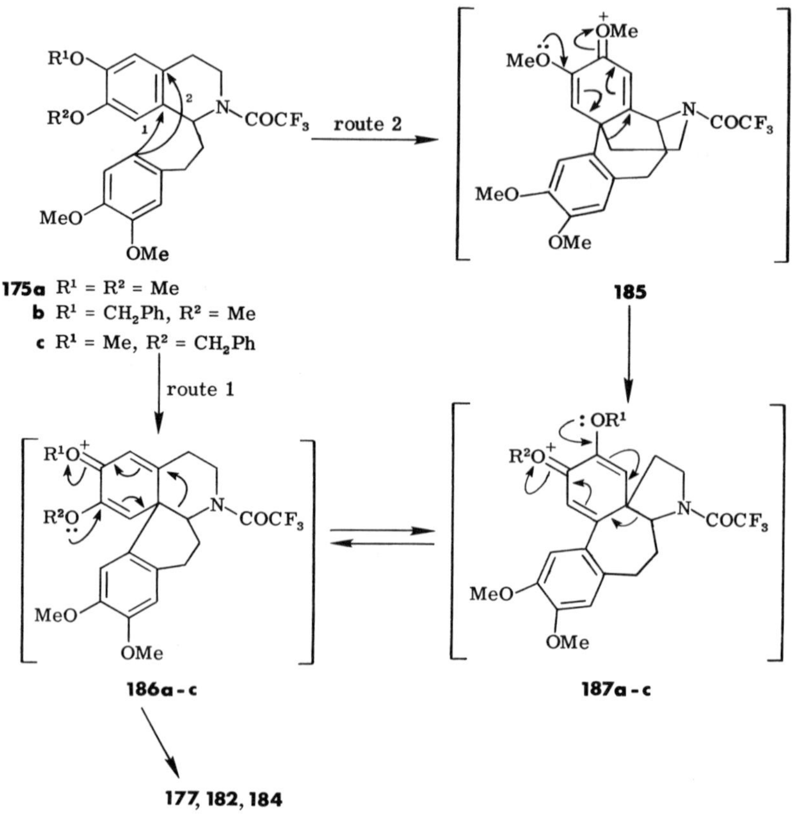

SCHEME 19

The homoproerythrinadienone-type intermediates (e.g., **186**) and homoneospirenedienone-type intermediates (e.g., **187**) exist in equilibrium in the reaction medium. In the oxidation of **175b** (Table II, expt. 6), cleavage of a benzyl group from the homoproerythrinandienone-type intermediate (**186b**) shifts the equilibrium toward the homoproerythrinadienone-type intermediate (**186b**), and thus homoproerythrinadienone **178a** is isolated as the major product. In the oxidation of **175c**, cleavage of the benzyl group from **187c** shifts the equilibrium toward homoneospirenedienone-type intermediate (**187c**) and homoneospirenedienone (**176a**) is isolated as the major product (Table II, expt. 9). Furthermore, when VOF_3–TFA:TFAA oxidation of **175a–c** is run for a longer period of time (Table II, expts. 4, 7, 10), the homoproerythrinadienone-type (**186**) and homoneospirenedienone-type intermediates (**187**) rearrange to give homoaporphines **177**, **182**, and **184**, respectively, in very high yield. (Scheme 19)

TABLE II

OXIDATION OF PHENETHYLTETRAHYDROISOQUINOLINES WITH VOF_3 in CH_2Cl_2–TFA:TFAA[a]

Expt.	Substrate	Temp (°C)	Time	Products	Yield (%)
1	**136**	−10	6 min	**137**	38
				138	30
2	**170**	−10	40 min	**171**	40
3	**175a**	−10	10 min	**176a**	64
				177	2
				178a	5
				179	22
4	**175a**	−10 → 30	3 hr	**177**	84
5	**175b**	−10	10 min	**178a**	50
				176b	42
6		−10 → 30	15 hr	**178a**	71
7		−10 → 30	24 hr	**182**	80
8	**175c**	−10	10 min	**176a**	60
				178b	3
9		−10 → 30	1 hr	**176a**	65
10		−10 → 30	24 hr	**184**	67.5
11	**188a**	−10	10 min	**189a**	40
				190a	18
12		−10 → 30	60 min	**189a**	77
13	**188b**	−10	6 min	**189b**	14
				190b	42
14		−10 → 30	60 min	**189b**	60
15	**188c**	−10	10 min	**189c**	46
				190c	4
16		−10 → 30	30 min	**189c**	54
17	**188d**	−10	10 min	**189d**	54
				190d	16
18	**192**	−10	10 min	**178a**	98

[a] Reprinted with permission from Kupchan *et al. J. Org. Chem.* **43,** 2521 and 4076. Copyright (1978) American Chemical Society.

C. MONOPHENOLIC SUBSTRATES

The potential of the VOF_3–TFA:TFAA coupling method for the synthesis of homoaporphines, homoproaporphines, and homoproerythrinadienones by the oxidation of monophenolic phenethyltetrahydroisoquinolines has also been evaluated[13] (Table II, expts. 11–17). Thus oxidation of **188a** yielded homoaporphine **189a** (40%) along with homoproaporphine **190a** (18%), and **188b** gave homoproaporphine **191b** (42%) along with homoaporphine **189b** (14%). Only one isomer of homoproaporphine **190a** or **190b** was obtained, in contrast to the diastereoisomeric mixture obtained

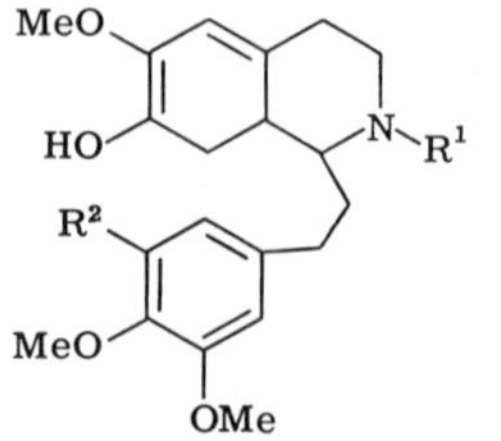

188a $R^1 = COCF_3$, $R^2 = H$
b $R^1 = Me$, $R^2 = H$
c $R^1 = COCF_3$, $R^2 = OMe$
d $R^1 = Me$, $R^2 = OMe$

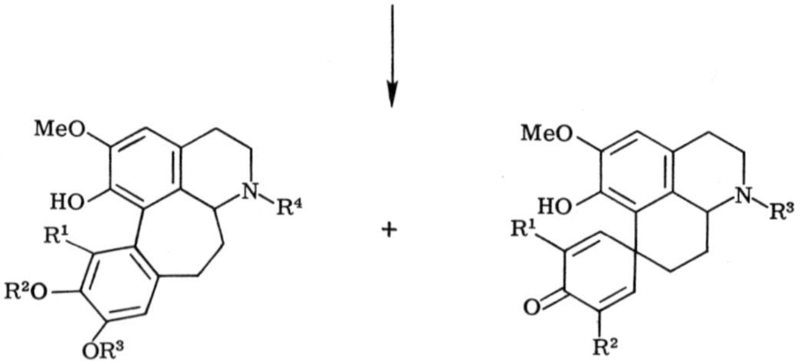

189a $R^1 = H$, $R^2 = R^3 = Me$, $R^4 = COCF_3$
b $R^1 = H$, $R^2 = R^3 = R^4 = Me$
c $R^1 = OMe$, $R^2 = R^3 = Me$, $R^4 = COCF_3$
d $R^1 = OMe$, $R^2 = R^3 = R^4 = Me$
e $R^1 = R^3 = H$, $R^2 = Me$, $R^4 = COCF_3$
f $R^1 = R^3 = H$, $R^2 = R^4 = Me$
g $R^1 = OMe$, $R^2 = H$, $R^3 = Me$, $R^4 = COCF_3$
h $R^1 = OMe$, $R^2 = H$, $R^3 = R^4 = Me$

190a $R^1 = OMe$, $R^2 = H$, $R^3 = COCF_3$
b $R^1 = OMe$, $R^2 = H$, $R^3 = Me$
c $R^1 = R^2 = OMe$, $R^3 = COCF_3$
d $R^1 = R^2 = OMe$, $R^3 = Me$

188a-d

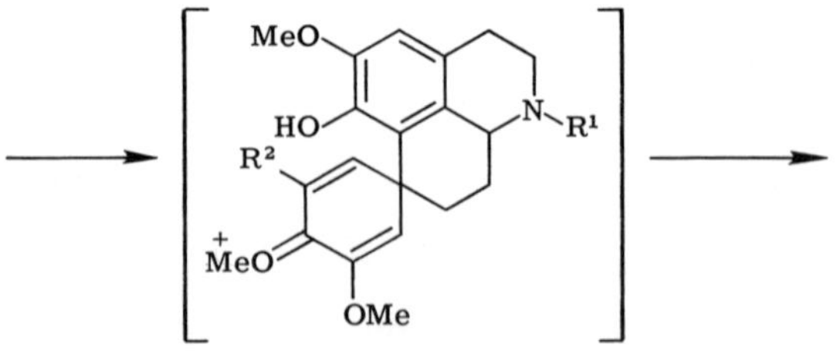

189a-d

191a-d

by the oxidation of the diphenolic precursor **136**. Similarly, oxidation of **188c** yielded homoaporphine **189c** (46%) along with homoproaporphine **190c** (4%), and **188d** gave homoproaporphine **190d** (54%) along with (±)-kreysigine (**189d**, 16%). Homoproaporphines **190a–d** undergo smooth dienone–phenol rearrangement upon treatment with boron trifluoride etherate to afford homoaporphines **189e** (87%), **189f** (70%), **189g** (87%), and **189h** [(±)-multifloramine, 72%], respectively. However, if **188a–c** are treated with VOF_3–TFA:TFAA for 30–60 min (Table II, expts. 12, 14, 16) so as to allow the complete rearrangement of homoproaporphine-type intermediates **191a–d**, homoaporphines **189a–d** are obtained in high yield.

Oxidation of 6-hydroxy-1-phenethyltetrahydroisoquinoline **192** with VOF_3–TFA:TFAA yields homoproerythrinadienone **178a** in 98% yield.

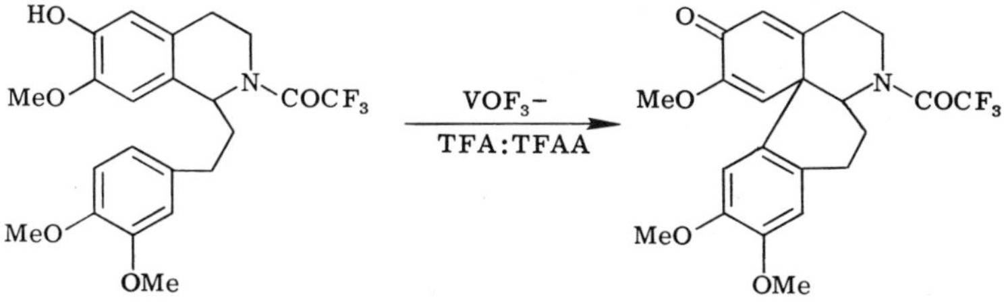

192 **178a**

Umezawa and his colleagues have pioneered the application of lead tetraacetate oxidation to the synthesis of various isoquinoline alkaloids.[19] The synthetic method using *p*-quinol acetate to prepare aporphines (Section II,C) has been extended to the synthesis of homoaporphines, homoproaporphines, and homomorphinandienones. Thus, treatment of *p*-quinol acetate **194a–c** with TFA gave homoaporphines **195a–c** in 52%, 54.3%, and 63% yield, respectively.

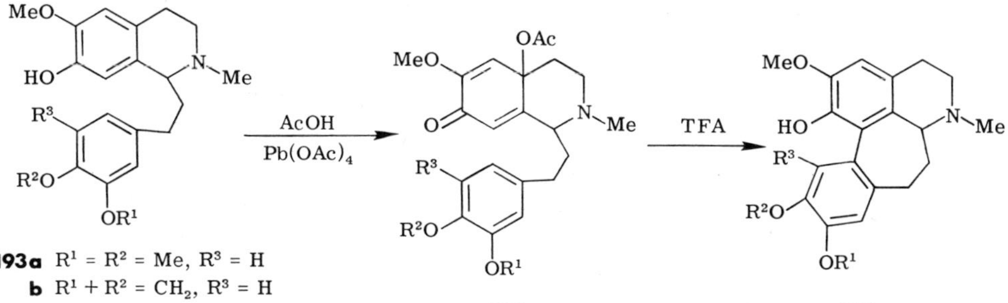

193a $R^1 = R^2 = Me$, $R^3 = H$
b $R^1 + R^2 = CH_2$, $R^3 = H$
c $R^1 = R^2 = Me$, $R^3 = OMe$

194a–c **195a–c**

TABLE III

OXIDATION WITH $Pb(OAc)_4$ IN TFA/CH_2Cl_2 [a]

Starting materials				Products yield (%)	
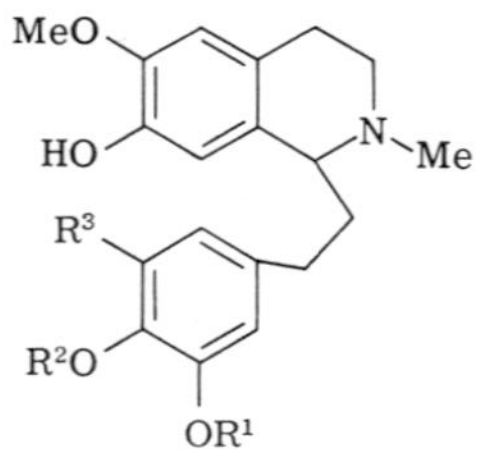 **193**			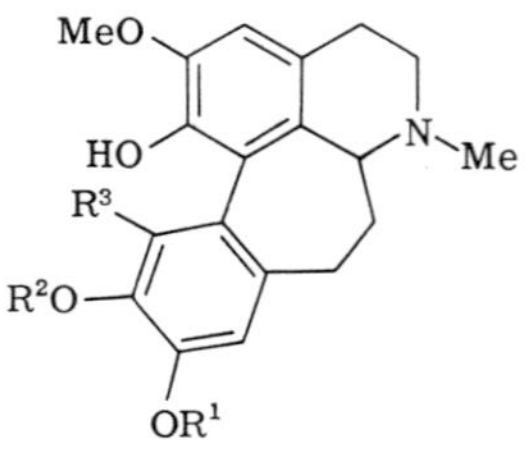 **195**	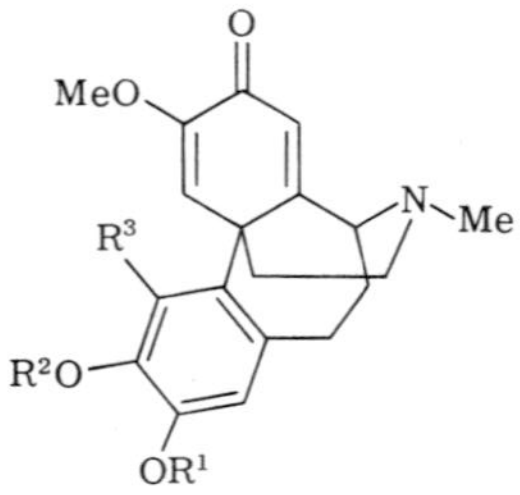 **196**	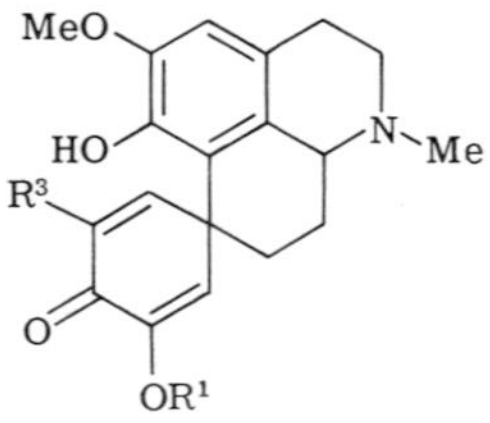 **197**
	R^1	R^2	R^3		
a	Me	Me	H	44.2	3.3	6.9
b	—CH_2—		H	55.0	—	—
c	Me	Me	OMe	35.2	23.1	31.3
d	Me	CH_2Ph	H	39.3	6.1	15.5
e	CH_2Ph	Me	H	29.6	0.3	11.4
f	Me	CH_2Ph	OMe	27.3	12.2	15.9

[a] Reprinted with permission from Umezawa and Hoshino,[19] courtesy of The Sendia Institute of Heterocyclic Chemistry.

On the other hand, if TFA in CH_2Cl_2 is employed as the cyclizing reagent, homomorphinandienone (**196**), homoproaporphine (**197**), and homoaporphine (**195**) are formed as shown in Table III. The yields of homomorphinandienones can be slightly improved (Table IV) by oxidizing **193a–e** with lead tetraacetate in trifluoroacetic acid. Homomorphinandienone (±)-*O*-methylandrocymbine (**196c**) has also been synthesized in 20% yield by thallium trifluoroacetate oxidation of **193c**.[73]

TABLE IV

OXIDATION WITH $Pb(OAc)_4$ IN TFA[a]

Starting materials **193**	Products yield (%) **195**	**196**	**197**
a	—	16	31
b	19	—	—
c	7.5	26	15
d	—	14	18
e	—	5	16
f	10	13	10

[a] Reprinted with permission from Umezawa and Hoshino,[19] courtesy of The Sendia Institute of Heterocyclic Chemistry.

IV. Oxidative Coupling of *N*-Benzylphenethylamines

In 1957, Barton and Cohen[5] suggested that norbelladine (**199a**), a *N*-benzylphenethylamine, represents a single precursor for the three main types of Amaryllidaceae alkaloids. They are crinine-type alkaloids by para–para coupling, galanthamine-type alkaloids by para–ortho coupling, and lycorine-type alkaloids by ortho–para coupling. Extensive tracer experiments[116,117] have verified that alkaloids of this type are biosynthesized from L-tyrosine via the phenolic oxidative cyclization of *O*-methylnorbelladine (**199b**). Synthetic approaches to these three types of compounds have been extensively investigated.[72,118] Biogenetic-type syntheses of crinine- and galanthamine-type alkaloids have appeared recently, utilizing newly

[116] W. C. Wildman, H. M. Fales, and A. R. Battersby, *J. Am. Chem. Soc.* **84**, 681 (1962).
[117] D. H. R. Barton, G. W. Kirby, J. B. Taylor, and G. M. Thomas, *J. Chem. Soc.*, 4545 (1963).
[118] R. V. Stevens, *in* "The Total Synthesis of Natural Products" (John ApSimon, ed.), Vol. 3, p. 439. Wiley, New York, 1977.

developed coupling reagents for diphenolic, monophenolic, and nonphenolic substrates.[21,119–123]

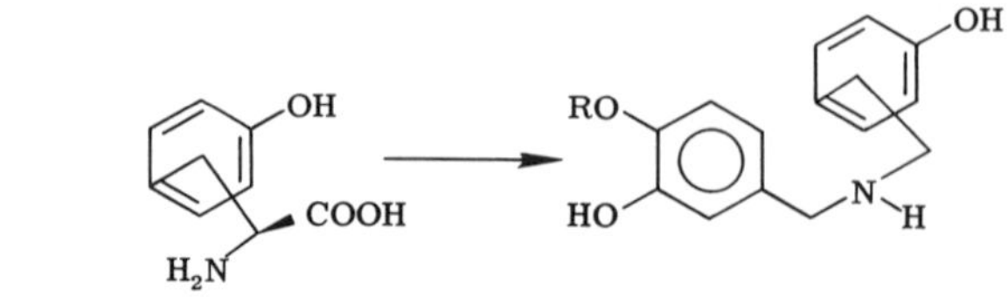

198 **199a** R = H
b R = Me

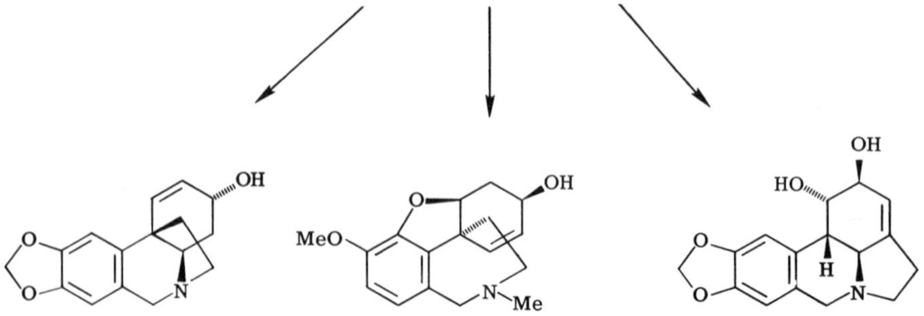

(−)-Crinine **200** (−)-Galanthamine **201** (−)-Lycorine **202**

A. DIPHENOLIC SUBSTRATES

Of the three alkaloid types obtained by phenolic oxidation of *O*-methylnorbelladine (**199b**), (−)-galanthamine (**201**) was the first to be synthesized via a biogenetic route. Thus, Barton and Kirby[124] oxidized *O*,*N*-dimethylnorbelladine (**203**) with ferricyanide and obtained (±)-norwedine (**204**) in 1.4% yield. Norwedine was resolved and (±)-norwedine reduced to a mixture of (±)-galanthamine (**201**) and (±)-epigalanthamine (**205**) with lithium aluminum hydride.

---

[119] M. A. Schwartz and R. A. Holton, *J. Am. Chem. Soc.* **92**, 1090 (1970).
[120] E. Kotani, N. Takeuchi, and S. Tobinaga, *Tetrahedron Lett.*, 2735 (1973).
[121] T. Kametani, T. Kohno, S. Shibuya, and K. Fukumoto, *Tetrahedron* **27**, 5441 (1971).
[122a] S. Yamada, K. Tomioka, and K. Koga, *Tetrahedron Lett.*, 57 (1976).
[122b] K. Tomioka, K. Koga, and S. Yamada, *Chem. Pharm. Bull.* **25**, 2681 (1977).
[123] K. Shimizu, K. Tomioka, S. Yamada, and K. Koga, *Chem. Pharm. Bull.* **26**, 3765 (1978).
[124] D. H. R. Barton and G. W. Kirby, *J. Chem. Soc.*, 806 (1962).
[125] T. Kamitani, K. Yamaki, H. Yagi, and K. Fukumoto, *Chem. Commun.*, 425 (1969).

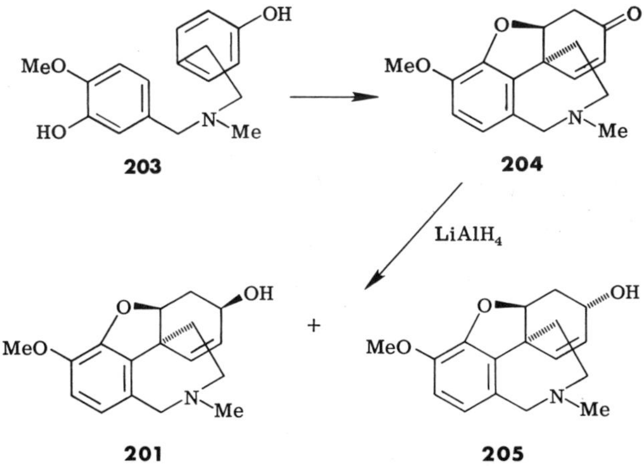

Kametani *et al.* has synthesized (±)-galanthamine (**201**) by oxidizing amide **206** with ferricyanide to give norwedine-type enone **208** (40%) followed by lithium aluminum hydride reduction.[125]

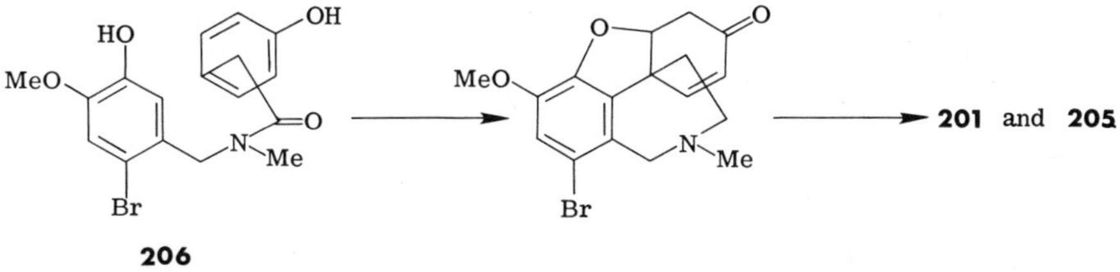

Recently K. Shimizu and co-workers[123] have reported a biogenetic-type asymmetric synthesis of (+)- and (−)-galanthamine. Oxidation of the *N*-benzylphenethylamine derivative **208** with manganic tris(acetylacetonate) in acetonitrile gave the norwedine-type enone **209** in 49% yield. Reaction of **209** with diethyl phosphorochloridate in the presence of triethylamine afforded the phosphate (+)-(**210**) in 81% yield.

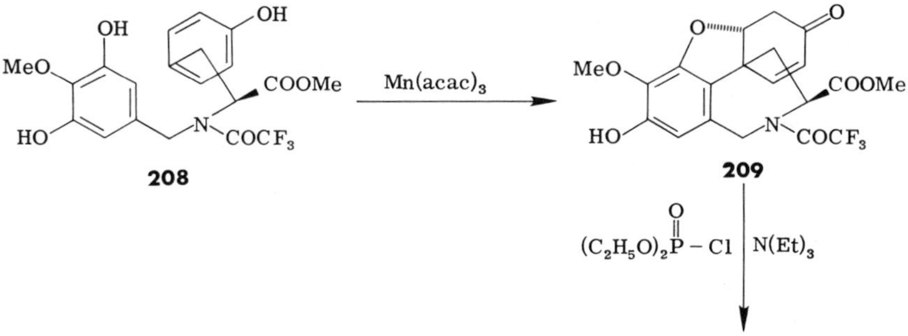

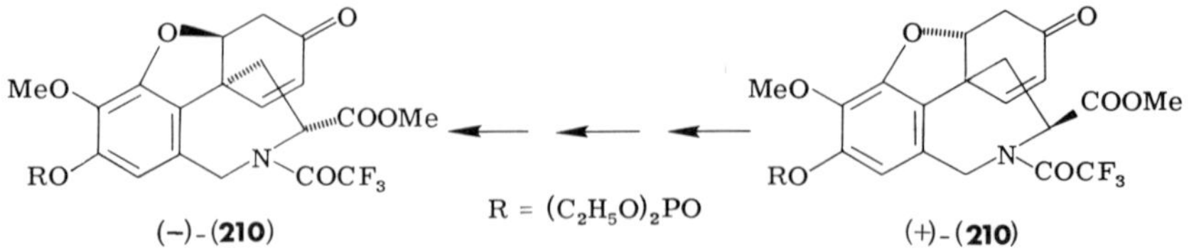

(−)-(**210**) (+)-(**210**)

The compound (+)-**210** was transformed into (+)-galanthamine [(+)-**201**] via the route shown in Scheme 20. The norwedine-type enone (+)-**210** could also be transformed into its antipode [(−)-**210**] by utilizing optical instability

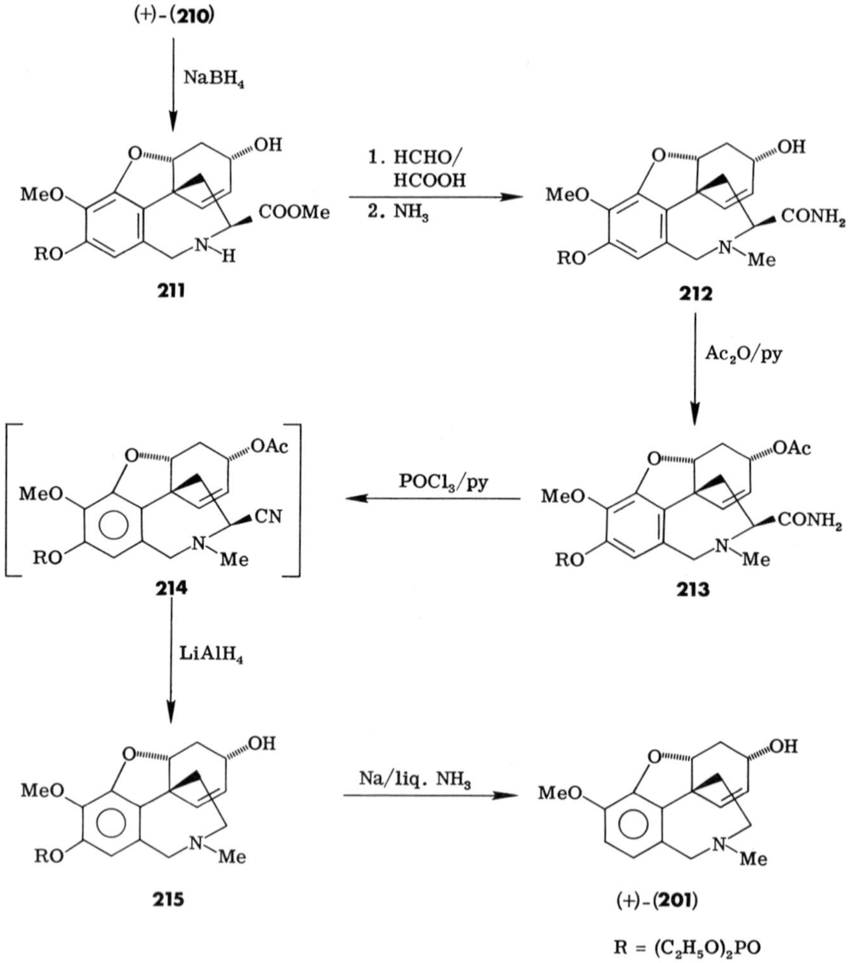

SCHEME 20. (Reprinted with permission from Shimizu *et al.*,[123] courtesy of The Pharmaceutical Society of Japan.)

of norwedine-type enone skeleton[124] and thus constitutes the formal total synthesis of (−)-galanthamine [(−)-**201**].

Maritidine (**219**), a representative of the 5,10b-ethanophenanthridine class of Amaryllidacae alkaloids, has also been synthesized via a biogenetic-type route. Oxidation of *N*-trifluoroacetyl-*O*-methylnorbelladine (**216**) with VOF_3–TFA:TFAA gave the dienone **217** in 71% yield.[126] Schwartz has reported a yield of 24% for **217** using $VOCl_3$ in ether as the coupling reagent.[119] Kotani and Tobinaga[120] have reported a yield of 35% for **217** using dimethylformamide–ferric chloride complex. Hydrolysis of **217** resulted in spontaneous cyclization to give the phenolic enone **218** which was converted into (±)-maritidine (**219**) by borohydride reduction followed by methylation and epimerization at C-3.

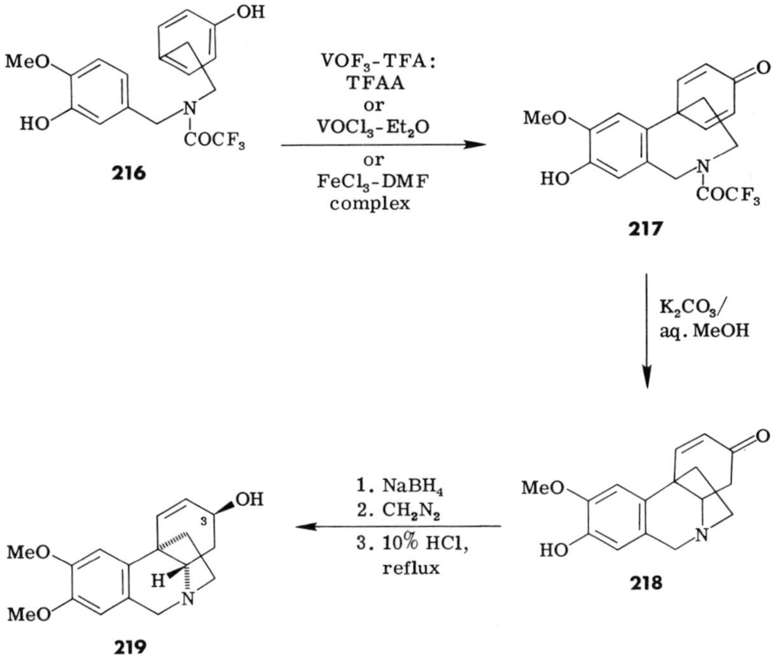

A biogenetic-type asymmetric synthesis of (+)-maritidine has recently been reported by Yamada and co-workers.[122] Oxidation of the diphenolic *N*-benzylphenethylamine derivative **220** with ferric chloride–DMF complex gave the spirodienone **221** in 14% yield. Methylation of **221** with methyl iodide and *tert*-BuOK in DMF afforded the dienone **222** in 25% yield.

[126] Om P. Dhingra, Ph.D. Thesis, Univ. of Virginia, Charlottesville, 1977.

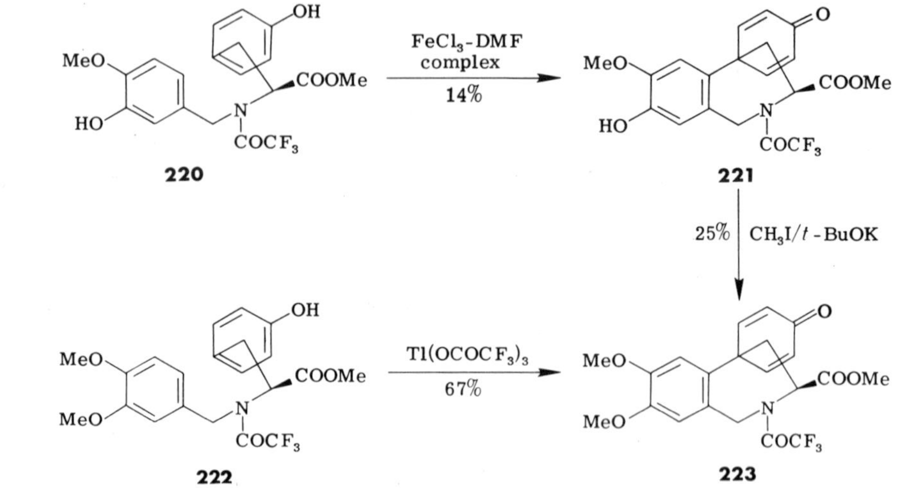

Transformation of the dienone **222** to (+)-maritidine was achieved via the route shown in Scheme 21.

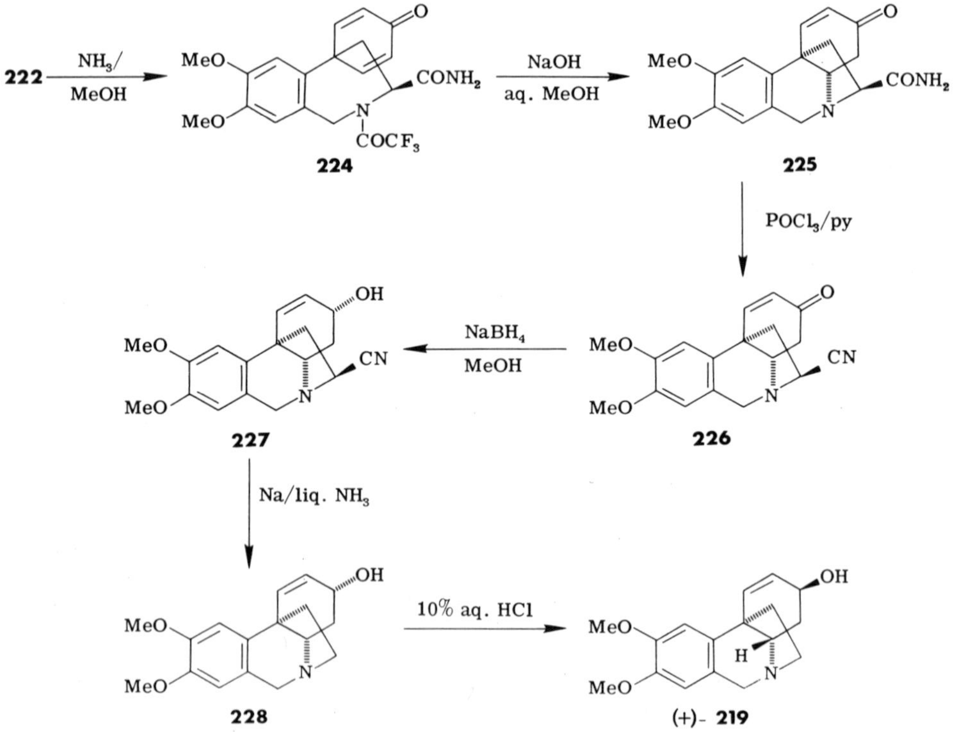

SCHEME 21. (Reprinted with permission from Yamada *et al.*,[122a] courtesy of Pergamon Press.)

Schwartz and co-workers[16] have recently reported on $VOCl_3$-induced oxidative coupling of the diphenolic acetamide **229**. The dienone **230** was obtained in 50% yield when **229** was treated with 3.5 eq of $VOCl_3$ in acetonitrile.

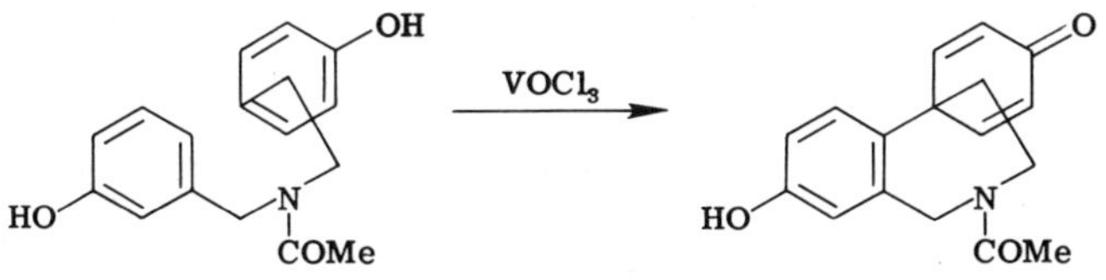

229 **230**

B. MONOPHENOLIC SUBSTRATES

The spirodienone **222**, obtained above in overall yield of 3.5% from the diphenolic substrate **220**, was synthesized in 67% yield when monophenolic compound **223** was oxidized with thallium trifluoroacetate in acetonitrile containing small amount of trifluoroacetic acid.[123]

The alkaloid (±)-oxocrinine (**233a**) has been obtained in excellent yield from the monophenolic compound **231a**. Thus, oxidation of **231a** with VOF_3–TFA:TFAA for 10 min gave the dienone **232a** in 89% yield.[13] Schwartz[73] has reported a yield of 19% for **232a** using thallium trifluoroacetate as the coupling reagent. Hydrolysis of the dienone **232a** with aqueous alkali afforded (±)-oxocrinine (**233a**) in 95% yield.

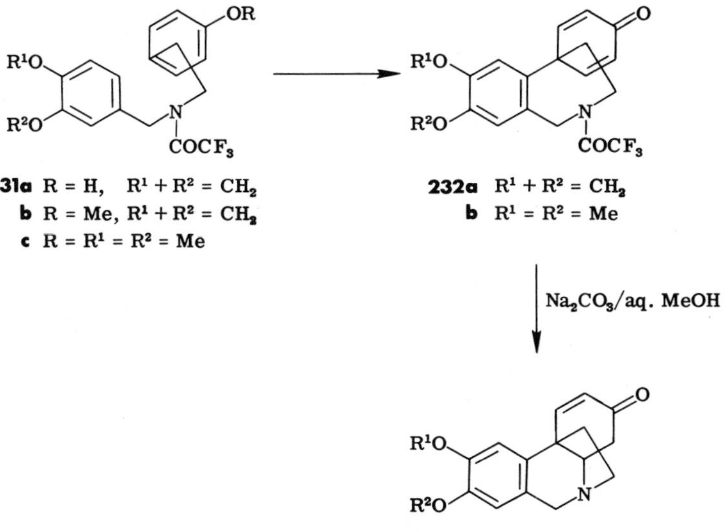

231a R = H, $R^1 + R^2 = CH_2$
b R = Me, $R^1 + R^2 = CH_2$
c $R = R^1 = R^2 = Me$

232a $R^1 + R^2 = CH_2$
b $R^1 = R^2 = Me$

233a $R^1 + R^2 = CH_2$
b $R^1 = R^2 = Me$

C. NONPHENOLIC SUBSTRATES

Recently, the syntheses of (±)-oxocrinine (**233a**) and (±)-oxomaritidine (**233b**) by anodic oxidation of nonphenolic *N*-benzylphenethylamines have been reported.[21] Tobinaga converted the *N*-benzylphenethylamines **231b** and **231c** to the dienones **232a** and **232b** in 62% yield in both cases. Hydrolysis of the dienones **232a** and **232b** with aqueous alkali afforded (±)-oxocrinine (**233a**) and (±)-oxomaritidine (**233b**), respectively.

Anodic oxidation of the amides **234** and **236** in acetonitrile, containing sodium perchlorate as supporting electrolyte, at an anode potential of ~1.1 V, has been shown by Sainsbury and Wyatt[127] to give the dibenzazonine derivatives **235** and **237** in 45% and 60% yield, respectively.

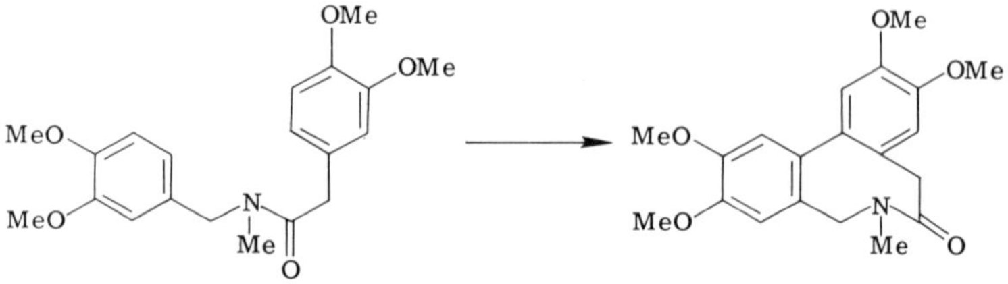

234 **235**

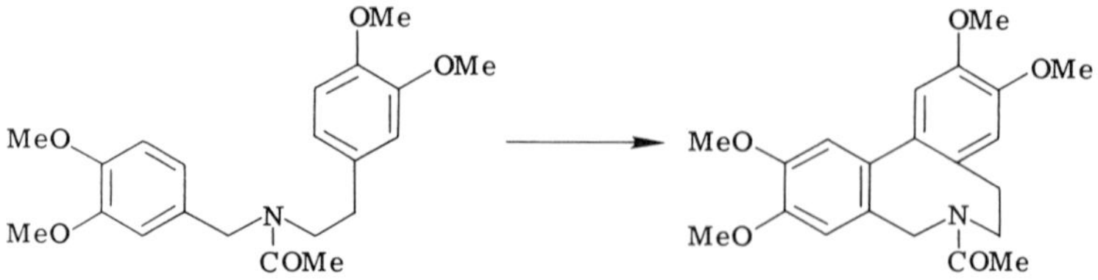

236 **237**

V. Oxidative Coupling of Diarylalkanes

A. DIPHENOLIC SUBSTRATES

In 1969, Schwartz[52] first reported on the intramolecular oxidative coupling of a diphenolic 1,3-diarylpropane **238** using a variety of coupling reagents. Oxidation of **238** with a number of one-electron oxidants such as $K_3Fe(CN)_6$, $FeCl_3$, or Ag_2CO_3 gave low yields of the intramolecularly coupled dienone **240** (TableV). On the other hand, oxidation of **238** with $VOCl_3$ in ether gave the dienone **240** in 76% yield. The high yield with this

[127] M. Sainsbury and J. Wyatt, *J. Chem. Soc. C*, 661 (1976).

TABLE V

OXIDATIVE COUPLING OF 1-*m*-HYDROXYPHENYL-3-*p*-HYDROXYPHENYLPROPANE WITH VARIOUS REAGENTS[a]

Oxidant	Solvent	Temp (°C)	Yield (%)	
			240	**238**
$K_3Fe(CN)_6/Na_2CO_3$	$H_2O/CHCl_3$	25	4	—
$FeCl_3$	H_2O/t-BuOH	50	3	66
$AgNO_3$/NaH	CH_3CN/Et_2O	−60;80	0	16
Ag_2CO_3-celite	C_6H_6	80	2	72
$Mn(acac)_3$	CH_3CN	81	10	16
$MoOCl_4$	CCl_4/Et_2O	−30;25	18	26
Pt anode	H_2O/CH_3CN	25	25	5

[a] Reprinted with permission from Schwartz *et al.*, *J. Am. Chem. Soc.* **99**, 2571. Copyright (1977) American Chemical Society.

reagent results from oxidation via a species such as **239**. A result of this mechanism would be a diminished tendency for intermolecular coupling. Similarly, oxidation of the diphenol **241** with $VOCl_3$ in ether gave 69% yield of the dienone **242**.[16]

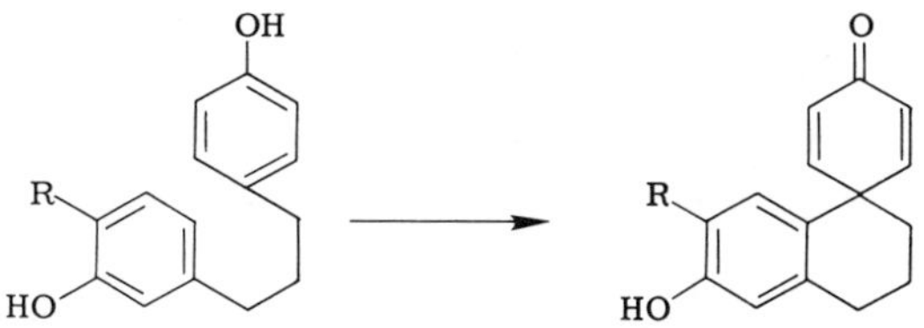

238 R = H
241 R = OCH_3

240 R = H
242 R = OCH_3

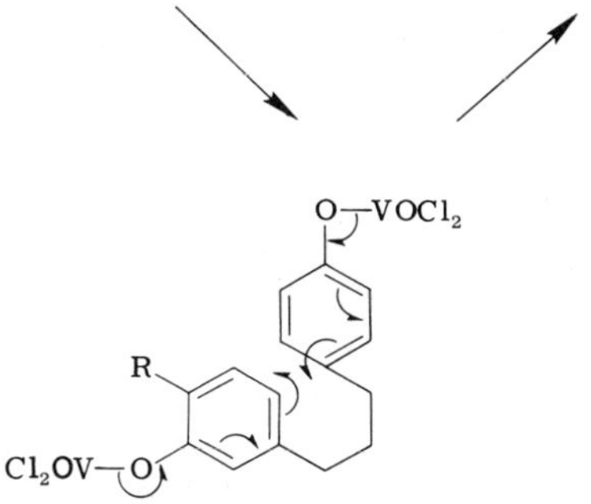

239

Recently, Tobinaga and co-workers[18] have discovered that ferric chloride and dimethylformamide form a very stable complex that is a good reagent for intramolecular coupling reactions of diphenolic compounds. Thus oxidation of the diphenolic 1,3-diarylpropanes **243a–d** with the above reagent gave spirodienones **244a–d** in good yield (Table VI). The spirodien-

TABLE VI

ANODIC AND $FeCl_3$–DMF OXIDATION OF 1,3-DIPHENYLPROPANES

	Starting materials					Products yield (%)	
	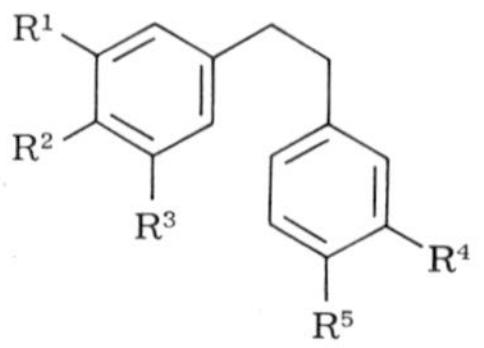 **243**					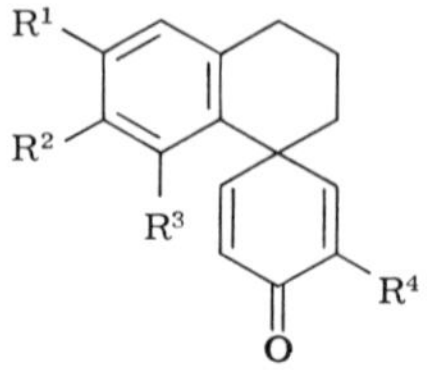 **244**	
	R[1]	R[2]	R[3]	R[4]	R[5]	$FeCl_3$–DMF oxidation	Anodic oxidation
a	OH	OMe	OMe	OMe	OH	91	60
b	OH	OMe	H	OMe	OH	67	
c	OH	H	H	OMe	OH	39	
d	OH	OMe	OMe	H	OH	16	
e	OMe	OMe	H	H	OMe		90
f	OCH_2O		H	H	OMe		95
g	OMe	OMe	OMe	H	OMe		96
h	OMe	OMe	OMe	OMe	OH		80

one **244a** has been utilized by Tobinaga *et al.*[24] in developing a very short and efficient route to colchicine (**136**) (Scheme 22).

The dienone **244a**, obtained in 91% yield by oxidation of **243a** with $FeCl_3$–DMF complex, was converted to a mixture of epimeric dienols **245** by methylation with diazomethane, followed by sodium borohydride reduction. Treatment of **245** with Simmons–Smith reagent afforded the cyclopropanated product **246** which was oxidized with Jones' reagent to yield the α,β-unsaturated cyclopropyl ketone **247**. Treatment of **247** with Ac_2O–H_2SO_4 at room temperature for 2 hr afforded desacetamidoiso-colchicine (**248**) directly. The transformation of **248** to colchicine is already known in literature.

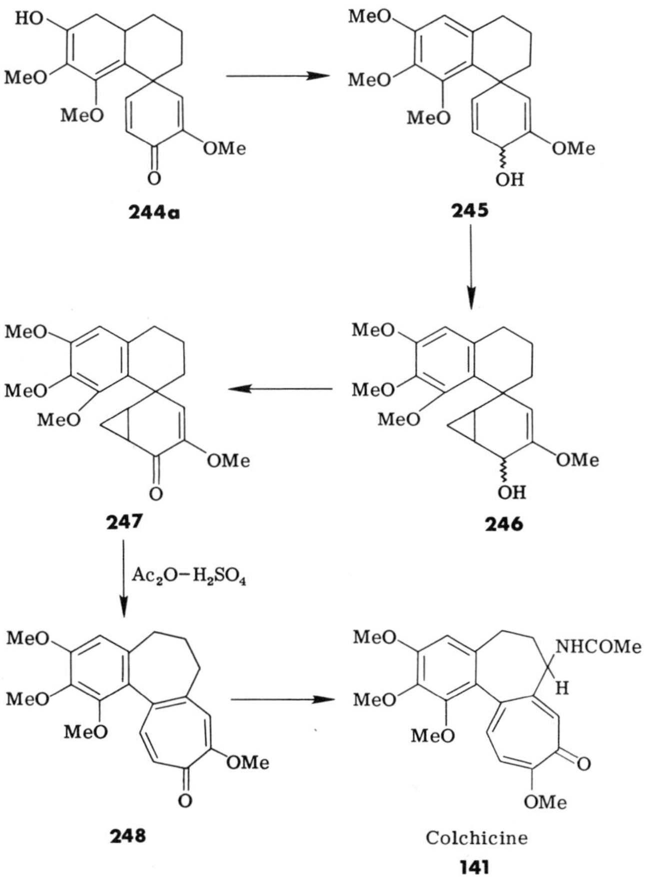

SCHEME 22

B. NONPHENOLIC SUBSTRATES

The remarkable success achieved in the anodic oxidation of (±)-laudanosine (**86**) to morphinandienone (**87**) prompted Miller and co-workers[128] to attempt to form the A, B, and C ring of the steroids with the proper angular methyl group by oxidizing the bibenzyl **249**. However, anodic oxidation of **249** unfortunately led to an efficient synthesis of dihydrophenanthrone **253** via the route shown in Scheme 23. The dihydrophenanthrone **253** can also be obtained in very good yield by VOF_3–TFA oxidation of the bibenzyl **249**.[11]

[128] J. R. Falck, L. L. Miller, and F. R. Stermitz, *J. Am. Chem. Soc.* **96**, 2981 (1974). For a detailed discussion of the mechanisms of anodic intramolecular coupling reactions see: J. B. Keri, T. C. Jempty, and L. Miller, *ibid.* **101**, 7338 (1979).

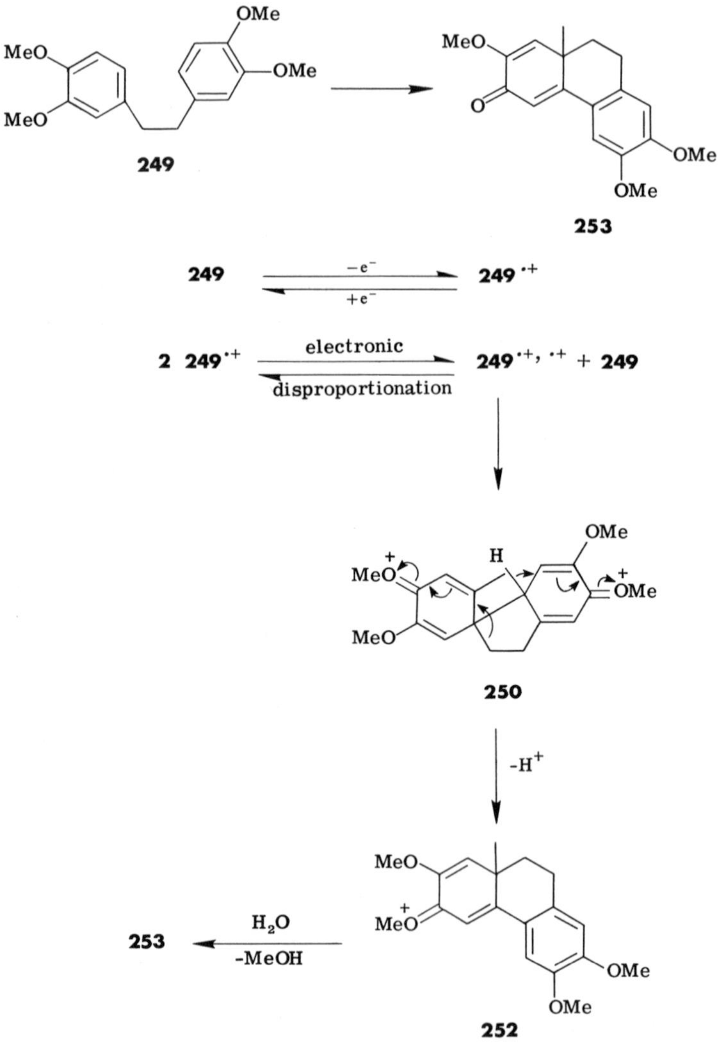

SCHEME 23. [Reprinted with permission from Falck *et al.*, *J. Am. Chem. Soc.* **96**, 2981. Copyright (1974) American Chemical Society.]

Parker and Rolan have extensively studied the oxidative coupling of nonphenolic diarylalkanes.[17,22] In the case of anodic oxidation of symmetrically substituted diarylalkanes (**254**-*n*), the mode of coupling depends on the value of *n* (Scheme 24). When *n* is 4 or smaller, intramolecular cyclization giving the bridged biphenyl (**256**-*n*) is the exclusive reaction pathway.

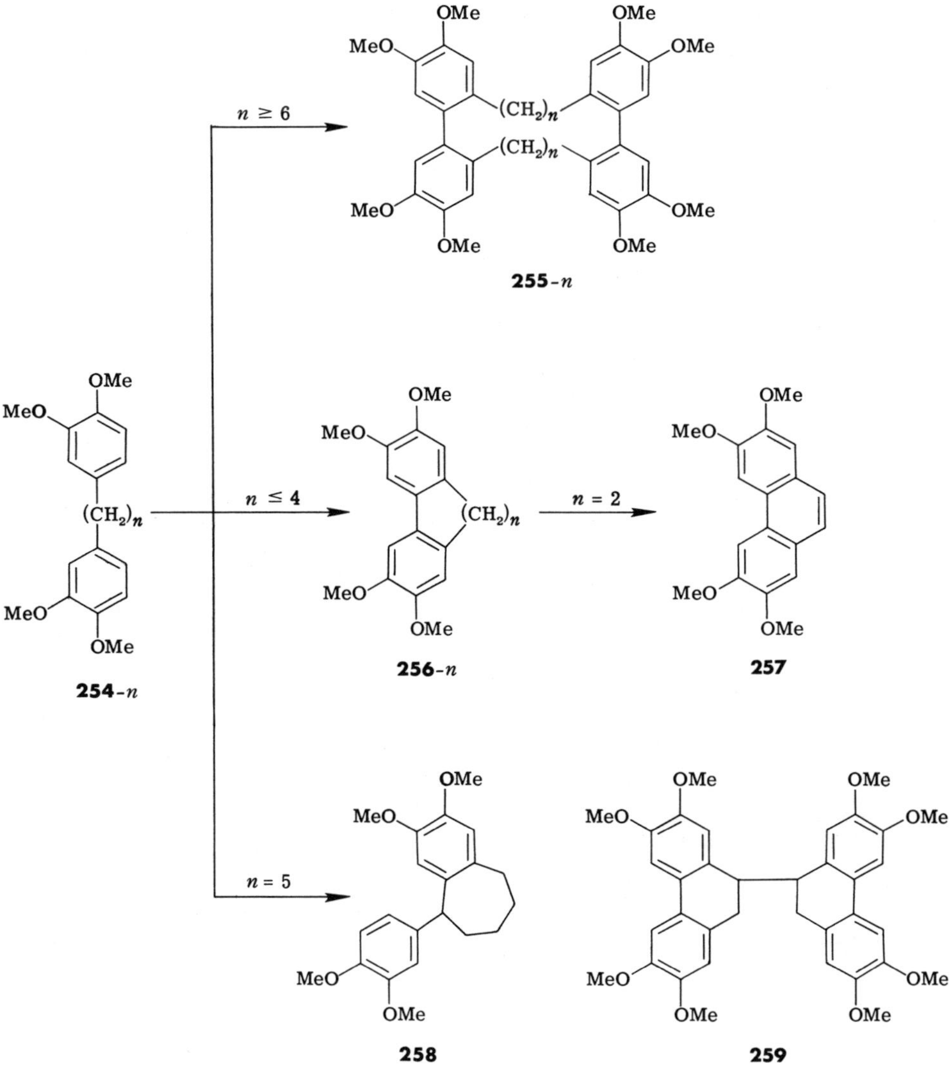

SCHEME 24. [Reprinted with permission from Rolan and Parker, *J. Org. Chem.* **39**, 1014. Copyright (1974) American Chemical Society.]

In the specific case of $n = 2$, the phenanthrene **257** is the product isolated. The case when $n = 5$, cyclization occurs between one ring and the position α to the other ring to give **258**. When n is 6 or larger, large ring compounds (**255**-n) are formed. The yields of anodic oxidation were compared with those obtained by manganic tris(acetylacetonate) (MTA) oxidation (Table VII).

TABLE VII

OXIDATION OF BIS(3,4-DIMETHOXYPHENYL)ALKANES (**254-*n***) in TFA–CH_2Cl_2[a]

Substrate	Anodic oxidation products yield (%)	MTA oxidation products yield (%)
254-1	**256-1** (95)	**256-1** (45)
254-2	**257-2** (95)	**256-2** (45)
		257 (19)
254-3	**256-3** (94)	**256-3** (60)
254-4	**256-4** (93	**256-4** (90)
254-5	**258** (38)	**258** (25)
254-6	**255-6** (4)	Only polymers
254-7	**255-7** (12)	Only polymers
254-8	**255-8** (43)	**255-8** (8)
254-9	**255-9** (42)	**255-9** (7)
254-10	**255-10** (38)	**255-10** (5)
254-16	**256-16** (16)	**256-16** (4)
	255-16 (8)	

[a] Reprinted with permission from Rolan and Parker, *J. Org. Chem.* **39**, 1014. Copyright (1974) American Chemical Society.

In general, the yields of cyclized products were found to be lower when MTA was the oxidant. Another difference in the anodic and MTA oxidation was seen for $n = 2$. Anodically **254-2** is oxidized cleanly to **257** while MTA oxidation gives the intermolecular dimer **259**. Thallium trifluoroacetate and the dimethylformamide–ferric chloride complex were found to be inferior to MTA. In the case of oxidation of unsymmetrically substituted nonphenolic diarylalkanes **260** and **262**, $n = 1$–4, intramolecularly coupled products were obtained.

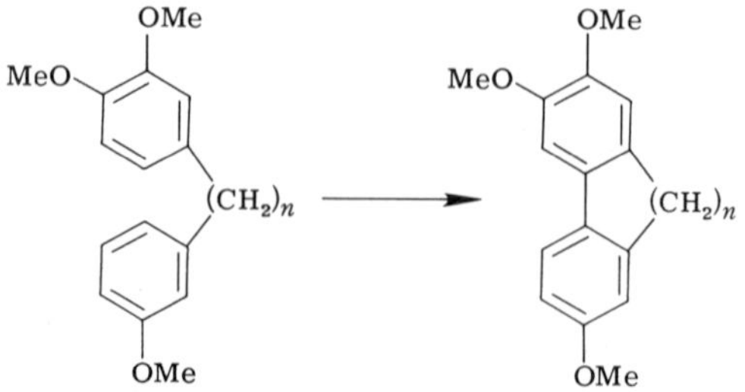

260 **261** $n = 3$

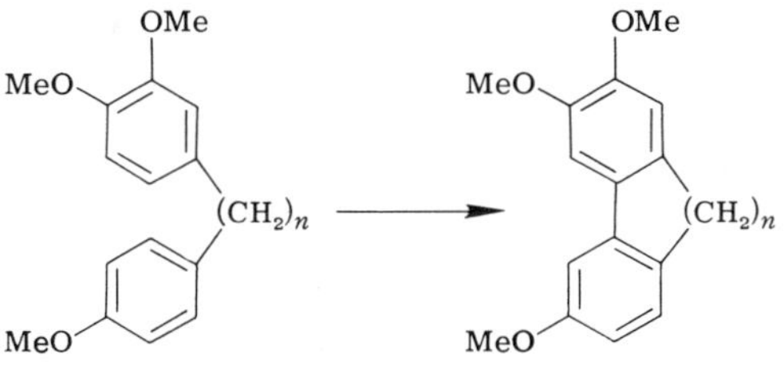

262 **263** $n = 3$

Anodic oxidation of nonphenolic 1,3-diarylpropanes **243e–g** in acetonitrile containing fluoroboric acid as the supporting electrolyte has been shown by Tobinaga *et al.*[21,25] to give spirodienones **244e–g** in excellent yield (Table VI). The dienone **244f** can also be obtained in good yield (70%) by $VOCl_3$ oxidation of **243f.**[16]

Recently, Liepa and Summons[129] have reported on the VOF_3-TFA induced oxidative coupling of 1,2-diarylethylenes. It was found that VOF_3 in TFA–CH_2Cl_2 at 0°C readily converted the stilbene derivatives **264a–c** into the corresponding phenanthrenes **265a–c**, in 75%, 85%, and 91% yield, respectively. This cyclization procedure was equally applicable to the preparation of the benzophenanthrenes **266** (68%) and **267** (88%) as well as methylpicene 14-carboxylate **268** (47%), from the corresponding 1,2-diarylethylene precursors. Anodic oxidation of 1,2-diarylethylenes has also been reported. Tobinaga *et al.*[25] subjected **264a** and **d** to anodic oxidation in acetonitrile containing Et_4NClO_4 as the supporting electrolyte to yield phenanthrenes **265a** and **d** in 50 and 45% yield, respectively.

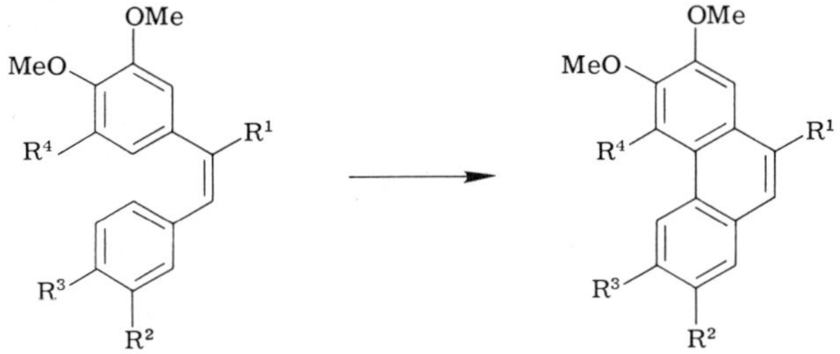

264a $R^1 = CO_2Me$; $R^2 = R^3 = OMe$; $R^4 = H$
b $R^1 = CO_2Me$; $R^2 = R^3 = —OCH_2O—$; $R^4 = H$
c $R^1 = CN$; $R^2 = R^3 = —OCH_2O—$
d $R^1 = CO_2Me$; $R^1 = R^3 = —OCH_2O—$; $R^4 = OMe$

265a–d

[129] A. J. Liepa and R. E. Summons, *J. Chem. Soc. Chem. Commun.*, 826 (1977).

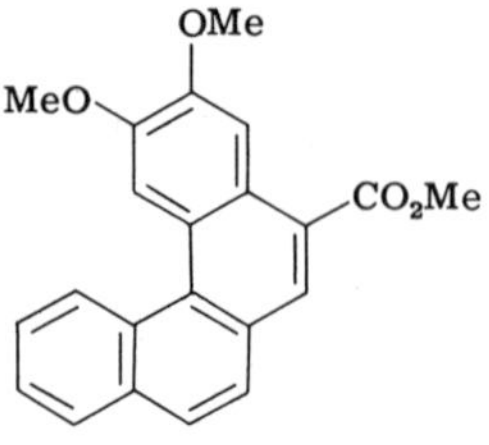

266

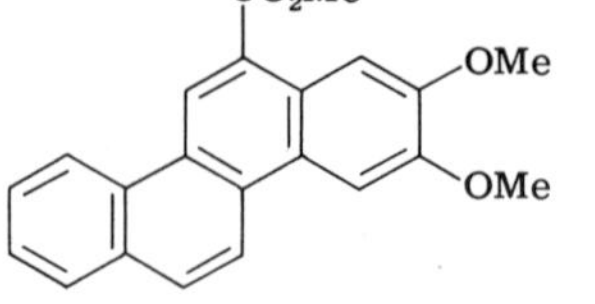

267

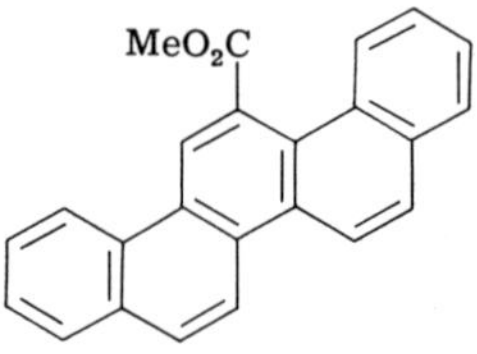

268

This oxidative cyclization procedure was utilized in developing a short and efficient synthesis of (±)-tylophorine (**271**), a member of the phenanthroindolizidine group of alkaloids. Thus, oxidation of **269** with VOF_3 in TFA–CH_2Cl_2 at 0°C resulted in ring closure with simultaneous dehydrogenation to **270** (59%). Reduction of **270** with diborane gave (±)-tylophorine (**271**) in 85% yield.

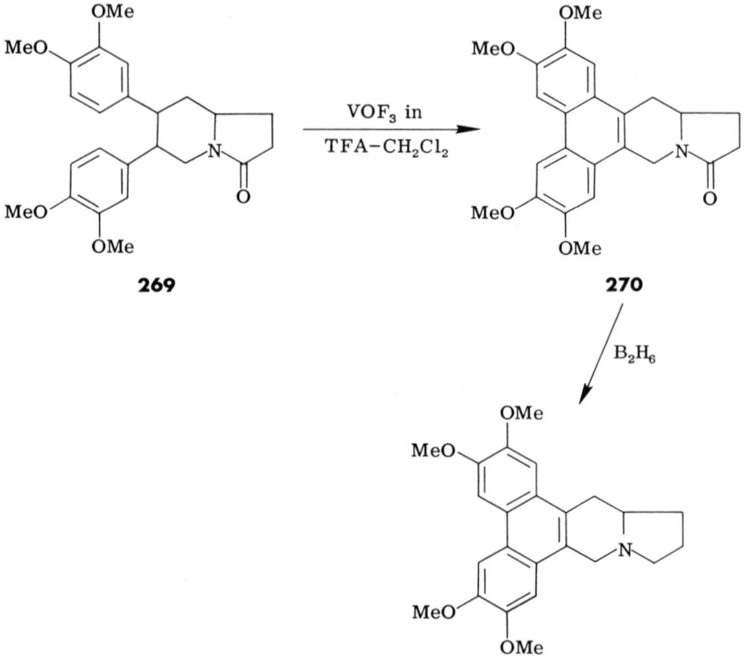

269 **270** **271**

[130] N. R. Farnworth, N. K. Hart, S. R. Johns, J. A. Lamberton, and W. M. Messmer, *Aust. J. Chem.* **22**, 1805 (1969).
[131] E. Krmpotic, N. R. Farnworth, and W. M. Messmer, *J. Pharm. Sci.* **61**, 1508 (1972).
[132] J. M. Paton, P. L. Pauson, and T. S. Stemens, *J. Chem. Soc. C*, 1309 (1969).
[133] J. M. Quante, F. R. Stermitz, and L. L. Miller, *J. Org. Chem.* **44**, 293 (1979).

(±)-Cryptopleurine (**278**), which has the phenanthroquinolizidine ring system, possesses interesting medicinal properties, such as cytotoxic[130] and antiviral activities.[131] In 1969, Pauson *et al.*[132] reported a biogenetic-type synthesis of (±)-cryptopleurine (**278**). The oxidative coupling reaction by manganese dioxide of the phenol **272** to give the spirodienone **273** (20%) was a key step for the construction of phenanthrolizidine nucleus in their synthesis.

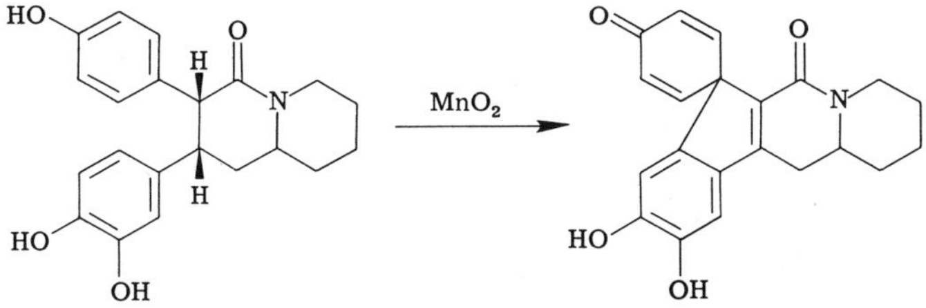

272 **273**

Tobinaga *et al.*[25] recently reported an efficient synthesis of (±)-cryptopleurine (**278**) via anodic oxidation of the nonphenolic quinolizidinone **274**. Oxidation of **274** in acetonitrile containing fluoroboric acid as the supporting electrolyte gave spirodienone **275** and ketocryptopleurine (**276**) in 60% and 31% yield, respectively. The former was converted into the latter by acid-catalyzed dienone–phenol rearrangement to **277**, followed by hydrolysis and methylation. Reduction of **276** with lithium aluminum hydride afforded (±)-cryptopleurine (**278**) in quantitative yield.

Stermitz and Miller[133] have recently reported on benzophenanthridine syntheses through chemical and electrochemical oxidative coupling of

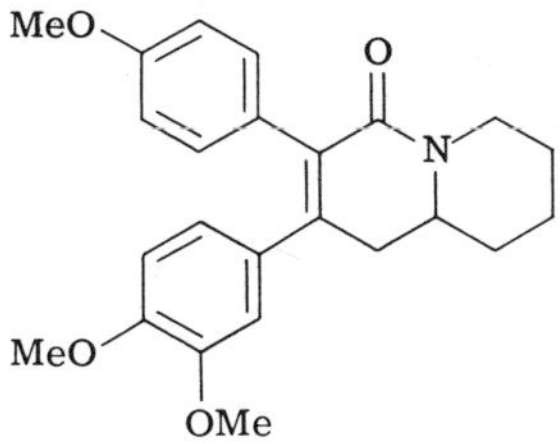

274

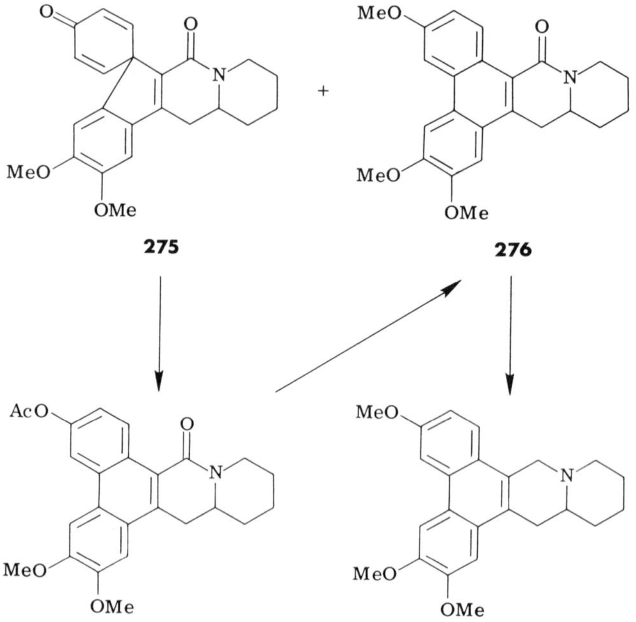

277 **278**

napthylamines and napthylimines. Oxidation of napthylimine **279** with VOF_3 in TFA–CH_2Cl_2 at 25°C gave 30% yield of the benzophenanthridine **280**. However, when **279** was oxidized with VOF_3 in TFAA–CH_2Cl_2, benzophenanthridine **281** was obtained in 50% yield.

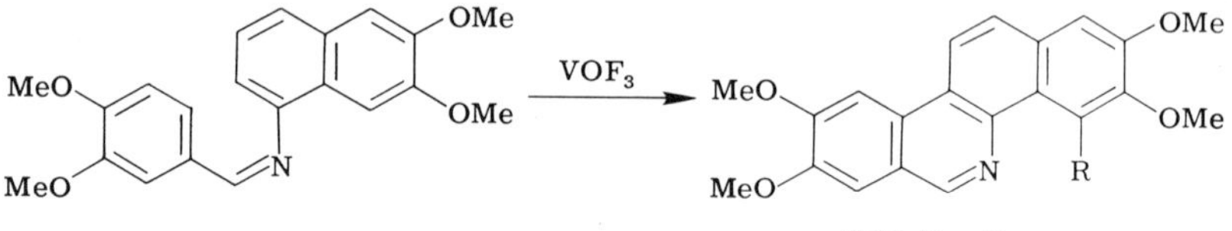

279

280 R = H
281 R = $COCF_3$

On the other hand, anodic oxidation of **279** did not yield any cyclized product. When the napthylamine **282** was oxidized anodically, dimer **283** was isolated in 42% yield.

The ability of VOF_3 in TFA–CH_2Cl_2 to couple two nonphenolic aromatic rings intramolecularly has been beautifully utilized by Kende[14a] in developing a synthesis of (±)-steganacin. Steganacin is a member of a rare group of lignans characterized by a dibenzocyclooctadiene skeletal structure and

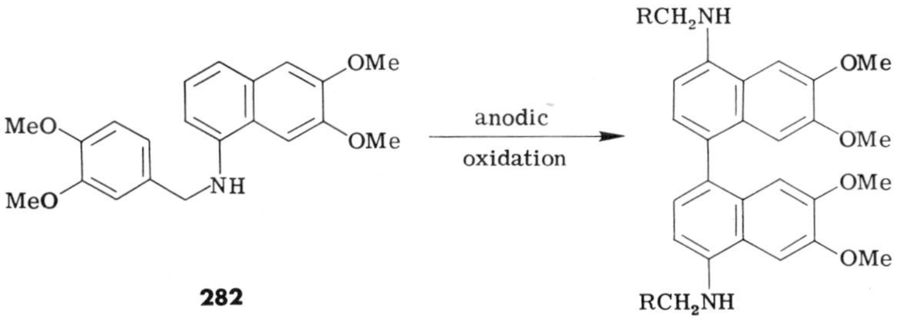

282

R = 3,4-Dimethoxyphenyl

283

by a varied medicinal activity which includes, *inter alia*, a significant anti-leukemic activity *in vivo*.[134] Kende oxidized the biarylbutane **284** with VOF_3 in CH_2Cl_2 containing TFAA to yield the coupled product **285** (45%) which was then transformed into (±)-steganacin (**286**) by a series of steps. Similarly, Schlessinger[15a] has synthesized isostegnane **288** by oxidizing **287** with VOF_3 in TFA–CH_2Cl_2.

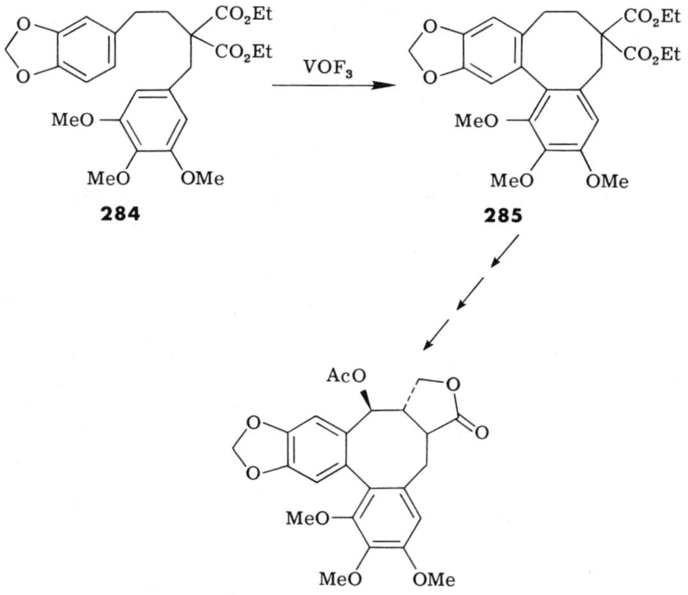

284 **285**

286

[134] S. M. Kupchan, R. W. Britton, M. F. Ziegler, C. T. Gulmore, R. J. Restivo, and R. F. Bryan, *J. Am. Chem. Soc.* **95**, 1335 (1973).

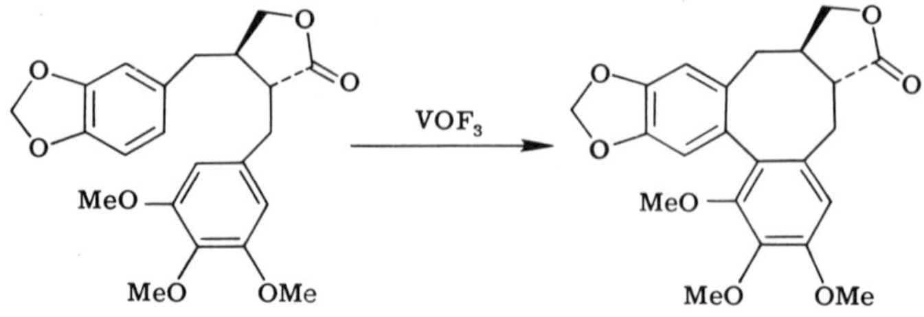

287 **288**

C. MONOPHENOLIC SUBSTRATES

Parker and Rolan have reported fully on the anodic intramolecular coupling of a series of monophenolic diarylalkanes.[23] When oxidation of **289** is carried out in acetonitrile containing *n*-tetrabutylammonium tetrafluoroborate as supporting electrolyte, dienones are obtained in high yield (75–85%) for $n = 3$; no cyclization is found for $n = 2$ or 4. However, if instead of $CH_3CN/n\text{-}Bu_4{}^+NBF_4{}^-$, CH_2Cl_2-trifluoromethanesulfonic acid (TMS) is used as electrolysis medium, dienone–phenol rearrangement products **291a–d** and **292** are obtained in good yield. Oxidation of **289** with thallium trifluoroacetate in CH_2Cl_2 also gives dienones in good yield for $n = 3$, starting material is recovered for $n = 2$ or 4 (Table VIII).

TABLE VIII

ANODIC AND TTFA OXIDATIONS OF THE DIARYLALKANES **289a–i**[a]

Substrate **289**	Anodic oxidation: Acetonitrile, Products yield (%)	Anodic oxidation: CH_2Cl_2-TFMS, Products yield (%)	TTFA oxidation, Products yield (%)
a	**290a** (85)	**291d** (90)	**290a** (80)
b	**290b** (80)	**291c** (94)	**290b** (90)
c	**290c** (80)	unidentified	**290c** (85)
d	**290d** (75)	**291a** (93)	**290d** (95)
e	**290e** (80)	**291b** (96)	**290e** (95)
f	**290f** (80)	**292** (60)	**290f** (90)
g	Polymers	Polymers	**290g** (80)
h	**290h** (80)	Polymers	**290h** (75); **i** (15)
i	**290j** (40); Polymers	Polymers	**290j** (60); **k** (40)

[a] Reprinted with permission from Palmquist *et al.*, *J. Am. Chem. Soc.* **98**, 2571. Copyright (1976) American Chemical Society.

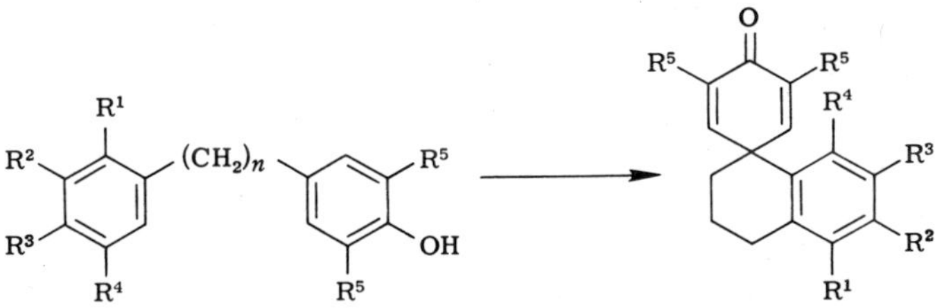

$n = 3$

289 a $R^1 = R^4 = H$, $R^2 = R^3 = R^5 = OMe$
b $R^1 = R^4 = H$, $R^2 = R^3 = OMe$, $R^5 = Me$
c $R^1 = Me$, $R^3 = R^4 = OMe$, $R^2 = R^5 = H$
d $R^1 = R^4 = R^5 = H$, $R^2 = R^3 = OMe$
e $R^1 = R^4 = R^5 = H$, $R^2 + R^3 = —OCH_2O—$
f $R^1 = R^4 = H$, $R^2 = R^3 = OMe$, $R^5 = t\text{-Bu}$
g $R^1 = R^5 = H$, $R^2 = R^3 = R^4 = OMe$
h $R^1 = R^2 = R^3 = H$, $R^4 = OMe$, $R^5 = t\text{-Bu}$
i $R^1 = R^2 = R^3 = R^5 = H$, $R^4 = OMe$

290 a $R^1 = R^4 = H$, $R^2 = R^3 = R^5 = OMe$
b $R^1 = R^4 = H$, $R^2 = R^3 = OMe$, $R^5 = Me$
c $R^1 = Me$, $R^3 = R^4 = OMe$, $R^2 = R^5 = H$
d $R^1 = R^4 = R^5 = H$, $R^2 = R^3 = OMe$
e $R^1 = R^4 = R^5 = H$, $R^2 + R^3 = —OCH_2O—$
f $R^1 = R^4 = H$, $R^2 = R^3 = OMe$, $R^5 = t\text{-Bu}$
g $R^1 = R^5 = H$, $R^2 = R^3 = R^4 = OMe$
h $R^1 = R^2 = R^3 = H$, $R = OMe$, $R^5 = t\text{-Bu}$
i $R^1 = R^2 = R^3 = H$, $R^4 = OMe$, $R^5 = t\text{-Bu}$
j $R^1 = R^3 = R^4 = R^5 = H$, $R^2 = OMe$
k $R^1 = R^2 = R^3 = R^5 = H$, $R^4 = OMe$

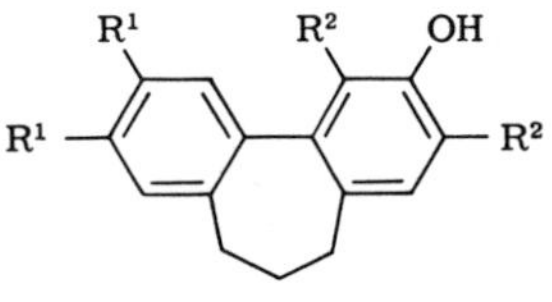

291 a $R^1 = OMe$, $R^2 = H$
b $R^1 = —OCH_2O—$, $R^2 = H$
c $R^1 = Me$, $R^2 = Me$
d $R^1 = R^2 = OMe$

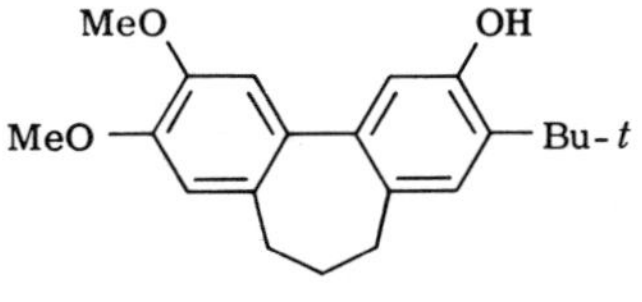

292

Oxidation of the monophenolic substrate **293** with TTFA or Ag_2O in TFA or $VOCl_3$ in CH_2Cl_2 has been shown to give dienone **294** in excellent

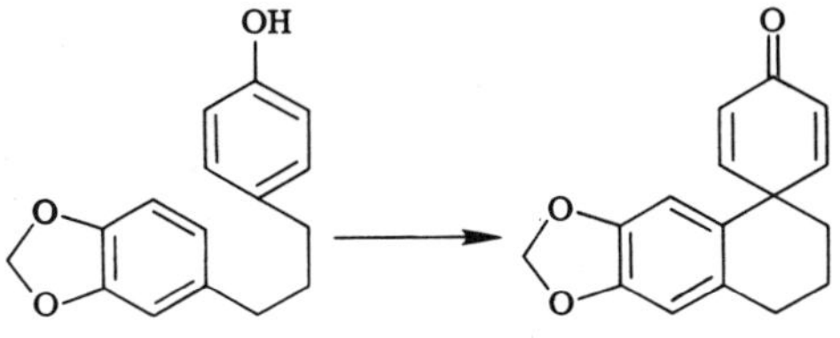

293 **296**

yield[16] (88–97%). Anodic oxidation of **243h** affords spirodienone **244h** in 80% yield[135] (Table VI).

Acknowledgments

The author is greatly indebted to his teachers in organic chemistry, Professors S. Morris Kupchan (deceased), F. A. Carey, and J. D. White for their invaluable guidance and encouragement. In particular, the author wishes to express his deepest gratitude and appreciation to the late Professor S. Morris Kupchan whose exciting endeavors in alkaloid chemistry are in large part responsible for some of the most interesting work described in this chapter. Many thanks to Professor A. T. Sneden, Virginia Commonwealth University, for reading the manuscript and making valuable comments. Thanks also to Mrs. Alexis Massa for typing the manuscript.

[135] S. Tobinaga, *Bioorg. Chem.* **4**, 110 (1975).

CHAPTER **IV**

Oxidation of Coal

RYOICHI HAYATSU, ROBERT G. SCOTT, RANDALL E. WINANS

I. Introduction 279
II. Coal Samples 280
A. General References 280
B. Petrographic Classification 281
C. Rank Classification 281
D. Pretreatment of Coal Before Oxidation 282
E. Methylation of Coal 283
III. Extraction, Separation, and Identification of Oxidation Products 284
A. Extraction of Organic Acids from Oxidation Products 284
B. Separation and Identification 285
IV. Oxidation 291
A. General Comments 291
B. Chromic Acid 292
C. Potassium Permanganate 304
D. Alkaline Cupric Oxide 314
E. Alkaline Nitrobenzene 321
F. Hydrogen Peroxide in Acetic Acid 324
G. Hydrogen Peroxide–Trifluoroacetic Acid 327
H. Sodium Hypochlorite 330
I. Nitric Acid 334
J. Molecular Oxygen 340
K. Miscellaneous 348
V. Conclusion 351
VI. Glossary 352

I. Introduction

Oxidative degradation procedures have been widely used in the elucidation of the complex structures of carbonaceous materials such as coal, kerogen, and humic substances for many decades. Such studies have provided information which cannot be obtained directly due to the chemical nature of these

Oxidation in Organic Chemistry, Part D

ISBN 0-12-697253-2

substances. The organic material in a coal consists of a complex array of insoluble, heterogeneous macromolecules with lesser amounts of volatile or solvent-extractable compounds trapped within the pores of the macromolecules. To obtain structural information, oxidative degradation procedures have been used to degrade coal to lower molecular weight compounds which are soluble, can be characterized readily, and related to the original coal structure. The oxidation of coal has also been studied as a relatively simple method for converting the coal into useful chemical products such as terephthalic acid.

Commonly used oxidants are nitric acid, permanganate, chromic acid, hydrogen peroxide, hypohalites, and oxygen. Perchloric acid, cupric oxide, nitrobenzene, and ozone have also been used. The yield and nature of the products vary widely with the oxidants and conditions. In past studies, many common oxidants and conditions led to a high conversion of coal but with rather drastic degradation to simple, stable compounds such as carbon dioxide, short-chain aliphatic carboxylic acids, and benzenecarboxylic acids. In addition, the oxidation products were in general studied by limited classical methods for separation and identification (see reviews by van Krevelen[1] and Lowry[2,3]). Therefore, these results gave only very limited structural information. On the other hand, mild oxidation with certain oxidants generally produces a high yield of high molecular weight acidic materials, (so-called humic acid or humic acid-like materials) and also does not give direct information on structural units of coal.

However, several selective oxidation methods have been developed recently to characterize the aliphatic nature and aromatic structural units of coal, and modern analytical techniques have been used for separation and identification of the oxidation products.

In this chapter, the emphasis will be on the selective oxidative degradation procedures as tools for characterization of coal structure. For comparison, oxidations of lignin, humic acid, and kerogen (Section VI), which are related to the structural study of coal, will be discussed. However, production of chemicals by oxidation of coal will not be discussed in detail.

II. Coal Samples

A. GENERAL REFERENCES

Since a number of reviews and books on the basic science of coal have been published, it is recommended that these works be consulted for de-

[1] D. W. van Krevelen, "Coal." Elsevier, Amsterdam, 1961.
[2] H. H. Lowry, ed., "Chemistry of Coal Utilization," Vols. 1 and 2. Wiley, New York, 1945.
[3] H. H. Lowry, ed., "Chemistry of Coal Utilization." Suppl. vol. Wiley, New York, 1963.

tailed accounts of the geological and chemical aspects (van Krevelen,[1] Lowry,[2,3] Murchison and Westoll,[4] Eglinton and Murphy,[5] Manskaya and Drozdova,[6] Swain,[7] Tingey and Morrey,[8] Karr,[9] Tissot and Welte,[10] Berkowitz,[11] Pitt and Millward,[12] Davidson,[13] Hayes and Swift[14]).

B. PETROGRAPHIC CLASSIFICATION

Coal macerals are analogous to the minerals in inorganic rocks. The petrographic classification of these coal macerals provides useful information on the origin of coal. Macerals are divided into three groups based on color or reflectance: vitrinite, exinite (or liptinite), and inertinite; each group is further subdivided on the basis of morphology. Typically, vitrinite is the most common maceral comprising approximately 80% of the American coals.

C. RANK CLASSIFICATION

Rank classification is based on the diagenesis of coal which is called coalification. The degree of coalification progressively increases from lignite through sub-bituminous coal, bituminous coal to anthracite. The carbon content and aromaticity increase with decreasing oxygen and hydrogen contents, whereas the chemical reactivity generally decreases. Lignite and sub-bituminous coals are referred to as low rank coals, and bituminous and anthracite coals are called high rank coals,[11] however, this distinction is not rigid. In addition, one must be careful about using elemental analyses

[4] D. Murchison and T. S. Westoll, eds, "Coal and Coal-Bearing Strata." Oliver & Boyd, Edinburgh, 1968.

[5] G. Eglinton and M. T. J. Murphy, eds., "Organic Geochemistry: Methods and Results." Springer-Verlag, Berlin and New York, 1969.

[6] S. Manskaya and T. V. Drozdova. "Geochemistry of Organic Substances." Pergamon, Oxford, 1968.

[7] F. M. Swain, "Non-marine Organic Geochemistry." Cambridge Univ. Press, London, 1970.

[8] G. L. Tingey and J. R. Morrey, "Coal Structure and Reactivity," Battelle Energy Program Rep. Battelle, Richland, Washington, 1973.

[9] C. Karr, Jr., ed., "Analytical Methods for Coal and Coal Products," Vols. 1 and 2. Academic Press, New York, 1978.

[10] B. P. Tissot and D. H. Welte, "Petroleum Formation and Occurrence." Springer-Verlag, Berlin and New York, 1978.

[11] N. Berkowitz, "An Introduction to Coal Technology." Academic Press, New York, 1979.

[12] G. J. Pitt and G. R. Millward, eds., "Coal and Modern Coal Processing: An Introduction." Academic Press, New York, 1979.

[13] R. M. Davidson, "Molecular Structure of Coal." Rep. No. ICTIS/TR08. IEA Coal Research, London, 1980.

[14] M. H. B. Hayes and R. S. Swift, *in* "The Chemistry of Soil Constituents" (D. J. Greenland and M. H. B. Hayes, eds.), p. 179. Wiley, New York, 1978.

of coals because of errors resulting from the mineral content in the coals[15] and the fact that most oxygen contents are determined by difference. The elemental analysis values reported here are on a moisture and ash-free (maf) basis.

D. PRETREATMENT OF COAL BEFORE OXIDATION

Organic acids and humic acid-like materials as well as many nonacidic materials such as aliphatic and aromatic hydrocarbons, terpenoids, waxes, etc., are known to be trapped in coal, and in particular, low rank coal.[1,3,5,16–25] Therefore, the mode of preparation of the starting material is of particular importance in oxidation studies. Upon oxidation, raw coals, or insufficiently extracted samples, would yield oxidation products not only of the macromolecular material of interest, but also of the trapped material, resulting in a misleading characterization of coal structure, particularly for the study of the aliphatic structure of coal.[26–28] In many oxidation studies, the starting material was not always subjected to exhaustive extraction methods before oxidation. Weathered coals also may yield misleading information upon further oxidation.

The following procedures are recommended for the removal of trapped organic materials: (1) Coal sample (10–200 μm particle size) is treated with 1–2 *N* HCl ($\geq 50°$C for 24 hr with stirring under N_2 atmosphere) to remove mineral carbonate and to convert salts of acids to free acids. It should be noted that the use of drastic conditions may cause changes in coal structures. In addition ferric ion, Fe(III), may form during HCl treatment of coal, particularly in the presence of air. Although ferric ion is a mild oxidant, it still may cause alteration of chemical structures of coal macromolecules.

[15] C. P. Rao and H. J. Gluskoter, Illinois State Geological Survey, Circular 476, 1973.
[16] J. D. Brooks, K. Gould, and J. W. Smith, *Nature* (*London*) **222**, 257 (1969).
[17] J. D. Brooks and J. W. Smith, *Geochim. Cosmochim. Acta* **33**, 1183 (1969).
[18] K. Ouchi, *J. Fuel Soc. Jpn.* **56**, 779 (1977).
[19] R. Hayatsu, R. E. Winans, R. G. Scott, L. P. Moore, and M. H. Studier, *Fuel* **57**, 541 (1978).
[20] R. Hayatsu, R. E. Winans, R. G. Scott, L. P. Moore, and M. H. Studier, *Nature* (*London*) **275**, 116 (1978).
[21] T. Yoshii and Y. Satou, *J. Fuel Soc. Jpn.* **58**, 783 (1979).
[22] T. Yoshii and Y. Satou, *J. Fuel Soc. Jpn.* **58**, 791 (1979).
[23] T. Yoshii and Y. Satou, *J. Fuel Soc. Jpn.* **58**, 797 (1979).
[24] M. Schnitzer and J. A. Neyroud, *Fuel* **54**, 17 (1975).
[25] L. B. Alemany, S. R. King, and L. M. Stock, *Fuel* **57**, 738 (1978).
[26] R. Hayatsu, R. E. Winans, R. G. Scott, and R. L. McBeth, *Fuel* **60**, 158 (1981).
[27] R. Hayatsu, R. E. Winans, R. G. Scott, R. L. McBeth, and G. F. Vandegrift, *Fuel* **60**, 161 (1981).
[28] R. Hayatsu, R. E. Winans, R. G. Scott, R. L. McBeth, and L. P. Moore, *Fuel* **60**, 77 (1981).

Indeed $FeCl_3$–HCl oxidation of Aleksinac kerogen concentrate[28a] altered its H/C and O/C ratios and IR spectra. After filtration, the residue is washed with distilled water until the filtrate is neutral. This procedure also removes the major portion of trapped volatile hydrocarbons. (2) The residue is then treated with 0.5–1.0 *N* aqueous NaOH (25°–50°C, 24 hr with stirring). After separating the insoluble residue by centrifugation, the residue is washed with water and methanol repeatedly until the supernatant solution is neutral and clear. By this treatment, acidic material and acids originally present as salts or esters can be removed. (3) Finally, the acid and alkali insoluble residue is refluxed with a mixture of benzene–methanol (3:1–4:1, 24–48 hr) which is commonly employed to extract organic compounds in organic geochemical studies. The residue obtained is dried at 80°–100°C for 1–2 days under vacuum. By these treatments, trapped organic material is found to account for approximately 0.5–12 wt % of coal, the yields varying with rank. In general, high rank coals do not contain much trapped material.

E. METHYLATION OF COAL

Coal contains hydroxyl groups on aromatic rings which account for approximately 30–70% of the organic oxygen.[8 29–32] In general, organic oxygen content decreases progressively from low rank coal (~18–30%) to high rank coal (~2–17%).[11,12] Except for those oxidations that do not attack phenols, such as alkaline cupric oxide, it is necessary to protect aromatic rings with free hydroxyl groups from degradation by the oxidant. This is particularly true for low rank coals. As a general procedure, 1.0 g of a solvent-extracted coal is methylated with 0.8 g dimethyl sulfate and 1.0 g K_2CO_3 in 25 ml of anhydrous acetone. The reaction mixture is refluxed for 2–3 days with stirring, and then the mixture is filtered and washed with water, acetone, and methanol. Finally, it is dried at 80°–100°C under vacuum. Although this technique may not methylate the coal sample quantitatively, it is found to be much more useful than diazomethane, in particular for methylation of hindered acidic materials.[33] In the authors' laboratory, d_6-dimethyl sulfate is also used for methylation of coal. In general, low rank coal contains aromatic hydroxyl and methoxy groups; therefore, oxidation

[28a] D. Vitorovic, V. D. Krsmanovic, and P. A. Pfendt, *in* "Advances in Organic Geochemistry 1979" (A. G. Douglas and J. R. Maxwell, eds.), p. 585. Pergamon, Oxford, 1980.

[29] G. A. Vaughan and J. J. Swithenbank, *Analyst* (*London*) **95**, 890 (1970).

[30] Coal Mining Research Center, Japan, Special Report (November), 1973.

[31] Z. Abdel-Baset, P. H. Given, and R. F. Yarzab, *Fuel* **57**, 95 (1978).

[32] R. F. Yarzab, Z. Abdel-Baset, and P. H. Given *Geochim. Cosmochim. Acta* **43**, 281 (1979).

[33] L. F. Fieser and M. Fieser, "Reagents for Organic Synthesis," Vol. 1 (see also Vols. 2–7). Wiley, New York, 1968.

of d_3-methylated low rank coal yields d_3-methoxy and regular methoxy aromatic acids, and the OCD_3/OCH_3 ratio can be determined by gas chromatography–mass spectrometry, gas chromatography–microwave plasma detection, and high resolution mass spectrometry[34–37] (see Section IV, B, Fig. 2).

III. Extraction, Separation, and Identification of Oxidation Products*

A. EXTRACTION OF ORGANIC ACIDS FROM OXIDATION PRODUCTS

Since a complex mixture of organic acids is generally produced from coal and coal products by oxidation, a single solvent may not be sufficient for extracting all organic acids produced; refluxing the products in the solvent may be necessary. Duty *et al.*[38] have reported that methanol incompletely extracted products obtained from the aqueous $Na_2Cr_2O_7$ oxidation of solvent-refined lignite (SRL). Additional organic acids were extracted from the remaining residue with benzene–ether (1:1) and chloroform. As shown in Table I, it is obvious that whereas naphthalene and phenanthrenecarboxylic acids can be sufficiently extracted with methanol, methanol is a poor solvent for extraction of benzenepolycarboxylic acids. Methyl ethyl ketone may be used also as an extraction solvent for these acids. Quantitative extraction of very water-soluble or amphoteric acids may be difficult from oxidation products that contain inorganic compounds. In such cases, it is recommended that the products be esterified by the method of Eisenbraun *et al.*[39] for the preparation of methyl esters of very water-soluble acids. Oxidation products can then be easily extracted as their methyl esters with appropriate solvents.

Insoluble residues obtained after oxidation, as indicated in Fig. 1, are sometimes referred to as nonoxidized or "unoxidized" coal in many refer-

* General Procedure, see Fig. 1.

[34] R. Hayatsu, R. G. Scott, R. L. McBeth, and R. E. Winans. Symp. on Polymers and Coal, **43**, 401 (Div. of Org. Coatings and Plastics Chem., Am. Chem. Soc.), Las Vegas, August, 1980.

[35] R. E. Winans, R. Hayatsu, R. G. Scott, L. P. Moore, and M. H. Studier, Preprint 24 (No. 2), p. 196 (Div. of Fuel Chem., Am. Chem. Soc.), Honolulu, April, 1979.

[36] R. E. Winans, R. G. Scott, R. L. McBeth, and R. Hayatsu, Preprint 25 (No. 2), p. 262 (Div. of Pet. Chem. Am. Chem. Soc.), Houston, March, 1980.

[37] R. E. Winans, R. Hayatsu, R. L. McBeth, R. G. Scott, L. P. Moore, and M. H. Studier, *Adv. Chem. Ser.*, No. 192, p. 161, 1981.

[38] R. C. Duty, R. Hayatsu, R. G. Scott, L. P. Moore, R. E. Winans, and M. H. Studier, *Fuel* **59**, 97 (1980).

[39] E. J. Eisenbraun, R. N. Morris, and G. Adolphen, *J. Chem. Educ.* **47**, 710 (1970).

TABLE I

RELATIVE ABUNDANCE MOLE (%) OF EXTRACTABLE ORGANIC ACIDS OBTAINED FROM AQUEOUS $Na_2Cr_2O_7$ OXIDATION OF SRL[a,b]

Organic acid	Methanol extract	Benzene–ether (1:1) and chloroform extracts of methanol extract residue
Benzenedicarboxylic acid	7.2	10.1
Benzenetricarboxylic acid	0.9	14.4
Benzenetetracarboxylic acid	—	11.6
Benzenepentacarboxylic acid	—	4.2
Naphthoic acid	7.3	0.8
Naphthalenedicarboxylic acid	12.9	6.8
Naphthalenetricarboxylic acid	0.6	2.5
Phenanthrenemonocarboxylic acid	9.7	1.1
Dibenzothiophenemonocarboxylic acid	7.3	0.9
Dibenzofuranmonocarboxylic acid	1.0	0.3
Dibenzofurandicarboxylic acid	0.3	0.7

[a] Prepared using data from Duty *et al.*[38]
[b] Only major products are shown.

ences, including some in this chapter. This designation may not be correct because such residues can consist of both nonoxidized coal and partially oxidized coal.

Humic acid-like materials are frequently produced together with organic acids from the oxidation of coal. The yield varies with oxidants and conditions. These materials are soluble in alkaline aqueous solution and some of them are soluble in polar solvents such as methanol. They are not soluble in acids or nonpolar solvents. The isolation method is very similar to that for soil humic acids which has been reviewed in detail by Eglinton and Murphy[5] and Hayes and Swift.[14] Since ease or difficulty of extraction of organic acids is influenced by oxidation reagents and conditions, it is recommended that original papers and reviews on oxidation studies of coal and coal products be consulted for detailed accounts of extraction methods.

B. SEPARATION AND IDENTIFICATION

1. GAS CHROMATOGRAPHY AND MASS SPECTROMETRY*

Mass spectrometry (MS) is an exceedingly useful technique for the elucidation and identification of oxidation products of coal. A combination

* High resolution mass spectrometry will be discussed in Section III,B,2.

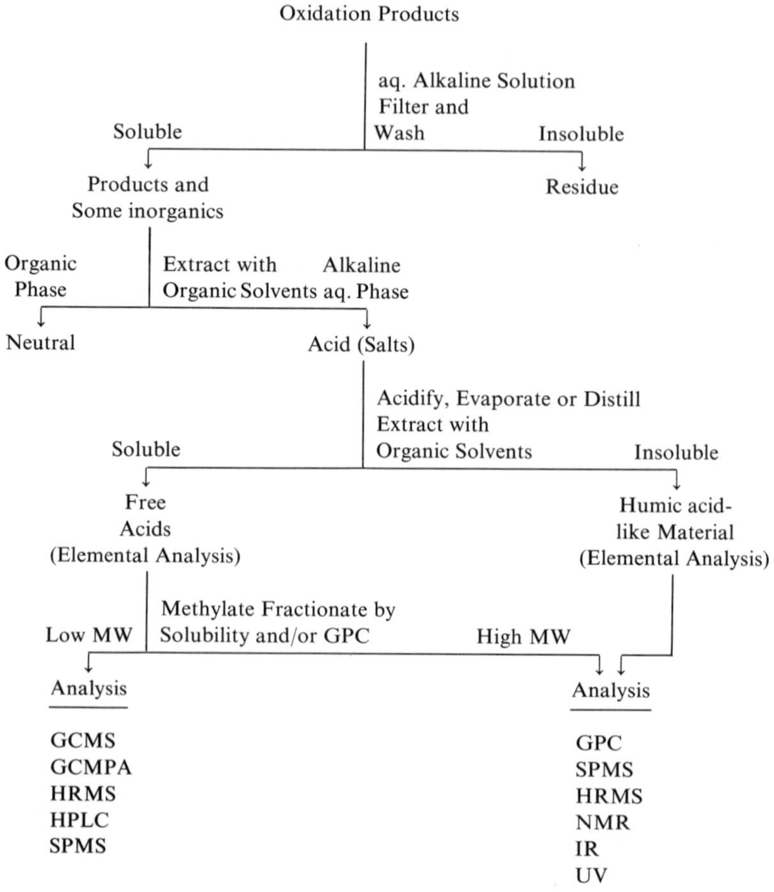

FIG. 1. Generalized scheme for separation and characterization of a coal oxidation product. Abbreviations: GPC—gel permeation chromatography, GCMS—gas chromatography–mass spectrometry, GCMPA—gas chromatography–microwave plasma analysis, HRMS—high resolution mass spectrometry, HPLC—high performance liquid chromatography, SPMS—solid probe mass spectrometry, NMR—nuclear magnetic resonance, IR—infrared, and UV—ultraviolet spectroscopy.

of MS with gas chromatography (GC) results in a system with greatly increased analytical capability. For analysis of the methyl esters of a complex mixture of organic acids isolated from oxidation products, a GCMS system is essential. As derivatives of organic acids for GCMS analysis, methyl esters are superior to other derivatives such as ethyl and trimethylsilyl esters because of: (1) ease of preparation; (2) easier interpretation of mass spectra; (3) much lower probability of producing misleading by-products.

Diazomethane is the most common reagent for esterification. The major artifacts produced by the reaction of diazomethane with coal oxidation products are pyrazoline compounds that are formed from the reaction of unsaturated compounds with diazomethane. Indeed, these compounds are found in the peracetic acid oxidation of lignin[40,41] and nonbuffer-controlled permanganate oxidation of lignite coal.[26] It is obvious that some drastic oxidations result in the degradation of benzene and phenolic rings to produce unsaturated compounds such as maleic and muconic acids. For example:

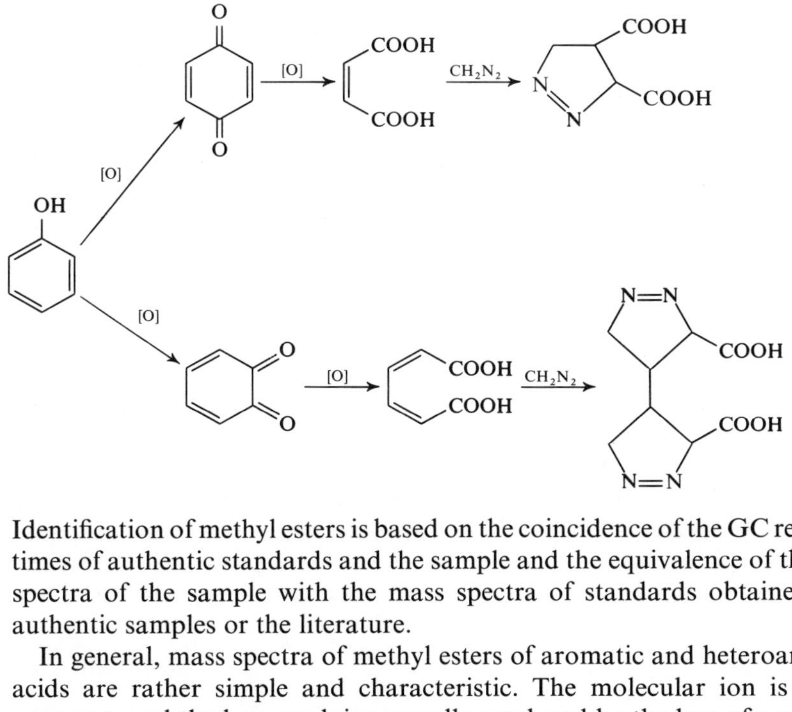

(1)

(2)

Identification of methyl esters is based on the coincidence of the GC retention times of authentic standards and the sample and the equivalence of the mass spectra of the sample with the mass spectra of standards obtained from authentic samples or the literature.

In general, mass spectra of methyl esters of aromatic and heteroaromatic acids are rather simple and characteristic. The molecular ion is always apparent, and the base peak is generally produced by the loss of a methoxy group $(M - OCH_3)^+$. In addition, a sizable ion $[(M - OCH_3) - CO]^+$ is also seen. In particular, some polynuclear aromatic acid methyl esters (e.g., methyl-1-naphthoate) produce this ion as a prominent peak. Ring-substituted methyl or polymethyl derivatives of aromatic acid methyl esters can be identified in the same manner. If the nuclear methyl group is substituted ortho to the ester group, it can be easily distinguished by the elimination of CH_3OH from the molecule (ortho effect).

[40] K. V. Sarkanen and J. Suzuki, *Tappi* **48**, 459 (1965).

[41] H. M. Chang and G. G. Allan, *in* "Lignin" (K. V. Sarkanen and C. H. Ludwig, eds.), p. 433. Wiley (Interscience), New York, 1971.

Individual identification of possible isomers of aromatic or heteroaromatic acid methyl esters cannot be accomplished by mass spectral interpretation alone. For example, naphthalenedicarboxylic acid dimethyl ester has 10 theoretical isomers, and mass spectra of six of these isomers were found to show similar fragmentation patterns in the authors' laboratory. Therefore, for individual identification of isomers, it is necessary to compare the GC retention times of the sample with those of authentic standards.

The mass spectra of aliphatic and alicyclic acid methyl esters show rather complex patterns, and differ markedly from those of the methyl esters of aromatic and heteroaromatic acids. In general, the molecular ion of these groups appears, if at all, only as a very small peak. However, identification of short-chain α,ω-dicarboxylic acid methyl esters (except methyl oxalate) is relatively easy, because the spectra of these esters are distinguished by intense peaks at M —31, M —59, or M —71. Identification of alicyclic acid methyl esters may be the most difficult because of complex mass spectra and ambiguous information on the fragmentation patterns. In such instances, high resolution mass spectrometry is required.

2. High Resolution Mass Spectrometry (HRMS)

Another very useful tool for studies of coal oxidation products is the high resolution mass spectrometer. An essential feature of this instrument is its capability of measuring masses with an accuracy of 10 ppm or better. The significance of this capability is especially evident when identification of heteroatomic species in the oxidation products is required. For example, mass spectra of the following esters all show an intense peak at *m/e* 211 which is due to the loss of a methoxy group from the parent molecule $(M—OCH_3)^+$.

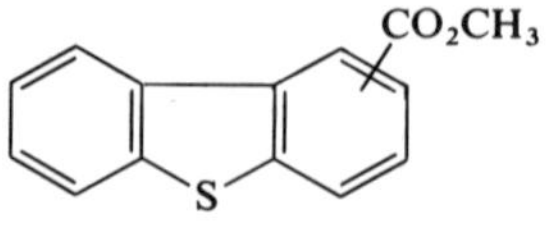

(1)

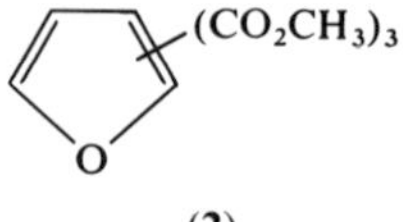

(2)

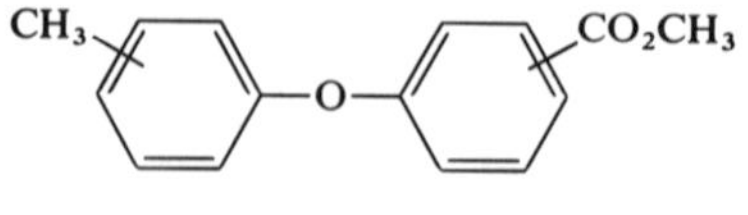

(3)

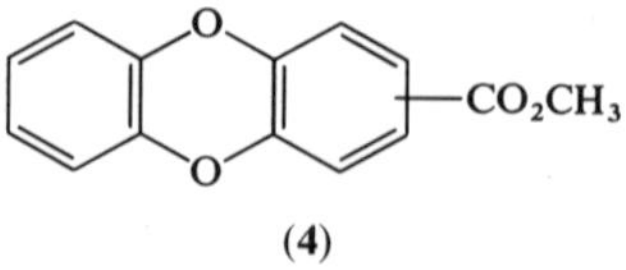

(4)

However, these species can be distinguished by HRMS since, as shown below, they have different precise masses resulting from their different elemental compositions. Indeed, these particular methyl esters have actually

Compound	$(M - OCH_3)^+$	Precise mass*
1	$C_{13}H_7OS$	211.0218
2	$C_9H_7O_6$	211.0243
3	$C_{14}H_{11}O_2$	211.0759
4	$C_{13}H_7O_3$	211.0395

* It should be noted that these precise masses are for molecular species containing only the most abundant isotope of each constituent element.

been found in coal oxidation products (after esterification) by GCMS and HRMS.[19,42,43]

Since the oxidation of coal always produces a complex mixture of organic acids, it is essential for definite identification that the methylated oxidation products be analyzed by high resolution mass spectrometry as well as by GCMS. Also, because these complex mixtures contain compounds with different volatilities, it is best to take many scans while increasing the source temperature. This results in partial fractionation and may also give several measurements on the same peaks.

3. Gas Chromatography–Microwave Plasma Detector-Computer

A gas chromatography (GC)–microwave plasma detector (MPD)-computer system has recently been developed for analyzing complex mixtures resulting from degradation reactions on coal and other fossil fuels by Winans *et al.*[36] The main feature of this system is an Applied Chromatography Systems MPD-850 which is an emission spectrometer with a helium plasma source directly connected to a gas chromatograph.[44,45] This instrument is set up to simultaneously detect the elements C, H, D, S, F, Cl and O or N. An Antek 300 GC with temperature programming and a modified injector is used with both support-coated open tubular (SCOT) and short (10 m) wall-coated open tubular (WCOT) glass columns. The output from the emission spectrometer is amplified and transferred to a central computer facility via a CAMAC interface. To obtain quantitative data, accurate peak areas must be calculated. This presents a problem when several hundred compounds are present in a sample. With the appropriate

[42] R. Hayatsu, R. E. Winans, R. G. Scott, L. P. Moore, and M. H. Studier, *in* "Organic Chemistry of Coal" (J. W. Larsen, ed.), p. 108. Am. Chem. Soc. Symp. Ser. 71, 1978.

[43] R. Hayatsu, R. G. Scott, L. P. Moore, and M. H. Studier, *Nature* (*London*) **257**, 378 (1975).

[44] W. R. McLean, *in* "Recent Analytical Developments in the Petroleum Industry" (D. R. Hodges, ed.), p. 139. Wiley, New York, 1974.

[45] K. S. Brennar, *J. Chromatogr.* **167**, 365 (1978).

standards, however, quantitative data are obtained and empirical formulae can be determined. Analysis of samples with both the GCMPD and the GCMS using the same type of columns consititutes a very powerful analytical system for complex mixtures.

The system has been applied to the analysis of organic acids (as methyl esters) obtained from oxidation of coals.[36] In particular, it has been found to be a useful technique for analysis of the following cases: (1) identification of a series of thiophenecarboxylic acids (as methyl esters) produced from oxidation of coals (because of their small mass defect they cannot be separated from the reference peaks in the normal HRMS); (2) determination of the ratio of methoxybenzene acids/d_3-methoxybenzene acids as their methyl esters.

4. Analysis of the Nonvolatile Oxidation Products

Polynuclear aromatic polycarboxylic acids, polycyclic heteroaromatic acids, or humic acid-like materials that are also produced by selective or mild oxidation of coal are difficult to analyze by GCMS because even their esters are not volatile enough for GC analysis.

In general, nonvolatile organic acid methyl esters can be analyzed by variable-temperature solid probe MS[46] and HRMS. If these techniques are combined with high performance liquid chromatography (HPLC), more secure identification can be obtained. Humic acid-like materials are still higher molecular weight nonvolatile acids, and mass spectral analysis of these acids reveals only degradation product fragments at high temperatures (150°–300°C), and hence, does not give direct information on coal structure. However, while coal macromolecules are insoluble materials, coal derived humic acid-like materials are known to be soluble in alkaline solution. Some of them are also soluble in polar organic solvents such as methanol. This may ease the difficulty of structural analysis. For example, these materials can be analyzed by LC, GPC, LCMS, NMR, IR, UV, etc. (see Fig. 1). Indeed, a great number of chromatographic and spectral studies of soil humic acid have been reported by Hayes and Swift[14] and Ruggiero *et al.*[47] However, to obtain detailed information on the structures of humic acid-like materials derived from coal, it is necessary to further oxidize these materials to simpler acids that are more readily interpreted in terms of coal structure.

[46] M. H. Studier, R. Hayatsu, and R. E. Winans, *in* "Analytical Methods for Coal and Coal Products" (C. Karr, Jr., ed.), Vol. 2 p. 43. Academic Press, New York, 1978.

[47] P. Ruggiero, F. S. Interesse, L. Cassidei, and O. Sciacovelli, *Geochim. Cosmochim. Acta* **44**, 603 (1980).

5. General References

Those readers interested in the chromatographic and spectral analyses of the oxidation products and related organic compounds may wish to consult the following papers and books which are only a limited representation of the many excellent references available in the literature. MS and GCMS (Eglinton and Murphy,[5] Biemann,[48] McFadden,[49] Stenhagen *et al.*,[50] Roboz and Chait,[51] Studier *et al.*,[46] Tam[52]), LC and LCMS (Scott,[53] Coleman *et al.*[54]), and NMR (Chang and Allan,[41] Ivin,[55] Karr,[9] Ruggiero *et al.*,[47] Hatcher *et al.*[56]).

IV. Oxidation

A. General Comments

Because of the diversity of the structures in coal, varying not only with rank but within the same rank, there are several difficult problems involved in its oxidative degradation.

One of these problems is establishing a standard procedure specifying such variables as reaction temperature, time, ratio of oxidant to coal sample, etc. To determine optimum conditions, more than one run for each oxidation of coal is required. For example, because of the uncertainty of the chemical structure of coal, stoichiometric considerations for the ratio of oxidant/coal sample cannot be directly applied to the oxidation of coal. This means that single-run coal oxidations can result in either insufficient oxidation or extensive degradation of the macromolecular material. Therefore, the characterization of coal structure based on the interpretation of oxidation products

[48] K. Biemann, *in* "Elucidation of Organic Structures by Physical and Chemical Methods" (K. W. Bentley and G. W. Kirby, eds.), 2nd ed., Vol. 4, Part 1, p. 377. Wiley (Interscience), New York, 1972.

[49] W. H. McFadden, "Techniques of Combined Gas Chromatography–Mass Spectrometry: Applications in Organic Analysis." Wiley (Interscience), New York, 1973.

[50] E. Stenhagen, S. Abrahamsson, and F. W. McLafferty, "Registry of Mass Spectral Data," Vols. 1–4 Wiley, New York, 1974.

[51] J. Roboz and E. Chait, *in* "Physical Methods of Chemistry" (A. Weissberger and B. W. Rossiter, eds.), Vol. 1, Part VI. Wiley, New York, 1977.

[52] S. W. Tam, *in* "The Chemistry of Acid Derivatives" (S. Patai, ed.), Suppl. B, Part 1, p. 121. Wiley, New York, 1979.

[53] R. P. W. Scott, "Contemporary Liquid Chromatography," Techniques of Chemistry (A. Weissberger, ed.), Vol. XI. Wiley, New York, 1976.

[54] M. W. Coleman, D. L. Woaton, H. C. Dorn, and L. T. Taylor *Anal. Chem.* **49**, 533 (1977).

[55] K. J. Ivin, "Structural Studies of Macromolecules by Spectroscopic Methods." Wiley, New York, 1976.

[56] P. G. Hatcher, D. L. VanderHart, and W. L. Earl, *Org. Geochem.* **2**, 87 (1980).

from single-run experiments may be very misleading if the parameters of the oxidation have not been established previously.

The presence of mineral matter (all forms of inorganic material) has to be considered in coal oxidation. For example, HCl-treated coal is found to be more easily oxidized by aqueous sodium dichromate oxidation than untreated coal.[28] In addition, minerals in coal may contain transition metals such as iron. From the standpoint of a structural characterization study, these metals may cause an unfavorable oxidation. For example, as will be discussed later (Section IV,F and G), H_2O_2–ferrous salt may lead to hydroxylation of phenolic rings known to be present in coal, and particularly in low rank coal. These hydroxylated phenols readily undergo ring degradation.

Since a great number of oxidation studies of coal have been reported, dating from about 1800, it is recommended that the following review papers and books be consulted in order to obtain a broad background of information.[1–3,8,57]

B. CHROMIC ACID

Essentially, this section has been divided into three parts that discuss results of: acidic dichromate oxidation which is nonselective and drastic, the more selective and useful aqueous sodium dichromate oxidation, and finally stepwise chromic acid oxidation.

Despite the fact that chromic acid is one of the most versatile oxidants, only a few chromic acid oxidation studies of coal had been made[1,57] before the aqueous sodium dichromate oxidation was successfully applied to the degradation of coal in 1975 by Hayatsu *et al.*[43] Kinney[58] found that oxidation of raw coal (1 g) using a mixture of concentrated nitric acid (25 ml) and potassium dichromate (20 g) and refluxing for 0.5 hr, yields acetic acid which may be directly correlated with the amount of methane produced from the pyrolysis of the same raw coal at 500°C. It was suggested that the methane produced resulted from thermal degradation of aliphatic and alicyclic groups of the coal and that these groups were responsible for the acetic acid formed during the oxidation. Indeed no acetic acid was produced from the oxidation of coal which had been previously pyrolyzed at 500°C.

The oxidizability of coals was studied using a solution of potassium dichromate in concentrated phosphoric acid by Honda and Hirose.[59] The

[57] A. E. Bearse, J. L. Cox, and M. Hillman, "Production of Chemicals by Oxidation of Coal," Battelle Energy Program Rep., Columbus, Ohio, 1975.

[58] C. Kinney, *J. Am. Chem. Soc.* **69**, 284 (1947).

[59] H. Honda and Y. Hirose, *Fuel* **37**, 323 (1958).

reactivity of various rank coals (from lignite to anthracite coal) was measured by the weight of carbon dioxide evolved in 150 min at 109.5°C. The resistivity of the coals to the oxidation seemed to be hardly affected by the internal surface area of the coals, but showed a close correlation with the mechanical properties such as hardness (or Young's modulus). Therefore, it was suggested that the oxidizability of coals depends on the persistence or decay during coalification of the bridge linkages or cross-linkages that connect the coal structural units.

However, it should be realized that because of the previously mentioned diversity of structures in coals, a meaningful determination of comparative oxidizability may be very difficult to achieve using only the results of a single condition potassium dichromate-phosphoric acid oxidation.

While the common oxidizing agents (HNO_3, $KMnO_4$, H_2O_2, O_2, NaOCl, etc.) are known to destroy polynuclear aromatic rings, it is also known that the amount of ring degradation which occurs in dichromate oxidation of polynuclear aromatic compounds increases with increasing acid or alkali concentration. A highly efficient oxidation procedure, aqueous sodium dichromate at elevated temperature, first appeared in the patent literature for the oxidation of toluene, chlorotoluene and xylenes to their corresponding carboxylic acids. This procedure was also applied to the oxidation of polynuclear aromatics such as acenaphthalene[60] and several methyl derivatives of benzo[*c*]phenanthrenes.[61] The oxidation has been further studied in detail by Friedman *et al.*[62]

The main virtue of this procedure is that it results in the oxidation of side chains on polynuclear aromatic rings to their corresponding carboxylic acids with minimum degradation of ring systems. However, Friedman *et al.*[62] and Hayatsu *et al.*[28] have shown that phenols and some of the heterocyclics such as furans and pyrroles are partially or extensively degraded by the aqueous $Na_2Cr_2O_7$ oxidation. Therefore, to isolate and identify these compounds from the oxidation of coal, different oxidants and conditions must be used; for example, CuO–NaOH for phenols, and $CH_3COOH–H_2O_2$ or 15% HNO_3 for furans.

As Friedman *et al.*[62] have discussed, oxidation of a methyl group of an aromatic hydrocarbon to the corresponding carboxylic acid requires 1 mol of $Na_2Cr_2O_7$. That is,

$$Ar—CH_3 + Na_2Cr_2O_7 \rightarrow Ar—COONa + NaOH + Cr_2O_3 + H_2O \quad (3)$$

[60] S. Kato, H. Hashimoto, and H. Sugiyama, *J. Synth. Org. Chem. Jpn.* **14**, 723 (1956).

[61] M. S. Newman and H. Boden, *J. Org. Chem.* **26**, 1759 (1961).

[62] L. Friedman, D. L. Fishel, and H. Shechter, *J. Org. Chem.* **30**, 1453 (1965).

Sodium hydroxide is formed during the conversion and slows the oxidation. To minimize such retardation, it is advantageous to use up to 50% excess $Na_2Cr_2O_7$ as a buffer:

$$Ar—CH_3 + 1.5\ Na_2Cr_2O_7 \rightarrow Ar—COONa + Na_2CrO_4 + Cr_2O_3 + 1.5\ H_2O \quad (4)$$

Under such conditions, it was found that the initial pH of the reaction medium is 4.5–5.0 and the final pH between 7 and 8, and little or no ring degradation occurs. Indeed, 22 mono- and polyalkyl substituted polynuclear aromatic hydrocarbons were oxidized in excellent yield to their corresponding carboxylic acids. In the same study, it was observed that, whereas considerable ring destruction occurs in the oxidation of monomethoxy derivatives of alkyl benzene, these compounds are oxidized in good yield to the corresponding acids in the presence of an acidic buffer, NaH_2PO_4. However, Hayatsu *et al.*[28] have found that dimethoxytoluenes were oxidized to the corresponding acids in poor yield (9–12%) even in the presence of NaH_2PO_4.

Recently, the aqueous sodium dichromate oxidation has been applied to the degradation of coals,[19] coal products,[37,38,42] and meteoritic polymer.[63] Further, to determine optimum conditions, the oxidation has been examined in considerable detail using various rank coals, coal products, model compounds, and polymers.[28,*]

In general, a coal sample is oxidized with 0.4–1.4 *M* aqueous $Na_2Cr_2O_7$ in an autoclave at about 250°C for 40–72 hr. From 50 to 90% of the starting material is converted to organic acids, humic acid-like materials, and carbon dioxide. Usually coal products are more easily oxidized than their feed coals. Highly condensed macromolecules such as graphitic material and some high rank coals are oxidized only slightly or not at all. As pointed out by Hayatsu *et al.*,[28] the yield and nature of the products are found to be affected not only by coal rank but also by several experimental factors such as the ratio of dichromate to sample, use of a buffer, oxidation time and temperature, type of autoclave, and sample pretreatment with acid before oxidation.**

* Duty and Sines[63a] have recently reported that sodium dichromate has only a limited ability to oxidize coals and coal products. However, this is a conclusion that seems unjustified by their report which has been shown to contain incorrect data (involving simple mathematical calculations and carbon contents of the samples), misleading statements, and evidence of misunderstandings.[63b]

** In addition, methanol treatment may be necessary for some samples before the oxidation. The purpose of this treatment is to wet the sample so that good contact between the sample and the oxidant is achieved. Such a consideration also applies to the acid treatment of geological samples for the removal of minerals.[5]

As shown in the GC of Fig. 2a, with numbered peaks identified in Table II, the major oxidation products from lignite coal are identified as benzenepolycarboxylic acids.[34] The sample was methylated with d_6-dimethyl sulfate before oxidation. It is obvious that d_3-methoxy acids resulted from benzene rings which had free phenolic groups in the coal structure, as shown in the examples below.

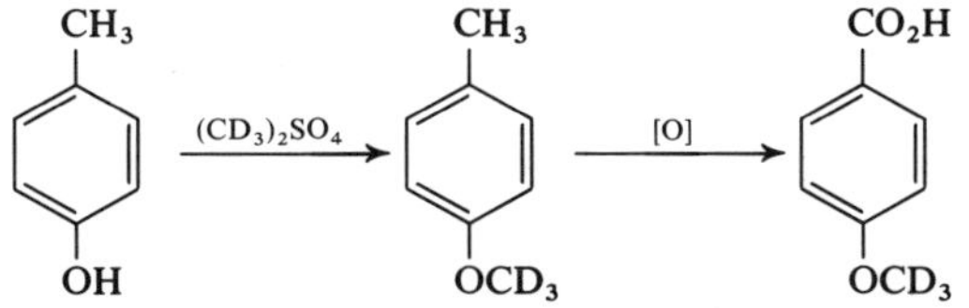

(5)

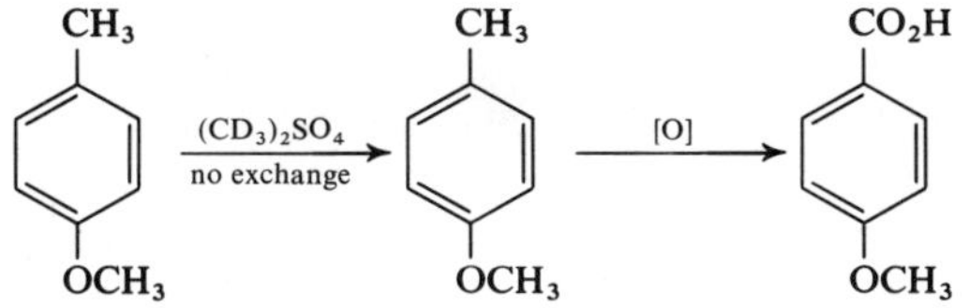

(6)

In the gas chromatograms (Fig. 2b) of organic acid methyl esters obtained from the oxidation of d_3-methylated high volatile bituminous coal, sizable peaks due to polynuclear aromatic and heteroaromatic carboxylic acid methyl esters are seen. The amounts of such carboxylic acids in the oxidation products generally increase with increasing carbon content of the coals.[19,42,46] For example, phenanthrenecarboxylic acids are identified in relatively large amounts in the oxidation products of anthracite coal (C = 91.3%, maf), and the relative abundance of phenanthrenes produced is nearly equal to that of benzenes identified as their carboxylic acids in the oxidation products. Condensed-ring heterocyclics (e.g., dibenzothiophene, benzonaphthothiophene, acridine, and benzoquinoline carboxylic acids) that are not seen or are detected in only very small amounts in the oxidation products of low rank coals, are produced in relatively large amounts from the oxidation of the same anthracite.

The aqueous $Na_2Cr_2O_7$ oxidation has also been used for characterization of aromatic and heteroaromatic units in the structures of coal products such as solvent refined lignite coal (SRL),[38] solvent refined coal (SRC),[42]

---

[63] R. Hayatsu, S. Matsuoka, R. G. Scott, M. H. Studier, and E. Anders *Geochim. Cosmochim. Acta* **41**, 1325 (1977).

[63a] R. C. Duty and R. Sines, *Fuel* **60**, 534 (1981).

[63b] R. Hayatsu, R. G. Scott, and R. E. Winans, to be published.

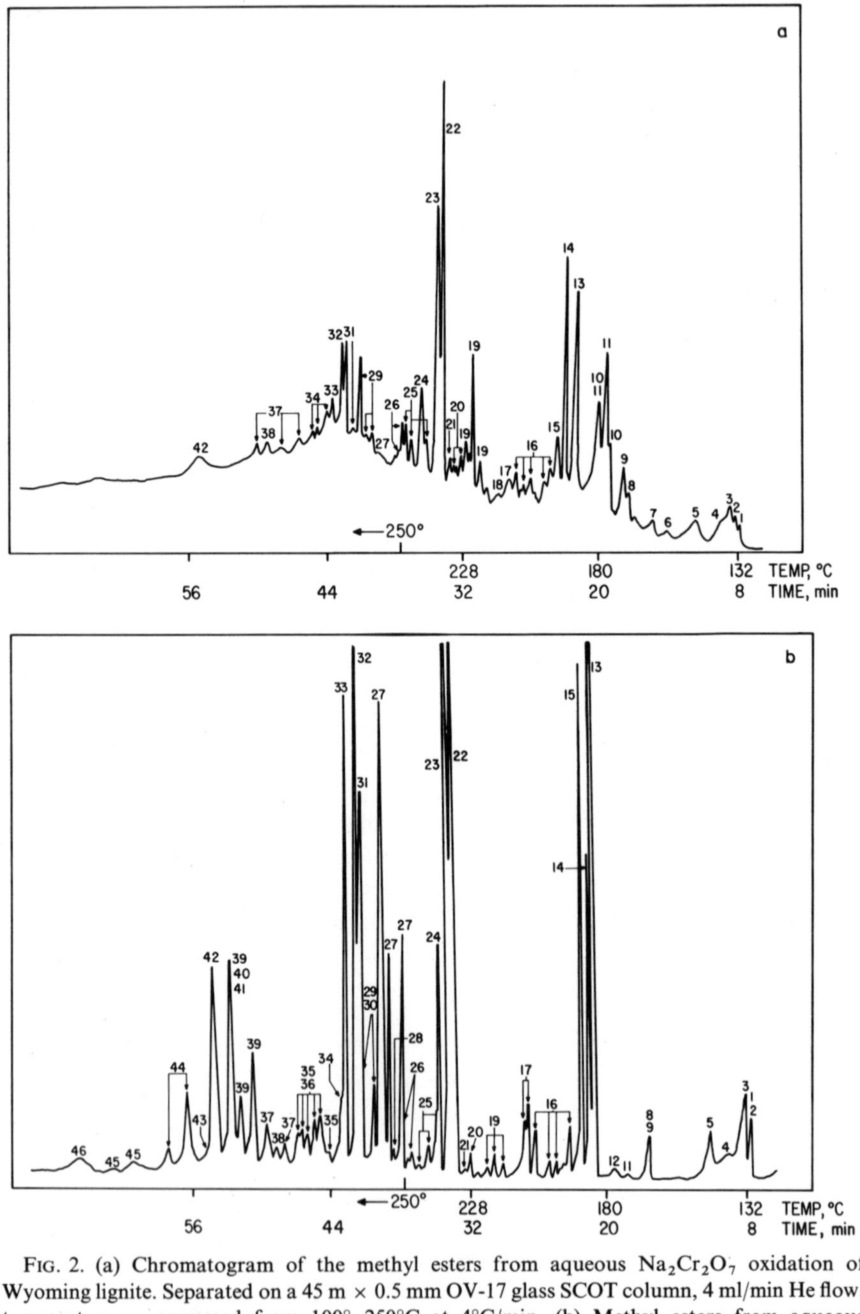

FIG. 2. (a) Chromatogram of the methyl esters from aqueous $Na_2Cr_2O_7$ oxidation of Wyoming lignite. Separated on a 45 m × 0.5 mm OV-17 glass SCOT column, 4 ml/min He flow, temperature programmed from 100°–250°C at 4°C/min. (b) Methyl esters from aqueous $Na_2Cr_2O_7$ oxidation of an Illinois No. 2 bituminous coal. Same conditions as (a). (Peaks identified in Table II.)

TABLE II

ORGANIC ACIDS FOUND IN THE AQUEOUS Na_2Cr_2O- OXIDATION OF LIGNITE AND BITUMINOUS COALS IDENTIFIED AS METHYL ESTERS BY GCMS AND HRMS[a]

GC peak	Compound
1	Succinic acid
2	Methylsuccinic acid
3	Benzoic acid
4	Methylfurancarboxylic acid
5	Methylbenzoic acid, Glutaric acid
6	Adipic acid
7	Pimelic acid
8	Methoxybenzoic acid
9	d_3-Methoxybenzoic acid
10	Methoxytoluic acid
11	d_3-Methoxytoluic acid
12	Methylfurandicarboxylic acid
13	1,2-Benzenedicarboxylic acid
14	1,4-Benzenedicarboxylic acid
15	1,3-Benzenedicarboxylic acid
16	Methylbenzenedicarboxylic acid
17	Naphthoic acid
18	Pyridinedicarboxylic acid
19	Methoxy- and d_3-methoxybenzenedicarboxylic acid
20	Dimethylfurandicarboxylic acid (T)[b]
21	Diphenylmonocarboxylic acid
22	1,2,4-Benzenetricarboxylic acid
23	1,2,3-Benzenetricarboxylic acid
24	1,3,5-Benzenetricarboxylic acid
25	Methylbenzenetricarboxylic acid
26	Dibenzofuranmonocarboxylic acid
27	Naphthalenedicarboxylic acid
28	Pyridinetricarboxylic acid
29	Methoxy- and d_3-methoxybenzenetricarboxylic acid
30	Fluorenonemonocarboxylic acid
31	1,2,4,5-Benzenetetracarboxylic acid
32	1,2,3,4-Benzenetetracarboxylic acid
33	1,2,3,5-Benzenetetracarboxylic acid
34	Methylbenzenetetracarboxylic acid
35	Phenanthrenemonocarboxylic acid
36	Dibenzothiophenemonocarboxylic acid
37	Xanthonemonocarboxylic acid
38	Anthraquinonemonocarboxylic acid
39	Naphthalenetricarboxylic acid
40	Methylxanthonemonocarboxylic acid

(*Continued*)

TABLE II (*cont.*)

GC peak	Compound
41	Dibenzofurandicarboxylic acid
42	Benzenepentacarboxylic acid
43	Benzoquinoline-/acridine-monocarboxylic acid
44	Phenanthrenedicarboxylic acid
45	Carbazolemonocarboxylic acid
46	Xanthonedicarboxylic acid

[a] Nonvolatile esters which cannot be analyzed by GCMS were identified (not shown in this table) by HRMS and solid probe MS.
[b] (T) Identification tentative.

and hydrolytically solubilized coal (HSC).[35,37] As discussed and suggested by many workers,[64–68] the SRL and SRC processes involve the thermal cracking of C—C, C—O, and C—H bonds and partial hydrogen transfer from hydroaromatic structures of coal. Therefore, SRL and SRC have lower H/C ratios and higher aromaticities than their feed coals. From the aqueous $Na_2Cr_2O_7$ oxidation study, a greater degree of aromatic condensation of the SRL and SRC compared to their feed coals is indicated since naphthalene- and phenanthrenecarboxylic acids are much more abundant in the oxidation products of the SRL and SRC.

Sodium dichromate–acetic acid oxidation has also been applied to the degradation of SRC.[42] The oxidation was carried out below 40°C for 70–72 hr. About 97% of the sample was oxidized, and humic acid-like material (49.4 wt% of the oxidation products) and an aromatic acid fraction (26.5 wt%) were isolated from the oxidation products. The aromatic acid fraction showed the presence of aromatic and heterocyclic carboxylic acids similar to those found in the aqueous $Na_2Cr_2O_7$ oxidation products of the same sample. In addition, diphenyl ether mono- and dicarboxylic acids were identified. This is an interesting finding because the diphenyl ether system has been considered to be present in coal and coal products, but there has been so far little direct evidence. This oxidation procedure has not been applied to the degradation of coal itself.

[64] R. J. Baltisberger, K. J. Klabunde, V. I. Stenberg, N. F. Woolsey, K. Saito, and W. Sukalski, *in* "Organic Chemistry of Coal" (J. W. Larsen, ed.), p. 294. Am. Chem. Soc. Symp. Ser. 71, 1978.
[65] M. Farcasiu, *Fuel* **56**, 9 (1977).
[66] R. G. Ruberto, D. C. Cronauer, D. M. Jewell, and K. S. Seshadri *Fuel* **56**, 25 (1977).
[67] Y. Kamiya, *J. Fuel Soc. Jpn.* **57**, 12 (1978).
[68] B. M. Benjamin, V. F. Raaen, P. H. Maupin, L. L. Brown, and C. J. Collins, *Fuel* **57**, 269 (1978).

Winans *et al.*[35,36] have found through characterization of aqueous $Na_2Cr_2O_7$ oxidation products of hydrolytically solubilized coal (HSC), that the hydrolytic reaction of coal with glycol/KOH involves the reduction of polynuclear aromatics and the cleavage of aryl ethers. This is also confirmed by ^{13}C-NMR. It is interesting to note that this solubilization process showed some effects directly opposite to those of the SRL and SRC processes; that is, the oxidation of HSC produced polynuclear aromatic and heteroaromatic acids in much lower amounts than are produced from the feed coal.

It has been shown recently, that coals can be solubilized in a donor solvent at >400°C with a short contact time (SCT). This SCT product does not appear to be as highly aromatic as the product that results from the normal SRC process. An SCT product has been produced from an Illinois No. 6 bituminous coal (77.7% C, 5.4% H, 1.4% N, and 4.1% S total on an maf basis) by treatment with 425°C tetralin for several minutes.[69] The benzene–methanol (3:1) soluble portion, which typically accounts for 30–39% of the SCT product, has been oxidized using buffered aqueous sodium dichromate after the extract has been methylated with dimethyl sulfate. This portion of the product has been chosen for its ease of oxidation and for direct comparison with the benzene–MeOH extracts of SRC, SRL, and HSC. Also, the feed coal has been oxidized in the same manner for direct comparison.

In Fig. 3 the chromatograms for three microwave plasma analyzer channels (carbon, hydrogen, and sulfur) are shown for the methyl esters from the oxidation of the SCT product. The mixture was separated on a glass SCOT column coated with OV-17. A GCMS analysis was performed with a similar column under the same conditions. The identification of prominent peaks which are listed in Table III has been made using a combination of MS, MPA and retention time data.

An important aspect of the MPA results is the capability of directly determining the distribution of sulfur in the volatile esters. Table IV gives the breakdown of the data for the sulfur compounds from Fig. 3 and from the analysis of the oxidation products of the whole coal.

It is interesting to note the identification of a naphthalene sulfonate that probably results from the oxidation of an aryl or alkyl naphthalene sulfide or thiol. This is the first sulfonic acid that we have found in an oxidation product.

As is typical in the dichromate oxidation of bituminous coal, benzene is the most abundant aromatic unit isolated. The oxidation products of SCT material differ from those obtained from the oxidation of an SRL where a significant amount of polycyclic aromatics are found. However, naphthalene

[69] R. E. Winans, R. Hayatsu, R. L. McBeth, and R. G. Scott, *Q. Prog. Rep.*, Fossil Energy, U.S. Dept. Energy (submitted), January–March, 1980.

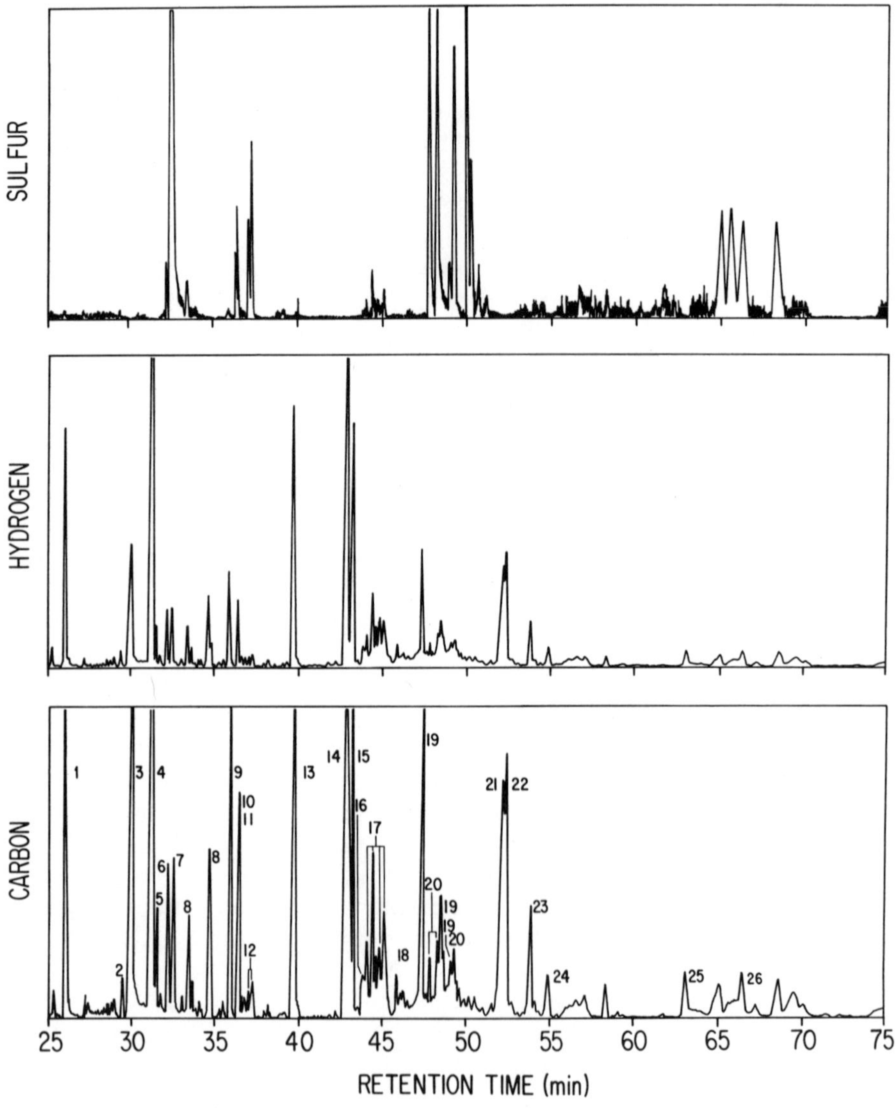

FIG. 3. Chromatograms from three channels, C, H, and S, of the GCMPA of methyl esters from the aqueous sodium dichromate oxidation of an SCT coal product from Illinois No. 6 bituminous coal. The separation was performed on a 50 m × 0.5 mm OV-17 glass SCOT column, 4 ml/min He flow, temperature programmed from 70°–250°C at 4°C/min. (Peaks identified in Table III.)

TABLE III

CARBOXYLIC ACIDS OBTAINED FROM THE AQUEOUS $Na_2Cr_2O_7$ OXIDATION OF AN SCT COAL PRODUCT IDENTIFIED AS METHYL ESTERS

GC peak	Compound
1	Methyl methoxybenzoate
2	Methyl biphenylcarboxylate
3	Methyl hydroxybenzenecarboxylate
4	Methyl 1,2-benzenedicarboxylate
5	Methyl 1,4-benzenedicarboxylate
6	Methyl 1,3-benzenedicarboxylate
7	Methyl thiophenedicarboxylate
8	Methyl methylbenzenedicarboxylate
9	Methyl 1-naphthoate
10	Methyl 2-naphthoate
11	Methyl naphthalenesulfonate
12	Methyl benzothiophenecarboxylate
13	Methyl methoxybenzenedicarboxylate
14	Methyl 1,2,4-benzenetricarboxylate
15	Methyl 1,2,3-benzenetricarboxylate
16	Methyl 1,3,5-benzenetricarboxylate
17	Methyl methylbenzenetricarboxylate
18	Methyl dibenzofurancarboxylate
19	Methyl naphthalenedicarboxylate
20	Methyl benzothiophenedicarboxylate
21	Methyl 1,2,4,5-benzenetetracarboxylate
22	Methyl 1,2,3,4-benzenetetracarboxylate
23	Methyl 1,2,3,5-benzenetetracarboxylate
24	Methyl methylbenzenetetracarboxylate
25	Methyl benzothiophenetricarboxylate
26	Methyl naphthalenetricarboxylate

carboxylic acids are abundant in the oxidation products from the SCT material. A comparison has been made between these products and those obtained from the oxidation of the raw coal. A portion of the chromatograms of the carbon channel from the GCMPA of both mixtures is shown in Fig. 4. The numbered peaks are identified in Table III. The naphthalene monocarboxylates are much more abundant in the SCT derived products. There are two possible sources of these compounds. One is from loss of hydrogen from tetralin originally attached to the coal macromolecule. The other source is from attachment of the solvent-derived naphthalene to the coal products. This second method has been shown to occur by using ^{14}C-labeled tetralin.[70]

[70] C. J. Collins, B. M. Benjamin, V. F. Raaen, P. H. Maupin, and W. H. Roark, *in* "Organic Chemistry of Coal" (J. W. Larsen, ed.), p. 165. Am. Chem. Soc. Symp. Ser. 71, 1978.

TABLE IV

DISTRIBUTION OF SULFUR IN THE BUFFERED $Na_2Cr_2O_7$ OXIDATION PRODUCTS OF ILLINOIS NO. 6 BITUMINOUS COAL AND ITS SCT PRODUCT

Methyl ester [$Ar(CO_2H)_n$]		% Sulfur	
Ar	n	Coal	SCT
Thiophene	1	3.5	4.6
	2	67.0	47.9
Benzothiophene	1	6.0	5.6
	2	11.0	26.2
	3	11.0	14.4
Naphthalene sulfonate		1.4	1.5

The aqueous $Na_2Cr_2O_7$ oxidation procedure was also successfully applied to the degradation of polymeric material in the Murchison carbonaceous meteorite; 15 aromatic and heteroaromatic ring systems as the corresponding acids were identified.[63]

Recently, the aqueous $Na_2Cr_2O_7$ oxidation in the presence of acidic buffer NaH_2PO_4 has been applied to the characterization of the structure of methylated cannel-like coal.[28] Petrographic analysis of a coal (Ohio PSOC #297 C, 80.5% maf) used in this oxidation shows that this coal has a high content of exinite (19.2 v%) and a relatively low content of vitrinite (63.0 v%). Only 21.7% of the coal was oxidized; however, almost all of the oxidation product was found to consist mainly of straight-chain aliphatic dicarboxylic acids (C_{3-20}, maximum at C_9). On the other hand, nonbuffered dichromate oxidation of the same coal indicated that the oxidant attacked the major portion of the coal macromolecules (vitrinite) and produced various aromatic, heteroaromatic, and phenolic acids together with very little short-chain dicarboxylic acids. This demonstrates that the buffer-controlled oxidation may preferentially attack aliphatic-rich material in the exinite macerals.

The aqueous $Na_2Cr_2O_7$ oxidation possesses an important advantage over most other oxidations because even though its results are semiquantitative rather than quantitative, most polynuclear aromatic and heteroaromatic structures are preserved rather than destroyed.

Successive chromic acid oxidation that was applied to the characterization of kerogen structure[71,72] has been used for the degradation of low rank

[71] A. L. Burlingame and B. R. Simoneit, *Nature* (*London*) **222**, 741 (1969).
[72] B. R. Simoneit and A. L. Burlingame, *Geochim. Cosmochim. Acta* **37**, 595 (1973).

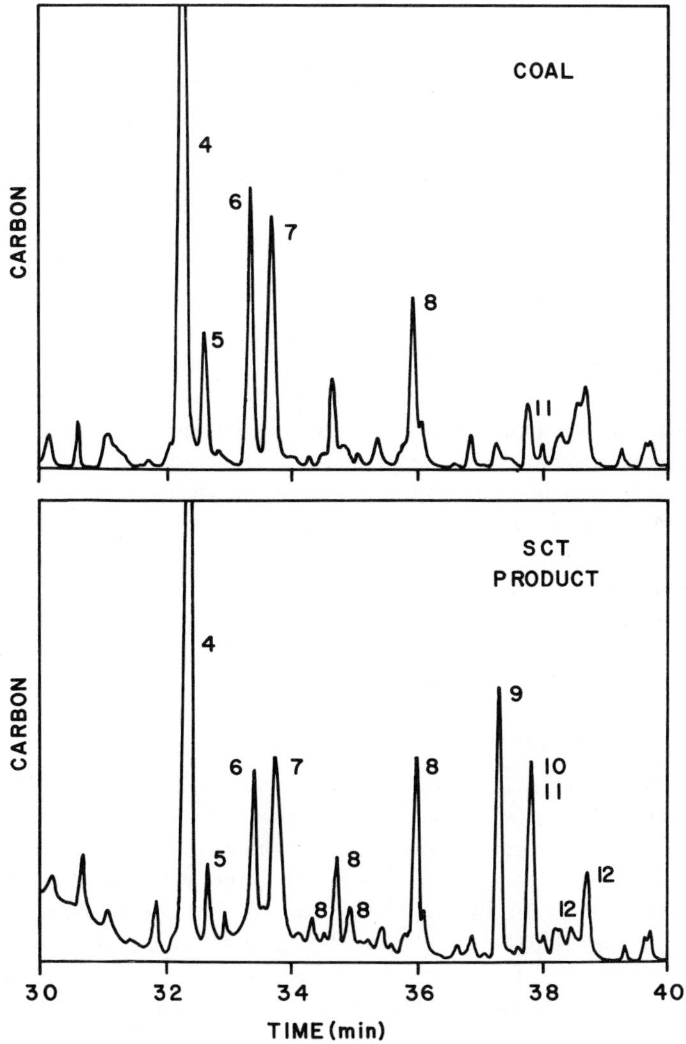

FIG. 4. A comparison of the methyl esters from the aqueous $Na_2Cr_2O_7$ oxidation of SCT product and the raw Illinois No. 6 bituminous coal. Only the carbon channel from the GCMPA is shown and the conditions are the same as for Fig. 3. (Peaks identified in Table III.)

coals.[27] For example, methylated lignite coal was oxidized at 80°C for 16 hr with 0.3 *M* chromic acid and 1.5 *M* H_2SO_4 (first-step oxidation). Organic acids (9.2 wt% of the sample) and humic acid-like material (9.6 wt%) were obtained. Major oxidation products in the organic acid fraction were identified as unbranched aliphatic dicarboxylic acids (C_3–C_{20}, maximum at

C_9) and benzenecarboxylic acids (mono–hexa). Branched dicarboxylic acids and cyclic acids were also found as minor products. The ratio of aliphatic acid to aromatic acid was shown to be 1.92. The nonoxidized residue from the first-step oxidation was further oxidized an additional 16 hr. In the second-step oxidation, the major products were benzenecarboxylic acids. Short-chain aliphatic dicarboxylic acids ($C \leq 8$) were also identified in relatively high abundance; however, long-chain dicarboxylic acids ($C \geq 9$) were found only in low concentration. The ratio of aliphatic acid to aromatic acid was given as 0.63. For comparison with coal oxidation, Green River kerogen concentrate (76.0% C; 9.95% H; 2.4% N; 1.4% S; 2.0% ash) was also oxidized under the same conditions. The kerogen is thought to have a central nucleus that is a highly cross-linked aliphatic structure. Indeed, the major oxidation products (first-step) were aliphatic dicarboxylic acids and the ratio of aliphatic acid to aromatic acid was shown to be 20.3. The oxidation of bituminous coal resulted in the production of large amounts of benzenecarboxylic acid with lesser amounts of short-chain dicarboxylic acid (C_3–C_8).

It is interesting to note that the first-step chromic acid oxidation of the lignite coal preferentially attacks an aliphatic-rich material present in that coal, and yields a distribution pattern similar to that of the first-step oxidation products of the kerogen. A more detailed discussion on the selective oxidation of the aliphatic-rich material in coal is presented in Section IV, C on buffer-controlled $KMnO_4$ oxidation.

C. POTASSIUM PERMANGANATE

Permanganate oxidation has been extensively used in the study of coal, lignin, humic substances, and kerogen. A large body of literature is available dealing with various applications and techniques of the oxidation and readers are referred to reviews by Lowry[2,3] and van Krevelen[1] for coal; Chang and Allan[41] for lignin; Hayes and Swift[14] for humic substances; Stewart,[73] Fieser and Fieser,[33] Lee,[74,75] and Ogliaruso and Wolfe[76] for general background information on the permanganate oxidation.

The first systematic study of the alkaline permanganate oxidation of coal was made during the 1920s and 1930s and these results have been reviewed in detail by van Krevelen.[1] Rates of alkaline permanganate oxidation of bituminous coal and condensed cyclic compounds including heterocyclics

[73] R. Stewart, *in* "Oxidation in Organic Chemistry" (K. B. Wiberg, ed.), Part A, Chap. 1. Academic Press, New York, 1965.

[74] D. G. Lee, *in* "Oxidation" (R. L. Augustine, ed.), Vol. 1, p. 1. Dekker, New York, 1969.

[75] D. G. Lee, Chapter 2, this volume.

[76] M. A. Ogliaruso and J. F. Wolfe, *in* "Chemistry of Acid Derivatives" (S. Patai, ed.), Suppl. B, Part 1, p. 267. Wiley, New York, 1979.

were reported by Ward *et al.* [77] A similar but more extensive investigation using a number of model compounds lignin, and coal-derived humic acids was undertaken by Yokokawa *et al.*[78] Amounts of oxalic acid and carbon dioxide produced from each sample were determined. Mononuclear phenols were shown to be easily oxidized to yield about an equal molar quantity of oxalic acid, but their methoxy derivatives were found to be stable to the oxidation. However, Freudenberg *et al.*[79,80] have noted in the permanganate oxidation study of lignins that stabilization of phenolic rings with methylation was only successful to a certain degree and nuclear degradation still occurred as evidenced by the formation of succinic acid.

It is interesting that succinic acid is formed. We think that it could be produced by either or both of the following pathways for example:

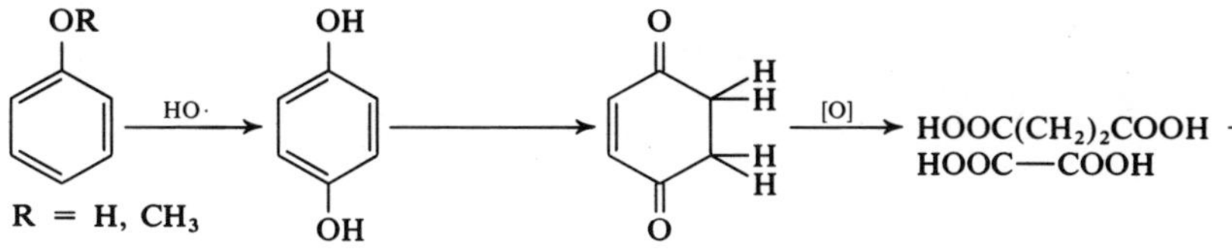

(7)

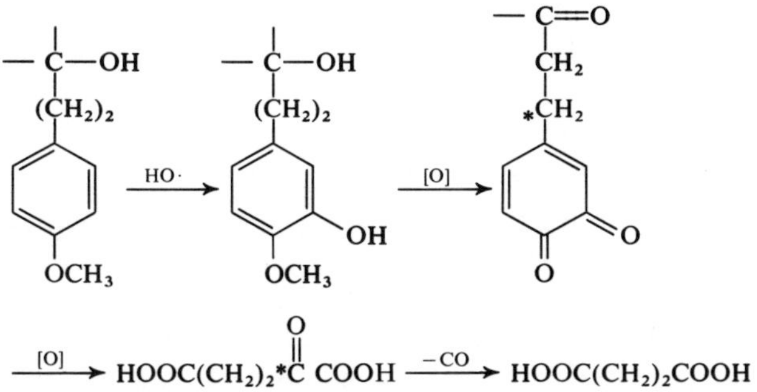

(8)

The intermediate valence states of manganese (Mn^{4+}) will form hydroxyl radicals from water. These radicals will react with phenols to form a dihydroxy compound which could tautomerize to a small extent to the quinone form (7). This intermediate then would be oxidized to the saturated acid. In the second pathway the succinic acid would be derived from the saturated aliphatic side chain (8).

Lignin and humic acid extracted from lignite are found to be more reactive and to consume more oxidant than humic acid* derived from bituminous

* Prepared by mild oxidation with 1 *N* HNO_3.

[77] J. J. Ward, W. R. Kirner, and H. C. Howard, *J. Am. Chem. Soc.* **67**, 246 (1945).

[78] C. Yokokawa, Y. Watanabe, S. Kajiyama, and Y. Takegami, *Fuel* **41**, 209 (1962).

[79] K. Freudenberg and C. Chen, *Chem. Ber.* **93**, 2533 (1960).

[80] K. Freudenberg, C. Chen, and G. Cardinale, *Chem. Ber.* **95**, 2814 (1962).

TABLE V

ALKALINE PERMANGANATE OXIDATION OF LIGNIN AND HUMIC ACIDS[a,b]

Sample	Carbon content %	Oxidant consumption 10^{-2} mol/g sample	Oxalic acid (C distribution %)	CO_2 (C distribution %)
Lignin	64.3	3.2	25.6	N.D.[c]
Humic acid (L)[d]	60.3	2.3	14.8	N.D.
Methylated humic acid (L)	65.5	1.6	10.8	18.1
Humic acid (B)[e]	65.4	1.1	6.2	16.9
Methylated humic acid (B)	65.8	1.3	6.0	15.8
Cinnamic acid	—	1.6	14.1	6.15
p-Hydroxycinnamic acid	—	4.5	50.3	24.4

[a] Prepared using data from Yokokawa *et al.*[78]
[b] At 25°C, 3 hr. with 1.25 *N* $KMnO_4$ in 1 *N* KOH.
[c] N.D.—no data shown.
[d] (L)—lignite.
[e] (B)—bituminous coal.

coal.[78] Methylated humic acid from lignite is more resistant to the oxidation than is nonmethylated. Table V shows the oxidant consumption ratios, and the carbon distribution percentages of oxalic acid and carbon dioxide produced from the oxidation of these samples. It is interesting to note that methylation was not effective in the case of oxidation of bituminous coal derived humic acid. This apparently indicates that this humic acid contains few phenolic OH groups.

A Russian group[81] has oxidized methylated humic acids, prepared from several sources ranging from soil to weathered bituminous coal, with alkaline permanganate (0.5 *N* $KMnO_4$–1% KOH 1:1; refluxing). The yield of methoxybenzoic acids (*p*-methoxybenzoic, veratric, 3,4,5-trimethoxybenzoic acids, and methoxybenzenedi- and tricarboxylic acids) is shown to decrease with increasing coalification. These oxidation products obtained from humic acids of different origins (soil, peat, lignites, and brown coals) are similar to the permanganate oxidation products obtained from lignin. Short-chain aliphatic dicarboxylic acids are also found in the oxidation products of these humic acids. However, it is interesting that aliphatic acids have not been found in the oxidation products of humic acids derived from brown coal of the fusain–xylain type, or weathered bituminous coal. Benzenecarboxylic acids are the major oxidation products from these two samples.

Alkaline permanganate oxidation has also been applied to the degradation

[81] D. B. Maximov, V. E. Schapovalov, and T. V. Shvets, *Fuel* **51**, 185 (1972).

of methylated humic acid extracted from Spanish lignite.[82] Normal alkanes, ranging from C_{14} to C_{35}, were found in abundance and constituted about 90% of the total products. As minor products, fatty acids, aliphatic dicarboxylic acids, methyl alkyl ketones, benzenecarboxylic acids, and phenolic acids were found. Some unusual oxidation products such as dehydroabietic acid, bromohydroxybenzoic acids, and bis(2-ethylhexyl) phthalate were also identified. However, on the basis of work by another group[24,83] (see also Section II,D of this chapter), we believe that most of the products identified, and in particular the *n*-alkanes, were probably physically adsorbed on and in the humic acid and are not likely to be structural constituents of that humic acid. Indeed, the supposed structural constituent oxidation products, extracted with ethyl acetate, represented only 12% of the humic acid starting material.

Allan[84] has systematically investigated stepwise alkaline permanganate oxidation of exinite macerals (carbon content ≈75–87%) separated from English and Scottish bituminous coals. A sample (5 g) of extracted and saponified exinite residue was oxidized with $KMnO_4$–1.6% KOH aqueous solution (100 ml) at 80°C for each step (sample/oxidant ratio ranged from 4 to 1). After reduction of each portion of reagent, the remaining exinite and MnO_2 were filtered off and treated with a fresh aliquot of alkaline permanganate solution. Successive oxidation steps were continued until the permanganate was not fully reduced after 16 hr. This procedure was first applied to oxidation of kerogen of ancient sediments by Djuricic *et al.*[84a]

The yields of organic acid and humic acid-like materials were shown to vary with the exinite sample and the conditions used (number of steps, ratio of sample/$KMnO_4$). For example, organic acids (8.2–24 wt %) and humic acid-like material (2.8–32.4 wt %) were produced from the seven sporinite (an exinite maceral) samples after stepwise oxidation. The major organic acids produced were polar* and aromatic and accounted for about 80–92 wt % of the total organic acid fractions, but only the benzenecarboxylic acids were identified. Aliphatic monocarboxylic acids (ranging from C_7–C_{34}, 0.2–1.4 wt % of the organic acid fractions) and dicarboxylic acids (C_5–C_{28},

* Although no explanation is given in the original paper, "polar acids" are probably aromatic and cyclic acids. Recently, the same group isolated a polar acid fraction from the oxidation of a different maceral, alginite. The fraction consists of benzene carboxylic acids and unidentified compounds which have been suggested to be aliphatic chain or cyclic compounds. They have also been suggested to be polyfunctional.[87a]

[82] F. Martin and C. Saiz-Jimenez, *Fuel* **57**, 353 (1978).

[83] S. M. Griffith and M. Schnitzer, *Soil Sci.* **122**, 191 (1976).

[84] J. Allan, Ph.D. Thesis, Univ. of Newcastle-upon-Tyne, 1975.

[84a] M. V. Djuricic, D. Vitorovic, B. D. Andresen, H. S. Hertz, R. C. Murphy, and K. Biemann, *in* "Advances in Organic Geochemistry 1971" (H. R. Gaertner and H. Wehner, eds.), p. 305. Pergamon, New York, 1972.

7.3–18.7 wt %) were also identified. From the relatively large amounts of aromatic and polar acids produced in this study, it appears that the chemical structures of the exinites oxidized may be similar to those of type II kerogen[10] (cf. oxidation results obtained from Aleksinac, Kimmeridge and Torbanite,[84a,85] Tasmanite,[72] and Green River kerogens[86,87]). Young *et al.*[88] have also isolated aliphatic dicarboxylic acids (C_5–C_{18}) together with benzene di- and tricarboxylic acids from a single alkaline permanganate oxidation of a North Dakota lignite coal (C, 69.3%; H, 4.87% maf). Methanol-soluble organic material constituting about 46% of the original weight of the coal was isolated. After esterification, however, only heptane soluble esters were analyzed by GC and many of these GC peaks were unidentified. In this study, no aliphatic monocarboxylic acids were detected.

In order to characterize the nature of aliphatic structures in low rank coal, Hayatsu *et al.*,[26] more recently, have developed a two-step buffer-controlled permanganate oxidation. Before oxidation, each coal sample (Wyoming lignite coal and Illinois No. 2 high volatile bituminous coal) was exhaustively extracted with 2 *N* HCl, 0.5 N NaOH, benzene–methanol, and finally chloroform to remove trapped organic material completely. Then, in order to protect aromatic rings with free phenolic groups from destruction by oxidation, the samples were methylated. Each methylated sample (3 g) was oxidized at 80°C with buffer-controlled permanganate (300 ml of 4.3% aqueous $KMnO_4$ and 11.5 g of $MgSO_4 \cdot 7H_2O$) with stirring, until the reagent was consumed. The final pH of the oxidation medium was neutral. The nonoxidized residue was further oxidized under the same conditions as in the first step. The following yields based on weight percent of the starting coal were obtained:

	Step	Organic acids (wt %)	Humic acid-like materials (wt %)
Lignite	1	31.3	15.3
	2	20.0	8.7
Bituminous coal	1	27.0	21.0
	2	23.0	24.0

[85] M. V. Djuricic and D. Vitorovic, *Org. Geochem.* **1**, 79 (1978).

[86] M. Djuricic, R. C. Murphy, D. Vitorovic, and K. Biemann *Geochim. Cosmochim. Acta* **35**, 1201 (1971).

[87] D. K. Young and T. F. Yen, *Geochim. Cosmochim. Acta* **41**, 1411 (1977).

[87a] J. Allan, M. Bjoroy, and A. G. Douglas, *in* "Advances in Organic Geochemistry, 1979" (A. G. Douglas and J. R. Maxwell, eds.), p. 599. Pergamon, Oxford, 1980.

[88] D. K. Young and T. F. Yen, *Energy Sources* **3**, 49 (1976).

As shown in Fig. 5a and Table VI straight-chain aliphatic dicarboxylic acids (C_4–C_{18}, HRMS detected up to C_{21}) were identified in high abundance with lesser amounts of branched dicarboxylic acids (C_5–C_{10}) and tricarboxylic acids (C_6–C_8) in the organic acid fraction obtained from the first-step oxidation of the lignite. The ratio of aliphatic acid to aromatic acid produced was 3.1. In the second-step oxidation, the major products were aromatic acids and the ratio of aliphatic acid to aromatic acid was 0.49. Long-chain dicarboxylic acids ($C > 8$) were found again, but in much lower concentration. On the other hand, in the first-step oxidation products from the bituminous coal, large amounts of aromatic acids were identified together with smaller amounts of aliphatic dicarboxylic acids (C_3–C_8). No long-chain acids were detected by GCMS.

From these results, the authors have suggested that the lignite coal contains at least two types of organic macromolecular material differing in chemical structure and perhaps origin. One of these is an aliphatic-rich kerogen-like material which produced the major portion of the long-chain aliphatic dicarboxylic acids resulting from the oxidation, and the other is an aromatic-type material which is generally thought of as "coal macromolecule." While it seems less likely, it is perhaps also possible that the long-chain dicarboxylic acids were derived from aromatic-type material which consists of aromatic ring clusters interconnected by long aliphatic cross-links. Most of the aromatic acids identified are considered to be derived from the aromatic-type material. From a calculation on the basis of the carbon analysis and GC data, about 15% of the lignite macromolecular material appears to be the kerogen-like material. It is interesting to note that petrographic analysis of the lignite coal oxidized apparently shows a very low content of exinite ($\approx 1\%$). Therefore, this kerogen-like material may not be directly related to exinite macerals. In contrast, Allan[84] found long-chain aliphatic mono- and dicarboxylic acids in the alkaline permanganate oxidation products of sporinite (an exinite) macerals separated from whole coals. Perhaps the chemical structure of the kerogen-like material is different from those of the exinite macerals. From the oxidation results, the kerogen-like material is thought to be more susceptible to the permanganate oxidation. Perhaps the kerogen-like material structure has more branched chains, aromatic moieties, ethers, and activated methylene groups that may contribute to a greater oxidation rate. Indeed, most of the kerogen-like material in the lignite is degraded in the first-step buffer-controlled permanganate oxidation. As coalification progressed, such kerogen-like material might have been degraded under the existing geological conditions. In support of this theory, bituminous coal did not produce long-chain acids when subjected to the oxidation.

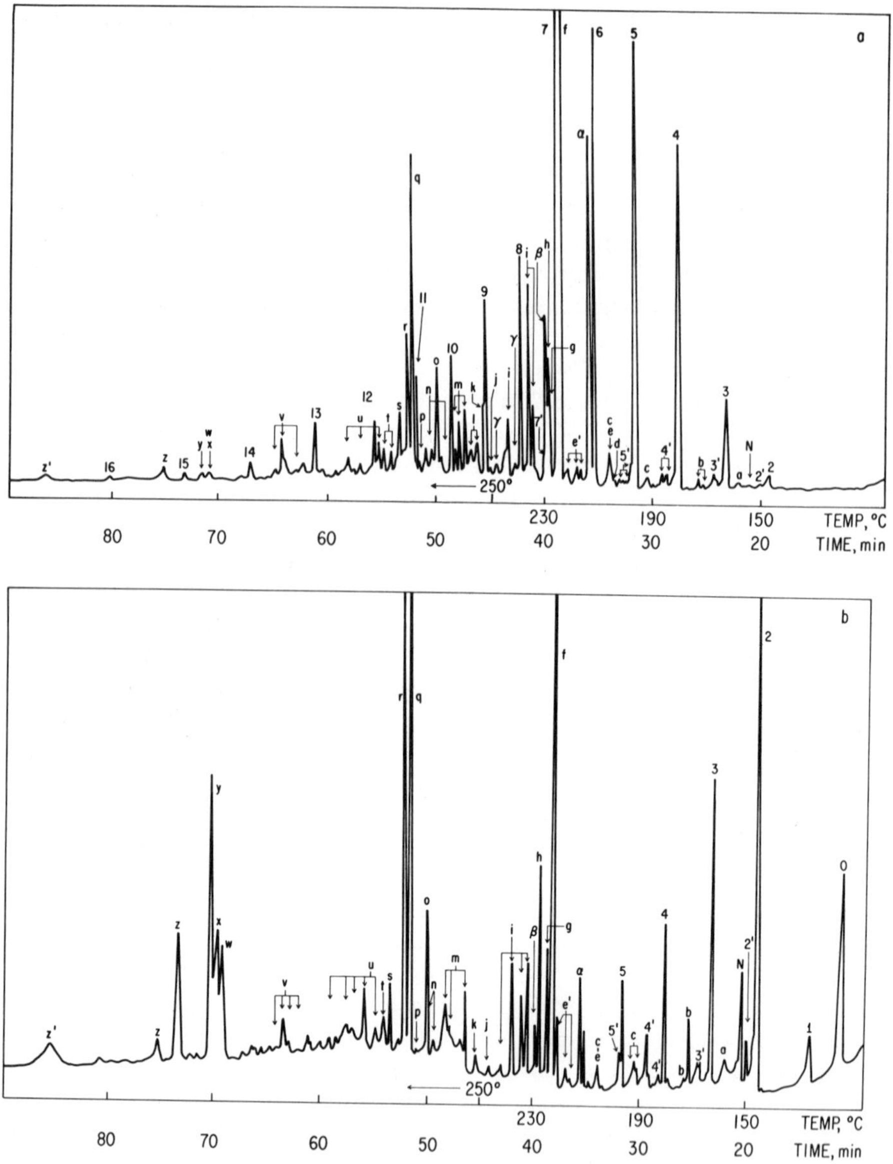

FIG. 5. Gas chromatogram of methyl esters of the organic acid fraction obtained from the first-step buffer-controlled permanganate oxidation of (a) lignite, (b) the non-buffered oxidation of lignite, and (c) the first-step buffered oxidation of bituminous. The separation was made on a 45.9 m × 0.5 mm glass SCOT column coated with OV-17 and temperature programmed from 70°–250°C at 4°C/min. (Peaks identified in Table VI.)

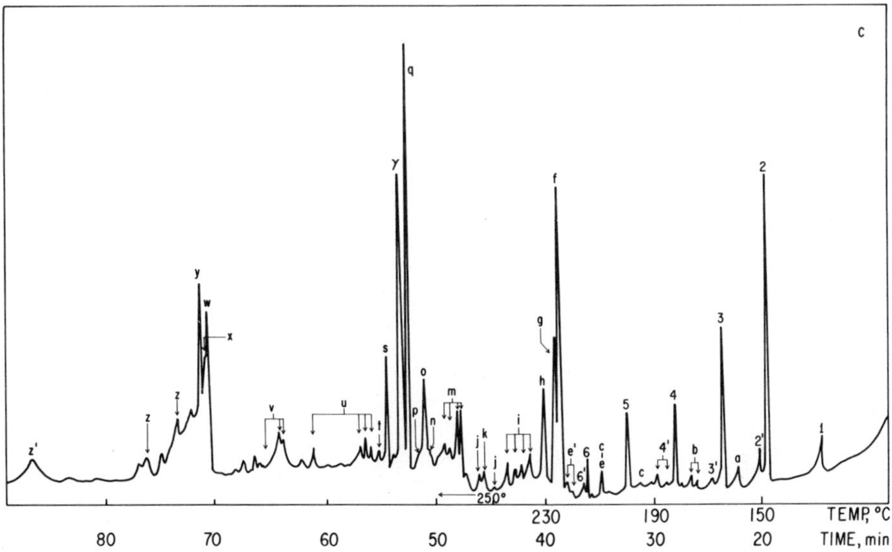

FIG. 5. (*continued*)

For comparison, a methylated sample (3g) of the same lignite was oxidized at 80°C under nonbuffered conditions (300 ml of 4.3% $KMnO_4$; the final pH of the reaction medium was 9.0–9.5). As shown in Fig. 5b, aromatic acids were produced in even greater quantities. Short-chain dicarboxylic acids (C_2–C_8) were found in significant amounts; however, long-chain acids ($C > 8$) were almost nonexistent. These facts show that under nonbuffered permanganate oxidation conditions as well as the alkaline permanganate oxidation, initial oxidation products (long-chain acids) are further oxidized to more stable short-chain acids and CO_2. Oxalic acid and pyrazoline compounds (see Section III,B,1), nearly absent from the buffer-controlled oxidation products of the same coal, are found in the nonbuffered oxidation product. They are believed to result, in part, from the degradation of benzene and phenolic rings even though the benzene ring is resistant to the oxidation. The greater abundance of succinic acid (Fig. 5b peak 2) is also due in part to the degradation of such aromatic rings as shown in the oxidation studies of lignin.[41]

In contrast with these observations, little or no short-chain dicarboxylic acids (C_2–C_5) are found in the alkaline permanganate oxidation products of lignite[88] and exinite macerals[84] already discussed. It is rather surprising that these short-chain dicarboxylic acids are not produced as a result of (1) aromatic ring degradation, (2) oxidation of short methylene chains or alicyclics probably present in the structures of the starting materials, and (3)

TABLE VI

ORGANIC ACIDS OBTAINED FROM BUFFERED PERMANGANATE OXIDATION OF LIGNITE IDENTIFIED AS METHYL ESTERS BY GCMS AND HRMS

GC peak	Carbon number	n	Compound
			Aliphatic dibasic acids $(CH_2)_n(COOH)_2$
0	2	0	Oxalic acid
1	3	1	Malonic acid
2	4	2	Succinic acid
3	5	3	Glutaric acid
4	6	4	Adipic acid
5	7	5	Pimelic acid
6	8	6	Suberic acid
7	9	7	Azelaic acid
8	10	8	Sebacic acid
9	11	9	Nonane-1,9-dicarboxylic acid
10	12	10	Decane-1,10-dicarboxylic acid
11	13	11	Brassylic acid
12	14	12	Dodecane-1,12-dicarboxylic acid
13	15	13	Tridecane-1,13-dicarboxylic acid
14	16	14	Thapsic acid
15	17	15	Pentadecane-1,15-dicarboxylic acid
16	18	16	Hexadecane-1,16-dicarboxylic acid
			Aromatic acids
a			Benzoic acid
b			Methylbenzoic acid
c			Dimethylbenzoic acid
d			*o*-Hydroxybenzoic acid[c]
e			*p*-Hydroxybenzoic acid
e′			Methylhydroxybenzoic acid
f			1,2-Benzenedicarboxylic acid
g			1,4-Benzenedicarboxylic acid
h			1,3-Benzenedicarboxylic acid
i			Methylbenzenedicarboxylic acid
j			Dihydroxybenzoic acid
k			3,4-Dihydroxybenzoic acid
l			Methylhydroxybenzofuranmonocarboxylic acid (T)[a]
m			Dimethylbenzenedicarboxylic acid
n			Hydroxybenzenedicarboxylic acid
o			4-Hydroxy-1,2- (and/or 1,3-) benzenedicarboxylic acid
p			Methylhydroxybenzenedicarboxylic acid
q			1,2,4-Benzenetricarboxylic acid
r			1,2,3-Benzenetricarboxylic acid
s			1,3,5-Benzenetricarboxylic acid

TABLE VI (*cont.*)

GC peak	Carbon number	*n*	Compound
t			Hydroxybiphenylmonocarboxylic acid (T)
u			Methylbenzenetricarboxylic acid (dimethylhydroxybenzenedicarboxylic acid, minor)
v			Hydroxybenzenetricarboxylic acid
w			1,2,4,5-Benzenetetracarboxylic acid
x			1,2,3,4-Benzenetetracarboxylic acid
y			1,2,3,5-Benzenetetracarboxylic acid
z			Methylbenzenetetracarboxylic acid
z′			Benzenepentacarboxylic acid
			Methyl dibasic acids
2′	5	2	
3′	6	3	
4′	7	4	
5′	8	5 (n = 4, dimethyl)	
6′	9	6 (n = 5, dimethyl)	
7′	10	7 (T)	
			Aliphatic tribasic acids
α	6		Propane-1,2,3-tricarboxylic acid
β	7		Butanetricarboxylic acid
γ	8		Pentanetricarboxylic acid
N			Pyrazole compounds[b]

[a] (T) Identification tentative.

[b] Pyrazoline compounds are formed from the reaction of diazomethane with oxidative degradation products from benzene rings.

[c] *m*-Isomer may be present in very small amounts.

further oxidation of initial products (longer chain aliphatic acids) under the alkaline permanganate oxidations used. Furthermore, since all of the starting samples were not methylated before the oxidations, aromatic rings with phenolic OH groups present in the coal originally, must have been degraded for the most part to short-chain aliphatic acids and CO_2 even at room temperature.[41] Matsuda and Schnitzer[89] have shown the importance of premethylation for permanganate oxidation of soil humic acids. However, even methylated lignins[79,80] and methylated humic acids[90] are known to be degraded to some degree by alkaline permanganate oxidation. To obtain a milder permanganate oxidation, potassium hydroxide is omitted[91] in some

[89] K. Matsuda and M. Schnitzer, *Soil Sci.* **114**, 185 (1972).

[90] M. Schnitzer and J. G. Desjardins, *Soil Sci. Soc. Amer. Proc.* **34**, 77 (1970).

[91] D. H. R. Barton and M. Schnitzer, *Nature* (*London*) **198**, 217 (1963).

experiments with soil humic acids. However, as Hayatsu *et al.*[26] have already shown, the final oxidation medium becomes alkaline (pH 9.0–9.5) even using aqueous $KMnO_4$ solution, and such a condition results in the destruction of more aromatic rings and longer chain aliphatics.

These facts suggest that the oxidation results given by Allan[84] and Young *et al.*[88] may be inadequate for quantitative evaluations. Furthermore, it is unfortunate that mass spectrometric techniques were not used for the identification of the oxidation products which make up these complex mixtures since even a sharp GC peak is often found by MS to consist of two or more components. Therefore, GC analysis aloue, as used by these two groups, may not permit definite identification of individual compounds.

In some studies, aromaticity of humic acids, kerogens, or coals was determined from calculations based on the percentages of aromatic acids produced in permanganate oxidative degradations. This was justified because the permanganate oxidation was thought to be a selective procedure that cleaved branched chain or activated chain structures, leaving aromatic and straight-chain aliphatic structures nearly intact. However, when the permanganate oxidation, particularly in alkaline medium, is used for structural characterization of coal, humic substances, or other polymeric materials, the following must be kept in mind. (1) Different conditions (ratio of substrate/oxidant, temperature, time) give different yields and kinds of products; in particular prolonged reaction with excess permanganate in alkaline solution results in greater degradation of organic material. (2) Polynuclear aromatic structures as well as phenols are easily degraded in the oxidation, for example, 1,4-dimethylnaphthalene gives 1,2,3,4-benzenetetracarboxylic acid.[41,74,76] (3) Heterocyclic compounds are also oxidized.[73,76] (4) Initial oxidation products are further oxidized to more stable products such as oxalic acid, acetic acid, and carbon dioxide; e.g., for a 3-hr alkaline permanganate oxidation of dodecanedioic acid, product recovery amounted to only 28%.[92]

D. ALKALINE CUPRIC OXIDE

The alkali solutions of metal oxides (copper(II), mercury(II), silver(I), and cobalt(II) oxides) are known to selectively oxidize lignins (for review, Chang and Allan,[41] Adler[93]), fulvic and humic acids (for review, Hayes and Swift[14]), land-derived marine sediments (Hedges *et al.*[94]), and coals (Hayatsu *et al.*[95–97]) without degradation of aromatic rings, and in particular, phenolic

[92] R. Hayatsu, R. E. Winans, R. L. McBeth, and R. G. Scott, unpublished results.

[93] E. Adler, *Wood Sci. Technol.* **11**, 169 (1977).

[94] J. I. Hedges and P. L. Parker, *Geochim. Cosmochim. Acta* **40**, 1019 (1976).

[95] R. Hayatsu, R. E. Winans, R. G. Scott, R. L. McBeth, L. P. Moore, and M. H. Studier, *Nature (London)* **278**, 41 (1979).

nuclei. Although there are some similarities between alkaline nitrobenzene and alkaline metal oxide oxidations, the latter oxidations have a significant advantage because undesirable by-products are not formed by them (see Section IV, E) Therefore, the alkaline metal oxides, and in particular, cupric oxide which is milder than the other metal oxides, have been used more extensively. Unlike the alkaline nitrobenzene oxidation of lignin which is a two-electron transfer process, as suggested by Chang and Allan,[41] alkaline cupric oxide oxidation is known to involve only a one-electron transfer. For example, vanillin is formed via a quinone methide from vanillyl

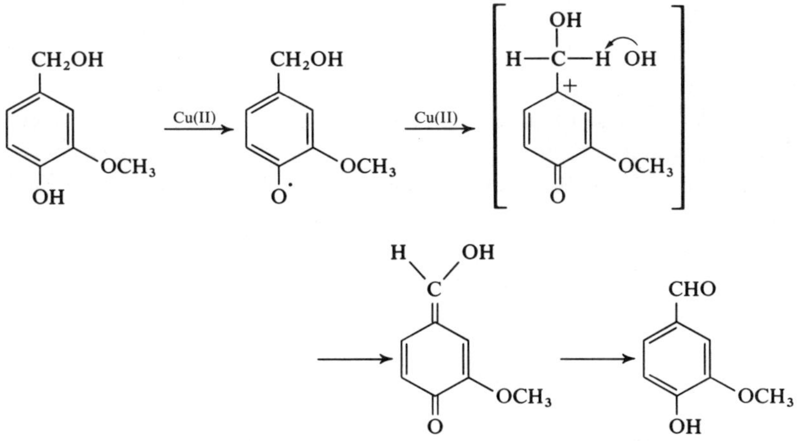

(9)

alcohol. Before one phenoxy radical couples with another, a second electron is transferred to another cupric oxide preventing the coupling reaction and allowing the formation of the intermediate.

To characterize the structure of lignin, five depolymerization techniques have been used and evaluated by Morohoshi *et al.*[98]: sodium in liquid ammonia, thioglycolic acid degradation, thioacetic acid degradation, alkaline nitrobenzene oxidation, and alkaline cupric oxide oxidation. Among the five depolymerization techniques evaluated in combination with permanganate oxidation, with regard to their capabilities for generating detailed analytical structural information on lignin polymeric materials, the alkaline cupric oxide oxidation appeared to be most useful. Optimum

---

[96] R. Hayatsu, R. E. Winans, R. L. McBeth, R. G. Scott, L. P. Moore, and M. H. Studier, Preprint **24** (No. 1), p. 110 (Div. of Fuel Chem., Am. Chem. Soc.) Honolulu, April, 1979.

[97] R. Hayatsu, R. E. Winans, R. L. McBeth, R. G. Scott, L. P. Moore, and M. H. Studier, *Adv. Chem. Sec.*, No. 192, p. 133, 1981.

[98] N. Morohoshi and W. G. Glasser, *Wood Sci. Technol.* **13**, 165 (1979).

[99] M. Schnitzer and M. I. O. DeSerra, *Can. J. Chem.* **51**, 1554 (1973).

oxidation conditions of lignin with the alkaline cupric oxide were determined. In the oxidation temperature range from 100 to 174°C, the yields of oxidation products as methyl esters increased approximately sixfold from 13 to 76 mg/100 mg of lignin.

A number of different common plants and marine sediments have also been oxidized with alkaline cupric oxide.[94] Characteristic lignin oxidation products (*p*-hydroxybenzene derivatives, vanillyl, and syringyl groups) are identified in the oxidation products of vascular land plants and nearshore marine sediments.

The oxidation of humic and fulvic acids was investigated in detail by Schnitzer's group[83,99–101] and the major products were found to be phenolic, benzenecarboxylic, and *n*-fatty acids. Lignin oxidation products, *p*-hydroxybenzoic acids (**5**), 3,4-dihydroxybenzoic acid (**6**), 3,4,5-trihydroxybenzoic acid (**7**), and 4-hydroxybenzene-1, 3-dicarboxylic acid (**8**) were isolated and identified as their methyl esters. Hydroxybenzenepolycarboxylic acids that are considered to be lignin degradation products were identified as well.

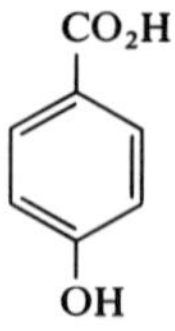

(**5**)

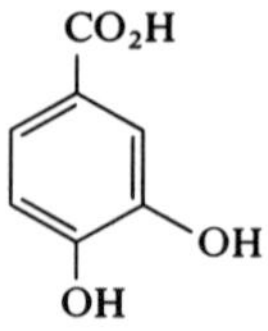

(**6**)

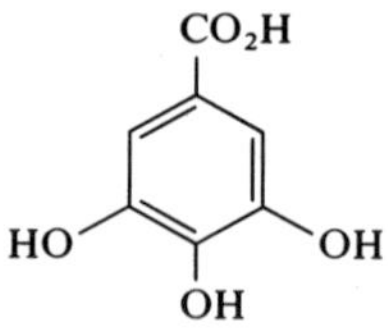

(**7**)

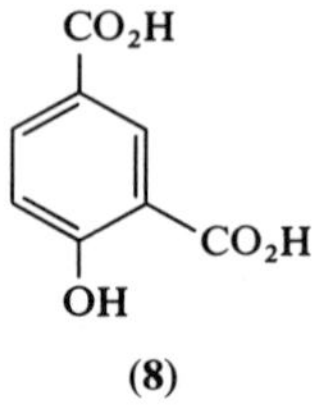

(**8**)

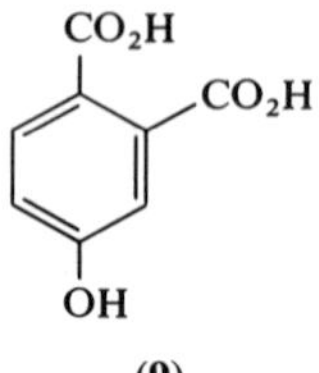

(**9**)

Lignins have long been regarded as major constituents of vascular land plants from which coals are derived; however, their roles in the coalification process and resulting coal structures were not defined. In an effort to clarify these roles, alkaline cupric oxide oxidation was applied to coals for the first time by Hayatsu *et al.*[95–97] Seven various rank coals from lignite to anthracite were oxidized. In general, lignin, fulvic acids, humic acids, and marine sediments were oxidized at 170°C for 2–4 hr. However, this treatment was found to be too mild to oxidize coals, particularly, high rank coals. Therefore, coal samples were oxidized at 200°C for 8–10 hr.

[100] J. A. Neyroud and M. Schnitzer, *Can. J. Chem.* **52**, 4123 (1974).
[101] J. A. Neyroud and M. Schnitzer, *Soil Sci. Soc. Amer. Proc.* **38**, 907 (1974).

TABLE VII

CuO–NaOH OXIDATION OF MODEL COMPOUNDS AT 200°C FOR 8 HOURS

Compound	Type of reaction	Major product[a]	Yield mol %
4-Methylbenzyl alcohol	$—CH_2OH \rightarrow —COOH$	4-Methylbenzoic acid	93
3-Phenylpropan-1-ol	$—(CH_2)_2—CH_2OH \rightarrow —COOH$	Benzoic acid	68[b]
Benzylmethylcarbinol	$—CH_2—CH(OH)—CH_3 \rightarrow COOH$	Benzoic acid	70
β-Naphthylmethanol	$—CH_2OH \rightarrow —COOH$	2-Naphthoic acid	77
Benzaldehyde	$—CHO \rightarrow —COOH$	Benzoic acid	95
4-Methoxybenzaldehyde	$—CHO \rightarrow —COOH$	4-Hydroxybenzoic acid	89[c]
2,6-Dimethoxyphenol	$—OCH_3 \rightarrow —OH$	1,2,6-Trihydroxybenzene	2.0[d]
Syringaldehyde	$—CHO \rightarrow —COOH$	3,4,5-Trihydroxybenzoic acid	4.5[d]
Dibenzoylmethane	$—CO—CH_2— \rightarrow —COOH$	Benzoic acid	76
1,4-Naphthoquinone	$—CO—C \rightarrow —COOH$	Phthalic acid	91
Flavone	$—CO—C{=}C— \rightarrow COOH$	Benzoic acid	37
		Salicylic acid	41
Phenylacetic acid	$—CH_2—CO— \rightarrow —COOH$	Benzoic acid	68
Diphenyl ether	$—O— \rightarrow —OH$	Phenol	67
Dibenzyl ether	$—CH_2—O—CH_2— \rightarrow —COOH$	Benzoic acid	81
4,4′-Dihydroxydibenzyl ether	$—CH_2—O—CH_2— \rightarrow —COOH$	*p*-Hydroxybenzoic acid	79
Dibenzyl sulfide	$—C—S—C— \rightarrow —COOH$	Benzoic acid	64[e]
Dibenzyl disulfide	$—C—S—S—C— \rightarrow —COOH$	Benzoic acid	72[e]
Diphenylmethane	$=C—C— \rightarrow —COOH$	Benzoic acid	11
2,4-Dimethoxytoluene	$—C—C— \rightarrow —COOH$	2,4-Dihydroxybenzoic acid	5[f,g]
3,4-Dimethoxytoluene	$—C—C— \rightarrow —COOH$	3,4-Dihydroxybenzoic acid	7[f]
Poly(4-methoxystyrene)	$—C—C— \rightarrow COOH$	4-Hydroxybenzoic acid	6[h,i]
Poly(2-vinylnaphthalene)	$—C—C— \rightarrow —COOH$	2-Naphthoic acid	4[h,i]
1-Methylnaphthalene	$—C—C— \rightarrow —COOH$	1-Naphthoic acid	3[i]
β-Naphthol	$=C—OH \rightarrow =C=O$	Naphthoquinone	≤3[i]
1-Methylphenanthrene	$—C—C— \rightarrow —COOH$	Phenanthrene-1-carboxylic acid	≤2[g,i]

[a] All products were derivatized with [d_6]-dimethyl sulfate or diazomethane, and analyzed by GCMS.

[b] Phenylacetic acid and hydrocinnamic acid were also obtained as minor products.

[c] 45% of the product was found to be demethylated.

[d] The product consisted mainly of polymerized material and of extensively degraded material. Milder oxidation (170°C for 4 hr) also gave similar results.

[e] As minor product, dibenzoyl disulfide was tentatively identified from both the samples.

[f] Nonoxidized compounds were amply obtained under the oxidation conditions of 170°–200°C for 4–8 hr. They consist of: (1) starting material (recovery 40–45 wt %), (2) monomethoxy compound (32–37 wt %), and (3) dihydroxy compound (16–20 wt %).

[g] The oxidation was also carried out in the presence of lignite coal. It was found that the coal mineral did not produce any hydroxylation, ring destruction, or other side reactions.

[h] Yield is shown as wt %.

[i] In most cases the remaining compound is starting material.

To obtain detailed information on the CuO–NaOH oxidation at elevated temperatures (170°–200°C for 3–8 hr), control experiments with a total of 25 compounds and polymers containing a variety of functional groups were carried out. As shown in Table VII, the alkaline cupric oxide was found to be a selective oxidant. The results are generalized as follows: (1) The treatment results in hydrolytic oxidation of ether linkages in the aliphatic side chains of aromatics; (2) the oxidation is not effective for cleavage of C—C bonds such as aryl, methyl, and methylene chains; (3) there is effective oxidation of aliphatic chains which contain oxygen or sulfur functional groups such as —OH, —C(O), —COOH, —S—, —S—S—; (4) there is no hydroxylation of aromatic rings; (5) no ring destruction occurs in polynuclear aromatics; (6) no ring degradation of mono- or dihydric phenols occurs.

Although trihydroxybenzenes and their derivatives have been isolated from the oxidation products of lignins, humic acids, and land-derived marine sediments, it is found that 2,6-dimethoxyphenol and syringaldehyde

TABLE VIII

SUMMARY OF CuO–NaOH OXIDATION PRODUCTS

Coal sample No.[a]	1	2	3	4	5	6	7
	Weight %[b]						
Extract							
Benzene–ether	35.3	19.6	17.1	7.0	14.7	1.3	—[c]
Methanol	54.3	61.0	46.2	9.3	39.5	69.5	—
Only aqueous alkali soluble	≦1.0	≦1.5	16.1	51.0	20.2	2.0	7.5
Nonoxidized coal[d]	11.0	21.0	22.2	30.4	26.5	32.0	95.0
Identified acids[e]							
Phenolic	66.6	54.1	14.6	41.0	8.3	5.8	—
Benzenecarboxylic	26.0	36.2	69.4	50.2	62.7	76.7	—
Naphthalenecarboxylic	2.0	3.3	6.2	5.5	22.9	10.6	—
Heterocyclic[f]	—	1.6	3.5	2.0	2.0	3.2	—
Aliphatic dibasic	2.6	1.0	1.4	—	≦1.0	≦1.0	—
Others	2.8	3.8	4.9	1.3	3.5	3.0	—
Ratio of phenol/benzene acid	2.56	1.49	0.21	0.82	0.13	0.07	—

[a] (1) Wyoming lignite (66.4%, C), (2) Ill. No. 2 bituminous (73.9%, C), (3) Ill. No. 6 bituminous (77.7%, C), (4) PSOC 297 bituminous (80.5%, C), (5) Pittsburgh No. 8 bituminous (82.0%, C), (6) PSOC 258 bituminous (86.5%, C), and (7) PSOC 85 anthracite (91.0%, C); all carbon analysis on an maf basis.

[b] Wt % was obtained from each coal sample on an maf basis.

[c] — = Not found.

[d] Small amounts of insoluble Cu salts are present.

[e] Wt % was obtained from each benzene–ether extract. Determination was made from the gas chromatograms of their methyl esters.

[f] Thiophene and pyridine carboxylic acids.

(4-hydroxy-3,5-dimethoxybenzaldehyde) are quite susceptible to the CuO–NaOH oxidation. The conditions used (170°–200°C for 3–8 hr) seem to be somewhat drastic for trihydric phenol derivatives. A summary of the oxidation products obtained from seven coal samples is shown in Table VIII. As seen in the chromatograms in Fig. 6*, the abundant oxidation products identified are shown to be *p*-hydroxybenzoic and 3,4-dihydroxybenzoic acids for low rank coals (lignite and high volatile C bituminous coals). Two isomers of hydroxybenzenedicarboxylic acids (**8**, **9**) among the six possible isomers are also identified in large amounts. Phenolic tricarboxylic acids are found as minor products. All these phenolic acids are considered typical lignin degradation products. In general, phenolic di- and polycarboxylic acids have not been found in the alkaline cupric oxide oxidation products of lignins, land plants, and marine sediments. It is interesting to note that, whereas phenolic acids are found in the oxidation products of fulvic and humic acids together with considerable amounts of fatty acids, little or no fatty acids are observed in the coal oxidation products. According to Schnitzer *et al.*[24] and Griffith *et al.*,[83] most of the fatty acids released by the alkaline cupric oxide oxidation from fulvic and humic acids are esterified to OH groups of phenolic acids. For coal, Hayatsu *et al.*[95–97] have suggested that the phenolic acids found are derived mainly from the cleavage of ether-linked aromatic systems present in the structures of coals as shown in the following example:

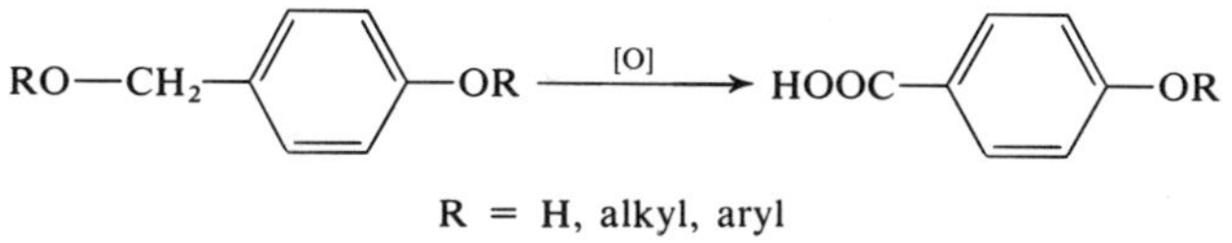

(10)

R = H, alkyl, aryl

Finally, the same authors have concluded that: (1) The alkaline cupric oxide oxidation of coals shows that lignin-like polymers have been incorporated in the macromolecules of coals, and are still identifiable in lower rank coals. Evidence for this is the identification of *p*-hydroxy- and 3,4-dihydroxybenzoic acids (**5**, **6**) and hydroxybenzenedicarboxylic acids (**8**, **9**), which are known lignin oxidation products. (2) As coalification progresses, cross-linking increases and the lignin-derived polymers become more aromatic. Phenolic polycarboxylic acids and hydroxynaphthalene dicarboxylic acids that were identified are not found in the CuO–NaOH oxidation products of lignins and plant materials. (3) Extensive transformation of lignin-like polymers occurred perhaps through reactions such as demethylation, demethoxylation, isomerization, dehydrogenation, oxidation, cleavage of ring structures, recondensation, etc. during coalification from lignite to

* For identification of these chromatogram peaks see Hayatsu *et al.*[95]

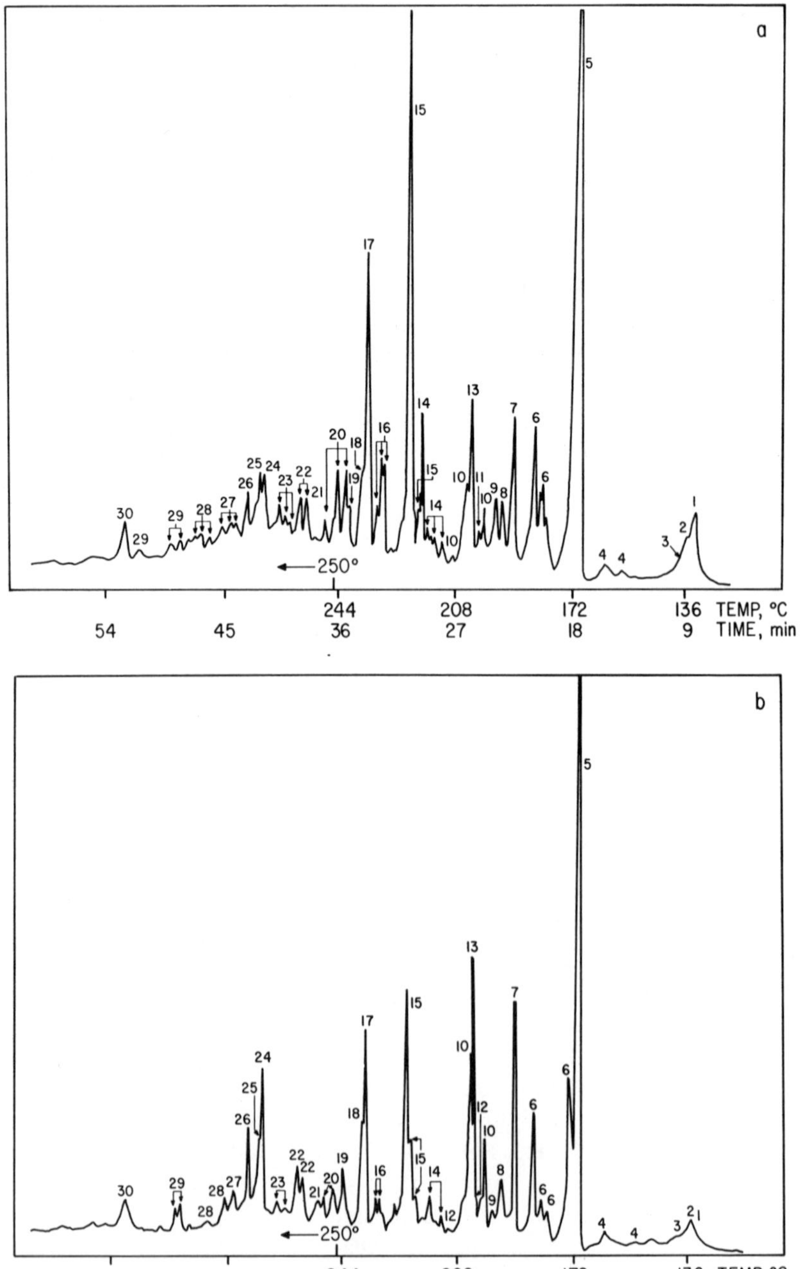

FIG. 6. Gas chromatograms of the methyl esters derived from the organic acid fraction of the CuO–NaOH oxidation of (a) Wyoming lignite and (b) Illinois No. 2 bituminous coal. The separation was made on a 15 m × 0.5 mm OV-17 SCOT column temperature programmed from 100°–250°C at 4°C/min with 4 ml/min He flow. Peaks 5, 13, and 15 have been identified as *p*-hydroxybenzoic acid, 3,4-dihydroxybenzoic acid, and hydroxybenzenedicarboxylic acids, respectively. (Peaks identified in reference 95.)

bituminous and anthracite coals. This conclusion is supported by the following observations: (1) Lower rank coals gave higher yields of the CuO–NaOH oxidation product; (2) a higher phenolic acid/benzenecarboxylic acid ratio is shown for the lower rank coals; (3) very little or no dihydroxy benzoic acid is identified in the oxidation products of higher rank coals; (4) more naphthalene- and heterocarboxylic acids are found in the oxidation products of higher rank coals; (5) major parts of lignin-like polymers are transformed into nonlignin-type polymers at a later stage of coalification. Indeed, the later coalification product, anthracite coal, did not yield any phenolic acid as a result of the oxidation.

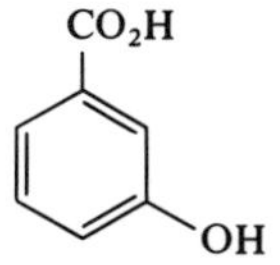

(10)

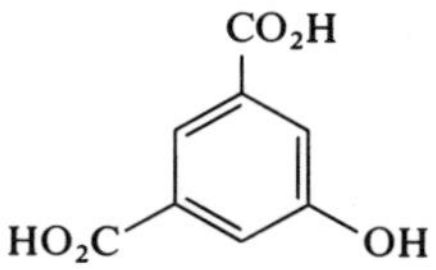

(11)

In sharp contrast, meteoritic organic polymer yields mainly 3-hydroxybenzoic (**10**) and 3-hydroxy-1,5-benzenedicarboxylic (**11**) acids by the alkaline cupric oxide oxidation.[102] These phenolic acids are thought to be derived from phenolic ethers in the polymer, which in turn probably were formed from CO and H_2 by catalytic Fischer–Tropsch type reactions in the solar nebula. Apparently, the polymer in the meteorite has been shown to be unrelated to lignin which is biogenic in origin.

E. ALKALINE NITROBENZENE

Since the pioneering work of Freudenberg's group,[103] alkaline nitrobenzene oxidation has been applied widely to the characterization of structure of lignins, humic substances, marine sediments, coals, and coal products. In this procedure (2 *N* NaOH and nitrobenzene, heated at 170°–180°C for 2–2.5 hr) 20–75% of lignin is converted into identifiable aromatic compounds that are principally a mixture of aldehydes comprising *p*-hydroxybenzaldehyde (**12**), vanillin (**13**), and syringealdehyde (**14**).[103–106]

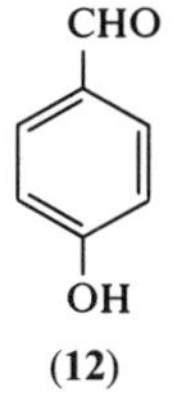

(12)

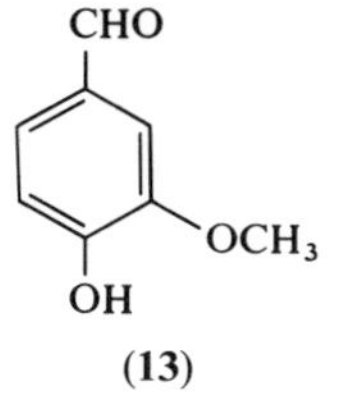

(13)

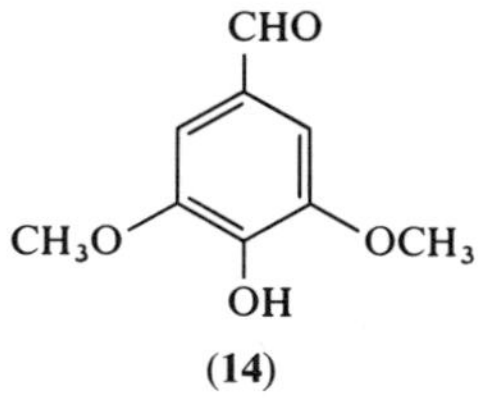

(14)

[102] R. Hayatsu, R. E. Winans, R. G. Scott, R. L. McBeth, L. P. Moore, and M. H. Studier, *Science* **207**, 1202 (1980).

These three compounds are considered characteristic lignin oxidation products. Leopold,[107] Chang and Allan,[41] and Hayes and Swift[14] have discussed the following mechanism proposed for oxidation of isoeugenol (**15**) to vanillin (**13**):

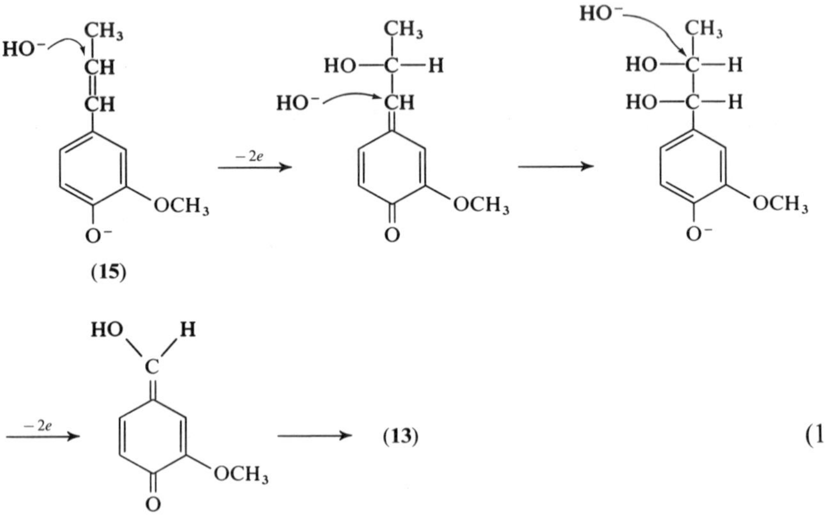

(11)

The oxidation of humic substances yields the three aldehydes (**12–14**) and three acids which are further oxidation products of the aldehydes.[108–110] These workers also indicated that the yields of aldehydes decrease in the order: plant material > peat > mineral soil.

Under the oxidation conditions, nitrobenzene acts as a two-electron acceptor and undergoes stepwise reduction through nitrosobenzene and phenylhydroxylamine to aniline. Secondary reactions between these reduction intermediates also occur in alkaline media, and give azoxybenzene, azobenzene, and hydroxy (*ortho* and *meta*) azobenzenes. All of these compounds have been isolated from lignin oxidation products. As already

---

[103] K. Freudenberg, W. Lautsch, and K. Engler, *Chem. Ber.* **73B**, 167 (1940).
[104] B. Leopold, *Acta Chem. Scand.* **4**, 1523 (1950).
[105] K. R. Kavanagh and J. M. Pepper, *Can. J. Chem.* **33**, 24 (1955).
[106] J. G. Bicho, E. Zavarin, and D. L. Brink, *Tappi* **49**, 218 (1966).
[107] B. Leopold, *Srensk Kem Tid.* **64**, 18 (1952).
[108] R. I. Morrison, *J. Soil Sci.* **9**, 130 (1958).
[109] R. I. Morrison, *J. Soil Sci.* **14**, 201 (1963).
[110] R. E. Wildung, G. Chesters, and D. E. Behmer, *Plant Soil* **32**, 221 (1970).

pointed out by many workers,[14,94] these by-products not only interfere with analysis of the oxidation products but also may lead to misinterpretation of the origins of the compounds produced (see Section IV,D).

A Turkish group[111–113] has oxidized coals with alkaline nitrobenzene at 190°–240°C for 2–5 hr, and on the basis of IR, UV, NMR and spot tests, reported the presence of aromatic aldehydes and ketones with phenolic OH and aliphatic substituents in the oxidation products. However, the individual compounds produced were not separated and identified. Recently, the same group reported the characterization of the structure of reductively ethylated products, (with an average molecular weight of a few thousand) obtained from lignite and coking coal using alkaline nitrobenzene oxidation.[114] The oxidation products were analyzed by GCMS and a few hundred GC peaks were observed. A great number of compounds were identified that have molecular weights of less than 300, and that were classified by groups: (1) aliphatic and aromatic compounds not containing oxygen such as aliphatic, alicyclic, aromatic, and hydroaromatic hydrocarbons, with or without alkyl and methyl sulfide substituents; (2) aromatic compounds with carbonyl groups such as aldehydes, ketones, and carboxylic acids; (3) aromatic compounds containing oxygen with ether linkages or hydroxyl groups such as alcohols, phenols, benzyl ether types, and furan derivatives; (4) aliphatic compounds containing oxygen such as aldehydes, ethers, ketones, esters, and carboxylic acids.

However, the identification of many of these compounds is uncertain for the following reasons.

(1) Gas chromatograms and/or mass spectral data are not shown for identification of a great number of compounds of various types.

(2) Packed columns using nonpolar silicone liquid phases are used. It is unlikely that these columns would give sufficient separation and resolution for all groups listed above; this is especially true for analyses of polar compounds (alcohols, phenols, and carboxylic acids) and nonpolar compounds (hydrocarbons).

(3) Standard samples or published data for GC retention times and mass spectra may not be available for many of the compounds identified. Some of these compounds are complex and quite uncommon[114] and, therefore, mass

[111] A. Olcay, H. Evliya, A. F. Gaines, and J. Homer, *Fuel* **52**, 20 (1973).
[112] B. Bozkurt, A. F. Gaines, A. Ogut, Y. Yurum, and A. Olcay, *Fuel* **52**, 81 (1973).
[113] K. Burgur, A. F. Gaines, A. Gulec, S. Guyer, Y. Yurum, and A. Olcay, *Fuel* **52**, 115 (1973).
[114] R. Dogru, A. Gaines, A. Olcay, and T. Tugrul, *Fuel* **58**, 823 (1979).

spectral identification cannot be made reliably on the basis of general principles alone as was done by the authors.

(4) No confirmation of identification was made by means of high resolution mass spectrometry.

F. HYDROGEN PEROXIDE IN ACETIC ACID*

Even though hydrogen peroxide–acetic acid has been used widely for oxidation studies of lignin, this oxidant has been applied rarely to the oxidation of soil humic acids and coal. The peracetic acid oxidation of phenols and their ether derivatives (as lignin model compounds) was extensively studied in the 1950–1960s (e.g. see review by Chang and Allan[41]). In general, the oxidation of phenolic compounds leads to ring hydroxylation in available *ortho*- or *para*-positions, and then to the formation of *ortho*-quinones or *para*-quinones. Finally, by oxidative ring opening, muconic acid (**16**) is formed from *ortho*-quinone while maleic (**17**) and fumaric acid (**18**) are formed from *para*-quinone. A muconic acid derivative (**19**) is isolated from

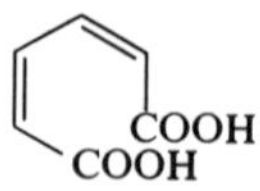

(**16**)

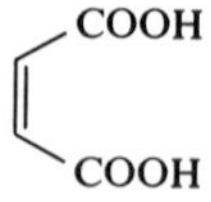

(**17**)

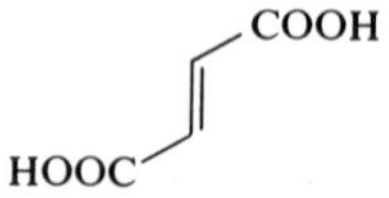

(**18**)

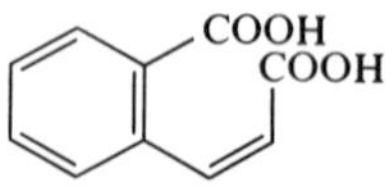

(**19**)

the oxidation of 2-methoxynaphthalene via 1-hydroxy-2-methoxynaphthalene. Methoxybenzene (anisole) is hydroxylated to form 2- and 4-methoxyphenols by either Fenton's reagent [H_2O_2 and Fe(II)] or H_2O_2 catalyzed by Fe(III) and catechol.[115,116] Under different conditions, peracetic acid oxidation is found to result in the demethylation of methoxybenzene derivatives;[117] for example:

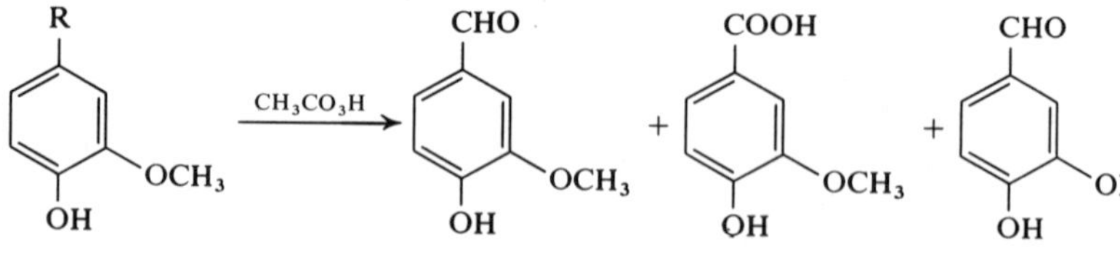

(12)

R = $-CH_2OH$, $-CH_2OCH_2CH_3$, $-C(O)CH_3$, $-CH(OH)CH_3$

* As Fieser and Fieser[33] noted, this solution contains peracetic acid, and is commonly called peracetic acid; however, it differs substantially in composition and properties from commercial 40% peracetic acid and anhydrous peracetic acid.

In general, peracetic acid oxidation of lignin results in the oxidative rupture of the aromatic rings and the formation of dibasic acids. Similarly, phenolic compounds did not survive mild peracetic acid oxidation of peat humic acid.[118] However, Schnitzer *et al.*[119] have successfully oxidized soil humic and fulvic acids under Meneghel's mild conditions (H_2O_2–CH_3COOH, 40°C for 8 days), and have identified phenolic acids, benzenecarboxylic acids, aliphatic acids, and methyl-3,4-furandicarboxylic acid as the oxidation products. Soon after, this procedure was applied to the oxidation of lignite and bituminous coals.[19,42] As a general procedure, a 2 g sample of coal which had been solvent-extracted previously, was stirred in 250 ml of glacial acetic acid with 10 ml of isopropyl alcohol at 40°–60°C. Four 20 ml aliquots of 30% aqueous H_2O_2 were added dropwise at 24-hr intervals. After 96 hr, the mixture was filtered and the filtrate evaporated under vacuum. The resulting solids were taken up in methanol and methylated with diazomethane. The conversion of the solid coals was 80% for lignite and 77% for the bituminous coal with a weight yield for soluble acids of 96% and 127%, respectively. The esterified organic acid fraction of the lignite oxidation products was analyzed by GCMS and HRMS. Phenolic acids (22.1% of the fraction), benzenecarboxylic acids (43.7%), furancarboxylic acids (15.9%), and aliphatic dibasic acids (18.2%) were identified. For the bituminous coal the phenolic acids were found to be half as abundant. These results suggest that the lignite has twice the phenolic content of the bituminous coal. However, it is necessary to consider the fact that the H_2O_2–CH_3COOH oxidation is somewhat similar to the H_2O_2–CF_3COOH oxidation, which produces phenolic and furancarboxylic acids as artifacts (see Section IV, G). This means that some of the phenolic and furancarboxylic acids produced abundantly from the coals by the H_2O_2–CH_3COOH oxidation may be secondary products. However, the results from the oxidation of model compounds, higher rank coals, and the oxidation conditions used, suggest that the formation of these artifacts is much lower than for the H_2O_2–CF_3COOH oxidation.[120] One of the model compounds studied was 2,6-dimethylnaphthalene (**20**), which under typical reaction conditions with H_2O_2–CH_3COOH, was oxidized cleanly to a quinone (**21**). With the addition of a

[115] G. A. Hamilton and J. P. Friedman, *J. Am. Chem. Soc.* **85**, 1008, (1963).
[116] G. A. Hamilton, J. P. Friedman, and P. M. Campbell, *J. Am. Chem. Soc.* **88**, 5266 (1966).
[117] H. Ishikawa and T. Oki, *J. Japan TAPPI* **18**, 477 (1964).
[118] R. Meneghel, C. Petitsarlotte, and J. M. Block, *Bull. Soc. Chim. Fr.*, 2997 (1972).
[119] M. Schnitzer and S. I. M. Skinner, *Can. J. Chem.* **52**, 1072 (1974).
[120] R. E. Winans, R. L. McBeth, R. G. Scott, and R. Hayatsu, unpublished results.

small amount of lignite coal, the reaction proceeds to the benzene carboxylic acids (preliminary results were shown by Hayatsu *et al.*[42] and Studier *et al.*[46]).

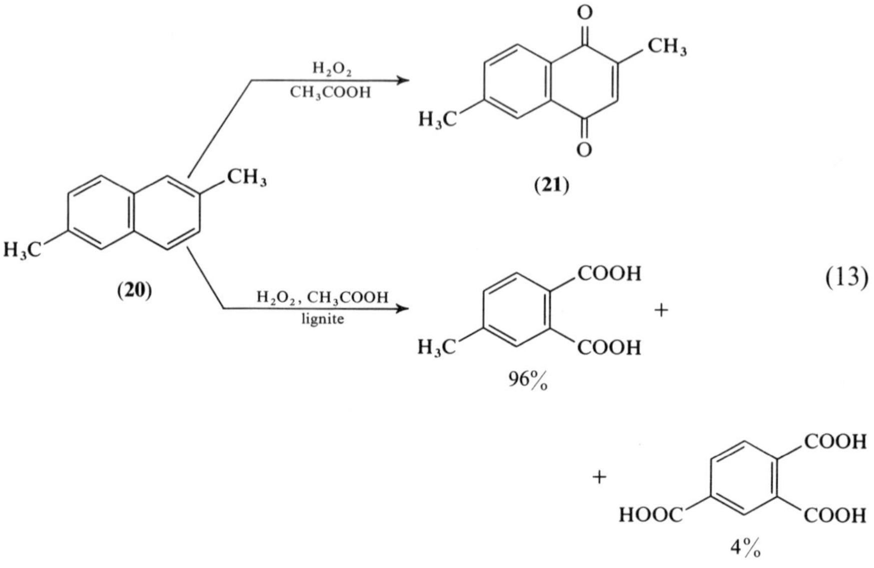

It should be noted that no hydroxylated compounds or furans were isolated. The whole coal appears to be acting as a catalyst. In addition, the fact that the abundance of phenols in the oxidation products of the coals depends on the oxygen content is consistent with minimal hydroxylation of the coals by the reagent.

The H_2O_2–CH_3COOH oxidation is found to be effective for isolating thiophene carboxylic acids. It has been noted that thiophenes are more resistant to oxidation than sulfides.[121] These observations have been confirmed because under these relatively mild conditions, no thiophene sulfoxides or sulfones are detected. In the oxidation of a high volatile C bituminous coal rich in organic sulfur [C, 78.9%; H, 5.9%, N, 1.6%; S, 4.01% (organic)], a series of thiophenecarboxylic acids and benzothiophenecarboxylic acids were identified. The chromatograms for the C, H, and S channels from the GCMPD resulting from the separation of the methyl esters derived from this oxidation are shown in Fig. 7. Tentatively, several methyl benzenesulfonates have also been identified by Winans *et al.*[120]

[121] S. Gronowitz, *in* "Advances in Heterocyclic Chemistry" (A. R. Katritzky, ed.), Vol. 1, p. 2. Academic Press, New York, 1963.

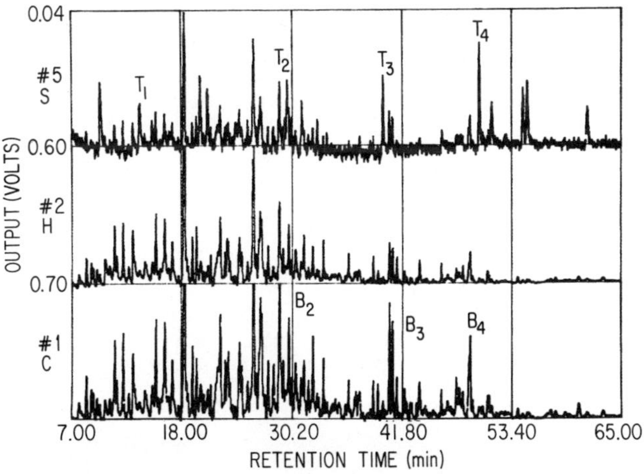

FIG. 7. Output from three channels, C, H, and S for GCMPA of the methyl esters from the H_2O_2/CH_3COOH oxidation of PSOC 244-HVC bituminous coal. Separated on a 45 m × 0.5 mm OV-17 glass SCOT column, with 4 ml/min He flow, temperature programmed from 70°–250°C at 4°C/min. T_n, methyl thiophenecarboxylates (n = number of $COOCH_3$ groups); B_n, methyl benzenecarboxylates.

G. HYDROGEN PEROXIDE–TRIFLUOROACETIC ACID (TFA)*

Pertrifluoroacetic acid (30% H_2O_2–CF_3COOH with or without the addition of 96% H_2SO_4) was first applied to the oxidation of lignite and bituminous coals by Deno *et al.*[122,123] Using this method, this group found that whereas all aromatic ring systems were destroyed, most of the original aliphatic structures of coal were preserved. The nature of the oxidation was illustrated using model compounds such as toluene, ethylbenzene, *n*-propylbenzene, and isopropylbenzene.[124] The major oxidation products from these compounds were acetic, propionic, butyric, and isobutyric acids, respectively. In contrast, commonly used oxidants such as HNO_3, O_2, Mn(VII), and Cr(VI) attack the benzylic hydrogen yielding benzoic acid for all four of these aromatics. In general, H_2O_2–TFA–H_2SO_4 oxidation of alkylbenzenes leads to degradation of the benzene ring to a carboxyl group and over 70% preservation of the aliphatic structure. From these model

* See the series, "Reagents for Organic Synthesis" by Fieser and Fieser.[33]

[122] N. C. Deno, B. A. Greigger and S. G. Stroud, Preprint **23** (No. 2), 54 (Div. of Fuel Chem., Am. Chem. Soc.) Anaheim, March, 1978.

[123] N. C. Deno, B. A. Greigger, and S. G. Stroud, *Fuel* **57**, 455 (1978).

[124] N. C. Deno, B. A. Greigger, L. A. Messer, M. D. Meyer, and S. G. Stroud, *Tetrahedron Lett.*, 1703 (1977).

compound results, it appeared that this might be a good technique for examining the aliphatic structure in coals.

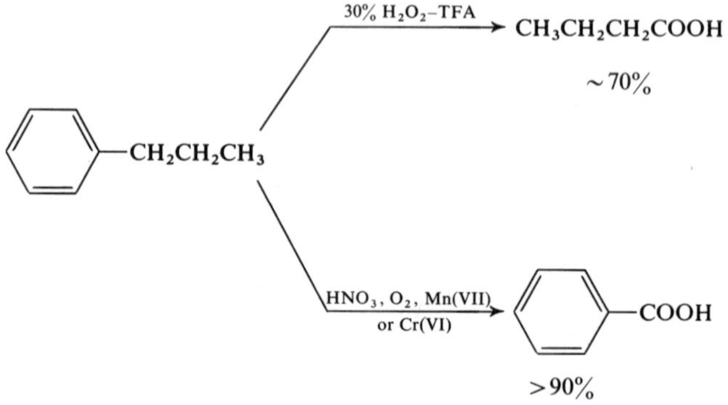

(14)

The oxidation procedure is simple. For example, 0.8 g of a coal is oxidized with a mixture of 10 ml of 30% H_2O_2 and 8 ml of TFA (in some cases, 8 ml of 96% H_2SO_4 is added). Completion of the oxidation is indicated by a colorless or nearly colorless solution. The lignite coals were stirred without additional heating and bituminous coals were refluxed. Typical reaction times were 1–4 hr.* The resulting acids and their methyl esters were analyzed by NMR and GC. It is interesting to note that of the sample's original hydrogen, the amount that appeared in the oxidation products varied from 9–34% for bituminous coals and was 55% for a lignite, and about three-fourths of this hydrogen was present in acetic, propionic, succinic, glutaric acids, and methanol. The identification of acetic acid, propionic acid, and methanol in coal oxidation products seems to indicate that arylmethyl, ethyl, and methoxy groups were present originally in the coals. Since methanol was found only in the oxidation products of lignite coal, Deno *et al.*[122,123] have suggested that no methoxyaryl compounds are present in the structures of bituminous coals. Glutaric acid could arise from 1,3-diarylpropane or indan structures, or possibly tetralin structures substituted at the 1-position. Succinic acid which accounts for 4.4–13.4% of the total hydrogen in the coals studied may be derived from Ar—CH_2—CH_2—Ar and indans. It was claimed that no tetralin unsubstituted on the reduced ring or aryl carboxylic acids could be present in the coals since neither cyclohexane-1,2-dicarboxylic acid anhydride or aromatic acids were identified.

* In a later paper by the same group the oxidation conditions were shown to be somewhat different.[125]

[125] N. C. Deno, B. A. Greigger, A. D. Jones, W. G. Rakitsky, K. A. Smith, and R. D. Minard, *Fuel* **59**, 694 (1980).

From these findings and results, it seems that the H_2O_2–TFA (or with H_2SO_4) oxidation is a very useful technique for characterization of aliphatic structures in coal. However, the same group using the same oxidation methods has now found[125] (reviewed by Davidson[13]) that benzenecarboxylic acids are produced abundantly from the Monterey bituminous coal used in the previous study. This is rather surprising because the H_2O_2–TFA oxidation is supposed to destroy all the aromatic ring structures of coal.[122,123] These benzenecarboxylic acids identified in the oxidation products may derive from dihydrobenzene ring systems such as 5,12-dihydronaphthacene, but not from polyalkylbenzenes and polynuclear aromatics. For example, the products from the H_2O_2–TFA oxidation of 5,12-dihydronaphthacene duplicate the majority of products from Monterey bituminous coal, and the relative amounts are comparable. Major products from both samples are 1,2-benzenedicarboxylic, 1,4-benzenedicarboxylic, 1,2,4-benzenetricarboxylic, and 1,2,4,5-benzenetetracarboxylic acids. Benzoic acid and epoxide derivatives of aliphatic acids are also found. From these facts, Deno *et al.*[125] have concluded that Monterey bituminous coal is a polymer composed of fused benzene rings in which dihydrobenzene units are interspersed at frequent intervals. Two coals and their liquefaction products were also degraded using this procedure by Deno *et al.*[126,127]

Benzenecarboxylic acids are produced from naphthalene, anthracene, chrysene, and benzanthracene in model experiments by Deno *et al.*[125] Hayatsu *et al.*[34] have confirmed recently that alkyl-substituted aromatic rings yield aromatic acids as a result of H_2O_2–TFA oxidation. For example, oxidation of poly(2-vinylnaphthalene) at room temperature gave large amounts of phthalic acid, 1,2,4-benzenetricarboxylic acid, and 2-naphthoic acid with smaller amounts of short-chain aliphatic mono- and dicarboxylic acids. Hydroxybenzenedicarboxylic acids were also isolated. Additional secondary products, a series of furan di-, tri-, and tetracarboxylic acids, were identified as their methyl esters by GCMS. Perhaps these compounds were formed thermally during GC analysis from epoxides or glycols produced during the oxidation.

The oxidation of several lignite and bituminous coals and kerogen was also examined. For example, a Wyoming lignite was subjected to a step-wise oxidation with H_2O_2–TFA–H_2SO_4. The first-step oxidation, at room temperature for 5 hr, yielded organic acids (22.3 wt % of the starting material) and humic acid-like material (14.1 wt %). About 47% of the starting material

[126] N. C. Deno, B. A. Greigger. A. D. Jones, W. G. Rakitsky, K. A. Smith, and R. D. Minard, *Fuel* **59**, 699 (1980).

[127] N. C. Deno, B. A. Greigger, A. D. Jones, W. G. Rakitsky, D. D. Whitehurst, and T. O. Mitchell, *Fuel* **59**, 701 (1980).

was recovered. The organic acid fraction was found to consist of benzenecarboxylic acids (58.5%), polynuclear aromatic acids (1.8%), phenolic acids (4.2%), heterocyclic acids (9.7%), aliphatic acids (17.9%), and other acids (7.9%). These results show that aromatic moieties in the polymer and coals can survive under mild oxidation conditions. If the oxidation were carried out at higher temperatures ($\geq 50°C$) however, the aromatic moieties would be oxidized. In addition, such higher temperature conditions may result in further oxidation of any long methylene chain or alicyclic compounds produced initially from the oxidation of coal or kerogen. As discussed in Section V, this is one of the examples showing that the optimum condition for oxidation of a simple model compound may not be directly applicable to the oxidation of a polymer or coal. Recently, Liotta and Hoff[128] have found that no ring degradation occurs in the H_2O_2–TFA–H_2SO_4 oxidation of nitrogen and sulfur heterocyclics. Venier *et al.*[128a] have reported that aromatic rings directly attached to sulfur can survive H_2O_2–TFA oxidation at 60°C for 5 hr; for example, Ph—SH or Ph—S—S—Ph yielded Ph—SO_3H (92–94%), Ph—S—Ph and Ph—S—CH_2—Ph were oxidized to Ph—SO_2—Ph and Ph—SO_2—CH_2Ph in excellent yield, respectively. This procedure has also been applied to the determination of the nature and amounts of organic sulfur functional groups in coal.

H. SODIUM HYPOCHLORITE*

Chakrabartty *et al.*[129–132] have studied the aqueous sodium hypochlorite oxidation of coals and model compounds in detail, and have concluded that this oxidation is a highly selective procedure which discriminates between aromatic (sp^2) and aliphatic (sp^3) carbon and oxidizes only the aliphatic carbon. The hypohalite oxidation process is well known as the haloform reaction. In general, the substrate stirred in a nitrogen atmosphere at 60°–70°C and 1.5–1.6 *M* sodium hypochlorite solution was then added in small portions over a period of 2–3 hr until no more oxidant was consumed. The reactivity of different coals was found to be related to the amount of oxygen in the coal. Low rank coal has a higher percentage of oxygen and

* For background information, see Chakrabartty.[128b]

[128] R. Liotta and W. S. Hoff, *J. Org. Chem.* **45**, 2287 (1980).

[128a] C. G. Venier, T. G. Squires, Y. Y. Chen, J. C. Shei, R. M. Metzler, and B. F. Smith, Preprint Vol. 26 (No. 1), p. 20 (Div. of Fuel Chem. Am. Chem. Soc.) Atlanta, March, 1981.

[128b] S. K. Chakrabartty, *in* "Oxidation in Organic Chemistry" (W. H. Trahanovsky, ed.), Part C, p. 343. Academic Press, New York, 1978.

[129] S. K. Chakrabartty and H. O. Kretschmer, *Fuel* **51**, 160 (1972).

[130] S. K. Chakrabartty and H. O. Kretschmer, *J. Chem. Soc. Perkin 1*, 222 (1974).

[131] S. K. Chakrabartty and H. O. Kretschmer, *Fuel* **53**, 132 (1974).

[132] S. K. Chakrabartty and N. Berkowitz, *Fuel* **53**, 240 (1974).

reacted vigorously with hypochlorite, whereas bituminous coal had to be activated by nitration or acylation to cause it to react. Oxidation studies on a number of model compounds indicated that unactivated aromatic, hydroaromatic, alicyclic, and alkane systems were not oxidized by hypochlorite unless an activating group permitted formation of a carbanion, or unless a tertiary sp^3 center was present in alicyclic systems. Whereas cyclohexane was not reactive with sodium hypochlorite, cyclohexanecarboxylic acid and cyclohexanone could be oxidized to dicarboxylic acids as well as CO_2. Hypochlorite oxidation of adamantane-1-carboxylic acid gave extensively chlorinated products and carbon dioxide possibly through the pathway shown below.[129] Heteroaromatic ring systems such as pyridine and quinoline

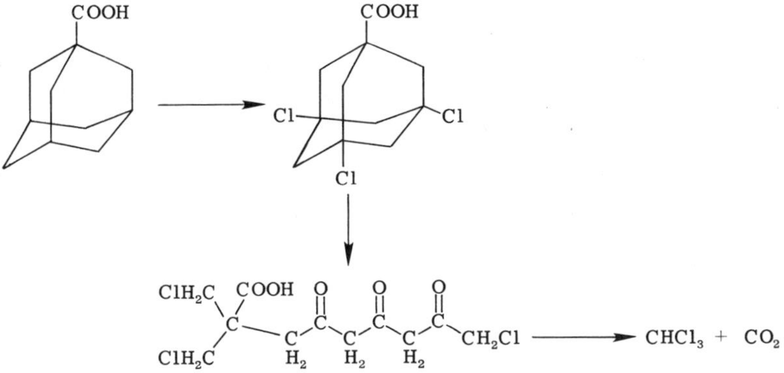

(15)

were readily cleaved; for example, carbon dioxide and ammonia were obtained from pyridine and quinoline gave isophthalic acid, dichloro- and trichloroquinoline-2-one. However, 2-methylpyridine and 2,6-dimethylpyridine yielded fair amounts of pyridine-2-carboxylic acid and 6-methylpyridine-2-carboxylic acid, respectively.

From the hypochlorite oxidation of coals containing 76 to 90% carbon, the estimation of sp^3 carbon abundance in the coals was based on the yield of carbon dioxide and the carboxyl acidity of the products. Major organic acid products were benzenepolycarboxylic acids; short-chain aliphatic acids were found as minor constituents. If the contribution of heteroaromatic structures were subtracted, the estimate of sp^3 carbon would be 55–60% of the total carbon.

All of these results and observations led to the conclusion that the greater part of the molecular skeleton in coal structure is made up of a bridged tricycloalkane configuration analogous to a variously extended polyamantane (for structural information see Fig. 8, Chakrabartty *et al.*[132] and Berkowitz[11]).

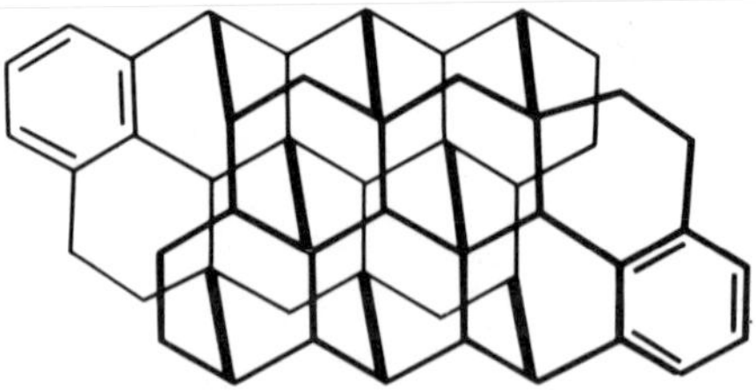

FIG. 8. Regular polyamantane structure from Berkowitz.[11]

However, this work and conclusion have been severely criticized by many other investigators. Whereas no evidence for aromatic rings other than benzene derivatives was found in the aqueous sodium hypochlorite oxidation products, Hayatsu *et al.*[43] demonstrated in an aqueous $Na_2Cr_2O_7$ oxidation study the presence of polynuclear aromatic ring systems in coal and found no support for the proposed tricycloalkane or polyamantane structure. Chakrabartty and Berkowitz[133] then claimed that the aqueous $Na_2Cr_2O_7$ oxidation at elevated temperature may result in pyrolytic cleavage and major structural rearrangements, and that about 65% of the total carbon in coal exists in the sp^3 state based on a calculation using the results of the oxidation by Hayatsu *et al.*[43] However, these claims were immediately replied to by Hayatsu *et al.*[134] and rejected for the following reasons. Detailed pyrolysis studies of the same coal used for the oxidation preclude the possibility that major structural rearrangements and pyrolytic formation of polynuclear aromatics occur during aqueous $Na_2Cr_2O_7$ oxidation, and since the aromatic units obtained from the oxidation do not represent the total aromatic content of the coal, the results are consistent with a structure that is predominantly aromatic (see Section IV, B).

These findings have been further confirmed by their later work.[19,28,38] The same group has oxidized 2,6-dimethylnaphthalene, solvent-refined coal, and meteoritic polymer under the conditions described by Chakrabartty *et al.*[130–132] The two latter materials have highly condensed aromatic structures.[38,42,63] No naphthalene or other polynuclear aromatic acids were detected in the oxidation products of these samples. Benzenecarboxylic acids were the major products. This apparently shows that polynuclear aromatic systems were destroyed by the hypochlorite oxidation. Landolt[135] showed that 1- and 2-naphthols, 1-hydroxyanthracene, and 2,3-naphthalenedicarboxylic acids were readily oxidized to benzenecarboxylic acids with aqueous

[133] S. K. Chakrabartty and N. Berkowitz, *Nature* (*London*) **261**, 76 (1976).
[134] R. Hayatsu, R. G. Scott, L. P. Moore and M. H. Studier, *Nature* (*London*) **261**, 77 (1976).
[135] R. G. Landolt, *Fuel* **54**, 299 (1975).

hypochlorite, under conditions analogous to those reported by Chakrabartty *et al.*[132] Cleavage of phenolic rings was also demonstrated. Mayo also reported[136] that 2-naphthol and 2-naphthoic acid reacted vigorously with aqueous hypochlorite at 65°C to give phthalic acid. Ghosh *et al.*[137] concluded from the oxidation study of naphthol–formaldehyde polymer and fluorescein that hypochlorite oxidation cannot always differentiate between sp^2 and sp^3 carbon. Although Chakrabartty *et al.*[138] addressed these disagreements, their arguments are not very convincing. Landolt's group[139] further studied the oxidation of naphthalene and its derivatives and showed that the efficiency of naphthalene oxidation by hypochlorite is markedly pH-dependent; that is, naphthalene is most effectively oxidized to phthalic acid at pH 8.5–9, and at pH > 11 yields of phthalic acid are significantly reduced. It was also observed that whereas 2-methylnaphthalene exhibits a reactivity to hypochlorite similar to that of naphthalene, 2-methyl-1-nitronaphthalene has a substantially reduced susceptibility to the oxidation. A similar observation was made by Hayatsu *et al.*[42] on the basis of hypochlorite oxidation of nitrated 2,6-dimethylnaphthalene; only 26% of the substrate was converted to 4-methylphthalic acid, 1,2,4-benzene tricarboxylic acid, and their nitro derivatives. These results and observations are not in agreement with those of Chakrabartty *et al.*[129–131] who reported that the hypochlorite is ineffective in oxidizing coal until the aromatic rings are activated by nitration.

Recently, Mayo and Kirshen[140] reported on an in-depth study of the effects of pH on the yields of products for a mild 30°C sodium hypochlorite oxidation of coals. The majority of the experiments were performed on an Illinois No. 6 bituminous coal which had been extracted with pyridine. They obtained both humic acid-like material that they called black acids, and water-soluble low molecular weight acids. The ratio of these two types of acids was found to be dependent on both pH and coal particle size. At a high pH (13) most of the coal was converted to the black acids. At lower pH, the benzenepolycarboxylic acids or aliphatic acids and CO_2 were the main products. They observed also that at the higher pHs, although the aromaticity of the humic acids was less than that of the starting coal, they still contained more aromatic than aliphatic carbon.

From these data they concluded that two mechanisms are responsible for the action of sodium hypochlorite on coals. At the higher pH levels, the enolizable compounds and some others are attacked and this is called the

[136] F. R. Mayo, *Fuel* **54**, 273 (1975).
[137] G. Ghosh, A. Banerjee, and B. K. Mazumdar, *Fuel* **54**, 294 (1975).
[138] S. K. Chakrabartty and N. Berkowitz, *Fuel* **55**, 362 (1976).
[139] J. L. Angert, S. L. Gratton, M. T. Reilly, and R. G. Landolt *Fuel* **56**, 224 (1977).
[140] F. R. Mayo and N. A. Kirshen, *Fuel* **58**, 698 (1979).

"high-pH mechanism." At lower pH levels, an unselective reaction occurs which will attack most organic groups ("low-pH mechanism"). They used these two mechanisms as a possible explanation of why nitration activates the coal. Initially, some groups activated by nitration are oxidized which lowers the pH so that the low pH mechanisms will operate and degrade the coal. In addition, they suggested that nitration may increase the surface area or porosity of the coal and result in an increase in the rate of oxidation.

Unlike most oxidants, the hypochlorite at high pH preserved the aliphatic portion of the coal, and Mayo and Kirshen believed that this humic-like material would be interesting to study. They suggested that because the aliphatic H/C ratio is so low for the black acids, the coal must have a large percentage of condensed aliphatic rings in its structure. While this conclusion supports the original conclusion of Chakrabartty *et al.*[132] and Berkowitz[11] that bituminous coal contains many polyamantanes, Mayo and Kirshen pointed out that these ring structures would most likely be highly irregular, rather than regular as is the case for polyamantanes. Perhaps, as a result of the further investigations needed in this area, these conclusions can be reconciled with the oxidation results obtained with other oxidants.

I. NITRIC ACID*

Since Friswill's nitric acid oxidation study of coal in 1891, many workers have extensively investigated this reaction, as well as the oxygen or air–alkali reactions (reviewed by van Krevelen[1] and Bearse[57]) in an effort to elucidate not only coal structure but also possible reaction routes for the commercial production of chemicals. The major products resulting from the nitric acid oxidation of coal are generally benzenecarboxylic acids, short-chain aliphatic acids and humic acid-like materials. Their yields depend on the substrate and the oxidation conditions.

Since coal as a substrate possesses aromatic ring systems, nitric acid acts not only as an oxidant but also as a nitrating agent. Indeed, the contamination of the oxidation products by nitrogen-containing compounds (mainly nitro compounds) is one of the major disadvantages of this reaction. Although in past studies the presence of such nitrogen compounds in the oxidation products of coals has been recognized, definite identification of individual nitrogen compounds has not been made except in a few instances. For example, picric acid was produced (11 wt %) besides CO_2 (20%), water-soluble acids (22%), and water-insoluble acids (26%) from the 40% HNO_3 oxidation (90°C, 40 min, then refluxing for 10 hr) of coal. The yields of

* For background information, see Ogata.[141]

[141] Y. Ogata, *in* "Oxidation in Organic Chemistry" (W. S. Trahanovsky, ed.), Part C, p. 296. Academic Press, New York, 1978.

picric acid, CO_2, and water-soluble acids increased with reaction time and acid concentration.[142] Grosskinsky[143] also isolated picric acid in high yield (9–20%) together with aliphatic acids and benzenecarboxylic acids from the nitric acid oxidation of coal. However, considering the nature of the nitric acid oxidation and the rather high yields of picric acid reported, its identification and quantitative determination may be viewed with some suspicion.

Aqueous nitric acid oxidations on model compounds have been studied in an attempt to understand the action of nitric acid on material such as lignin. Sobolev[144] found that 4-methylguaiacol (**22**), a lignin model compound, is easily oxidized with 2.6 *N* HNO_3 at 70°C for 2 hr to form oxalic acid, other organic soluble products, and gaseous products such as N_2O, NO, NO_2, CO, and CO_2. He concluded that the reaction gives rise to the following sequence of intermediates:

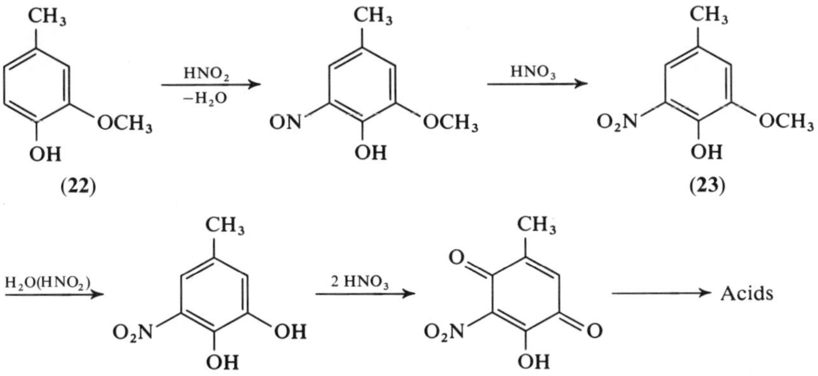

(16)

Nitrosonium (NO^+) ion mechanisms for the nitration of (**22**) and demethylation of (**23**) are proposed to account for the observed catalysis by nitrous acid. As Dense[145] has reviewed in detail, nitric acid cleaves alkyl aryl ethers by a solvolytic reaction that results in the liberation of phenolic hydroxyl groups. Such phenols are highly susceptible to oxidative degradation by nitric acid. Indeed, Yokokawa *et al.*[78] have shown that 1 *N* HNO_3 oxidations of mono- and dihydric phenols (refluxed for 5 hr) produce up to 44% and 22% of their carbon content as carbon dioxide and oxalic acid, respectively.

Hayatsu *et al.*[34,42] have recently studied 15% nitric acid oxidation of Wyoming lignite (C, 66.4% maf), high volatile Illinois No. 2 bituminous

[142] K. Ota and L. Hirao, Kyushu Inst. of Tech. Res. Rep. No. 15, p. 49, 1965.
[143] O. Grosskinsky, *Glueckauf* **88**, 376 (1952).
[144] I. Sobolev, *J. Org. Chem.* **26**, 5080 (1961).
[145] C. W. Dense, *in* "Lignin" (K. V. Sarkanen and C. H. Ludwig, eds.), Chapter 10. Wiley (Interscience), New York, 1971.

(C, 73.9% maf) and Pennsylvania anthracite (C, 91.0% maf) coals. Originally this procedure was found to selectively oxidize dialkylbenzenes to alkyl benzoic acids by Ferguson and Wims.[146] The preferential oxidation of alkyl groups increases with increasing alkyl chain length except for the *tert*-butyl group; e.g., the oxidation of *p-n*-propylethylbenzene results in *p*-ethylbenzoic acid.

As a general procedure, 3.5 g of each coal sample, which had already been extracted with solvents to remove trapped organic material, was oxidized with 15% HNO_3 (100 ml) under refluxing and stirring for 12 hr, and then with an additional 15% HNO_3 (40 ml) under the same conditions for 10 hr. Table IX contains a summary of the oxidation products. The gas chromatograms of the organic acid fractions (as methyl esters) from the lignite and bituminous coals are shown in Figs. 9a and 9b. The numbered peaks are identified in Table X. As shown in these tables and figures, benzenecarboxylic acids were the most abundant among the aromatic acids identified. Nitrobenzene di-, tri- and tetracarboxylic acids were also identified. Solid

TABLE IX

SUMMARY OF 15% HNO_3 OXIDATIONS OF WYOMING LIGNITE, ILLINOIS BITUMINOUS, AND PENNSYLVANIA ANTHRACITE COALS

	Lignite	Bituminous	Anthracite
		Weight %	
Product[a]			
Organic acids	58.2	74.5	—
Humic acid-like materials	38.9	28.0	1.5
Nonoxidized coals	16.6	18.3	98.0
Organic acids[b]			
Benzenecarboxylic	71.3	58.5	—
Nitrobenzenecarboxylic	7.3	16.7	—
Heterocyclic	3.6	11.0	—
Aliphatic[c]	11.4	7.5	—
Others[d]	6.4	6.3	—

[a] Wt % was obtained from the sample on a dry ash-free basis.

[b] Wt % was obtained from the calculation based on the GC, solid probe MS and HRMS data.

[c] Aliphatic dicarboxylic acids (ranging from C_4 to C_{11}) were produced in high yield when the lignite coal was oxidized with 15% HNO_3 at 80°C for 14 hr.

[d] Polynuclear aromatic carboxylic acids and unidentified compounds.

[146] L. N. Ferguson and A. L. Wims, *J. Org. Chem.* **25**, 668 (1960).

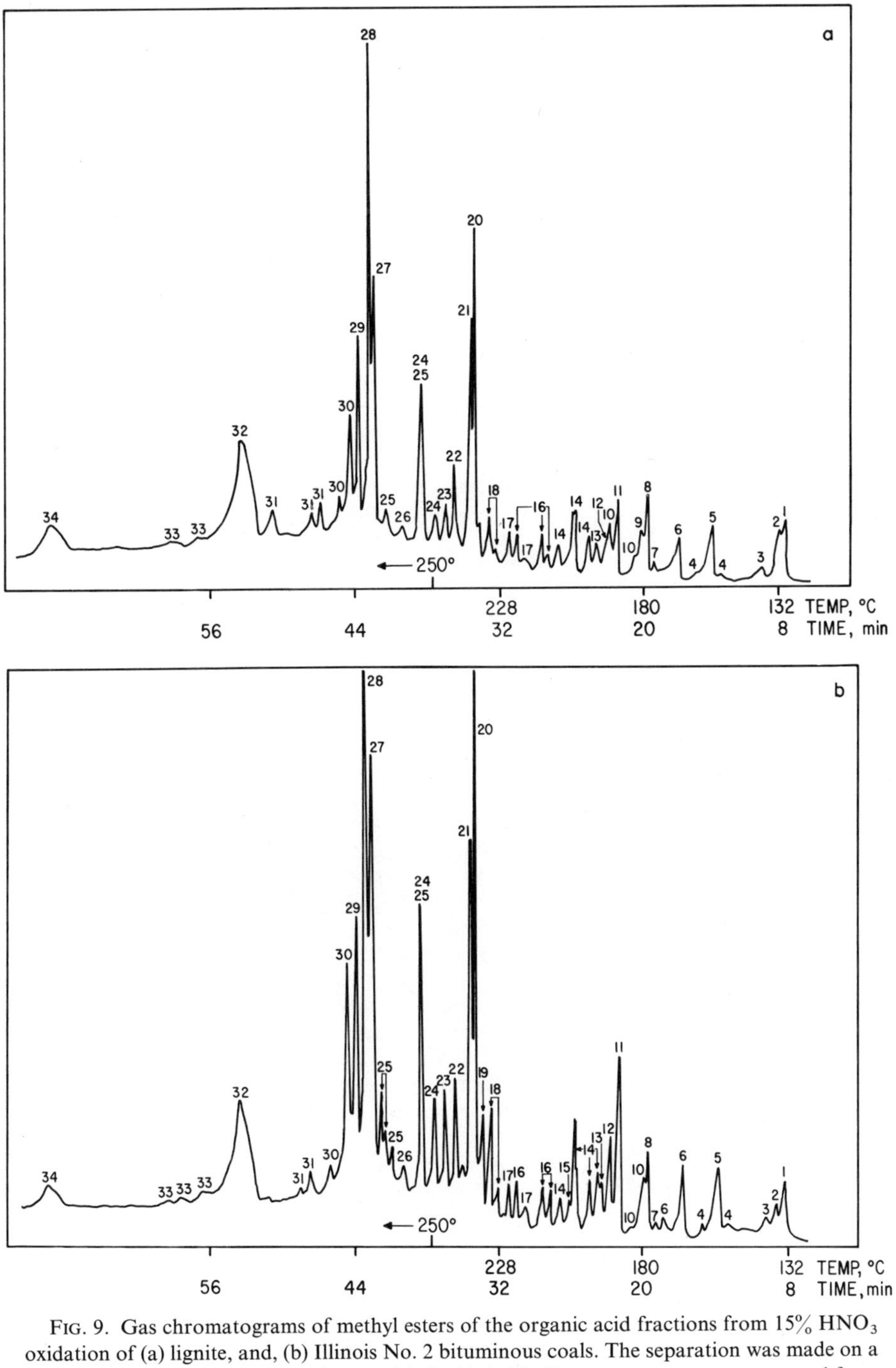

FIG. 9. Gas chromatograms of methyl esters of the organic acid fractions from 15% HNO_3 oxidation of (a) lignite, and, (b) Illinois No. 2 bituminous coals. The separation was made on a 15 m × 0.5 mm OV-17 SCOT column, with 6 ml/min He flow, temperature programmed from 100°–250°C at 4°C/min (Peaks identified in Table X).

probe TOFMS and HRMS showed the presence of nitrobenzenepentacarboxylic acid as its methyl ester, which was not analyzed by GCMS because of its low volatility. It is interesting to note that 15% HNO_3 oxidations of the coals produced several furancarboxylic acids (peaks 7, 17, 23, and 26), thiophenetricarboxylic acid (peak 19) and pyridinedicarboxylic acid (peak 15). Since these compounds are also found in the products of other

TABLE X

ORGANIC ACIDS FROM 15% HNO_3 OXIDATIONS IDENTIFIED AS METHYL ESTERS BY GCMS AND HRMS

GC peak	Compound
1	Succinic acid
2	Methylsuccinic acid
3	Benzoic acid and glutaric acid
4	Methylbenzoic acid
5	Adipic acid
6	Pimelic acid
7	Methylfurandicarboxylic acid
8	Suberic acid
9	Propane-1,2,3-tricarboxylic acid
10	Methylsuberic acid
11	1,2-Benzenedicarboxylic acid
12	1,4-Benzenedicarboxylic acid
13	1,3-Benzenedicarboxylic acid
14	Methylbenzenedicarboxylic acid
15	Pyridinedicarboxylic acid
16	Dimethylbenzenedicarboxylic acid
17	Dimethylfurandicarboxylic acid
18	Nitrobenzenedicarboxylic acid
19	Thiophenetricarboxylic acid
20	1,2,4-Benzenetricarboxylic acid
21	1,2,3-Benzenetricarboxylic acid
22	1,3,5-Benzenetricarboxylic acid
23	Methylfurantricarboxylic acid
24	Methylbenzenetricarboxylic acid
25	Nitrobenzenetricarboxylic acid
26	Furantetracarboxylic acid
27	1,2,4,5-Benzenetetracarboxylic acid
28	1,2,3,4-Benzenetetracarboxylic acid
29	1,2,3,5-Benzenetetracarboxylic acid
30	Methylbenzenetetracarboxylic acid
31	Nitrobenzenetetracarboxylic acid
32	Benzenepentacarboxylic acid
33	Methylbenzenepentacarboxylic acid
34	Benzenehexacarboxylic acid

oxidative degradation reactions with coals (e.g. H_2O_2–CH_3COOH or aqueous $Na_2Cr_2O_7$), they are thought to be indigenous to the coals. Little or no nitrophenols were detected in the oxidation products. It was found that while 70% HNO_3 oxidized demineralized Pennsylvania anthracite coal (refluxing for 22 hr, 39% of the coal was oxidized; in 68 hr, 100% of the coal was oxidized) very little or no oxidation was observed for 15% HNO_3 (refluxing for 22 to 68 hr).[42,46]

It is interesting to note that aliphatic-rich macromolecules in coal are oxidized by concentrated HNO_3 without extensive degradation. For example, 70% HNO_3 oxidation of cannel-like coal, which has a high content of exinite, yields large amounts of aliphatic unbranched dicarboxylic acids. Recently, Hayatsu *et al.*[92] have oxidized methylated cannel-like high volatile A bituminous coal (Ohio PSOC 297; C, 80.5 maf; exinite content is about 19.2 vol %) with 70% HNO_3 at 110°C for 48 hr. Organic acids (31 wt % of the substrate) and humic acid-like material (58 wt %) were isolated, and approximately 47% of the organic acids were found to consist of aliphatic unbranched dicarboxylic acids ranging from C_4 to C_{13}. Perhaps these aliphatic dicarboxylic acids are derived mainly from the exinite which has a high H/C ratio and a highly cross-linked aliphatic structure. A similar observation was made by the same group when a petroleum residue from a catalytic hydrocracking of petroleum was oxidized with 70% HNO_3 at 110°C for 20 hr. The residue has a high hydrogen content (C, 85.0% H, 11.0%) and a low aromaticity with the fraction of aromatic carbon (f_a) equal to 0.32 (f_a in coals varies from 0.6 to 0.9). Over 69 wt % of organic acids obtained from the oxidation (yields of the oxidation products: organic acids, 48 wt % and high molecular weight acids, 53%) were identified as aliphatic and alicyclic acids such as $C_nH_{2n+1}COOH$ ($n = 13$–19), $C_nH_{2n}(COOH)_2$ ($n = 2$–8), $C_nH_{2n-1}(COOH)_3$ ($n = 3$–6), and $C_nH_{2n-2}(COOH)_2$ ($n = 6$–7). Recently Deno *et al.*[146a] have oxidized six coals with 40% HNO_3 at 60°C for 5 hr, and have found a homologous series of aliphatic dicarboxylic acids ranging from C_4 to C_9 or C_{10} (one of the coals, Utah Cannel coal, produced up to C_{23}). However, yields of these acids were low except for Utah cannel coal.

Under mild conditions (15% HNO_3 80°C for 2 hr) lignite coal produces nitrohumic acids in 84% yield.[147,148] Their molecular weights range from 900 to 1500, and one of the molecular formulas is $C_{40}H_{44}O_{10}(COOH)_5$-$(OH)_6(CO)_4(NO_2)_3(NOH)$. These nitrohumic acids are of interest as

[146a] N. C. Deno, K. W. Curry, A. D. Jones, K. R. Keegan, W. G. Rakitsky, C. A. Richter, and R. D. Minard, *Fuel* **60**, 210 (1981).

[147] K. Higuchi, T. Arai, M. Tsuyuguchi, and K. Asakawa, *J. Fuel Soc. Jpn.* **35**, 404 (1956).

[148] K. Higuchi, T. Arai, M. Tsuyuguchi, and K. Asakawa, *J. Fuel Soc. Jpn.* **35**, 640 (1956).

possible soil conditioners and organic fertilizers. Bituminous coal (C, 82.5%) also yielded humic acid containing 64% carbon upon refluxing with 1 *N* HNO_3. After 5 hr refluxing, a maximum yield (98% of coal on maf basis) of humic acid was obtained.[149] The equivalent weight of the humic acid was 190 by conductometric titration. The percentages of total oxygen (26%) present in different functional groups in the humic acid are carboxylic (50%), phenolic (7%), carbonyl (4%), and ether (39%).

Micheel *et al.*[150] identified 28 interesting compounds, including polynuclear aromatic hydrocarbons, phenol, biphenyl ether, xanthene. aniline, alkylpyridines, alkylquinolines, and benzoquinolines from oxidation of high volatile bituminous coal with HNO_3–air at 250°–300°C. Copper salts of the oxidation products were then decarboxylated at 250°–300°C in water and in polyethylene glycol monomethyl ether. Finally, zinc dust distillation was carried out to remove all oxygen groups from the decarboxylation products. The compounds obtained were separated by column and vapor-phase chromatography and were analyzed by IR and UV. However, there are several significant factors that represent potential problems with the identification of these compounds: (1) In general, thermal degradation of coal begins at temperatures over 250°C; (2) in particular, zinc dust distillation requires a very high temperature and, therefore, the reaction often leads to excessive and extensive degradations and rearrangements; (3) zinc dust distillation usually gives a very low yield; (4) modern analytical techniques such as GCMS and HRMS were not used for the identifications. Therefore, it may be necessary to reconsider whether or not the compounds identified are really indigenous to the coal oxidized.

As discussed by van Krevelen,[1] the yield of benzenehexacarboxylic acid (mellitic acid) from the nitric acid oxidation of coal or coal products may serve as a criterion for measuring the degree of condensation of aromatic nuclei that make up the coal or coal products. Indeed, on the basis of 70% HNO_3 oxidation studies of various rank coals, solvent-refined coal, char, and a polymer obtained from a Fischer–Tropsch type reaction (see Fig. 10), Hayatsu *et al.*[42] and Studier *et al.*[46] have shown that the total acid yields and the number of carboxylic acid groups per benzene ring seem to be correlated with the degree of condensation of the original material.

J. MOLECULAR OXYGEN

In 1920 Fischer and Schrader showed that coal is oxidized to soluble acidic materials by oxygen-alkali at 200°–250°C. This procedure was studied

[149] Y. Takegami, C. Yokokawa, S. Kajiyama, Y. Tsuji, and A. Nishigori, *Kogyo Kagaku Zasshi* **60**, 1013; C. A. **53**, 10704C (1957).

[150] F. Micheel, J. Bernsmann, G. Draeger, and J. P. Riebe, *Brennst. Chem.* **45**, 360 (1964).

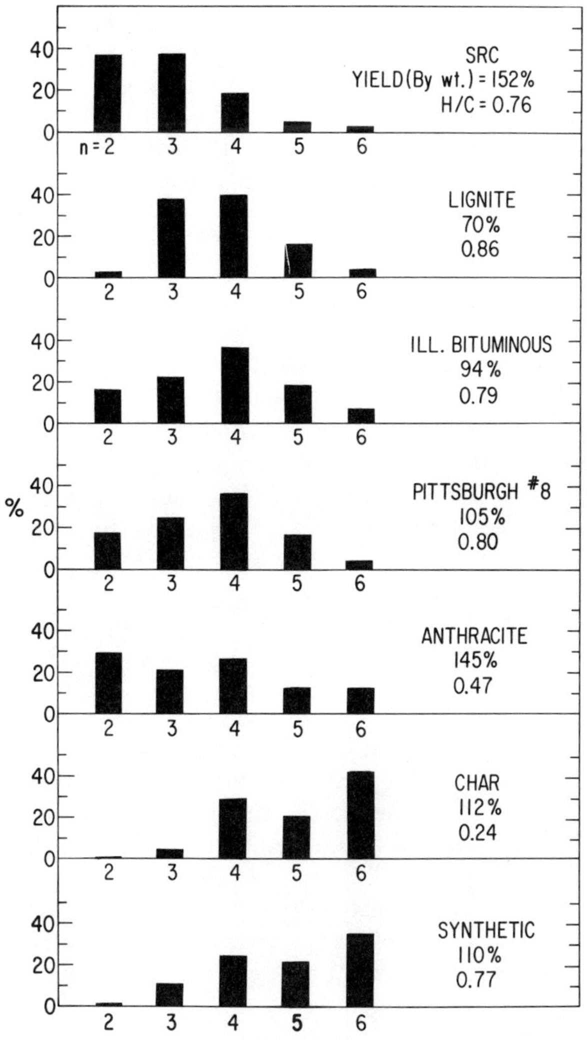

FIG. 10. Abundances (%) of benzenecarboxylic acids, determined as their methyl esters, produced by 70% nitric acid oxidation; n = number of ($COOCH_3$) groups per benzene ring.

extensively by several groups during the 1950s and 1960s not only for its application to the structural characterization of coal but also for the possibility of finding new methods of production for commercially useful chemicals (reviewed by van Krevelen,[1] Bearse *et al.*,[57] and Kamiya;[151] oxidative degradation of polymers with molecular oxygen has also been reviewed by Kamiya *et al.*[152]).

Generally, the coal to be oxidized is suspended in aqueous alkaline solution (Na_2CO_3, K_2CO_3, NaOH, or KOH) and oxidized with oxygen or air under pressure at 200° to 300°C. The conditions used varied with the coal.

At an early stage of the oxidation, initially formed humic acid-like materials are then further oxidized to water-soluble or organic solvent-soluble acidic materials in which benzenecarboxylic acids are found as the major products. Identification of some individual compounds in the oxidation products was achieved by Entel,[153,154] Holly *et al.*,[155,156] and Montgomery *et al.*[157–160] Entel[154] subjected the aromatic acids produced to decarboxylation and identified the following compounds in the products: benzene, naphthalene, biphenyl, biphenyleneoxide, phenanthrene, phenol, benzophenone, fluorenone, dibenzopyrone, pyridine, quinoline, α-, β-, γ-phenylpyridines, and β-naphthoquinoline. Holly *et al.*[155,156] investigated the acidic fraction obtained by the oxidation of an aqueous alkaline suspension of Pocahontas bituminous coal with oxygen at 270°C and 900 psig. Approximately 50% of the carbon of the coal could be converted into water-soluble acids and the balance into CO_2. The acidic fraction was identified as methyl esters by mass spectrometry and infrared spectrometry. A total of 39% of the acids were either positively or tentatively identified. The most abundant acid identified, was 1,2,4-benzenetricarboxylic acid which constituted 21% of the total. The three isomers of benzenedicarboxylic acid were also identified: 1,2-(11%), 1,3-(3.1%), and 1,4-(1.7%). Benzoic, benzenetetracarboxylic and naphthoic acids were found as minor products. The unidentified acids ($\approx$61%) were more highly condensed.

The same oxidation products were analyzed again by this group using a different method.[157–159] The acidic fraction was decarboxylated with anhydrous cupric sulfate and quinoline at a temperature below 365°C. The hydrocarbons resulting from the decarboxylation reaction were separated by fractional distillation and chromatography, and then identified primarily by mass spectrometry and infrared spectroscopy. Almost all of the aromatic rings found in the original acids obtained from the oxidation of the coal

---

[151] Y. Kamiya, *J. Fuel Soc. Jpn.* **58**, 148 (1979).

[152] Y. Kamiya and E. Niki, *in* "Aspects of Degradation and Stabilization of Polymers" (H. H. G. Jellinek, ed.), p. 79. Amer. Elsevier, New York, 1978.

[153] J. Entel, *J. Am. Chem. Soc.* **76**, 3646 (1954).

[154] J. Entel, *J. Am. Chem. Soc.* **77**, 611 (1955).

[155] E. D. Holly and R. S. Montgomery, *Fuel* **35**, 49 (1956).

[156] E. D. Holly, R. S. Montgomery, and R. S. Gohlke, *Fuel* **35**, 56 (1956).

[157] R. S. Montgomery, E. D. Holly, and R. S. Gohlke, *Fuel* **35**, 60 (1956).

[158] R. S. Montgomery and E. D. Holly, *Fuel* **36**, 63 (1957).

[159] R. S. Montgomery and E. D. Holly, *Fuel* **36**, 493 (1957).

[160] R. S. Montgomery and E. D. Holly, *Fuel* **37**, 181 (1958).

were found in this analysis, and about 23% of the original coal macromolecular material was represented by these identified aromatic rings.

Although many individual compounds were identified in the decarboxylation products, it was found that only eight make up more than 94% by weight of the total; they are: methylnaphthalene* (38%), benzene (35%), biphenyl (9.0%), naphthalene (4.0%), phenanthrene (3.3%), C_5-benzene (1.8%), benzophenone (1.7%), and toluene (1.5%). Terphenyl and phenylnaphthalene were also found in significant amounts.

However, we may have to be careful in evaluating these results because this work was done over 20 years ago and as these workers observed, the oxidation products consisted of a complex mixture. Therefore, the analysis of these products (either as methyl esters or hydrocarbons after decarboxylation) might be difficult using only direct inlet low resolution mass spectrometry and infrared spectroscopy. If these products were analyzed by modern techniques such as GCMS, HRMS, LCMS, etc., it seems likely that more detailed and reliable results would be obtained. Using model compounds such as naphthols, methylnaphthalene, naphthalenecarboxylic acids, anthracene, anthraquinones, phenanthrene, fluorenone, etc., Kamiya[151,161] has found that polynuclear aromatics are susceptible to the alkali–oxygen oxidation. Conversion of these model compounds into benzenecarboxylic acids was accomplished with oxygen in an alkaline medium (2.5 *N* NaOH or 3.7 *N* Na_2CO_3) at 240°–260°C for 1 hr. In particular, aromatic rings that were substituted with hydroxyl groups were found to be extensively degraded. However, Gartsman *et al.*[162] have identified naphthalene- naphthofuran- (or dibenzofuran-), fluorenone-, xanthone-, and anthraquinone- (or phenanthrene-9,10-dione-) carboxylic acids besides benzenecarboxylic acids in the oxidation products of coal (C, 86.27%; H, 5.68%). The coal was oxidized with oxygen in an aqueous alkaline medium at 260°–270°C for 1 hr. The products were analyzed as their methyl esters by GCMS. It is interesting to note that methoxybenzophenone- (mono-, di-, tri-) carboxylic acids were identified. This suggests the presence of diphenylmethane systems in the structure of the coal.

Many papers and reviews have discussed the determination of optimum conditions for the high temperature alkali–oxygen oxidation of coal. Conclusions can be drawn in three general areas.

* In a later paper, the same investigators pointed out that the presence of large amounts of this compound must be viewed with suspicion, because of an impurity in the quinoline used in the decarboxylation.[160]

[161] Y. Kamiya, *Fuel* **42**, 353 (1963).

[162] B. B. Gartsman, Z. A. Rumyantseva and N. N. Grishin, *Khim. Tverd. Topl.* (Solid Fuel Chem. Engl. transl.) **13**, 92 (1979).

1. Yield of Oxidation Products as a Function of Coal Rank

As shown in Fig. 11, the yield of water-soluble aromatic acids (benzenecarboxylic acids are major constituents) increases with increasing carbon content. In coals of lower carbon content, the oxidation results in a negligible amount of water-insoluble acids (humic acid-like material). This suggests that high molecular weight humic acids or humic acid-like materials formed as initial products in the early stages of the oxidation are finally oxidized to simple water-soluble acids. Low rank coals in general have high O/C ratios, which indicates that they contain more groups such as ethers, phenols, and carbonyls that are susceptible to the oxidation. The presence of these groups in the structure of coal promotes the rapid oxidative degradation of high molecular weight acids produced as intermediates. As an example, about 45% of the aromatic acids from the oxidation of Shikamachi bituminous coal (C, 87.3% maf) consist of benzenecarboxylic acids; this accounts for 26%

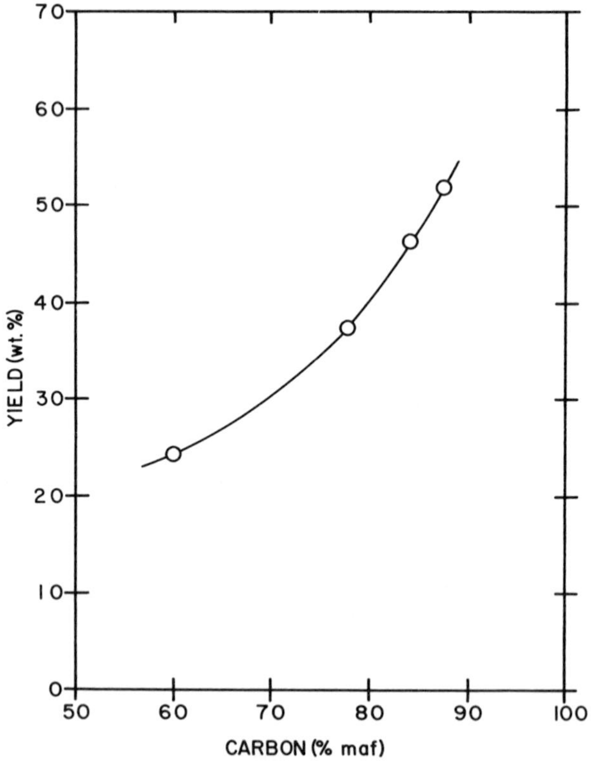

FIG. 11. Yields of water-soluble aromatic acids vs. carbon content of the coals. Prepared from Table I of Kamiya's review.[151]

of the coal. Benzenetri- (16.2%) and tetra- (18.3%) carboxylic acids were identified as major constituents.[163] In general, a tendency toward increasing yields of benzenepenta- and hexacarboxylic acids with increasing carbon content in coal is shown.

2. Temperature, Pressure, and Time

A temperature range of 260°–270°C appears to be the most suitable for the formation of water-soluble acids (the yield is about 50–60 wt % of the original coal). The yield of water-insoluble acids sharply decreases with increasing oxidation temperature; Shikamachi coal yields 25.3 wt % at 250°C and 3.9 wt % at 270°C.[151] In general, pressure and time are not dominant factors in the production of organic acids if the oxidation is run for 1–2 hr under partial oxygen or air pressure of approximately 4.6–8.0 megapascals (MPa). However, Pevzner *et al.*[164] have recently shown that the oxidation of low rank coals at 270°C in caustic soda solution for 2 hr, as compared with 1 hr, leads to a substantial decrease in the yields of benzene- and quinone-carboxylic acids.

3. Ratio of Alkali to Coal Sample

For a maximum yield of organic acids the alkaline concentration should be in the range of 2 to 3.5 *N*, with a ratio of alkali to coal of 1.5–3.5. Kamiya has shown the optimum conditions for the production of water soluble acids from Japanese bituminous coals using different alkali reagents (see Table XI).[151] In runs 3 and 4 (Table XI), to keep a high concentration of alkaline carbonate (e.g., $Na_2CO_3 + CO_2 + H_2O \rightleftarrows 2NaHCO_3$), carbon

TABLE XI

Effect of the Type of Alkali Used in Air Oxidation on Percentage Yields of Carboxylic Acids

Run no.	Ratio of alkali to coal	Water-soluble aromatic acids	Water-insoluble acids	Oxalic acid	Residue unoxidized coal
1	NaOH 3.4	61.0	6.5	12.1	1.2
2	Na_2CO_3 7.0	62.0	2.0	0.2	2.7
3	Na_2CO_3 3.0	60.8	3.8	0.2	3.1
4	K_2CO_3 3.0	65.8	2.1	2.8	1.4

[163] Y. Kamiya, *Fuel* **40**, 457 (1961).

[164] Z. I. Pevzner, Z. A. Rumyantseva, B. B. Gartsman, N. M. Karavaev, and B. I. Gizatova, *Khim. Tverd. Topl.* (Solid Fuel Chem. Engl. transl.) **12**, 25 (1978).

dioxide produced during the oxidation was continuously discharged from the reaction container. Nitrogen was found to be useful to expel the carbon dioxide. If such a technique is used, a ratio of alkaline carbonate/coal of over 2.0 gives a yield of over 60% for water-soluble aromatic acids.

Recently, Eisenberg and Solomon[165] have produced benzenepolycarboxylic acids by applying the alkali–oxygen oxidation to seven solid carbonaceous by-products (carbon contents 79.1–93.4% maf) obtained from coal gasification and liquefaction processes. The oxidation was carried out in 10% aqueous NaOH at 250°C and a total pressure of 8.27 MPa. Approximately 60% of the carbon in each of the original feed materials was oxidized to carbon dioxide. Yields of benzenecarboxylic acids in the organic extracts were 6.2–19.5%. For comparison, Illinois No. 6 bituminous coal was also oxidized using the same conditions. The yield for benzenecarboxylic acids was larger in the case of coal (23.6%). However, the percentage of highly substituted acids in the benzenecarboxylic acid components, i.e., benzene tetra-, penta-, and hexacarboxylic acids, was larger for the solid by-products than for coal itself. In the case of char obtained from COED process treatment of Illinois No. 6 feed coal, the percentage of higher acids was 89.5%, approximately 30% larger than the value for the feed coal. This indicates that the structures of these by-products exhibit a higher degree of aromatic ring condensation than their feed coals.

The high temperature oxidations we have just considered are interesting at the present time, primarily because they offer the high yields desirable for commercial production of chemicals. The following mainly low temperature oxidations, however, are much more likely to contribute to our understanding of oxidation mechanisms and coal structures. Some of them were carried out under nonalkaline conditions.

Ruschev *et al.*[166] have studied the kinetics of the oxidation of peat, lignites, and brown coal by molecular oxygen. They have shown that in the oxidation of peat, the maximum yield of the process is practically independent of the rate of sample stirring during the oxidation. For lignites, however, the maximum oxidation rates were clearly shown to be related to the rate of stirring. The oxidation rates increase with an increase in the stirring rate (from 300 to 630 rpm) by approximately a factor of 2. The oxidation temperature was also an important factor. For peat, in the 110°–150°C temperature range, the oxidation rate increased from 1.9 to 5.5 arbitrary units, for lignites from 1.0 to 2.4, and for brown coal from 0.5 to 1.7. In general, the rate of the oxidation processes of the low carbonification materials increases in this temperature range by approximately a factor of 3.

[165] W. C. Eisenberg and I. J. Solomon, *Fuel* **56**, 181 (1977).

[166] D. D. Ruschev, S. K. Ivanov, K. I. Markova, and L. D. Volkova, *Khim. Tverd Topl.* (Solid Fuel Chem Engl. transl.) **12**, 66 (1978).

Kucher *et al.*[167] have studied the kinetics of the formation of oxygen-containing groups and paramagnetic centers during the 150°C atmospheric oxygen oxidation of two coals and their vitrinite and fusinite concentrates. The primary stable products of the oxidation of medium rank coal (C, 83.0%) are peroxides. In lower rank coals (C, 78.5%), both peroxides and phenolic groups are observed. Phenol, carbonyl, and carboxy groups are formed in that order independently of the degree of coalification and the petrographic composition of the coal. Kinetic curves of accumulation of hydroxy, peroxide, carbonyl, and carboxy groups during the oxidation of coals and their vitrinites and fusinites are shown. It was found that the oxidation of vitrinites proceeds more rapidly than that of fusinites.

Mild oxidation of solvent extractable fractions of an Illinois No. 6 bituminous coal, unextractable residue, fractions of solvent refined coals, and model compounds has been investigated by Huntington *et al.*[168] Each sample was oxidized with oxygen at 50°C. Azobis-2-methylpropionitrile was used as an initiator, and in some experiments *tert*-butylhydroperoxide was added to accelerate oxidation, but this compound also appeared to act as an oxidizing and cleaving reagent. Rates of the oxidation, relative reactivities, and the oxidation products were determined and discussed in detail. Numerous coal model compounds were used by these authors in this investigation.[169]

Many of the oxidations were self-retarding, but refluxing with alcoholic KOH eliminates this retardation and effects considerable breakdown of the coal sample, even in the absence of oxygen.[168] The oxidation and base treatment of the pyridine-soluble fraction of Illinois No. 6 coal were shown in detail to result in more extensive degradation than occurred for the insoluble, extracted coal (corresponding to 85% of the original coal). The pyridine-soluble fraction (MW $\approx$ 1500) was broken down into fragments of average molecular weight of 700–800 by either refluxing with alcoholic KOH or gentle oxidation at 50°C or a combination of these, with only a moderate resultant increase in oxygen content. From the results of the cleavage reaction with alcoholic KOH (refluxing), the authors suggest that there may be a wide range of reactivities of ether linkages in Illinois No. 6 coal, and that a selective cleavage by alcohols and bases may be possible by choice of temperature.

However, it is well known that alkaline hydrolysis of coal macromolecules is ineffective at the low temperature used by the authors.[37,170–172] A similar observation has been made for lignin[41] and humic substances.[14] As discussed

[167] R. V. Kucher, L. F. Butuzova, V.A . Kompanets, L. V. Luk'yanenko, and N. M. Degtyar, *Khim. Tverd. Topl.* (Solid Fuel Chem. Engl. transl.) **12**, 17 (1978).

[168] J. G. Huntington, F. R. Mayo, and N. A. Kirshen, *Fuel* **58**, 24 (1979).

[169] J. G. Huntington, F. R. Mayo, and N. A. Kirshen, *Fuel* **58**, 31 (1979).

or reviewed by Eglinton *et al.*,[5] Djuricic *et al.*,[86] Schnitzer *et al.*,[24] Griffith *et al.*,[83] and Hayatsu *et al.*[102] (see also Section II, D), organic materials may be either loosely or tightly trapped in coal or related carbonaceous materials. Therefore, we may not rule out the possibility that the products yielded by the alcoholic KOH refluxing of the pyridine-soluble fraction or extracted coal might be derived from: (1) waxy material, (2) humic acid or humic acid-like material, or (3) material bonded to the coal macromolecules by labile linkages such as esters that can be readily hydrolyzed.

Oxidative ammoniation of coal is an unusual oxidation which produces neither aromatic acids nor humic acids. It has been known since the late 1920s that oxidation of coal with oxygen and ammonia at temperatures above 200°C yields black, infusible, and alkaline-insoluble materials which contain as much as 22–24% chemically combined nitrogen. This reaction has been considered as a possible route to coal-based nitrogen fertilizer.[11] According to Chakrabartty *et al.*,[173] the oxidatively ammoniated products are remarkably inert, being resistant to oxidation, reduction, deamination, acetylation, methylation, or pyrolysis. Only under severe hydrolytic conditions with HCl do they lose up to 80% of their nitrogen (as ammonia). These findings and results suggest that nitrogen in the product exists mainly in stabilized amidine or isoimidole configurations.

K. MISCELLANEOUS

1. PHOTOCHEMICAL

A number of photochemical cleavage reactions of polymers are known (Randy *et al.*[174]), but with few exceptions they do not appear to have been applied successfully to the degradation of bulk complex carbonaceous materials such as coals.

Lignin is degraded by oxygen in the presence of light and oxidative cleavage of methoxy groups has been noted particularly[175,176] (see also review by Chang and Allan[41]).

A British group[177] has reported photo-oxidation of coals in water, and found that no major structural changes occur. Recently, lignite, high volatile

---

[170] M. Makabe, Y. Hirano, and K. Ouchi, *Fuel* **57**, 289 (1978).

[171] K. Ouchi and M. Makabe, *J. Fuel Soc. Jpn.* **57**, 249 (1978).

[172] J. C. Pew and J. R. Withrow, *Fuel* **10**, 44 (1931).

[173] S. K. Chakrabartty and N. Berkowitz, *Fuel* **48**, 151 (1969).

[174] B. Randy and J. F. Rabek, "Photodegradation, Photooxidation and Photostabilization of Polymers." Wiley (Interscience), New York, 1975.

[175] G. F. Leary, *Tappi* **50**, 17 (1967).

[176] G. F. Leary, *Tappi* **51**, 257 (1968).

[177] R. Bent and J. K. Brown, Conf. on Sci. in the Use of Coal (Sheffield), p. A48. Inst. Fuel, London, 1958.

bituminous, and anthracite coals in 10% aqueous HCl solution have each been irradiated for 8 days with ultraviolet light from a high-pressure mercury lamp while air or oxygen was bubbled through the stirred HCl solution.[19,42] Almost no photo-oxidation was observed for anthracite, whereas lignite and bituminous coals were oxidized in yields of about 25 and 30%, respectively. About 35 to 40% of the oxidation products are high molecular weight acids which do not lend themselves to direct analytical identification. If the unreacted residue was further oxidized by the same procedure, 22–24% of the residue was degraded, again with the same distribution of products. Benzenecarboxylic acids are found to be most abundant among the oxidation products of bituminous coal. The products are: benzenecarboxylic acids (59.3 wt % of the oxidation products identified), chlorobenzenecarboxylic acids (17.1%), phenolic acids (<1.0%), polynuclear aromatic acids (2.8%), heterocyclic acids (11.3%), aliphatic acids (5.0%), and others (3.5%). The oxidation products of a lignite coal exhibit a distribution similar to that of the bituminous coal.[19,34]

Although this oxidation procedure destroys polynuclear aromatic rings such as naphthalene and phenanthrene, it is particularly attractive for the isolation and analysis of amphoteric nitrogen-containing acids produced (e.g., pyridinecarboxylic acids). This is due to the fact that nonvolatile water-soluble reagents, from which it is experimentally difficult to isolate the amphoteric acids, are not used. Model compounds including phenols, polynuclear aromatic hydrocarbons and heteroaromatics, and a poly(2-vinylpyridine)styrene polymer have been oxidized to elucidate the nature of the photo-oxidation.[63]

2. Perchloric Acid

Raj[178] has shown that perchloric acid oxidation of several coals at 50°C yields aromatic compounds which are related to lignin degradation products. Many of the oxidation products were identified as phenylpropanes (**24**) and phenylpropenes (**25**) with or without phenolic OH or OCH_3 groups. Vanillic

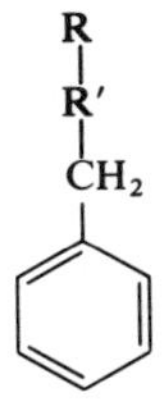

(**24**)

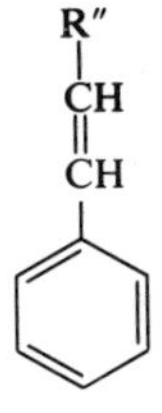

(**25**)

R = CH_2OH, CHO or COOH
R′ = CH_2, CO or C(H)OH

R″ = CH_2OH or COOH

and syringic alcohols, 2-methylcresol, and 2,5-dihydrobenzoic acid were also found (see Berkowitz[11]). These findings seem to be important evidence that many lignin structures and their related derivatives are present in the structure of the coal oxidized. However, more careful investigation may be needed in order to verify these results of perchloric acid oxidation. Since perchloric acid has been used rarely for the oxidation of organic material, little or no background information is available on its application to the oxidation of lignin compounds. The hot concentrated acid is often used, however, to destructively remove organic matter in geological materials. On the other hand, alkali salts of perchloric acid, either in acidic media or neutral, are well known as reagents for oxidation of phenols to quinones.[179,180] Therefore, the identification of mono-, di-, and trihydric phenols in the oxidation products may be open to question. Furthermore, as Dorofeenko *et al.*[181] and Fieser and Fieser[33] indicated in their reviews, perchloric acid is used widely as an acidic catalyst for many organic reactions. These facts suggest that the possibility of secondary reactions during the oxidation of coal with perchloric acid cannot be excluded.

3. Electrochemical

Only a few studies using electrochemical oxidation have been reported (see review by van Krevelen[1]). Using a copper anode, the electrolytic oxidation of coal suspended in 3 *N* NaOH solution at 80°C produces humic acid that is finally converted into polycarboxylic acids and CO_2 after 15 hr. The maximum yield of polycarboxylic acids was about 47% based on the coal carbon content.[182]

4. Ozone

The oxidation of coals and coal-derived humic acids with ozone has been studied by several workers (see review by van Krevelen[1]). However, identification of the products was ambiguous. Bitz *et al.*[183] oxidized a bituminous coal by an H_2O_2–O_3 combination technique and isolated acidic material

---

[178] S. Raj, Ph.D. Thesis, Penn State Univ., University Park, Pennsylvania, 1976.
[179] K. Schimmelschmidt, *Annalen der Chemie* **566**, 184 (1950).
[180] R. Kerber and W. Gestrich, *Chem. Ber.* **106**, 798 (1973).
[181] G. N. Dorofeenko, S. V. Krivun, V. I. Dulenko, and Yu. A. Zhdanov, *Russ. Chem. Rev.* (*Engl. Transl.*) **34**, 88 (1965).
[182] B. G. Gaberman and L. N. Yushkova, *Tr. Khim. Metall. Inst. Akad. Nauk Kaz. SSR* **5**, 3; *C. A.* **71**, 127089g (1969).
[183] M. C. Bitz and B. Nagy, *Anal. Chem.* **39**, 1310 (1967).

(131.6 mg/g of coal). Only benzene di-, tri, and tetracarboxylic acids and some aliphatic dicarboxylic acids were identified by preparative GC, IR, UV, and MS techniques. On the basis of these studies, the oxidation of coal with O_3 or O_3–H_2O_2 does not appear to be, at the present time, an attractive method for the structural study of coals.

5. Alkaline Hydrogen Peroxide

Yugoslavian lignite which contains ≈11% humic acid was oxidized with alkaline 1 *N* H_2O_2 solution at 20°C; the humic acid content increased from 11 to 42%. Increasing the oxidation temperature did not affect the yield of the humic acid substantially.[184]

V. Conclusion

Organic macromolecular material in coal is essentially a complex mixture of various types of rather simple compounds such as benzenes, phenols, naphthalenes, cycloalkanes, tetralins, furans, thiophenes, pyridines, etc. In the coal structures generally envisioned, these compounds are tied together by bridge linkages such as methylene chains, alicyclics, ethers, sulfides, disulfides, etc.

As we have already pointed out, if such a mixture of different type compounds is oxidized by only one procedure, while some of the compounds may indeed be oxidized selectively, others might be destroyed or might not be oxidized at all. Determining the carbon balance (carbon contents of oxidation products, nonoxidized residue, and CO_2) is an important factor in making quantitative determinations. However, such determinations cannot be made nor quantitative results obtained if only a single oxidation procedure is used. It would be very difficult for example, to estimate what percent of CO_2 formed during an oxidation was derived from aliphatics and what percent was derived from aromatics. Therefore, an oxidative degradation study of a coal should involve using a variety of oxidants and conditions, and comparing the nature and yield of the oxidation products to insure meaningful structural interpretation. In addition, in order to obtain background information, control experiments with a number of model compounds and polymers of known structure containing a variety of functional groups should be carried out under the same oxidation conditions used for the coal. However, it is necessary to keep in mind that even under the same conditions, model compounds in general are more easily oxidized than

[184] P. Petrovic and D. Vitorovic, *Agrohemija* **9**, **10**, 383 (1977).

polymers containing similar structures (e.g., 2-ethylnaphthalene and poly-2-vinylnaphthalene). It has been found[3] that whereas model compound oxidations produce stable, simple compounds, polymer or coal oxidations yield more complex products as well as simple compounds.

While there still are not enough data available to permit the comprehensive understanding necessary to provide detailed information concerning the origin, transformation, and structure of coal organic macromolecules, much work has been done, especially during the past decade, in developing the technique of oxidative degradation. The results of this work have led to a better understanding of the necessary precautions and limitations and also have contributed in the following ways to the characterization of the chemical structure of coal:

(1) Determination of the degree of condensation.
(2) Elucidation of aliphatic structures.
(3) Characterization of aromatic units.
(4) Determination of the chemical forms of heteroatoms such as nitrogen, oxygen, and sulfur.
(5) Provision of evidence for the presence of phenols and their polymeric materials.
(6) Indication that a coal may contain macromolecules of different chemical structures and perhaps origins.
(7) Improved insight into oxidizability and the kinetics of oxidation.

Although we have endeavored to make our review thorough and complete, we may not rule out the possibility that we have inadvertently overlooked some interesting papers. It is hoped that this chapter will stimulate work that will lead not only to reinvestigation of established oxidation procedures but also to finding new oxidation procedures. It is obvious that the oxidative degradation method is one of the most important research tools available and should continue to provide basic information that will be very useful in programs for coal gasification, liquefaction, and other utilization.

VI. Glossary

Cannel Coal

Cannel coal is a sapropelic coal characterized by a high concentration of fungal spores or pollen. Sapropelic coals were formed under anaerobic conditions from fine grained organic muds in shallow ponds or lakes and are unstratified.[10,11] Cannel coal derives its name from the word "candle" because the coal burns with a long and steady flame.

Humic Acid

Humic acids and their related materials are defined as those components of peat that are extracted by basic solvents but not acidic solvents. They are dark, amorphous organic macromolecular materials ranging in molecular weight from 2000 to 300,000. Widely distributed, they are found in soils, lignites and other sedimentary accumulations. *Fulvic acid* is a humic substance soluble in either acidic or basic solvents. (For more information, see Eglinton and Murphy,[5] Swain,[7] Greenland and Hayes[185] and Kukharenko.[186])

Humic Acid-like Materials

Humic acid-like materials are defined here as relatively high molecular weight materials (MW $\approx$ 600 $\sim$ several thousand) resulting from mild or partial oxidation of coal. They are usually only soluble in basic solvents. Some of them are also soluble in polar solvents such as methanol and probably consist of lower molecular weight materials (see Lowry[3] and Raj[187]).

Kerogen

Kerogen is the organic macromolecular constituent of sedimentary rocks which is insoluble in most common solvents. It is divided into three types. The first type of kerogen is rich in aliphatic structures with a high H/C ratio and a relatively low O/C ratio. Such kerogen was derived from autochthonous (local) aquatic organic matter. The second type of kerogen, with a low H/C ratio, is rich in polyaromatic nuclei and oxygen groups. This type was derived from higher plants and terrestrial humic material. The third type is an intermediate with a relatively high H/C ratio and a low O/C ratio, and was probably derived from both algal and higher plant materials (Eglinton and Murphy,[5] Tissot and Welte,[10] Philp *et al.*,[188] and van de Meent *et al.*[189]).

Lignin

Lignin consists of organic macromolecular materials formed from the polymerization of C_6–C_3 units. It is present in all woody tissue in amounts of

[185] D. J. Greenland and M. H. B. Hayes, eds., "The Chemistry of Soil Constituents." Wiley, New York, 1978.

[186] T. A. Kukharenko, *Khim. Tverd. Topl.* (Solid Fuel Chem. Engl. trans.) **13**, 1 (1979).

[187] S. Raj, Preprint 25 (No. 4), p. 58 (Div. of Fuel Chem. Am. Chem. Soc.), Las Vegas, August, 1980.

[188] R. P. Philp, M. Calvin, S. Brown, and E. Yang, *Chem. Geol.* **22**, 207 (1978).

[189] D. van de Meent, S. C. Brown, and R. P. Philp, *Geochim. Cosmochim. Acta* **44**, 999 (1980).

up to 35% or more and is divided into three broad classes: (1) softwood lignin (derived mainly from coniferyl alcohol); (2) hardwood lignin (derived from coniferyl and sinapyl alcohols); and (3) glass lignin (derived from *p*-coumaryl, coniferyl and sinapyl alcohols).

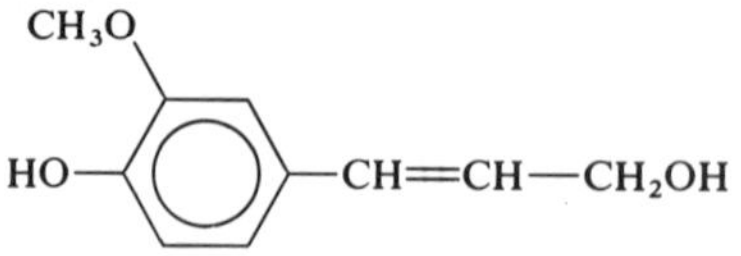

Coniferyl alcohol

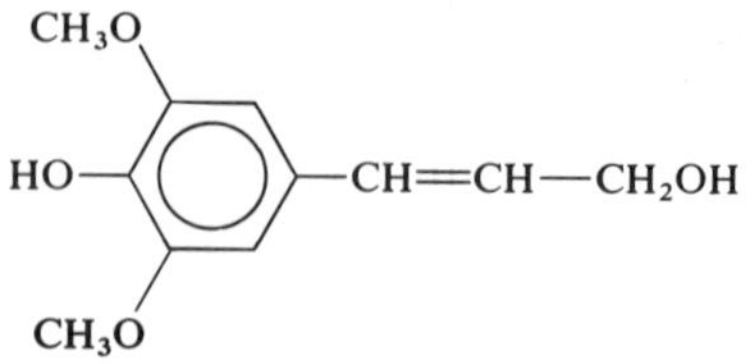

sinapyl alcohol

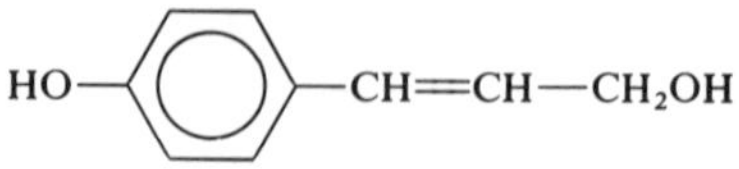

p-coumaryl alcohol

Recently Sakakibara[190] proposed a structural model for softwood lignin which has a formula of $C_{252}H_{222}O_{67}(OMe)_{26}$ and a molecular weight of 5124. More information may be obtained from Adler,[93] Freudenberg and Neish,[191] Sarkanen and Ludwig,[192] and Grimshaw.[193]

ACKNOWLEDGMENTS

We would like to express our thanks and appreciation to Drs. E. H. Appelman, R. C. Duty, G. R. Dyrkacz, Y. Kamiya, D. S. Newman, K. Ouchi, and L. M. Stock for carefully reading this chapter and offering us their helpful comments and suggestions. This work was performed under the auspices of the Office of Basic Energy Sciences, Division of Chemical Sciences, United States Department of Energy.

[190] A. Sakakibara, *Wood Sci. Technol.* **14**, 89 (1980).

[191] K. Freudenberg and A. C. Neish, "Constitution and Biosynthesis of Lignin." Springer-Verlag, Berlin and New York, 1968.

[192] K. V. Sarkanen and C. H. Ludwig, eds., "Lignin." Wiley (Interscience), New York, 1971.

[193] J. Grimshaw, *in* "Rodd's Chemistry of Carbon Compounds" (S. Coffey, ed.), Vol. 3, Part D, p. 203. Amer. Elsevier, 1976.

Index

A

Acetates, from olefins, 140–141
Alcohols
  from alkanes, 148, 190–192
  cyclic ethers from, 11–27
  fragmentation of, *see* Alcohols, oxidative cleavage of
  lead tetraacetate oxidation of, 2–27
  oxidative cleavage of, 5–27
  permanganate oxidation of, 150, 193–199
Aldehydes, 76, 77, 126
  from alcohols, 3–27
  from diols, 27–36
  formation of, 132
  permanganate oxidation of, 201
Aldimines, from amines, 89–98
Alkaloids, 56, 93–95, 208, 209
  (±)-*N*-acylnorlaudanosines, 228, 251–253
  Amaryllidacae, 261–262
  androcymbine, 242–243
  aporphines, 210–240, 250
  autumnaline, 242, 243
  biosynthesis of, 208
  boldine, 217
  (±)-bracteoline, 240
  bulbocapnine, 211
  cephlotaxine, 243–248
  (−)-coclaurine, 215
  codamine, 239, 240
  colchicine, 242, 243, 266, 267
  corydine, 216–218
  (±)-corytuberine, 226
  crinine, 257, 258
  crotonosine, 215
  (±)-cryptopleurine, 273
  demethoxy-*o*-demethylandrocymbine, 248–249
  deoxyharringtonine, 244
  dibenzazonine, 220–239
  dicentrine, 217, 218
  O,*N*-dimethylnorbelladine, 258, 259
  (±)-epigalanthamine, 258, 259
  erybidine, 233, 236, 237
  erythraline, 221, 222
  Erythrina, 221–224, 244–248, 250
  N-formylnorlaudanosine, 227–228
  galanthamine, 257–261
  glaucine, 217, 218, 228
  harringtonine, 243, 244
  homoaporphine, 241
  homoaporphines, 249–255, 257
  Homoerythrina, 241, 243, 244, 250
  homoharringtonine, 244
  (±)-homolaudanosine, 248, 249
  homomorphinandienones, 241, 257
  homoproaporphines, 241, 242, 253–255, 257
  homoproerythrinadienones, 253–255
  homoorientaline, 242
  homoreticuline, 243
  isoboldine, 211, 213
  isococlaurine, 215
  isocorytuberine, 216
  isoharringtonine, 243, 244
  isoquinoline, 209–240, 255–257
  isostegnane, 275, 276
  isothebaine, 216
  kreysiginone, 242
  laudanosine, 226, 227, 267
  laudanosoline, 212
  laureline, 214
  liriodenine, 209–210
  lycorine, 257, 258
  magnofluorine, 211
  maritidine, 261, 262
  mecambroline, 215
  (±)-*o*-methylandrocymbine, 257
  o-methylflavinantine, 226–229, 232, 233, 235
  O-methylnorbelladine, 257, 258
  (±)-*o*-methylthalisopravine, 229, 230
  N-mesylnorisobolding, 213
  N-mesylnorreticuline, 213
  morphinandienone, 267

morphine, 209–240, 248–252
norbelladine, 257, 258
norprotosinomenine, 217, 218
(+)-norprotosinomenine, 221, 222
norreticuline, 223
(±)-norwedine, 258–261
nuclferine, 216
orientaline, 216, 242
orientalinol, 216
orientalinone, 216, 242
(±)-oxocrinine, 263, 264
(±)-oxomaritidine, 264
phenethylisoquinoline, 241–257
(±)-predicentrene, 236, 237
proerythrinadienones, 217, 218
protosinomenine, 217
protostephanine, 211, 220, 221
pseudocorydine, 216
(±)-reticuline, 210, 211, 213, 217, 226
roemerine, 214
schelhammeridine, 243–248
(±)-steganacin, 274, 255
dl-tetradrine, 210
thalicarpine, 209, 210, 229–232
(±)-thalicmidine, 236, 237, 239, 240
N-trifluoracetylhomonorlaudanosine, 249, 250, 252
N-trifluoroacetyl-*o*-methylnorbelladine, 261
(±)-*N*-trifluoroacetylnorcodamine, 238, 239
(±)-*N*-trifluoroacetylwilsonirine, 239, 240
(±)-tylophorine, 272

Alkanes
lead tetraacetate oxidation of, 122–127
permanganate oxidation of, 148, 150, 190–192

Alkenes
from 1,2-dicarboxylic acids, 81–89
lead tetraacetate oxidation of, 131–140
permanganate oxidation of, 149, 168–181
strained, preparation of, 118

N-Alkoxyaziridines, from alkoxyamines, 91, 92

Alkylbenzenes, permanganate oxidation of, 100, 186–190

Alkynes
formation of, 142
permanganate oxidation of, 150, 182–186

Allenes, 133

Aluminum organometallics, *see* Organoaluminum compounds

Amides

Amides
from amines, 204–206
lead tetraacetate oxidation of, 98–101

Amines
lead tetraacetate oxidation of, 89–98
permanganate oxidation of, 204–206

α-Amino acids, lead tetraacetate, oxidation of, 76, 77

1,2-Amino alcohols, lead tetraacetate oxidation of, 35, 36

Ammonium salts, quaternary, 151, 153, 156–158, 160–162, 164–167, 169–171, 173–176, 179–181, 184–195, 197–205

Anthracite, 281

Aromatic hydrocarbons, lead tetraacetate oxidation of, 127–131

Arsines, 101

Arsinimines, 101

Arsonium salts, quaternary, 151, 156, 194

Azepines, 115

Azides, 138, 139
allyl, formation of, 139

α-Azido ketones, 139

Aziridines, 111–114, 117
from amines, 90–92
lead tetraacetate oxidation of, 94

1-*H*-Azirines, 116

2-*H*-Azirines, 116

Azo compounds, from hydrazines, 101–105

Azomethines, lead tetraacetate oxidation of, 118–122

B

Benzocyclobutadiene, 145

Benzophenones, 58

Benzyltetrahydroisoquinolines, oxidative coupling of, 209–240

Benzyne, from 1-aminobenzotriazole, 107–110

Biphenylenes, 109

Bituminous coal, 281

Boron organometallics, *see* Organoborane compounds

Bromine, 15
N-Bromosuccinimide, 17
with lead tetraacetate, 127
Brown coals, 306

Cupric oxide, 280, 283
Cupric oxide oxidation, of coal, 314–321
Cycloadditions, 85, 102–104, 107–117, 144, 145
Cyclobutadiene, 144, 145
Cyclobutadieneiron tricarbonyl, 144, 145

C

Cannel coal, definition of, 352
Carbamates, from amides, 98–101
Carboxylic acids
from alcohols, 193–199
from aldehydes, 201, 202
from alkanes, 150
from alkenes, 149, 150, 168–181
α-alkoxy acids, lead tetraacetate oxidation of, 75, 76
from alkynes, 182–186
from arenes, 150, 186–190
decarboxylation of, 61–78
from halides, 203, 204
α-hydroxy, lead tetraacetate oxidation of, 75
from oxidation of coals, 284–352
Ceric ammonium nitrate, 144
Ceric sulfate, 227
Chalcones, 58
Chromic acid, 280
Chromic acid oxidation, of coal, 292–304
Coals
high rank, 281
low rank, 281, 283, 284
methylation, 283–284
oxidation of, 279–354
oxidative ammoniation, 348
polyamantane structure of, 331–334
pretreatment of, before oxidation, 282–283
petrographic classification, 281
rank classification, 281
structure of, 351, 352
yield of oxidation products as a function of coal rank, 344, 345
Coalification, 281
Colchicine, 209
Copper salts, 16, 31, 62–66, 69–74, 76, 77, 88
Coumarins, 58
(±)-Cryptoplurine, 209
Cuprous chloride, 226

D

Dialdehydes, from the permanganate oxidation of alkenes, 178–181
Diazenes, 101, 105, 111, 112, 116, 118
Diaziridines, 105
Diazo compounds, 121
Dicarboxylic acids, lead tetraacetate oxidation of, 81–89
Dichromate oxidation, 284, 285
of coal, 292–304
Dienone–phenol rearrangement, 214–226
Diketones, *see* Diones
Diols
from alkenes, 149, 173–181
cleavage of, 27–36
lead tetraacetate oxidation of, 27–36
Diones
from alkenes, 150, 178–181
from alkynes, 150, 182–186
Diselenides, 139, 140
Disulfides, 139, 140

E

Electrochemical oxidation, 209, 242, 248, 249, 264, 265, 267–271, 273–278
of coal, 350
Enamines, 94, 95
lead tetraacetate oxidation of, 45, 46
Enols, lead tetraacetate oxidation of, 37–55
Esters, from ethers, 199–201
Ethers
cyclic, from oxidation of alcohols, 11–27
permanganate oxidation of, 199–201
silyl, 158, 159
Exinite, 281, 307, 309

F

Ferric chloride, 208, 212, 242, 264, 265
Ferric chloride–dimethylformamide complex, 208, 209, 261, 266, 270
Ferric ion, 282, 283
Furans, 126, 128, 129, 293

G

Glycols, *see* Diols

H

Halides, permanganate oxidation of, 203, 204
Halodecarboxylation, 78–81
Humic acid, 279, 280
  definition of, 353
  oxidations of, 280–351
Humic acid-like materials, 282, 285, 290, 303, 304, 353
Hydrazines, lead tetraacetate oxidation of, 101–105, 111, 112, 115
Hydrazones, lead tetraacetate oxidation of, 118–121
Hydrocarbons, permanganate oxidation of, 148, 149, 168–193
Hydrofluoric acid, with lead tetraacetate, 132, 137, 138
Hydrogen peroxide, 280–293
Hydrogen peroxide oxidation of coal, 324–327, 351
Hydrogen peroxide–trifluoroacetic acid oxidation, of coal, 327–330
α-Hydroxy carbonyl compounds, oxidation of, 31–34
Hypohalites, 280–293
Hyponitrite formation, 92

I

Indoles, 126
Inertinite, 281
Iodine, with lead tetraacetate, 18–27, 42, 80, 99
Iron organometallics, *see* Organoiron compounds
Iron salts, 16
Isocyanates, from amides, 98–101
Isoquinoline, permanganate oxidation of, 150

K

Kerogen, 283, 302, 304, 307
  definition of, 353
  oxidations of, 280–351
Ketones, 76, 77
  from alcohols, 3–27, 150, 193–199
  from alkanes, 190–192
Ketols, from alkenes, 150, 173–181
Ketones
  from diols, 22–36
  formation of, 132
  from α-hydroxy carboxylic acids, 75

L

Lead(IV)acetate azides, 138
Lead–aryl compounds, 53, 60, 61
Lead-carbon bonds, 128, 130–134, 137, 141, 142, 144
Lead dioxide, 208
Lead tetraacetate, 209, 255–257
  oxidations with, 1–145
Lead tetrabenzoate, 47, 129
Lead tetracarboxylates, 131
Lead tetra(trifluoroacetate), 17, 18
Lignans, 274, 275
  definitions of, 353
  oxidations of, 280–351
Liptinite, *see* Exinite
Lignite, 281, 306
  solvent-refined, 284
LTA, *see* Lead tetraacetate

M

Macerals, 281
Manganic tris(acetylacetonate), 208, 259, 265, 269, 270
Manganese dioxide, 208, 212, 213, 273
Mercury salts, 15
Mercury organometallics, *see* Organomercury compounds

Meteoritic organic polymer, oxidation of, 294, 321
(±)-*O*-Methylflavinantine, 209
α-Methylene lactones, 63, 64
Micelles, 167
Molybdenum oxytetrachloride, 265
Monocarboxylic acids
  lead tetraacetate oxidation of, 61–78

N

Nitrenes, 5, 92, 93, 101, 105, 111–117
  from azides, 91
Nitric acid, 280–293
Nitric acid oxidation, of coal, 324–341
Nitric acid and potassium dichromate, 292
Nitriles, from amines, 89–94
Nitroacetates, 120, 121
Nitrobenzene, 280
Nitrobenzene oxidations of coal, 315, 321–324
Nitrogen-containing compounds, lead tetraacetate oxidation of, 89–122
Nitrosoacetates, 120, 121
Nitrosamines, 77, 78

O

Organoaluminum compounds, 140, 141
Organoborane compounds, 140
Organoiron compounds, 144, 145
Organomercury compounds, 141–143
Organometallics, lead tetraacetate oxidation of, 140–145
Organopalladium compounds, 143, 144
Organothallium compounds, 141
Organotin compounds, 142
Osmium tetroxide, 28, 31, 34
Oxazine, 92
Oxazolidones, 117
Oxazolin-5-ones, 76, 77
Oxidative coupling
  of aromatic substrates intramolecular, 207–278
  of *N*-benzylphenethylamines, 257–264
  of benzyltetrahydroisoquinolines, 209–240
  of diarylalkanes, 264–278
  electrochemical, 209
  of phenethyltetrahydroisoquinolines, 241–257
  photochemical, 208
Oxidative degradation of coal, 279–354
Oximes, lead tetraacetate oxidation of, 118, 120, 121
Oxygen, 280–293
Oxygen oxidation, of coal, 340–343
Ozonation, of coal, 350–351
Ozone, 280

P

Palladium organometallics, *see* Organopalladium compounds
Peat, 306
Peracetic acid oxidation, of lignin, 287
Perchloric acid, 280
Perchloric acid oxidation, of coal, 349, 350
Permanganate, 280–293
  benzyltriethylammonium, 151, 188, 189, 190–192, 199, 200
  tetrabutylammonium, 151, 195, 202
Permanganate oxidation, 147–206
  of coal, 304
  of lignite coal, 287
Phase transfer agents, 156–160
  adogen, 157, 158, 160, 161, 165, 169, 176, 179, 181, 184, 192, 193–195, 197, 201, 202, 204
  aliquat, 157, 158, 160, 161, 165, 169, 170, 185
  solvent polarity, 160, 161
  structure of, 162, 163
Phase transfer assisted permanganate oxidations, 147–206
Phase transfer mechanisms, 166, 167
Phase transfer reaction
  nature of the anion, 161, 162
  solvent effects, 163, 164
Phenethyltetrahydroisoquinolines, oxidative coupling, 241–257
Phenols
  intramolecular oxidative coupling of, *see* Oxidative coupling
  lead tetraacetate oxidation of, 55–61
Phosphonium salts, quaternary, 151, 156, 161–163, 165, 166
Photochemical oxidation, of coal, 348, 349

Polycyclic aromatic hydrocarbons, permanganate oxidation of, 192, 193
Polyethers, cyclic and acyclic, 151, 152, 158–161, 163, 165, 169–172, 176, 179–181, 185, 194, 195, 202
Polynuclear aromatic hydrocarbons, lead tetraacetate oxidation of, 124, 125, 127, 128
Polynuclear aromatic polycarboxylic acids, 290
Potassium ferricyanide, 208, 212, 220, 223, 226, 242, 243, 246–248, 258, 259, 264, 265
Potassium hexacycanoferrate, 216
Potassium permanganate, *see* Permanganate
Prostaglandins, 62
Pschorr reaction, 208
Purification of water, by treatment with permanganate, 148
Pyrroles, 126, 293

Q

Quinones, 60
  from arenes, 192–193
  from phenols, 196–199

S

Selenides, 140
Selenocarbonyl, compounds, 54, 55
Semicarbazones, lead tetraacetate oxidation of, 121, 122
Silver carbonate, 212, 264, 265
Silver nitrate, 265
Silver oxide, 277, 278
Silver salts, 15
Silver ethers, of enols, 46–48
Sodium hypochlorite oxidation, of coal, 330–334
Solvent systems, used in permanganate oxidations, 148–152
Steroids, 7, 20–22, 28, 39–43, 46, 49, 62, 67, 120, 136–139
Stilbenes, 58
Subbituminous coal, 281
Sulfenamides, 93
Sulfides, 49–51, 116, 139, 140
Sulfinamides, permanganate oxidation of, 150
Sulfonamides, 150
  lead tetraacetate oxidation of, 100, 101
Sulfones, from sulfides, 150, 201–203
Sulfonic acids, from sulfides, 201–203
Sulfoxides, 116
  from sulfides, 201–203
Sulfoximides, 116
Sulfur compounds
  in coal, 289–352
  permanganate oxidation of, 150, 201–203

T

Tellurium compounds, 101
Tetrabutylammonium tetrafluoroborate, 228–230
n-Tetrabutylammonium tetrafluoroborate, 275–278
Thallium organometallics, *see* Organothallium compounds
Thallium trifluoroacetate, 208, 257, 262, 263, 270, 277, 278
Thiadiazolidines, 102
Thiocarbonyl compounds, 54
Tin organometallics, *see* Organotin compounds
Titanium tetrachloride, 140
Trifluoroacetic acid, with lead tetraacetate, 122, 123, 130, 132, 138–141

V

Vanadium oxytrichloride, 208, 213, 222, 223, 224, 263–265, 271
  oxytrifluoride, 208, 223, 224
  in trifluoracetic acid, 227–240, 242, 247–249, 253–255, 261, 267, 268, 272, 274–276
Vitrinite, 281

ORGANIC CHEMISTRY

A SERIES OF MONOGRAPHS

EDITOR

HARRY H. WASSERMAN
Department of Chemistry
Yale University
New Haven, Connecticut

1. Wolfgang Kirmse. CARBENE CHEMISTRY, 1964; 2nd Edition, 1971
2. Brandes H. Smith. BRIDGED AROMATIC COMPOUNDS, 1964
3. Michael Hanack. CONFORMATION THEORY, 1965
4. Donald J. Cram. FUNDAMENTALS OF CARBANION CHEMISTRY, 1965
5. Kenneth B. Wiberg (Editor). OXIDATION IN ORGANIC CHEMISTRY, PART A, 1965; Walter S. Trahanovsky (Editor). OXIDATION IN ORGANIC CHEMISTRY, PART B, 1973; PART C, 1978; PART D, 1982
6. R. F. Hudson. STRUCTURE AND MECHANISM IN ORGANO-PHOSPHORUS CHEMISTRY, 1965
7. A. William Johnson. YLID CHEMISTRY, 1966
8. Jan Hamer (Editor). 1,4-CYCLOADDITION REACTIONS, 1967
9. Henri Ulrich. CYCLOADDITION REACTIONS OF HETEROCUMULENES, 1967
10. M. P. Cava and M. J. Mitchell. CYCLOBUTADIENE AND RELATED COMPOUNDS, 1967
11. Reinhard W. Hoffmann. DEHYDROBENZENE AND CYCLOALKYNES, 1967
12. Stanley R. Sandler and Wolf Karo. ORGANIC FUNCTIONAL GROUP PREPARATIONS, VOLUME I, 1968; VOLUME II, 1971; VOLUME III, 1972
13. Robert J. Cotter and Markus Matzner. RING-FORMING POLYMERIZATIONS, PART A, 1969; PART B, 1; B, 2, 1972
14. R. H. DeWolfe, CARBOXYLIC ORTHO ACID DERIVATIVES, 1970
15. R. Foster. ORGANIC CHARGE-TRANSFER COMPLEXES, 1969
16. James P. Snyder (Editor). NONBENZENOID AROMATICS, VOLUME I, 1969; VOLUME II, 1971
17. C. H. Rochester. ACIDITY FUNCTIONS, 1970
18. Richard J. Sundberg. THE CHEMISTRY OF INDOLES, 1970
19. A. R. Katritzky and J. M. Lagowski. CHEMISTRY OF THE HETEROCYCLIC *N*-OXIDES, 1970
20. Ivar Ugi (Editor). ISONITRILE CHEMISTRY, 1971
21. G. Chiurdoglu (Editor). CONFORMATIONAL ANALYSIS, 1971
22. Gottfried Schill. CATENANES, ROTAXANES, AND KNOTS, 1971
23. M. Liler. REACTION MECHANISMS IN SULPHURIC ACID AND OTHER STRONG ACID SOLUTIONS, 1971
24. J. B. Stothers. CARBON-13 NMR SPECTROSCOPY, 1972

25. Maurice Shamma. THE ISOQUINOLINE ALKALOIDS: CHEMISTRY AND PHARMACOLOGY, 1972

26. Samuel P. McManus (Editor). ORGANIC REACTIVE INTERMEDIATES, 1973

27. H. C. Van der Plas. RING TRANSFORMATIONS OF HETEROCYCLES, VOLUMES 1 AND 2, 1973

28. Paul N. Rylander. ORGANIC SYNTHESES WITH NOBLE CATALYSTS, 1973

29. Stanley R. Sandler and Wolf Karo. POLYMER SYNTHESES, VOLUME I, 1974; VOLUME II, 1977; VOLUME III, 1980

30. Robert T. Blickenstaff, Anil C. Ghosh, and Gordon C. Wolf. TOTAL SYNTHESIS OF STEROIDS, 1974

31. Barry M. Trost and Lawrence S. Melvin, Jr. SULFUR YLIDES: EMERGING SYNTHETIC INTERMEDIATES, 1975

32. Sidney D. Ross, Manuel Finkelstein, and Eric J. Rudd. ANODIC OXIDATION, 1975

33. Howard Alper (Editor). TRANSITION METAL ORGANOMETALLICS IN ORGANIC SYNTHESIS, VOLUME I, 1976; VOLUME II, 1978

34. R. A. Jones and G. P. Bean. THE CHEMISTRY OF PYRROLES, 1976

35. Alan P. Marchand and Roland E. Lehr (Editors). PERICYCLIC REACTIONS, VOLUME I, 1977; VOLUME II, 1977

36. Pierre Crabbé (Editor). PROSTAGLANDIN RESEARCH, 1977

37. Eric Block. REACTIONS OF ORGANOSULFUR COMPOUNDS, 1978

38. Arthur Greenberg and Joel F. Liebman, STRAINED ORGANIC MOLECULES, 1978

39. Philip S. Bailey. OZONATION IN ORGANIC CHEMISTRY, VOLUME I, 1978; VOLUME II, 1982

40. Harry H. Wasserman and Robert W. Murray (Editors). SINGLET OXYGEN, 1979

41. Roger F. C. Brown. PYROLYTIC METHODS IN ORGANIC CHEMISTRY: APPLICATIONS OF FLOW AND FLASH VACUUM PYROLYTIC TECHNIQUES, 1980

42. Paul de Mayo (Editor). REARRANGEMENTS IN GROUND AND EXCITED STATES, VOLUME I, 1980; VOLUME II, 1980; VOLUME III, 1980

43. Elliot N. Marvell. THERMAL ELECTROCYCLIC REACTIONS, 1980

44. Joseph J. Gajewski. HYDROCARBON THERMAL ISOMERIZATIONS, 1981